Comparative Virology

CONTRIBUTORS

Max Bergoin

D. E. Bradley

Jordi Casals

Purnell W. Choppin

Richard W. Compans

Samuel Dales

T. O. Diener

A. F. Graham

L. Hirth

M. David Hoggan

A. F. Howatson

Klaus Hummeler

André Lwoff

Carl F. T. Mattern

Stewart Millward

William T. Murakami

Erling Norrby

Bernard Roizman

Roland R. Rueckert

Kenneth M. Smith

Patricia G. Spear

Kenneth K. Takemoto

Paul Tournier

Comparative Virology

Edited by

KARL MARAMOROSCH
BOYCE THOMPSON INSTITUTE FOR PLANT RESEARCH
YONKERS, NEW YORK

EDOUARD KURSTAK
DEPARTMENT OF MICROBIOLOGY AND
IMMUNOLOGY, FACULTY OF MEDICINE
UNIVERSITY OF MONTREAL
MONTREAL, CANADA

 1971

ACADEMIC PRESS   *New York and London*

ACADEMIC PRESS, INC.
111 Fifth Avenue, New York, New York 10003

United Kingdom Edition published by
ACADEMIC PRESS, INC. (LONDON) LTD.
Berkeley Square House, London W1X 6BA

LIBRARY OF CONGRESS CATALOG CARD NUMBER: 70-137607

PRINTED IN THE UNITED STATES OF AMERICA

Contents

Chapter 1. *Remarks on the Classification of Viruses*

ANDRÉ LWOFF AND PAUL TOURNIER

Chapter 2. *Small DNA Viruses*

M. DAVID HOGGAN

Chapter 7.   *A Comparative Study of the Structure and Biological Properties of Bacteriophages*

D. E. BRADLEY

Chapter 8.   *Picornaviral Architecture*

ROLAND R. RUECKERT

Chapter 9.   *Arboviruses: Incorporation in a General System of Virus Classification*

JORDI CASALS

List of Contributors

Numbers in parentheses indicate the pages on which the authors' contributions begin.

MAX BERGOIN (169), Station de Recherches Cytopathologiqes, St. Christol-les-Ales (Gard), France

D. E. BRADLEY (207), Department of Zoology, University of Edinburgh, Edinburgh, Scotland

JORDI CASALS (307), Yale Arbovirus Research Unit, Yale University Medical School, New Haven, Connecticut, and The Rockefeller Foundation, New York, New York

PURNELL W. CHOPPIN (407), The Rockefeller University, New York, New York

RICHARD W. COMPANS (407), The Rockefeller University, New York, New York

SAMUEL DALES (169), Department of Cytobiology, Public Health Research Institute of the City of New York, New York, New York

T. O. DIENER (433), Plant Virology Laboratory, Agricultural Research Service, Department of Agriculture, Beltsville, Maryland

A. F. GRAHAM (387), Department of Biochemistry, McGill University, Montreal, Canada

L. HIRTH (335), Laboratoire des Virus des Plantes, Institut de Botanique, Strasbourg, France

M. DAVID HOGGAN (43), Laboratory of Viral Diseases, National Institute of Allergy and Infectious Diseases, Bethesda, Maryland

A. F. HOWATSON (509), Ontario Cancer Institute and Department of Medical Biophysics, University of Toronto, Toronto, Canada

KLAUS HUMMELER (361), The Children's Hospital of Philadelphia, Department of Pediatrics Medical School, University of Pennsylvania, Philadelphia, Pennsylvania

ANDRÉ LWOFF (1), Institut de Recherches Scientifiques sur le Cancer, Villejuif, France

CARL F. T. MATTERN (81), Laboratory of Viral Diseases, National Institutes of Health, Bethesda, Maryland, and Graduate Department of Biochemistry, Brandeis University, Waltham, Massachusetts

STEWARD MILLWARD (387), The Wistar Institute of Anatomy and Biology, Philadelphia, Pennsylvania

WILLIAM T. MURAKAMI (81), Laboratory of Viral Diseases, National Institutes of Health, Bethesda, Maryland, and Graduate Department of Biochemistry, Brandeis University, Waltham, Massachusetts

ERLING NORRBY (105), Department of Virology, Karolinska Institute, School of Medicine, Stockholm, Sweden

BERNARD ROIZMAN (135), Department of Microbiology, University of Chicago, Chicago, Illinois

ROLAND R. RUECKERT (255), Biophysics Laboratory and Department of Biochemistry, University of Wisconsin, Madison, Wisconsin.

KENNETH M. SMITH (479), Cell Research Institute, University of Texas at Austin, Austin, Texas

PATRICIA G. SPEAR (135), Department of Microbiology, University of Chicago, Chicago, Illinois

KENNETH K. TAKEMOTO (81), Laboratory of Viral Diseases, National Institutes of Health, Bethesda, Maryland, and Graduate Department of Biochemistry, Brandeis University, Waltham, Massachusetts

PAUL TOURNIER (1), Institute de Recherches Scientifiques sur le Cancer, Villejuif, France

Preface

During the last two decades virology has become an independent science. Several major treatises have been published on general virology (Luria, 1953; Luria and Darnell, 1967), on methods in virology (Maramorosch and Koprowski, 1967–1971; Harris, 1964; Habel and Salzman, 1969; and others), and on descriptions of viruses affecting various hosts, such as man and animals (Horsfall and Tamm, 1965; Fenner, 1968; Melnick, 1958–1969; Andrewes and Pereira, 1967; Rhodes and Van Rooyen, 1968; Fenner and White, 1970), plants (Bawden, 1964), insects (Smith, 1967; Maramorosch, 1968), and bacteria (Adams, 1959; Stent, 1963). However, none of these books has dealt with comparative virology, which is the aim of this volume.

Viruses have provided tools for molecular biologists, and their studies have aided in our understanding of the chemistry and structure of viruses. Early attempts at classifying viruses were usually made according to hosts affected, describing groups of viruses multiplying in warm-blooded vertebrates, in insects, in bacteria, or in plants. Attempts to unify viruses and to consider these disease agents as a single kingdom began many years ago when it was realized that certain viruses infect both plants and invertebrate animals. These attempts at unification gained impetus with the announcement of the Lwoff–Horne–Tournier classification in 1959.

The primary purpose of this treatise is to provide an integrated comparison of viruses, based on their chemical and morphological characteristics. These descriptions will not only give the reader a background but also a detailed analysis of the various groups. In some instances the groups are still host related, as in the case of bacteriophages and polyhedral insect viruses. In others, for instance in pox viruses, the group comprises viruses of vertebrates and invertebrates. The hosts of the bacilliform Rhabdovirales range from man and other warm-blooded vertebrates through invertebrate animals to plants. A special chapter is devoted to viruses devoid of protein —a group that is of great interest and that has only recently been recognized. Since there is historical and practical interest in ecologic groupings, such as arboviruses and oncogenic viruses, chapters on such groups have also been included. Although the subject areas may overlap to some extent, the groupings in these two chapters are necessarily arbitrary.

Nobel laureate André Lwoff, who stimulated the lively discussions at the First International Conference on Comparative Virology, consented to contribute the first chapter, which is more comprehensive and theoretical in scope than any of the others. It serves as a preamble to the treatise. Chapters dealing with DNA viruses and RNA viruses follow, and the ecologically and disease-oriented groups complete the volume.

Virology has reached the stage where a number of specialists have felt justified in attempting to integrate and compare viruses irrespective of the hosts affected by them. In September 1969, the First International Conference on Comparative Virology was organized by Karl Maramorosch and Edouard Kurstak, the editors of this treatise. Under the sponsorship of the University of Montreal, the Ministry of Helath of the Province of Quebec, the National Research Council of Canada, the Medical Research Council of Canada, and the Quebec Medical Research Council, 150 virologists from seventeen countries met for four days at Mount Gabriel Lodge in the Laurentian Mountains near Montreal, Canada, to discuss the latest information and ideas concerning comparative virology.

The major objectives of the conference were the integration of animal and plant viruses into a single kingdom, Viridae, and the comparison of these virus groups, primarily on the basis of their chemical and morphological characteristics. Prior to the meeting the forty-eight speakers at the conference were informed that their contributions would not be published and that their presentations were to be made with the specific purpose of enlightening other conference participants only. When the conference ended the possibility of compiling a book on comparative virology began to be seriously considered. Authors were invited from among the participants as well as from experts in their fields who could not attend. Each of the contributors is well known, and each has prepared a thoughtful and well-documented treatment of his subject.

There were many more who made important contributions to the conference but did not contribute directly to the book. Thanks are due to all those who consented to have published and unpublished work quoted by the authors of the present chapters. Personal interpretations and conclusions of the authors, as well as numerous illustrations, many of unpublished material, provide a large body of information and bring into sharp focus current findings and new directions of research.

It is our hope that "Comparative Virology" will help bring unity to the science of virology through the comparative approach that is not dependent on virus–host interactions. The worldwide growth of virological science is perhaps unique in the history of biology, and the combined efforts of eminent contributors to discuss and evaluate new information will hopefully benefit all who are interested in virology.

The chairmen of the First International Conference on Comparative Virology and editors of this book wish to express their sincere gratitude to the contributors for the effort and care with which they have prepared their chapters; to the Faculty of Medicine of the University of Montreal for help during various stages of the treatise; to Professor André Lwoff, Institut de Recherches Scientifiques sur le Cancer, and many others who gave moral support to the idea of organizing the Conference; to Dr. Maurice L'Abbé, Vice-Rector for Research of the University of Montreal, to Dr. Eugène Robillard, past Dean, and to Dr. Pierre Bois, present Dean of the Faculty of Medicine of the University of Montreal, for their support and continued interest in comparative virology; and last, but not least, to Academic Press for their part in editing, indexing, proofreading, and other aspects of production of the volume.

KARL MARAMOROSCH
EDOUARD KURSTAK

CHAPTER 1   *Remarks on the Classification of Viruses*

ANDRÉ LWOFF AND PAUL TOURNIER

To know, is to classify.
Stuart Mill

The coining and application
of any collective or generic term
represents an act of classification.
Ernst Mayr

1

I. Introduction

Only a few years ago, the viral world was a chaos of "small infectious particles." Consequently, it embraced, at the same time, viruses and bacteria. Today, the viruses are well defined by the sum of the distinctive traits of the virion. These distinctive traits are as follows:

1. Presence of a single nucleic acid.
2. Incapacity to grow and to divide.
3. Reproduction from the genetic material only.
4. Absence of enzymes for energy metabolism.
5. Absence of ribosomes.
6. Absence of information for the production of enzymes in the energy cycle.
7. Absence of information for the synthesis of the ribosomal proteins.
8. Absence of information for the synthesis of ribosomal RNA and transfer RNA.

It would appear that a correlation exists among all these characteristics: only one of them is, in fact, sufficient to establish that an infectious particle belongs to the viral world. In the group of viruses recognized as such by the virtue of a definition, it became evident that order was mandatory.

In fact, the remarkable homogeneity, attested to by the number and extent of the common characteristics, masks a no less remarkable diversity. The viral infectious particle presents, in fact, a great diversity in composition and structure. Order could be achieved only through a classification, which is a system of order. The goal of biological classification is to group together organisms presenting certain analogies and certain affinities and, if possible, to also bring out phylogenic relationships.

TABLE I

DISCRIMINATIVE CHARACTERS OF VIRUSES AND PROTISTS OR BACTERIA

	Virions	Protists
Nucleic acid	DNA *or* RNA	DNA *and* RNA
Reproduction from the sole genetic material	+	0
Growth	0	+
Division	0	+
Information for the synthesis of enzymes of the energetic metabolism	0	+
Presence of the enzymes	0	+
Information for the synthesis of transfer RNA	0	+
Information for the synthesis of ribosomal RNA	0	+
Presence of ribosomes	0	+

The conceptions relative to the methodology of taxonomy, which is the science of classification, are diverse. The divergence of the conceptions necessarily leads to discussions, the heuristic value of which is incontestable. Nevertheless, a good classification must allow predictions and must also pose problems.

A few years ago we (Lwoff and Tournier, 1966) presented a critical history of various classifications of viruses, so we will not go back to that topic again. However, the symposium devoted to the classification of micro-organisms, held by the "Society of General Microbiology" in 1962, was overlooked. In the corresponding volume are to be found remarkable general articles and vast amount of information. Wildy (1962b) presented an article on the classification of viruses: his conclusions were a restatement of those of Horne and Wildy published the previous year and no new classi-fication was proposed.

In the course of this meeting, Sneath (1962) discussed the structure of taxonomic groups. Sneath, as is known, is one of the principal champions of the numerical method. His study was devoted principally to bacteria, but one short paragraph dealt with viruses. There, it is seen that the numerical method has permitted recognition of three groups: the arboviruses, myxoviruses, and enteroviruses; that is not much. Pirie (1962), too, talked about the classification of viruses and discussed the criteria, but did not propose any system.

The general atmosphere of the symposium was pessimistic; it was held that the viruses could not be classified, owing to insufficient data. But if a certain caution is always necessary, an exaggeratedly critical attitude is sterilizing. It was necessary to go ahead and to act.

II. The LHT system

A. THE SYSTEM

Lwoff, Horne, and Tournier, from here on referred to as LHT, proposed a System of Viruses (Lwoff *et al.*, 1962a,b) (see Table II) and discussed the classification (Lwoff and Tournier, 1966).

Four characteristics were utilized: the nature of the genetic material, the symmetry of the capsid, the naked or enveloped nature of the nucleocapsid, and, finally, the number of capsomers for the virions with cubical symmetry, the diameter of the nucleocapsid for the virions with helical symmetry, the numerical data allowing for diversity. LHT had noted that other properties in addition to those adopted could be used for grouping viruses; among them were: the molecular weight of the nucleic acid, the proportion of nucleotides, the number of strands in the nucleic acid, the properties of the capsomers, the nature of the envelope, its origin, the antigenicity of the viral proteins, the host, and the virulence. Twenty-nine characteristics were thus enumerated.

In 1965, the provisional committee on nomenclature of viruses (Anonymous, 1965) proposed that the groups defined by the four characteristics of the LHT system be considered as families. Since 1962, the LHT system has become enlarged: some new families have found their place in the system without difficulty. Table III affords a general picture of the viral world as we understand it in 1969. Table IV shows the various properties of the virus families.

TABLE II

THE LHT SYSTEM IN ITS ORIGINAL 1962 FORM[a]

I Nucleic acid	II Capsid symmetry	III Naked (N) or enveloped (E)	IV Capsid diameter or no. of capsomeres	Viruses	Groups
RNA	H	N	100–130 Å	White clover mosaic, potato X, cactus potato F, pea stripe, wheat stripe	RHN 115
			170–200 Å	Tobacco mosaic, Vigna sinensis, cucumber 3	RHN 185
			250 Å	Barley stripe mosaic	RNH 250
		E	90–100 Å	Influenza, fowl plague	I myxo (1)
			170 Å	Parainfluenzae, mumps, sendai, Newcastle disease, rinderpest, Carré disease, measles	II myxo (1)
	C	N	32	Turnip yellow mosaic	RCN 32
			60	Bushystunt, polio (2)	—
			92	Wound tumor, REO III	REO
	H	E	90–100 Å	Vaccinia, infectious pustular dermatitis	Pox (3)
			12	φX174	DCN 12
			42	Polyoma, warts, SV40	Schizo (4)
			252	Adeno, infectious canine hepatitis	Adeno
DNA	C	N	812	Gal (5)	DCN 812
				Tipula	
		E	162	Herpes, varicella, pseudorabies	Herpes
	B	N	Prism	Phage T2	T pairs
			Icosahedron	Phage of *B. megaterium*	—

[a] The table contains abbreviations which are those of the phanerogram.

TABLE III

THE LHT SYSTEM IN 1969

Nucleic acid	Capsid symmetry	Naked (N) or enveloped (E)	Helical diameter or no. of capsomeres	
DNA	H	N	50 Å	Inoviridae
		E	?	Poxviridae
	C	N	12	Microviridae
			32	Parvoviridae
			42	Densoviridae
			72	Papilloviridae
			252	Adenoviridae
			812	Iridoviridae
		E	162	Herpesviridae
	B	N Urovirales		Phages with tail
RNA	H	N Rhabdovirales		
			90 Å	Myxoviridae
			180 Å	Paramyxoviridae
		E Sagovirales		Stomatoviridae
			?	Thylaxoviridae
	C	N Gymnovirales	32	Napoviridae
			92	Reoviridae
			?	Blue tongue virus (sheep)
		E Togavirales	?	Encephaloviridae

Most of the elements of this table have been known for many years and can be traced back to the work of Luria and Darnell (1968), Davis, Dulbecco, Eisen, Ginsberg, and Wood (1968), and Fenner (1968). We have completed the table with the help of recent data. Now follows a sequence of comments on the various families (presented in the same order as Table IV).

1. Inoviridae—Type Genus: *Inovirus*

This family brings together the filamentous phages, the adsorption of which depends on the existence of sexual pili. It includes, specifically phages fd, fl, and M13. The fact that the single-stranded DNA is circular implies a different structure for the virion from that of the tobacco mosaic virus. Two patterns were proposed. One, roughly cylindrical, with the DNA located in a central core, the other with two parallel cylinders, each containing half of the DNA. In both cases, the units of the structure (protein B) join together in a helical formation.

TABLE IV

THE CHARACTERISTICS OF VARIOUS FAMILIES OF VIRUSES

Viridae	Nucleic acid	Symmetry	Naked (N) or enveloped (E) nucleocapsid	Virions with cubic symmetry			Virions with helical symmetry		Mol. wt. of nucleic acid ($\times 10^6$)	Number of nucleic acid strand (5)
				Number of capsomers	Diameter of the nucleocapsid (Å)	Diameter of the envelope (Å)	Diameter and length of the nucleocapsid (Å)	Dimensions of the enveloped virions (Å)		
Ino-	D	H	N				5–6×760–850		1,7–3	1
Pox-	D	H?	E				?	2500 × 1600 3000 × 2300	160–240	2
Micro-	D	C	N	12	250				1,7	1
Parvo-	D	C	N	32	220				1,8	1
Denso-	D	C	N	42	200				1,6–2,2	1
Papilloma- (papova)	D	C	N	72	450–550				3–5	2
Adeno-	D	C	N	252	700				20–25	2
Irido-	D	C	N	812	1300				126	2
Herpes-	D	C	E	162	775				54–92	2
Uro-	D	BC	N	—	—	1500–2000	20 × 130			2
Rhabdo-	R	H	N				10 × 1250			1
Myxo-	R	H	E				90 ×?	1000	2–3	1
Paramyxo-	R	H	E				180 ×?	1200	7,5	1
Stomato- (rhabdo)	R	H	E				180 ×?	1750 × 680	6	1
Thylaxo-	R	H	E				?	1000	10	1
Napo-	R	C	N	32	220–270				1,1–2	1
REO	R	C	N	92	700				10	2
Cyano	R	C	N	32 or 42	540				—	2
Encephalo	R	C	E	?	?		600–800		2–3	1

2. Poxviridae—Type Genus: *Poxvirus*

The helical structure of the nucleocapsid of the Poxviridae remains entirely hypothetical: the helical tubules seen at the periphery of the virion are part of the envelope system but are not associated with the nucleic acid. The main viruses or diseases include: vaccinia, small pox, mouse ectromelia, myxoma and fibroma of Shope, bovine papular stomatitis (dermatitis), orf, avian small pox (avipoxviruses), and *Molluscum contagiosum*.

3. Microviridae—Type Genus: *Microvirus*

The DNA of phage ϕX174 and related phages S13 and ϕR is circular.

4. Parvoviridae—Type Genus: *Parvovirus*

The DNA of "minute virus of mice" (Crawford, 1966) and of the K virus of rats is single-stranded.

5. Densoviridae—Type Genus: *Densovirus*

This type is the densonucleosis virus of *Galleria mellonella* as described by Kurstak and Cote (1969). The structure of its nucleic acid is not yet strictly determined.* These authors postulated a capsid with 42 capsomers. The same figure was first proposed for the papilloma and polyoma viruses but agreement has now been reached on the figure of 72.

6. Papillomarividae (or Papovaviridae)—Type Genus: *Papillomavirus*

This family includes two genera: *Papillomavirus*, with a diameter of 55 mμ, whose circular DNA has a molecular weight of 5 million [main viruses: rabbit papilloma virus (Shope), the human wart, bovine, equine papilloma] and *Polyomavirus*, with a diameter of 45 mμ whose circular DNA has a molecular weight of 3 to 3.5 million (main viruses: polyoma virus, SV40 virus).

7. Adenoviridae—Type Genus: *Adenovirus*

The 31 human adenoviruses, the simian, bovine, avian, and canine adenoviruses (Rubarth hepatitis).

8. Iridoviridae—Type Genus: *Iridovirus*

Tipula iridescens virus, *Chilo*, and *Sericesthis* viruses.

* See Note Added in Proof, p. 42.

9. Herpesviridae—Type Genus: *Herpesvirus*

In man this genus includes: herpes simplex virus, varicella-zoster viruses, cytomegalic inclusion virus disease, infectious mononucleosis virus, viruses associated with Burkitt's disease, and rhinocarcinoma of the pharyn. (The last three viruses are, if not identical, at least closely related.)

In animals this genus includes: feline, bovine rhinotracheitis viruses, abortion in equidae, cytomegalic inclusion virus disease in guinea pigs, mice, etc.

10. Urovirales

The world of bacteriophages provided with a tail. They are grouped, not into a family, but into an order: the urovirales.

The structural diversity of these phages enables us to envision a classification that will take into consideration the following:

1. *The tail*:
   a. Whether or not it is contractile.
   b. The structure of the terminal plaque.
   c. The structure of the collar.
   d. The length.
   e. The existence of fibers.

2. *The head*: octahedral, icosahedral, or any other form.

3. *The nucleic acid*: its molecular weight, the sequence of its nucleotides, its homologies with the nucleic acid from other phages. Its eventual single- or double-stranded structure.

11. Rhabdovirales—Type Genus: *Rhabdovirus*

This order brings together at least fifty of the plant viruses, classified by Brandes and Berks into six different groups, the most extensively studied of which is the tobacco mosaic virus. The name *Rhabdovirus* was proposed in 1965 by the Provisional Committee for the Nomenclature of Veruses (PCNV) (see Anonymous, 1965).

12. Myxoviridae—Type Genus: *Myxovirus*

The 3 types of *Myxovirus influenzae* A, B, and C.

13. Paramyxoviridae—Type Genus: *Paramyxovirus*

Among the RNA viruses with helical symmetry, enveloped, and with a nucleocapsid 180 Å in diameter, the following forms have been separated:

subspherical virions, which are the Paramyxoviridae, and bullet-shaped virions, which are the Stomatoviridae. The genus type of Paramyxoviridae is *Paramyxovirus.*

Apart from the parainfluenzae viruses I, II, III, and IV, the group includes the viruses of mumps, canine distemper, and rinderpest.

14. Stomatoviridae—Type Genus: *Stomatovirus*

This family, as was previously stated, groups viruses of the bullet-shaped virions. The name *Stomatovirus* was proposed in 1965 by the PCNV. A few years later, a group of the International Committee for the Nomenclature of Veruses (ICNV) proposed the name Rhabdoviridae for the same group.

Now, as already mentioned, the name Rhabdoviridae was proposed by the PCNV for plant viruses. It is unfortunate, and even inadmissible, to see virologists appropriate a name already given to one group and to apply it to another. Such practices can only lead to confusion. They are forbidden by the codes of plant, animal, and bacterial nomenclature and it is to be hoped that this ruling will soon apply, also, to the code for the nomenclature of viruses.

The group comprises rabies, vesicular stomatitis viruses, and the *Drosophila* sigma virus.

15. Thylaxoviridae—Type Genus: *Thylaxovirus*

These represent the group of RNA oncogenic viruses. The structure of their nucleocapsid is still poorly elucidated. A few observations suggest that it could be of helical symmetry and of smaller diameter lower than the Myxoviridae. The name was proposed by a study group (Anonymous, 1966):

Virus of avian sarcomas and leukosis
Virus of murine sarcomas and leukosis
Virus of the mammary tumor of mice (Bittner).

In Fenner's treatise, the Thylaxoviridae are called leukovirus.

16. Napoviridae—Type Genus: *Napovirus*

This is an important family by virtue of the number of its representatives. It comprises:

1. The *Napoviruses*, of which the typical species is the yellow mosaic virus of the turnip and several plant viruses with cubical symmetry.

2. The small RNA bacteriophages: f2, MS2, R17, R23, Qβ.

3. The polioviruses, coxsackieviruses, echoviruses, rhinoviruses, and the foot-and-mouth virus. These viruses are often referred to under the name of "picornavirus"—but it is not known whether this name corresponds to a genus or to a subfamily (see Lwoff and Tournier, 1966).

17. Reoviridae—Type Genus: *Reovirus*

The characteristics of the *Reovirus* virions and those of the wound tumor virus are the same. Furthermore, the RNA is double-stranded. For this reason, these viruses are classified together (Lwoff and Tournier, 1966). The group comprises:

1. The three types of *Reoviruses* I, II, and III common in humans and other mammals.

2. The clover-wound tumor virus, the rice dwarf virus.

18. Cyanoviridae—Type Genus: *Cyanovirus*

The blue tongue virus in sheep has a different structure from that of the reovirus. It has 32 or 42 capsomers, and its ribonucleic acid is double-stranded. Possibly, it represents the type of a new family.

19. Encephaloviridae (and Arboviruses)—Type Genus: *Encephalovirus*

This is one of the families resulting from the breaking up of the arboviruses. These are defined, not by their structure, but by their mode of transmission. They represent, therefore, an ecological group. These viruses appear to belong to three distinct groups:

1. The *Encephalovirus* group. The icosahedral structure of the capsid has been demonstrated only for a few viruses of this extremely vast group, which includes several subgroups:

Group A: eastern equine encephalitis (EEE), western equine encephalitis (WEE), Venezuela equine encephalitis (VEE), and sindbis.

Group B: West Nile virus (WNV), Japanese B encephalitis virus (JBE), St. Louis encephalitis virus, yellow fever virus, tick-borne encephalitis virus.

Group C: Oriboca.

Outside of these three groups are phlebotomus fever virus and hemorragic fever viruses.

For reasons of structure, the following are not included in the "Encephaloviridae" although these are "arboviruses" in the ecological sense of the term: African horsesickness virus, blue tongue virus, and vesicular stomatitis virus.

2. The vesicular stomatitis group: RNA, helical symmetry, nucleocapsid, enveloped, bullet-shaped.

3. The blue tongue virus: RNA, cubical symmetry, no envelope, 32 or 42 capsomers.

Miscellaneous

A. It was proposed to group under the name *Coronaviruses* (Anonymous, 1968) avian infectious bronchitis viruses, mouse hepatitis, and certain strains isolated during respiratory tract infections in humans. These contain a ribonucleic acid and are surrounded by a lipid envelope, but the type of symmetry of the nucleocapsid is not yet known.

B. A virus present in a *Penicillium cyaneofulvum* strain has recently been described. It has a diameter of 32.5 mμ, appears to present a cubical symmetry, its nucleocapsid is naked, and contains a double-stranded RNA. The number of capsomers has not been determined.

C. The alphalpha mosaic virus is elongated. The rounded extremities are semiicosahedral cut perpendicularly to a ternary axis. The cylindrical part in the center is a flat hexagonal network. Although elongated, the virion therefore belongs to a cubical symmetry system (Hull *et al.*, 1969).

B. DISCUSSION

Let us bear in mind that the viruses belonging to a given family in the LHT system have the following characteristics in common:

A. The four obvious characteristics of the LHT system:

    1. Nature of the nucleic acid.
    2. Symmetry of the capsid.
    3. Presence or absence of an envelope.
    4. Number of capsomeres (cubical symmetry) or diameter of the nucleocapsid (helical symmetry).

B. Other characteristics:

    5. The molecular weight of the nucleic acid.
    6. The proportion of nucleic acid in the nucleocapsid.
    7. The single- or double-stranded structure of the nucleic acid.

8. The percentace of guanine + cytosine (within certain limits).
9. The pattern of the doublets.
10. The homologies of the genetic material (within certain limits).
11. The form of the nucleocapsid.
12. The dimensions of the nucleocapsid (within certain limits).
13. The form of the virion (with one exception).
14. The dimensions of the virion.

The correlation of characteristics 5–14 with the sum of characteristics 1–4 could not be accidental. The viruses belonging to the families of the LHT system present similarities and are probably biologically related. It has been said that the families of the LHT system were conceived "by intuition." Such is not the case. The principles of the classification used by LHT and the choice of the characteristics are derived from a rational analysis and numerous trials.

Be that as it may, a system must prove itself and it is not the greater or lesser role played by intuition that will determine its value.

The opponents of the LHT system should explain why the grouping is not satisfactory. It must be recognized that the opponents are in a privileged situation, since they have not proposed any general classification of viruses. It is certainly easier to criticize than construct.

Some virologists consider that a classification must comprise coefficients of similarity. Within the LHT system, assuming that the viruses have 4, 3, 2, or 1 characteristics in common, the coefficients will be 100, 75, 50, or 25%, respectively. If the coefficient is 100%, the viruses will belong to the same family. The fact that a computer is not needed to calculate the coefficients of similarity should not diminish their value.

Within the LHT system, the characteristics are placed in a hierarchical order. The justification of the hierarchy has been presented (Lwoff and Tournier, 1966) and it seems useless to go back on that point again.

In the absence of phylogenic data, any hierarchy, as noted by Lwoff and Tournier (1966), is necessarily arbitrary. However, a hierarchical system for viruses presents some advantage. It affords, at a glance, an overall picture of the viral world and provides, if not enlightenment, at least a certain order. And order, even though arbitrary, is better than confusion.

The unitary concept of the viral world evolved somewhat belatedly. For a long time, the virologists were only concerned with groups pertaining to their specific interest: plant viruses, animal viruses, and bacterial viruses. With regard to the animal viruses, specialization came more and more to the fore. The human viruses, vertebrate viruses, invertebrate viruses, and insect viruses were considered separately. The situation was naturally complicated by the fact that certain plant viruses were evolving in insects, and certain vertebrate viruses in diverse arthropods.

The separatism manifested itself as classifications dealing exclusively with one group of viruses, followed by the observation that the virions of certain plant, bacteria, insect, and vertebrate viruses were similar. Little by little, the idea imposed itself that it was preferable to classify viruses according to the characteristics of the virion rather than according to the nature of the host or the vector. The unitary concept of the viral world is, today, universally accepted. It becomes more and more difficult to accept classifications that adopt as a discriminative character the systematic position of the host—or sometimes, even of the vector.

At present, the LHT system is the only one that embraces the entire viral world. It permits the definition of taxons having one, two, three, or four characteristics in common. It so happens that taxons exhibiting four common characteristics correspond to groups of genera, that is to say, to families that are today recognized by the majority of virologists. The LHT system also permits the classification of plant viruses. Nevertheless, plant virology is strongly handicapped by a group of factors that are more closely related to the mental confusion of some people than to the viruses themselves.

We would like to add a remark here. When total disorder reigns within a given domain, any system of order that emerges is resented by many as a constat of deficiency; all the more so, of course, as it affords more clarity.

C. Remarks on Viruses without Capsids

In their studies on the spindle tuber potato virus, T. O. Diener and Raymer (1969) were able to prove that the terminal form of the cycle is a double-stranded naked RNA, and that there are no nucleocapsids, in other words no virions. The molecular weight of the RNA is from 100 to 200,000; in other words, the nucleic acid can most likely "code" for one single protein. This protein, if it is unique, could be nothing more than replicase.

Naturally, we must ask ourselves if this naked nucleic acid can be considered a virus. It is a question of definition. The virus may be defined as an infectious particle that has only one type of nucleic acid and reproduces itself from the single genetic material, and the viral infection as the introduction into the cell of the genetic material of a virus. If these definitions are accepted, then the spindle tuber potato RNA is a virus.

Does this virus represent a primitive form or a degraded evolved form? For the moment, it is impossible to answer this question; the problem of the origin of RNA viruses was discussed recently (A. Lwoff, 1969). However, the very existence of this naked viral RNA obliges us to provisionally divide the RNA viruses into two groups: one with a capsid and the other without a capsid. When we shall know the frequency of the doublet pattern and the percentage of guanine cytosine, and when cross-breeding experiences have been performed, perhaps it will be possible to relate this RNA to the RNA of some "normal" virus.

III. Pros and Cons of the LHT System

It was evident that the LHT system would be accepted by some and rejected or even contested by others.

A. PROS

A. Cohen (1969) includes the LHT system in his "Textbook of Medical Virology"; M. Frobisher (1968) reproduces it in his "Fundamentals of Microbiology" and writes: "Neither perfection nor immutability are claimed for the system; but being the first of its kind, it marks a milestone i₁ the science of virology."

The system is also reproduced in "General Virology" by S. Luria and J. Darnell (1968), who write: "A major advance came when a system was proposed that took as its basis the structure and composition of virions and could embrace all viruses... ."

In addition, we know that many virologists use the LHT system in their teaching.

B. CONS

The offensive first began at Cold Spring Harbor. During the discussion that followed the presentation of the system, Wildy (1962a) declared that the viruses form a heterogeneous collection of entities considered together by virtue of an arbitrary definition.

It is only necessary to refer to the title of the Marjory Stephenson Lecture "Concept of Viruses" (A. Lwoff, 1958) to realize that, for a long time, viruses were held as a concept. In nature, one encounters individuals, not species, types, or families. Nature does not know categories constructed by the human mind; nevertheless, there is nothing to prevent taxonomists from operating logically and rationally.

Let us state, first of all, that the virus group is no more heterogeneous than any other group. The animal kingdom and vegetable kingdom are heterogeneous. The vertebrates constitute a heterogeneous group, as do the mammals. A taxon, whichever it may be, is a gathering of various organisms, that is to say, a heterogeneous group; and the higher the taxon in the hierarchy, the greater the heterogeneity. The herring and man belong to the same taxon, as do the paramecium and the elephant. Do the viruses differ more among themselves than the various representatives of certain categories of the animal kingdom? The judgment arrived at will depends upon the idiosyncrasy of the individual. Some people are inclined to retain what separates, others what unites. We belong to the latter group. This is why the group of viruses appears to us to be remarkably homogeneous. At any rate, if others consider it to be heterogeneous, it is up to them to put an end to the controversy by professing an alternative definition of viruses that will be less arbitrary.

Gibbs wrote, in 1969, that the LHT classification resembles, as much in its principles as in its defects, the first classifications of plants and animals. Gibbs adds: "see review by Adanson (1763)." The same author, in the same article, also writes (p. 309): "There are many ways to organize groups, but most seem quite arbitrary and of little value. The hierarchy based on four properties of the virions, as proposed by LHT in 1962, is of this type, for there is no evidence that any of the properties used in their proposed hierarchy will cluster related groups of viruses."

It should be noted that, within the LHT system, only one category is defined by a single character, the others are defined by two, three, or four characteristics. But Gibbs forgets this, no doubt unintentionally.

It is on the strength of these four characters, the RNA, a naked capsid with cubical symmetry, and the 92 capsomeres, that the LHT system has united, in one group, what is today the family Reoviridae, the reoviruses, and the plant rumor viruses. Gibbs (1969) also states that these viruses are related, but omits mention that LHT arrived at the same conclusion 7 years earlier.

The LHT system has been criticized as being àrbitrary. As we have said repeatedly and will repeat again, all classifications are arbitrary in the sense that categories or taxons do not exist in nature. Categories and taxons are concepts, as are evolution, heredity, or allosteric interaction. And any hierarchy shares the arbitrary characteristic of the classification. To blame a classification or a hierarchy for its arbitrary characteristic, is like blaming a cube for not being spherical.

There is more. Virologists have recognized that a certain number of species present some affinities. The species have been grouped into genera. Genera showing common characters have been grouped into families. The four characters of the LHT system make it possible to define families. The opponents of the LHT system make a point of ignoring that this system makes it possible to group viruses according to their natural affinities. To dispute the value of the LHT system under these conditions is curious. In a more general way, to dispute the principles of the classification of viruses without proposing a classification and to pretend to ignore that viruses are already distributed into families, to say the least, is somewhat singular. It is evident that the LHT system is very embarrassing to those who affirm that it is impossible to classify viruses.

Bellett (1967b) has stated that the LHT system is not scientific. He has in fact adopted Poper's ideas concerning classifications—which we will now summarize. A scientific classification is based on a scientific theory that attempts to consider the properties of the entities and the distribution of these properties within the population. That is alright. Starting with these premises, Bellett affirms that the LHT classification does not pretend to be scientific, that it belong to a kind of system constructed merely to solve the practical problems of nomenclature and identification, and that, in fact, LHT did not take into consideration the natural affinities of the viruses. It is perfectly true that we did not assign a scientific character to our system —that would have been pretentious. We did not intend to solve problems of nomenclature and identification: in our publication (Lwoff *et al.*, 1962b), there is no question of nomenclature or of identification. Finally, here is what we wrote: "In other terms, we feel that the various viruses, when their essential integrants are established, will find their "natural" place in the system. By natural place, we mean they will fall in the same group as biologically related entities."

Therefore, contrary to what Bellett affirms, LHT were concerned with the natural affinities of viruses. After all, it matters little whether a classification is based on one principle or another. What is important for a classification of viruses is that it groups the viruses according to their affinities and, also, that it presents a clear synoptical picture of the viral world. If it succeeds in so doing, the classification will be right. If it does not succeed, it will be wrong, even if it is based on a so-called scientific theory.

IV. Phanerogram, Cryptogram, and Gymnogram

A. PHANEROGRAM*

Nothing prevents the name of a virus from being followed by an abbreviation corresponding to the four characteristics of the LHT system: D and R for DNA and RNA, C and H for cubical symmetry and helical, etc. —as was done by LHT in 1962. In this way, for each virus, we can establish a formula, which will be referred to as a phanerogram.

B. CRYPTOGRAM

In 1966, Gibbs, Harrison, Watson, and Wildy proposed the use of eight characteristics for identifying viruses: the type of nucleic acid, its molecular weight, the percentage of nucleic acid within the virion, the form of the particle, the form of the nucleocapsid, the host, and the vector. The various parameters are defined by abbreviations and the whole of the formula has been named cryptogram.

It is evident that phanerogram and cryptogram correspond exactly to the same approach, which is the selection of characters. It should be noted, however, that the International Committee for the Nomenclature of Viruses named only one "cryptogram commission." It is true that the characteristics chosen by LHT were not given Greek names.

C. COMPARATIVE VALUES OF THE CHARACTERISTICS OF THE PHANEROGRAM AND THE CRYPTOGRAM

1. *The Form*

It is now time to critically examine the value of the characters proposed by one or the other, and we will start with the form.

We can do no better than to cite the ideas expressed by Horne and Wildy (1961): "Size and shape of the virion have so far been the only morphological characters used to classify viruses. These attributes are both unreliable and misleading as criteria and we suggest that they be abandoned forthwith." The form of the virion is "misleading," add Horne and Wildy, because it results from the symmetry of the capsid and from the presence or absence of such structures as the envelope and tail.

---

* Phanerogram from phaneros, visible and from gramme, writing. The list of discriminative characters employed by LHT in their classification.

Our colleagues conclude by saying that we must abandon the form in favor of the symmetry of the capsid. The number of capsomeres will also be a useful characteristic for the classification of viruses with cubical symmetry. Finally, Horne and Wildy proposed a scheme of classification of viruses based on the symmetry of the nucleocapsid and on the nature of the genetic material. Nevertheless, when Horne, with Lwoff and Tournier, put into practice the principles he had formulated with Wildy himself, the latter, together with Gibbs, Harrison, and Watson abandoned his stand.

In fact, while LHT were in favor of the *symmetry* of the nucleocapsid, Gibbs *et al.* were utilizing its *form*. For purposes of definition, the virologist has the choice, in the cryptogram, between *essentially spherical*, on one hand, *elongated*, on the other hand (with some variations), and, finally, *complex*. In the example given by Gibbs *et al.*, they talk about a spherical nucleocapsid. However in 1969, Gibbs refers to isometrical nucleocapsids. Here, the reader is confused; since isometrical has two different meanings. In current language, isometrical means of equal dimensions: a sphere is isometrical. In the language of crystallography, isometrical is synonymous with cubical symmetry. When Gibbs talks of *isometrical* capsid, is he using current language or that of crystallography? Probably current language, since isometrical is opposed to anisometrical—but this is only an assumption. Anyhow, the cryptogram considers the form, not the symmetry.

2. *The Elongated Character*

Next to the spherical character of the nucleocapsid, we mention, in the cryptogram, the elongated nucleocapsid with parallel sides. It is likely that this category corresponds to the nucleocapsids with helical symmetry. Nevertheless, here again, the reader is confused. In fact, do we have the right to say that the nucleocapsid of the tobacco mosaic virus, for instance, has parallel sides? Gibbs *et al.* obviously thought of a cylinder since, in a cylinder, the hypothetical straight lines perpendicular to both terminal circles are in fact parallel; and while a nucleocapsid with helical symmetry is roughly cylindrical, it does not, in fact, have "parallel sides." Furthermore, "elongated" does not always correspond to helical symmetry. The elongated nucleocapsid of certain viruses has a cubical symmetry.

3. *The Envelope*

The phanerogram takes into consideration the naked or enveloped character of the nucleocapsid. In the cryptogram, one considers the form of the virion and that of the nucleocapsid, separately. But, the one and only difference that the virion has to offer in relation to the nucleocapsid is the presence of an envelope.

Now, let us suppose that a "spherical" nucleocapsid is surrounded by an envelope, which too is "spherical." It will be impossible, according to the formula of the cryptogram, to know if the nucleocapsid is naked or enveloped. No clear distinction has been made in the cryptogram between naked and enveloped nucleocapsids (Gibbs, 1969).

This being said, it becomes evident that in certain very precise cases the form of the virion is a useful characteristic. In this way, in our classification, the Paramyxoviridae (family including the measles and the mumps viruses) are separated from the family of the Stomatoviridae = Rhabdoviridae (which includes the rabies and the *Drosophila* sigma viruses) by the characteristic, whether spherical or bullet-shaped, of the envelope.

D. Other Characters of the Cryptogram

1. *The Molecular Weight of the Nucleic Acid*

This is undoubtedly an excellent character.

2. *The Percentage of Nucleic Acid within the Virion*

This character can be maintained, even though it does not seem to be of great usefulness in a classification. The percentage of nucleic acid within the nucleocapsid should be more significant.

3. *Hosts*

It is known that several viruses are liable to evolve in a great number of, sometimes quite remote, organisms. The characteristic of the host could, nevertheless, be useful at times.

4. *Vectors*

The existence or lack of a vector, and the nature of the vector, should not be utilized for defining categories higher than the species.

E. General Remarks

Examination of the comparative table of the phanerogram and the cryptogram (Table V) enables us to observe the similarities and differences. It is seen that the "form" characteristic of the cryptogram will, sooner or

TABLE V

PHANEROGRAM AND CRYPTOGRAM

Virus		Phanerogram	Cryptogram
Nucleic acid	DNA or RNA	+	+
	Single- or double-stranded nucleic acid	0	+
	Molecular weight	0	+
	Percentage in the virion	0	+
Virions	Presence or absence of an envelope	+	0
	Shape	0	$+^a$
Nucleocapsid	Symmetry	+	0
	Number of capsomers (cubical symmetry) or diameter (helical symmetry)	+	0
	Form	0	$+^b$
Hosts		0	+
Vectors		0	+

[a] The difference between virion and nucleocapsid is necessarily related to the presence of an envelope (see text).

[b] The shape of the nucleocapsid will evolve rapidly toward symmetry (see text).

later, evolve toward the "symmetry" characteristic of the phanerogram. We also note that the cryptogram characteristic "form of the virion" is equivalent to the phanerogram characteristic "presence or absence of an envelope." In approaching the phanerogram, the cryptogram will no doubt gain in precision. Is it really justified to present the cryptogram as an original model at the expense of the phanerogram?

F. GYMNOGRAM*

Under the heading of gymnoviruses, we designate the viruses without capsid and whose infectious phase is a naked nucleic acid (see Diener and Raymer, 1969). The gymnoviruses could be classified according to a certain number of characteristics:

1. Nature of the genetic material.
2. Number of strands.
3. Molecular weight of the nucleic acid.
4. Percentage of guanine + cytosine.

* Gymnogram from gymnos, naked. The list of characters proposed for the classification of naked viruses, without capsid.

5. Affinities such as result from the hybridization test.
6. Sequences of bases.
7. Number and nature of the viral proteins.

Each one of these characteristics will be represented by abbreviations and the formula will be called *gymnogram*.

V. Evaluation of Characteristics

In order to define taxons and classify viruses, should we use the "totality" of characteristics or, on the contrary be selective?

As mentioned before, LHT chose four characters. Gibbs *et al.*, 1966, write that for them it is not possible to determine the relative importance of the different characters. They conclude that it is essential to follow the principles of Adanson, who suggested "that all the data should be used and that all the characteristics should be considered to be of equal importance."

However, Mayr (1965a) says in his excellent article "Numerical Phenetics and Taxonomy": "It is thus clear that Adanson... did not propose that the taxonomist abandon his prerogative to evaluate the characteristics. Indeed, Adanson can be considered the father of the method of character evaluation as practiced in classical taxonomy. It is a gross injustice to Adanson to label the nonweighing of characters: the Adansonian method."

Mayr adds that "the careful weighing of the few available characters is an absolute necessity.... To dilute these few useful characters by large numbers of useless ones in order to acquire a false sense of quantitative security is a procedure that, quite rightly, was already ridiculed by Adanson, who was not that naïve."

Mayr finally adds that "the sooner the myth that it is a sin to weigh characteristics disappears, the better for the numerical method. The numerical method will prove most successful that solves the problem of weighing most efficiently."

Thus, Adanson was a supporter of evaluation. It is of little importance since, any way, we do not accept the dogma of the infallibility of Adanson. It is preferable to ask, objectively, where is the refusal to evaluate and the aforementioned phenetic or numerical method leading to, and where is the computer leading to? Sokal and Senath, the principal protagonists of the nonevaluation method, admit that the only thing that the computer can do is to determine the similarity among taxons. "The rest of the taxonomical task," Sokal and Sneath write (1963) "still needs the experience and the judgement of specialists in the field."

The evaluation, or the choice of a character, is, therefore, considered by Gibbs *et al.* to be arbitrary. What should be said, then, of the refusal to chose, of the arbitrary decision according to which all characters are of equal value?

As was rightly noted by Mayr (1965b) the pheneticist does not differentiate between the important characteristics and those without any value. The pheneticist ignores the existence of "marked discontinuities" among the groups. "The categories of the pheneticists are based on arbitrary levels of phenetic distances."

It is to be wondered, then, why the pheneticists do not want to attribute different values to the characteristics. Perhaps it is because they do not know how to assess their value. Perhaps, also, it is because the programming of information into a computer is easier if all the characters have the same value, that is, the same content of information. Whatever it may be, a method should be judged according to the results it affords.

Certain characters are, as we know, stable. One has never seen either the nature of the genetic material or the symmetry of the capsid modified by a mutation. One also knows perfectly well that other characteristics are unstable: the virulence is easily modified by mutations. All the characters, therefore, do not have the same value. To attribute the same value to all of them is to renounce reason and to sink willingly into error.

What can be expected of a machine that has been supplied with sixty characteristics of necessarily unequal values? We agree with Sokal and Sneath: "the computer will determine the similarities among taxons." However, since these similarities are founded on characters of unequal values, the numerical expression of similarities will not make much sense.

Therefore, Gibbs *et al.* consider that a classification must rely on at least sixty characteristics to which an equal value must be attributed. Let us see where the refusal to evaluate can lead to.

Virus A is an RNA virus with cubical symmetry.

Virus B differs from A by its virulence, a characteristic that is apt to vary under the influence of mutations.

Virus C differs from A by the nature of the nucleic acid, a stable characteristic.

Virus D differs from A by the symmetry of the capsid, an equally stable characteristic.

Therefore, B, C, and D differ from A, each of them by a single characteristic. Since we decided that all the characteristics have the same value, B,

C, and D are, all three of them, at the same phenetic distance from A. And since A, B, C, and D differ by only one of the sixty characteristics, they will in a *numerical* classification, be placed into the same group. In a *normal* classification, A and B will belong to the same family, C to a different family from A and D, and D to a different family from A and C.

We can go further and pose a few questions to the virologists.

A. It is reasonable to place within the same species, the same genus, or the same family, two viruses differing by:

1. The nature of the genetic material.
2. The symmetry of the capsid.
3. The number of capsomeres.
4. The naked or enveloped nature of the nucleocapsid?

B. Is it reasonable to bring together within the same species, the same genus or the same family, viruses differing by two, three, or four of the cited characteristics?

Once again, species, genus, and family are arbitrary categories, or concepts, in the same way as the taxon virus is a concept. To use the word group is an evasion and it does not alter in any way the problem. This being said, we are now awaiting an answer from the followers of nonevaluation.

It should be noted, however, that Gibbs, along with Harrison, Watson, and Wildy, having asserted categorically that all characteristics have an equal value and, no less categorically, that we do not have the right to assign a different value to them, suddenly discovered, in 1969, that certain of the characteristics are stable while others are not, and that the value of one and the other is different. This is a reversal of their previous stand.

Perhaps it would serve some purpose here, to examine the very concept of viruses as it presents itself in regard to the nonevaluation and the numerical method.

The virus concept was formulated with the help of a comparative analysis of the characteristics of the small infectious particles. It was recognized that some of them had a group of characteristics in common, that were absent in others. In view of these discriminative characters, the small infectious particles were separated into *viruses* and *nonviruses*. The virus concept was, therefore, based on a rational analysis and on the evidence of discriminative characters. The supporters of the numerical method will obviously say that the choice is arbitrary. Nevertheless, it does not prevent them from using the term virus in the sence given to it by a method we consider rational.

That is what Bellett (1967c) expressed: "I shall use the term 'viruses' to refer to a class of entities defined by the sum of the properties listed under 'viruses' in Table 4 of Lwoff 1968, that is, to infectious agents which contain one type of nucleic acid and do not grow and divide (as defined by Lwoff) or contain a Lipmann system."

It is obvious that one should define what one is talking about. However, some virologists discuss viral nomenclature and classification without saying what they mean by viruses: they are the irreducible followers of the numerical and the nonevaluation methods.

Mayr (1965b) insisted very strongly on the fact that the numerical taxonomist ignores the sharpness of discontinuities among taxons and that his categories are based on arbitrary levels of phenetic distances.

Let us assume that a supporter of the numerical method decides to classify the small infectious particles. Since he refuses to make a choice, he will use all the characteristics among which, of course, some are not discriminatory, as, for instance, dimensions, form, virulence, symptoms. He will thus arrive at the conclusion that, among viruses and nonviruses, there is a certain phenetic distance, but he will not be able to say in what way viruses differ from nonviruses. In order to classify viruses, therefore, the followers of the numerical method make use of principles that do not permit to define viruses. This is the death sentence of the numerical method and of the nonevaluation.

We can only define viruses by discriminatory characters. What is true for the taxon virus is true for hierarchically inferior viral taxons.

VI. Categories and Taxons, Nomenclature

A. NAMES, CATEGORIES, TAXONS

For certain primitive peoples, the name designating an individual possesses a magical value. To know the name of an individual, is to know him, himself. The name, insofar as it denotes one individual to the exclusion of all others, is a specific element combining itself to the body to constitute the total being. This is the *nama-rupa*, the "name and body" of the Buddhists.

Doubtlessly these propositions will not be wholly accepted, but it is quite obvious that a name can and must express a virus, without being necessarily "combined" to the virion. However, it is the properties of the virion that define its personality and make it possible to assign a name to it. We are not too far from the *nama-rupa*.

Names are symbols that make it possible to recognize objects and organisms. A nomenclature is a system of names. A biological classification is a distribution of organisms into categories of various hierarchical value. Phyla, classes, orders, families, genera, and species are categories. A category corresponds to a rank within the hierarchy.

When a classification is applied to clearly defined organisms, a category embraces a definite group of organisms united by reason of certain similarities: the category thus becomes a *taxon*. Bacteria, mollusks, mammals, and viruses are taxons.

Finally, let us add that taxonomy, which is etymologically the distribution of taxons, is the science of classification.

B. REMARKS ON THE TERM "GROUP"

A few virologists reject species, genus, and family, and use the word "group" exclusively. While species, genus, and family are categories of a definite hierarchical value, "group," or its English equivalent *cluster*, have no definite hierarchical value.

In fact, at times, the *group* unites strains and corresponds to the species; in other instances, it unites species and conforms to the genus; and at other times, it unites genera and is the equivalent of a family.

In any case, nature knows nothing but individuals, and any grouping, as we said before, and are repeating, and shall repeat again, is arbitrary and the group is no less arbitrary than the species, the genus, and the family. Its sole interest is the lack of precision.

C. RULES OF NOMENCLATURE

Since it is necessary to know of what we speak, there is no classification without nomenclature. It is necessary, here, to recall certain rules.

We will give the essential points of the present international rules, as were established by the International Committee for Nomenclature of Viruses (INCV). First, it should be noted that, since these rules apply to species and to genera, it was necessary to define the species and genus.

1. The species groups identical viruses.
2. The genus is a group of species having common characteristics.
3. The name of the genus must terminate in the suffix *virus*.

The INCV decided that an effort should be made toward binomial Latin nomenclature. By adding the suffix *virus*, the names of the genus are automatically Latinized. The recommendation of the Committee is lawful as regards the genus. It is noted that the termination corresponds to usage. Such names as *Adenovirus*, *Myxovirus*, and *Poliovirus* have long been in use.

4. Each genus must have a type species. This is an indispensable measure in order to assure the stability of the name of the genus and in order to avoid interminable and irrelevant discussions. The name of the genus will always remain linked to the type species, whatever classification is adopted.

5. The Committee did not define families that are and can only be groups of genera.

The names of families must terminate in *idae*.

For Bellett, a nomenclature does not need to be based on a classification. That is correct as regards the nomenclature of individual viruses. However, the uniting of strains into species and of species into genera are acts of classification. The function of binomial nomenclature is to give a name to genera and species, and no scientific nomenclature is possible in the absence of this classification, which is elementary but essential.

Gibbs *et al.* (1966) consider that "a Latin binomial system is based rigidly on chosen characteristics in order to determine the form and the hierarchy of taxons: division, order, family, etc."

Divisions, orders, and families, as said before, are categories and not taxons. Now, the name of the genus and that of the species is not based on characters selected in order to determine the form or the hierarchy of categories above the genus. *Herpesvirus simplex* is the name of one of the viruses in the "herpes group." This name designates a specific organism and it will be the same, whichever classification of viruses is chosen. The names of the genus and species, are, once again, independent, totally independent, of characteristics used to define categories of a hierarchical order above the genus. We see no foundation for the statement of Gibbs *et al.*

D. CLASSIFICATION AND NOMENCLATURE

Not only do virologists often confuse category and taxon, but they also confuse classification and nomenclature. For example, Gibbs *et al.* (1966) attributed to Lwoff *et al.* (1962b) the authorship of a nomenclature code. Here is what LHT had said: "...that the binary nomenclature might be applied to viruses is subject to controversy. The problem will not be discussed here." This is the only reference to nomenclature in this paper.

The persistent confusion between nomenclature and classification is really curious.

E. Opposition to Binomial Nomenclature

Among the opponents of binomial nomenclature are Gibbs *et al.*, 1966, who write that the nomenclature of viruses must be based on a general classification of viruses established with the use of Adanson's principles. A nomenclature can only be international. Now, the committees on nomenclature are not competent in the matter of taxonomy. An international classification of viruses does not, will not, and cannot exist. To aspire to found an international nomenclature of viruses on a classification that will never exist is to hinder, forever, any international nomenclature. Furthermore, who has the right to decide if Adanson's principles have the force of law.

Gibbs (1969) once again joined the battle against Latinized binomial nomenclature, declaring it to be retrograde. One of the arguments is that not one international periodical publishes the articles on viruses in Latin. Nevertheless, plant virus specialists use Latinized binomial nomenclature for the designation of hosts and vectors. And does there exist one international periodical devoted to plants and arthropods that is printed in Latin?

The other argument is that binomial nomenclature is based on the species; that is true. However, if the viral species are refused, the nomenclature of viruses will become a catalog of strains.

Let us add that the International Committee on the Nomenclature of Viruses has accepted the genus and the species, that the name of the genus is already Latinized, and that the virologist must comply with the international rules in force.

Fenner (1966) also joins the war against latinized binomial nomenclature. "The use of Latinized binomial names," he writes, "is not successful and should be discontinued." There are two different targets for Fenner's condemnation: binomial nomenclature, on the one hand, and Latinization of the other. Since Fenner accepts neither species nor genus, binomial nomenclature cannot have any sense for him, naturally. We note, however, that Fenner has used Latinized names for "groups": myxovirus, adenovirus, reovirus, etc. Since in each one of these groups there are subgroups, and since in each of these subgroups there are infra subgroups, it can be presumed that Fenner's groups are families, subgroups genera, and infra subgroups species.

Fenner obviously forgets that Burnet (1967) furnished a remarkable review of the Poxviridae group and used a Latinized binomial nomenclature: *Poxvirus variolae*, *P. officinalae*, *P. bovis*, *P. avis*, and *P. myxomatis*. These are excellent names. One cannot see in what way they are inferior to the nomenclature used by Fenner in 1968: group of poxviruses, subgroups of vaccinia, infra subgroups variolae."

Poxviridae and *Poxvirus variolae* appear to be much simpler and also easier to understand. Furthermore, with the suffix, one knows immediately which hierarchical category is concerned. Besides, binomial nomenclature has been used for a long time. When one says *Poliovirus I*, one makes use of binomial nomenclature whether one wishes to or not. *Poliovirus* designates the genus and I the species, that is to say "groups" of different hierarchical value.

F. COMPETENCE OF INTERNATIONAL COMMITTEES FOR NOMENCLATURE

In order to settle the problems of nomenclature, zoologists, botanists, microbiologists, and virologists created international committees for nomenclature. The competence of the latter is nomenclature. Taxonomy, grouping of species into genera and genera into families, grouping families into orders, etc., is left to the decision of each. Biologists have recognized that a classification is necessarily arbitrary and that each and everyone of us must be free to propose a system of his own choosing. No international body of nomenclature has the power to legislate in matters of taxonomy. This rule applies to the International Committee for the Nomenclature of Viruses.

However, in his excellent book: "The Biology of Animal Viruses," Fenner writes that the International Committee for the Nomenclature of Viruses rejected the classification suggested by the Provisional Committee for Nomenclature. Note: (1) that the classification suggested by this Committee is the LHT system; (2) that the International Committee does not have to accept or reject the LHT classification; (3) that if it had rejected it, the decision would have been null and void, because it is not competent in matters of taxonomy.

It is advisable that virologists wishing to discuss taxonomy and nomenclature should first assimilate a few elementary principles with regard to these disciplines. This would avoid considerable confusion.

VII. Lanni's System

Since viruses are reproduced from their sole genetic material, the whole virus is determined by its nucleic acid. The genotype of the virus is its genetic material. However, what is important in the nucleic acid, is not its nature, its molecular weight, the percentage of purine and pyrimidine bases, or the patterns of doublet frequency. The major element is the information in the biological sense, that is, the sequence of codons responsible for the structure of the proteins.

Therefore, Lanni proposed creating a system based on the sequence of the codons. He called his system: molecular. Here is what we had to say about this system (Lwoff and Tournier, 1966, p. 61).

> Let us now examine Lanni's "molecular" system. Since viruses are reproduced from their genetic material, it follows that the whole virus, including the virion, is determined by the base sequence of its nucleic acid. Provided the code is entirely deciphered,—as today it is—knowing the base sequence of the structural genes means knowing the amino acid sequence of the proteins. Yet, for the time being, it is impossible to deduce the tertiary and quaternary structures of a protein from its amino acid sequence. Moreover, the architecture of a protein may be modified by various ligands. Of course, we all hope that in a not-too-distant future, the knowledge of the primary structure of the viral proteins will permit the deduction of the symmetry, size, and organization of the virion. This is not yet the case.
>
> Finally, let us assume that we are able to reconstruct the phylogeny of a given virus. Mutations have lead to substitution of amino acids, and to antigenic alterations. The question is immediately raised: how many amino acid substitutions will be needed in order to consider that we are dealing with a new species or a new genus? Not only the number of substitutions will have to be taken into account, but also their nature. If a few hundred amino acids belonging to twenty species are involved, one can foresee great battles.
>
> In Lanni's system, the base sequence of the nucleic acid is selected because it determines the properties of the virus and of the virion itself. Yet, if it is admitted that the most convenient classification of viruses is based on the virion, is it not simpler to consider the virion than the base sequence? We are afraid that, at least for a few years, it will be easier to state that a virion possesses DNA, a cubical symmetry, that to consider a list of 10,000 nucleotides. Lanni's system is a statement of inapplicable principles, and what virologists need is a real system.

In 1969 Gibbs also adopted the idea of virus classification based on the nucleic acid and arrived at a conclusion similar to ours, although neither Lanni nor Lwoff *et al.* were mentioned in this report (Gibbs, 1969).

VIII. Bellett's System

A. THE SYSTEM

Bellett (1967a,b,c, 1969) proposed classifying the viruses on the basis of the molecular weight of the nucleic acid and the percentage of guanine + cytosine. Since it is impossible, in this system, to compare usefully the single- and double-stranded viruses, the viral world has been separated into two parts, according to whether the nucleic acid had one or two strands. Bellett rightly insisted on the fact that his classification should be considered a preliminary guide.

The data are fed into the computer. The results of the calculations are, on the whole, remarkable (see Figs. 1 and 2). However, it should be noted that isolated data provided by the computer will sometimes give rise to conclusions judged unacceptable by Bellett himself. For instance, RNA viruses, like the reoviruses, fall within the same group as the DNA viruses, like the group of papilloma viruses. In fact, Bellett accepted the results furnished by the computer only when they were in harmony with the data relative to the phenotypic characteristics of the virion.

That is why the Shope papilloma virus and coxsackievirus 10, for instance, which the computer had grouped together, were separated. Bellett considers, and quite rightly, that viruses that differ in their phenotypic characteristics cannot be grouped together. Let us add finally that, in certain cases, Bellett

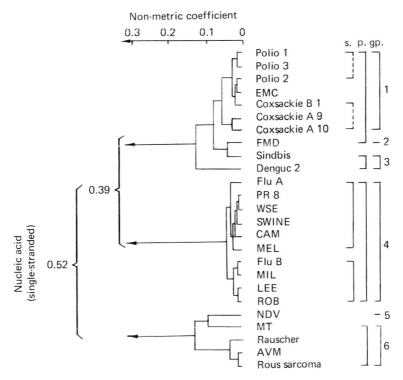

FIG. 1. Bellett's classification. Single-stranded viruses. S, Cross-breeding serological reactions; P, similar phenotypic properties. [With the author's permission, Bellett (1967c).]

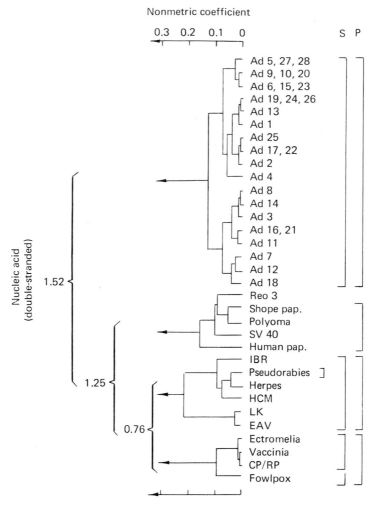

FIG. 2. Bellett's classification. Double-stranded viruses. S, Cross-breeding serological reactions; P, similar phenotypic properties. [With the author's permission, Bellett (1967c).]

used cross-breeding, which makes it possible to estimate the degree of homology of the nucleic acids.

Bellett's method, *supplemented by the utilization of phenotypic characters*, enables to rediscover fifteen of the "groups" of the LHT system. It should be recalled that the Provisional Committee for Nomenclature of Viruses assigned the value of families to these groups.

Within Bellett's classification, groups are defined by coefficients of similarity; that is the major defect of classification arising from the use of a computer. The computer is able to recognize that some viruses are related. But that will not suffice for the virologist, who will want to know the phenotypic characters of the virions. A mammal is not defined according to the molecular weight of its nucleic acid and according to the pattern of the doublets. It is defined by its phenotypic characteristics, which are and continue to be the basis of any classification.

Bellett then suggested that the theoretical basis for the classification of viruses could be furnished by the fact that its properties are determined by the base sequence of its nucleic acid. It is quite obvious, in fact, since viruses are reproduced from their genetic material, that the phenotype of the virion is determined by the genotype. It is noted, incidentally, that Bellett does not seem to be acquainted with Lanni's paper or with our remarks concerning this subject, which are cited here.

It is obvious that the similarity of the sequence of the codons will be important for establishing the phylogenic relationships of viruses. However, what is important in the genotype, is its expression. It is to be wondered whether the phenotype of the virion does not also furnish a major document on the functional value of the genetic information. These remarks in no way minimize the importance of Bellett's method or the results that he has obtained—and which confirm the validity of the LHT system.

B. NUCLEIC ACIDS VERSUS CAPSIDS: UROVIRUSES

The viruses provided with a "tail," whatever the structure and the dimensions of this appendage, are called uroviruses. Bellett's method, leads to classify uroviruses into three groups, that include, respectively:

1. Phages T3 and T7.
2. Phages lambda P22, T1 and phi 80.
3. Phages T2, T4, and T6.

In Bellett's index (1967c), groups 1 and 2 are related, but groups 2 and 3 are separated by a whole series of viruses: pseudorabies virus, herpesvirus, etc. This would mean that phage T3 is more closely related to herpesvirus than phage T2. The nucleic acids of these phages, as noted by Bellett, differ more by their molecular weight than by the percentage of their nucleic bases. Bellett, quite rightly, considered the grouping as temporary. Nevertheless, the molecular weight of DNA of two uroviruses can present great differences. At this point a discussion on the origin of bacteriophages would be appropriate.

Most virologists think that uroviruses have originated from bacterial

DNA. In fact, there are some common sequences in the genetic material of certain bacteriophages and in the bacterial chromosome. These are responsible for the coupling of temperate bacteriophages with the bacterial chromosome. It can be accepted that the procaryotic protista represent a monophyletic group. The Schizomycetes, however, have undergone considerable diversification, as much from the physiological as from the morphological point of view. The genetic material itself, has also evolved as evidenced by the variations of the percentage of guanine + cytosine.

Let us consider the hypothesis that uroviruses came into being late during the evolution of procaryotes. It is not difficult to accept that the tails of the uroviruses, in spite of their morphological and functional diversity, are homologous organelles. This tail represents a remarkable and unique structure. It is, therefore, possible to think that uroviruses probably originated from homologous sectors of DNA in bacteria. Homologous, but nevertheless, potentially capable of being diversified during evolution. Therefore, it is not surprising that the genetic material of phages exhibits some important differences. However, we repeat, the tail of uroviruses is an original organelle.

Within LHT's classification, all uroviruses are brought together in a group to which the Provisional Committee of Nomenclature of Viruses has assigned the value of an order, called Urovirales.

Should the need arise, it would be permissible to include in this order all the families whose creation is judged to be useful. It is not that, however, that is important, but the fact that all the uroviruses are grouped together into a single taxon.

It is not normal to see two groups of uroviruses separated by a series of animal viruses. Here, the evaluation of distances by numerical methods must yield to common sense: the tail transcends the nucleic acid.

A general remark is necessary here. The tail of urophages is a constituent of variable complexity. If one were to give a general definition of it, one could say that it is an organelle that is fixed to the nucleocapsid itself and enables the transfer of the nucleic acid through the bacterial wall. Naturally, it is to be wondered, how the viruses, other than urophages, can possibly inject their nucleic acid into the cytoplasm of the bacteria.

In the phage FD, the capsid contains two different proteins. One is represented in numerous examples and corresponds to capsomeres. The other one is represented by a single molecule that is responsible for the specific attachment on the *pili F* and assures penetration of the nucleic acid into their core from where it will reach the cytoplasm.

In addition to capsomeric proteins, there exists in certain RNA phages of cubical symmetry a single protein molecule which could play an analogous role. It is possible to envision a hypothesis, according to which the cycle of parasitic viruses of bacteria always require the presence of a specific structure responsible for the penetration of the nucleic acid into the cytoplasm. It is clear, however, that the single molecule of the phage FD can, under no circumstances, be considered as homologous to the tail or urophages.

IX. Miscellaneous Remarks

A. SYMMETRY

Certain capsids have a helical symmetry, others a cubical symmetry. For the sake of convenience, here, the corresponding capsomeres will be designated by "H" capsomeres and "C" capsomeres, respectively.

The structure of certain capsomeres thus involves a helical capsid architecture. Nothing is known concerning the tertiary structure that affects this property, except that it is controlled by the primary structure of the protein. Let us consider the facts.

We are now able to say that, within the primary structure of various "H" proteins, there are such sequences where the tertiary structure of capsomeres will force them to be arranged in helical sequences. The problem is to know whether or not the determining tertiary structures of all "H" capsomeres are bound to a common sequence of amino acids, that is, if the "H" symmetry of the capsomeres is bound to sequences of isosemantic codons on the structural gene of the capsomeric protein.

The same problem can arise as regards capsomeres with cubical symmetry or their subunits. The problem of the determinism of the specific properties of "H" and "C" capsomeres is not solved but merits consideration, since the hypothetical common sequences will not be found unless they are looked for.

B. RNA AND DNA VIRUSES

A virus may have evolved from a given sector of DNA within the host cell. It could also very well have derived from the corresponding RNA messenger that contains the same information, both qualitatively and quantitatively.

Therefore, theoretically, different viruses could have originated from nucleic acids of different but complementary nature. If this argument has any validity, then there should exist viruses whose phenotype would differ solely by the nature of the genetic material. Therefore, it would not be absurd to investigate whether the various nucleic acids of the viruses possessing an identical type of capsid may or may not be hybridized.

The Parvoviridae and the Napoviridae, as they are defined in our system, have a naked nucleocapsid with cubical symmetry and 32 capsomeres. Some are DNA and others are RNA. Hybridization merits consideration.

Should it evolve that viruses with an identical capsid and a different nucleic acid present complementary nucleotide sequences, it would be advisable to review the classifications and, in particular, to reconsider within the LHT system the hierarchical value of the nature of the nucleic acid.

C. SINGLE- AND DOUBLE-STRANDED VIRUSES

The nucleic acid of viruses is either single- or double-stranded. It is instructive to discuss the problem of the origin of these two types.

Viruses, as is generally accepted, derive from the nucleic acid of their host. RNA viruses could, thus, possibly have their origin in the RNA messenger. As a rule, the latter is double-stranded. It so happens that the RNA viruses which have been investigated, with the exception of the *Reoviridae* and the *blue tongue virus*, are single-stranded and thus correspond to the characteristics of the messenger. It is permissible to assume that the double-stranded RNA viruses originated as a consequence of an alteration of the replication system of the RNA.

The DNA viruses could derive from a segment of the DNA of their host. The great majority of DNA viruses are double-stranded, which corresponds to the characteristics of the genome. Thus, only the Inoviridae (filamentous phages), the Microviridae (ϕX174 group), and the Parvoviridae (Kilham virus and "minute virus of mice" group) are single-stranded and each one of these groups includes a very small number of species. It is permissible to assume that the single-stranded DNA viruses originate as a consequence of an alteration of the replication system of the DNA.

Although completely justified by practical considerations, the division of viruses, as practiced by Bellett, into two groups, one single-stranded and the other double-stranded, does not seem to correspond to the phylogeny of viruses. We do not think that single-stranded DNA viruses are more closely related to single-stranded RNA viruses than to double-stranded DNA viruses.

D. The Frequency of the Doublet Patterns

It is possible to characterize a nucleic acid in terms of the average frequency of the nucleic base doublets. This method was applied to the viruses (Subak-Scharpe *et al.*, 1966; Subak-Scharpe, 1967; Hay and Subak-Scharpe, 1968).

Nine mammalian viruses were studied. It happens that the doublet pattern of four small oncogenic DNA viruses, the SV40, polyoma virus, Shope papilloma virus, and human papilloma virus, bears a marked resemblance to the DNA patterns of mammals. Five large nononcogenic viruses, the herpesvirus, pseudorabies, equine abortion, vaccinia, and adenoviruses do not show this resemblance.

Bellett (1967b) studied the frequency of the doublet patterns. In this respect, the SV40 virus is closely related to the hamster. In the same group as the polyoma viruses, we find: man, rabbit, chicken, ox, mouse, and salmon. Bellett concludes that the similarity of the frequency of doublet patterns does not prove that two entities are phylogenetically closely related. In fact, it appears that the results of the pattern study should be interpreted with caution.

The identity of the code has given rise to the conclusion that the living kingdom is monophyletic. The vertebrates appear to descend from a relatively homogeneous group. But, if man and the salmon present closely related patterns, one can reach only to one conclusion: similar patterns are compatible with considerable phenotypic differences, that is, with marked functional differences of the genetic information.

In conclusion, therefore, it can be said that the doublet frequency may reveal a phylogenic relationship. It does not give any information on the possible structural similarities of viruses. But this should not, in any way, prevent us from provisionally accepting the conclusions of Subak-Scharpe and Bellett: viruses of the papilloma group probably originated in animals closely related to their present mammalian host, whereas the herpesviruses and the poxviruses could have originated from bacteria.

E. Doublet Frequency and Selection

The idea has been defended (Subak-Scharpe *et al.*, 1966) that the transfer RNA population belonging to an organism is adapted in an optimal way to the translation from the codon sequences of the RNA messenger.

It appears to be perfectly clear, in fact, that maximum economy must be achieved when the proportions of each of the transfer RNA's correspond to the frequency of incorporation of the corresponding amino acid into the proteins.

A virus, whose "pattern" is similar to that of a given organism, would therefore have its origin in an organism closely related to that organism. Viruses in which the pattern is different from the host's would be extrinsic. The pressure of selection would favor the viruses whose pattern is closely related to the host. Here, Bellett (1967c) has adopted Subak-Scharpe's conclusions. However, it should be noted that the vaccinia virus, despite the marked difference of pattern, multiplies quite well in mammals. It does not appear to have been selected against; quite the contrary.

X. Gibbs' Classification

Gibbs and Harrison (1968) reported that their classification method rediscovers several "intuitively constituted" groups of viruses.

The figures on the left-hand side of Fig. 3 correspond to the clusters in Table IV of Gibbs (1969). He took into consideration the shape, isometric or anisometric, of the capsid, the mode of transmission, the type of vector, the symptoms, and the accessory particles. The isometric particles of 25 to 30 mμ are divided into "rounded, angular, and squashy." In groups 1, 3, 4, and 5 we find viruses of both cubical and helical symmetry. It is impossible to know from the table whether the given virus is an RNA or a DNA virus, whether the symmetry of the capsid is cubical or helical, what is the number of capsomeres of the viruses with cubical symmetry or the dimensions of the viruses with helical symmetry. Gibbs' classification, could be cited as an example of nonfigurative virology.

XI. Classification of the Classifications

> *Moreover, some sort of arrangements is inescapable for organizing our facts about the bewildering diversity of living things; and a nomenclature is equally essential as a quick and accurate method of reference and communication.*
>
> A. J. Cain (1962)

In discussing the taxonomy of viruses, Pirie (1962) came to the conclusion that the first and essential step of a classification pertains to aes-

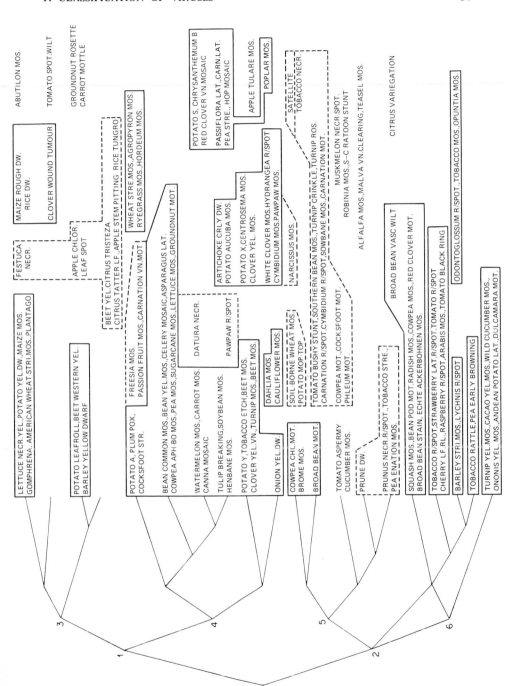

FIG. 3. Classification of plant viruses (modified from Gibbs, 1969).

thetics. It is a fact that the logic, symmetry, equilibrium, and harmony of a system pertain to aesthetics.

Now, let us try to fit the LHT system into the classification—necessarily arbitrary—of the classifications. For this purpose, we will follow closely the excellent article of Heslop-Harrison (1962). An artificial classification uses a small number of discriminatory criteria for subdivisions. A natural classification is based on the general resemblance and the maximum correlation of attributes. At first sight, since only the number of criteria is at stake, there is merely a difference of degree between artificial and natural classification.

In an artificial classification, when we use a small number of characteristics, each and everyone of the classes, regardless of its hierarchical level, is defined by the possession or absence of a single attribute. Each class, at each level, will be defined by attributes common to these elements. No individual will be "tolerated" in a category if it does not possess all of the diagnostic attributes of this category. "This," says Heslop-Harrison, "is an essential property of all logically constructed artificial classifications." The LHT system conforms to the definition of artificial classifications.

Within a natural classification, two individuals belonging to a group have more characteristics in common than an individual from this group has with an individual from another group. If a category was defined by the common possession of one or several characters, the category could be defined by these properties and the classification would then be artificial.

Within a numerical system of classification of viruses, since all of the characteristics have the same value, one may find within one family, or within the same group, two viruses that differ only by a single and unique characteristic, although it could be the nature of the genetic material or the symmetry of the nucleocapsid.

Naturally, the objection could be raised that, in a well-constructed numerical taxonomy, all of the viruses having the same nucleic acid, the same capsidal symmetry, etc., will be included in the same "group," species type, or family. If such were the case, this character could be used for the classification, which would then become "artificial."

As regards the taxonomic hierarchy of an artificial system, as well as the natural system, it has no bearing on phylogeny. As Heslop-Harrison writes, nobody would be so naïve as to think that taxons can be identified directly with the successive order of branching of a phylogenic tree. Heslop-Harrison adds that a state of chronic confusion exists between taxonomic tables and family tree. A classification, even a natural one, is in no way synonymous with phylogenic classification.

An artificial classification is the only one that can impart to the viral world an overall view that would enable us to visualize the structural and chemical characteristics of viruses. Since the architecture of the virion is controlled by the genetic information, the LHT system provides a picture of the functional value of the genetic information, that is, of the genotype.

XII. Conclusions

To reject the concept of viruses on the ground that the group is heterogeneous, to refuse any category and any hierarchy on the ground that there is no such thing in nature, to refuse a binomial nomenclature on the pretext that nature does not discriminate between species or genus, to systematically use the word "group," which does not correspond to any hierarchically defined category, to refuse discriminative characters on the ground that the value of characters cannot be ascertained, to accept, in spite of the evidence to the contrary, that all of the characters are of equal value, to feed these characteristics into a machine that will determine arbitrary distances, to claim that a table of coefficients of similarity is a classification, to cultivate imprecision and disorder, to call arbitrary any attempt at a universal, logical, and coherent system, to impose a Middle-Aged confusion on virology, this is what we are reproaching to the followers of the numerical method. A category exists only by the virtue of definition. A hierarchy implies a choice, a nomenclature demands a convention, a concept is the fruit of reasoning.

Science is a system of relations among facts and a body of concepts. To call arbitrary the logical procedures that are the very foundation of science is an antiscientific attitude.

REFERENCES

Anonymous. (1965). *Ann. Inst. Pasteur* **109**, 625–637.
Anonymous. (1966). *J. Nat. Cancer Inst.* **37** (No. 3), 395–397.
Anonymous. (1968). *Nature (London)* **220**, 650.
Bellett, A. J. D. (1967a). *J. Gen. Virol.* **1**, 583–585.
Bellett, A. J. D. (1967b). *J. Mol. Biol.* **27**, 107–111.
Bellett, A. J. D. (1967c). *J. Virol.* **1** (No. 2), 245–259.
Bellett, A. J. D. (1969). *Virology* **37**, 117–123.
Cain, A. J. (1962). *In* "Microbial Classification." Twelfth Symposium of the Society for General Microbiology, Cambridge Univ. Press, London, 1–13.
Cohen, A. (1969). "Textbook of Medical Virology." Blakwell, Oxford.

Crawford, L. V. (1966). *Virology* **29**, 605–612.

Davis, B., Dulbecco, R., Eisen, H., Ginsberg, H. S., and Wood, W. (1968). "Microbiology." Harper & Row. New York.

Diener, T. O. and Raymer, W. B. (1969). *Virology* **37** (No. 3), 351–366.

Fenner, F. (1968). "The Biology of Animal Viruses." Academic Press, New York.

Fenner, F. and Burnet, F. M. (1957). *Virology* **4**, 305–314.

Frobisher, M. (1968). "Fundamentals of Microbiology," 8th ed. Saunders, Philadelphia, Pennsylvania.

Gibbs, A. J. (1969). *Advan. Virus Res.* **14**, 263–328.

Gibbs, A. J., and Harrison, B. D. (1968). *Nature (London)* **218**, 927–929.

Gibbs, A. J., Harrison, B. D., Watson, D. H., and Wildy, P. (1966). *Nature (London)* **209** (No. 5022), 450–454.

Hay, J. and Subak-Scharpe, H. (1968). *J. Gen. Virol.* **2**, 469–472.

Heslop-Harrison, J. (1962). *In* "Microbial Classification." Cambridge Univ. Press, London.

Horne, R. W. and Wildy, P. (1961). *Virology* **15**, 348–373.

Hull, R., Hills, G. J., and Markham, R. (1969). *Virology* **37**, 416–428.

Kurstak, E. and Cote, J. R. (1969). *C. R. Acad. Sci. Paris* **268**, 616–619.

Luria, S. E. and Darnell, J. E. (1968). "General Virology," 2nd ed. Wiley, New York.

Lwoff, A. (1957). *Jl. Gen. Microbiology* **17**, 239–253.

Lwoff, A. (1969). *Bacteriol. Rev.* **33**, 390–403.

Lwoff, A. and Tournier, P. (1966). *Annu. Rev. Microbiol.* **20**, 45–74.

Lwoff, A., Horne, R. W., and Tournier, P. (1962a). *C. R. Acad. Sci. Paris* **254**, 4225–4227.

Lwoff, A., Horne, R. W., and Tournier, P. (1962b). *Cold Spring Harbor Symp. Quant. Biol.* **27**, 51–55.

Mayr, E. (1965a). *Systematic Zool.* **14** (No. 2), 73-97.

Mayr, E. (1965b). *Amer. Zoologist* **5**, 165, 174.

Pirie, N. W. (1962). *In* "Microbial Classification." Cambridge Univ. Press, London.

Rose, J. A., Hoggan, M. D., Koczot, F., and Shatkin, A. J. (1968). *J. Virol.* **2**, 999–1005.

Sneath, P. H. A. (1962). *In* "Microbial Classification." Cambridge Univ. Press, London.

Sokal, R. R. and Sneath, P. H. A. (1963). "Principles of Numerical Taxonomy." Freeman, San Francisco, California.

Subak-Scharpe, J. H. (1967). *Brit. Med. Bull.* **23** (No. 2), 161–168.

Subak-Scharpe, J. H., Burk, R. R., Crawford, L. V., Morrison, J. M., Hay, J., and Keir, H. M. (1966). *Cold Spring Harbor Symp. Quant. Biol.* **31**, 737–748.

Wildy, P. (1962a). *Cold Spring Harbor Symp. Quant. Biol.* **27**, 55.

Wildy, P. (1962b). *In* "Microbial Classification." Cambridge Univ. Press, London.

Note Added in Proof: We have learned that densonucleosis virus has a complementary single-stranded DNA, separately encapsidated. The MW of this DNA is 1.6×10^6 daltons (Kurstak *et al.*, *C. R. Acad. Sci.*, *Paris* **272**, 762, 1970), or 2.2×10^6 daltons (Barwise and Walker, *FEBS Letters* **6**, 13, 1970).

CHAPTER 2 *Small DNA Viruses*

M. DAVID HOGGAN

I. Introduction

The first of a rapidly growing list of small (18–25 nm) deoxyribonucleic acid (DNA)-containing viruses known as parvoviruses (Andrews, 1970) was first isolated about 10 years ago by Kilham and Olivier (1959) and was, at that time, thought to be a member of another group of slightly larger (35–45 nm) DNA viruses known to be tumorigenic, called papova–viruses. It was called rat virus or RV since it had been isolated in rat cell culture from a number of rat tumors. This was soon followed by the isolation of a similar but serologically distinct virus (H-1) from fractions of transplantable human tumors passed into rats and injected into newborn hamsters (Toolan *et al.*, 1960).

Since that time, similar viruses have been isolated from many diverse hosts. These include the hemadsorbing enteric virus (Haden) from cattle (Abinanti and Warfield, 1961; Storz and Warren, 1970); feline panleuko-penia virus of cats (Johnson, 1965; Johnson *et al.*, 1967); mink enteritis virus (Burger *et al.*, 1963; Johnson *et al.*, 1967; Gorham *et al.*, 1966); minute virus of mice (MVM) (Crawford, 1966); adeno-associated viruses (AAV) from humans and simians (Blacklow *et al.*, 1967b); porcine parvo-virus (Mayr and Mahnel, 1966; Mayr *et al.*, 1967, 1968; and Horzinek *et al.*, 1967); minute virus of canines (MVC) (Binn *et al.*, 1968); and the densonucleosis virus (DNV) of the insect *Galleria mellonella L.* (Meynadier *et al.*, 1964).

Although the bacteriophage ϕX174 (Sinsheimer, 1959) will not be con-sidered in detail in this report, it should be pointed out that it shares a number of properties with the other small DNA viruses under consideration (Payne *et al.*, 1964; Mayor and Melnick, 1966).

Since a number of the small DNA viruses have been the subject of recent reviews (Toolan, 1968; Rapp, 1969; and Hoggan, 1970) no attempt will be made to give an exhaustive survey of the literature, but I do hope to consider some of the early, as well as more recent findings, on these agents and show how these agents, even though quite distinct, have many common properties even though they come from widely divergent phylogenetic hosts.

II. Classification and Nomenclature: General Considerations

The detailed classification of the small DNA viruses has as yet not been finalized, but the generic name, parvovirus (Anonymous, 1965), has been approved by the Executive Committee of the International Committee on Nomenclature of Viruses (Andrewes, 1970), while the name, picodnavirus (Mayor and Melnick, 1966), has not.

Although no final classification system has been approved, many features of these agents are obvious to workers in the field. These include the follow-ing general characteristics: They contain DNA; they show icosahedral symmetry with a mean diameter between 15 and 30 nm; they are resistant to high temperature (e.g., 56°C for 30 minutes); they contain no essential lipid since they are resistant to ether and chloroform; and all show a high relative specific density in CsCl (1.38–1.46 gm/cm^3). Many of the basic ideas relating to the classification emanated from the report of the Provisional Committee for the Nomenclature of Viruses with Sir Christopher Andrewes as President (1965) and from a working subcommittee of the nomenclature of these small DNA viruses which is currently studying the problems and is assigned to make recommendations. I wish to emphasize that the thoughts and ideas presented here have in fact been influenced by the deliberations of this committee, but this report should not be construed as an official report by this committee, but is only an attempt by the author to better understand some of the interesting facts about these viruses and present a comparative study of these viruses in the light of current knowledge.

Some of the viruses considered have as yet not been sufficiently studied to be unequivocably placed into the parvovirus group. I wish to thank those who provided me with preprints and unpublished data, including Drs. Robert W. Atchison, L. Crawford, and J. Storz.

I should also point out that the excellent attempt to classify all vertebrate viruses by Wilner (1969) also influenced my thinking in approaching this present communication. It should be emphasized that the presentation will be somewhat biased and limited with most of the comparisons being made between the AAV and other potential members of the group. The author's personal laboratory experience has been limited to the four known types of AAV (Atchison et al., 1965; Hoggan et al., 1966a,b; Parks et al., 1967b), the prototype member of the parvovirus group, RV (Kilham and Olivier, 1959), H-1 (Toolan et al., 1960), Haden virus (Abinanti and Warfield, 1961; Storz and Warren, 1970), DNV (Meynadier et al., 1964), and MVM (Crawford, 1966).

III. Specific Properties of Various Parvoviruses and Parvovirus Candidates

A. Biological Properties

The biological properties of the various candidate members of the parvovirus group of viruses are not the primary consideration of this report and the interested reader is referred to the recent reviews by Toolan (1968), Rapp (1969), and Hoggan (1970). There are, however, a number of biological properties that should be considered in comparing these viruses. Perhaps the best known property of the RV and the H viruses which first brought these viruses to the attention of laboratory workers, is the osteolytic activity of some of these agents when injected into newborn hamsters (Toolan, 1960a,b; Kilham, 1961a,b). Attempts in our laboratory to induce such osteolytic activity with the various AAV have not been successful but whether such activity may be found with other new candidate members is not known. One property which seems to hold true for most parvovirus candidates, which may relate to the osteolytic activity of some of its members is that they all seem to require actively multiplying cells for replication (Toolan, 1968; Mayr *et al.*, 1968; Storz and Warren, 1970; Kilham, 1961a,b; Tennant *et al.*, 1969). Either nonconfluent actively dividing primary cultures or, in some cases, malignant cells which are not susceptible to contact inhibition have been shown to satisfy this requirement for a number of the parvovirus candidates.

One characteristic of the AAV subgroup, which on the surface sets it apart from other members of the group, is its dependence on adenovirus for the production of infectious virus (Atchison *et al.*, 1965; Hoggan *et al.*, 1966b; Smith *et al.*, 1966; Parks *et al.*, 1967a).

On the other hand, some of the nondefective parvovirus candidates are helped by adenoviruses when grown under restrictive conditions as described for the H-1 virus by Ledinko and Toolan (1968) and Ledinko *et al.* (1969). These workers showed that H-1 virus would produce infectious virus in human embryonic lung cells only in the presence of adenovirus. They showed, however, that H-1 virus was capable of producing virus-specific immunofluorescent (FA) antigen in these cells in the absence of adenovirus. A recent finding by Atchison (1970) has shown that herpes simplex and infectious bovine rhinotracheitis viruses (IBR) can elicit the production of AAV-1 FA antigen in cell culture but do not allow the production of infectious AAV. This finding was extended to include all the known human herpes viruses including human cytomegalovirus, EB virus, and herpes zoster (Blacklow *et al.*, 1970).

It was shown earlier by Chany and Brailovsky (1967) that some product of adenovirus replication helped partially inactivated RV to replicate. It is clear from this brief review of some of the biological properties of these viruses that even characteristics which can be used to distinguish between different candidate members can be shown to be similar under specific conditions.

B. Morphology

1. *Average Size of Parvovirus Candidates*

Although all of the potential members of the parvovirus group are small, there has been some variation in their size as reported by different workers.

Chandra and Toolan (1961) examined thin sections of H-1 virus-infected hamster tissue and found particles with 15 nm dense cores surrounded by a clear zone approximately 7.5 nm. From these findings they estimated the virus particles to be about 30 nm in overall diameter. Toolan *et al.* (1964), using negative staining with PTA on H-1 virus partially purified by adsorption to guinea pig erythrocytes found particles having a mean diameter of 24.5 nm. They occasionally saw "empty" particles having PTA filled centers measuring 13–15 nm.

Bernhard *et al.* (1963), using thin section techniques, found particles which measured about 15 nm in hamster fibroblasts infected with either RV or H-1 virus. Dalton *et al.* (1963) again using thin section techniques found numerous intranuclear particles approximately 13 nm in diameter in kidney stromal cells of hamsters infected with RV.

In studying the X14 virus, which is serologically indistinguishable from RV, by PTA negative staining, Payne *et al.* (1963) found considerable variation in size (18–24 nm) with a mean diameter of 22 nm. The particles they studied were purified either by adsorption to guinea pig erythrocyte ghosts or by centrifugation in CsCl gradients. Breese *et al.* (1964) found RV purified on potassium tartrate gradients to vary in size between 15 to 25 nm when stained with PTA, although they found a preponderance of 20 nm particles. Johnson *et al.* (1967) reported that feline panleukopenia virus and the serologically related feline ataxia virus were 20 nm in diameter when measured in the electron microscope and would pass through 45 nm, but not 39 nm filters. In earlier work, Johnson and Cruikshank (1966) found that the virus would not pass through an 80 nm APD membrane. Johnson *et al.* (1967) attributed this earlier finding to the fact that the virus was usually seen as large clumps in the microscope and care had to be taken to prevent aggregation.

Mayr and Mahnel (1966) reported seeing 20–22 nm particles in various hog cholera isolates which resembled RV. In a more detailed report, Mayr *et al.* (1967) using CsCl gradient factionation, found three classes of particles in such hog cholera preparations. The first class, reported to represent the hog cholera virus, was found at a density of 1.14–1.20 gm/cm³ and consisted of lipid enveloped particles 39–40 nm in diameter with 28–29 nm cores. The same fractions contained fewer but significant numbers of hexagonal appearing particles measuring 14–16 nm in diameter. The third class of particles found in the density range of 1.37 to 1.44 gm/cm³ measured 20–22 nm in diameter with a hexagonal silhouette. Similar findings were reported by Horzinek *et al.* (1967) who also reported that the 20–22 nm particles hemagglutinated various erythrocytes. Later Mayr *et al.* (1968)

made isolates of cytopathic viruses from uninoculated primary monolayer kidney cultures from normal 3-week-old piglets. These isolates measured between 20–22 nm when examined with PTA or silicatungstic acid (STA). These particles were ether and chloroform resistant and would pass through 44 nm, but not 39 nm Millipore filters.

The AAV were first noted in tissue culture harvests of various adenovirus stocks examined by negative staining techniques in the electron microscope (Archetti and Bocciarelli, 1964, 1965; Hoggan, 1965; Atchison *et al.*, 1965; Melnick *et al.*, 1965). They were noted because they were markedly smaller than the 70–80 nm adenovirions. Archetti and Bocciarelli (1964, 1965), found 24 nm particles in preparations of SV11. Atchison *et al.* (1965) found 24 nm particles in preparations of SV15 which could be separated from the adenovirions by passage through 50 nm Millipore filters. These later authors were the first to report in detail the defective nature of the AAV. Independently, Hoggan (1965) and Hoggan *et al.* (1966b) reported seeing 22 nm particles in various human adenovirus types, that the various AAV were defective and could be divided into various serological types. Melnick *et al.* (1965) and Mayor *et al.* (1965) reported seeing 20 nm particles in preparations of SV15 and various human adenovirus preparations and interpreted them as being adenovirus subunits. Smith *et al.* (1966) made careful measurements of a number of AAV strains and found them to be icosahedral in symmetry with diameter of 18 nm in one direction and 21 nm in the perpendicular plane.

Using thin section techniques, Archetti and Bocciarelli (1965) noted three size classes in cell cultures infected with SV11; 60–70 nm adenoviruses, 20–25 nm, and 8–13 nm particles. Atchison *et al.* (1966) using similar techniques saw crystalline arrays of particles up to 19 nm in diameter with 12 nm deep staining cores in cells infected with mixtures of SV15 and AAV-1. Mayor *et al.* (1967) reported similar findings with SV15 and AAV4.

The Haden virus was first isolated from young calves in 1959 by Abinanti and Warfield (1961). It was found to be ether and heat resistant and to be less than 30 nm in diameter as determined by Gradocol filtration. These authors felt it was not an enterovirus, but it was not until recently that it has been shown to have the characteristics of a parvovirus candidate (Storz and Warren, 1970). We have found similar results with Haden virus in which we found particles banding at 1.40–1.42 gm/cm^3 in CsCl and having an average diameter of 22 nm (M. D. Hoggan, G. F. Thomas, and C. E. Cox, 1969, unpublished data).

One candidate virus which has received considerable attention is DNV from *Galleria mellonella L.* This virus, first described by Meynadier *et al.*

(1964), was found in 1965 by Amargier *et al.*, to have a diameter of 21–23 nm. In sections of insect tissue (Vago *et al.*, 1966) the virus was reported to have a diameter between 19–20 nm. Kurstak and Cote (1969), found an average diameter of 20 ± 1.5 nm. Their analysis of the capsid structure will be discussed in more detail in a later section (III, B, 2, pp. 55–56).

Crawford (1966) described the isolation of a small DNA virus from preparations of mouse adenovirus. This agent, called the "minute virus of mice" (MVM), was reported to have a mean diameter of 25 nm. When close packed arrays of virus were supported on a carbon membrane and carefully measured, the center-to-center distance was found to be 26 nm. However, when particles were suspended between holes, the minimum was found to be 19 nm. They stained some preparations with 2% PTA and some with 1% uranyl acetate. In a later publication, Crawford *et al.* (1969) compared the size of MVM, ϕX174, and AAV-1. Their reported values for the average diameter of each were MVM = 28 ± 1.0 nm, ϕX174 = 30 ± 1.2 nm, and AAV = 28 ± 1.0 nm. Hall *et al.* (1959) had measured ϕX174 using both shadow length and a measure of the diameter with PTA. In the negative stained preparations ϕX174 was found to have an average diameter of 24.8 nm. McGeoch *et al.* (1970) compared the size of MVM = 28 ± 1.0 nm, RV = 28 ± 1.0, and H-1 = 30 ± 1.5 nm.

Karasaki (1966) made a comparative study of the size of four small DNA viruses, which was controlled by making specific aggregates of virus with antiserum. These included RV (Kilham and Olivier, 1959), H-1 virus (Toolan *et al.*, 1960), HT (Toolan, 1964), and H.B. (Toolan, 1964). He found that when the various virions were measured from side to side rather than from vertex to vertex, they fell into two size groups. The first group included the serologically identical RV and H-3 viruses and the serologically distinct HB virus. These viruses measured 19.0 ± 1.5, 19.1 ± 1.2, and 18.9 ± 1 nm, respectively. The H-1 and HT viruses which are serologically closely related to each other, were larger and measured 21.5 ± 1.6 and 21.5 ± 2.3 nm. Karasaki used 2% PTA at pH 7 for his tests and was careful to use areas for study that were lightly stained. Similar results have been found in our laboratory when we compared RV and H-1 with AAV types 1, 2, and 3 (Hoggan and Thomas, unpublished data, 1969). Using 2% PTA at pH 7.2, we found purified RV showed an average diameter of 19.1 while H-1 had an average diameter of 21.7 nm. The latter was indistinguishable fro AAV 1, 2, or 3. We have recently compared Haden, DNV, and MVM and found them all to have average diameters between 21.5–22.5 nm.

The variation in measurement reported from different laboratories is not surprising and can be accounted for in a number of ways. As has been pointed out by Hosaka (1965) and Karasaki (1966), the viruses are not truly spherical but icosahedral, and measurement of the diameters of various particles is dependent upon their orientation. In an icosahedral model, the relative length of the maximum to minimal diameter is 1.00–0.75 (Karasaki, 1966). With this fact in mind, Karasaki made his measurements from side to side on particles viewed along a three- or fivefold axis. In our laboratory, particles are measured from side to side in a similar manner. Smith *et al.* (1966) when studying the AAV made two measurements on each particle: one in its smallest dimension and one in its largest dimension. The average minimum dimension they found was 18 ± 1.1 nm and the maximum average dimension was 21 ± 1.8 nm. Other factors which must be considered is the well-known fact that the stain used may affect the apparent size. For example, we have found that when 1% uranyl acetate is used on small DNA viruses they appear 12 to 15% larger than when replicate preparations are stained with 2% PTA. We have found this to be true for RV, H-1, Haden, DNV, MVM, and AAV types 1, 2, 3, and 4.

2. *Fine Structure of Parvovirus Candidates*

Because of their small size and the probability that they have a very tight, close-packed capsid (suggested by their extreme resistance to various chemical and physical agents), the definition of the fine structure of parvovirus candidates has proved very difficult to resolve with any degree of certainty. The results and interpretations have varied greatly from laboratory to laboratory, and although these results have been stimulating, they have not as yet led to a final conclusion of just what the fine structure is and, therefore, cannot as yet be used unequivocally as a criterion for grouping.

The similarities of size and physical properties of the parvovirus candidates and the bacteriophage ϕX174 have been noted and since it was the first small (under 30 nm in diameter) DNA virus to be examined it shall be discussed first, even though it may not be ultimately designated a parvovirus.

Hall *et al.* (1959) examined ϕX174 using specialized shadowing techniques and negative staining with PTA. Their shadowed preparations suggested a particle with one apical "knob" surrounded by 5 peripheral "knobs" with the assumption of a similar group of 6 "knobs" on the underside of the particle. The authors interpreted these results as showing that ϕX174 contained a shell of 12 "knobs" arranged regularly as on the faces of a regular dodecahedron. Their results with negatively stained preparations were less convincing as they pointed out "the outline of negatively stained virus is consistent with the outlines of the type of polyhedron postulated although the resolution is not adequate to establish the shape unequivocally."

A further study of the ϕX174 by Tromans and Horne (1961) further supported the concept that the viral capsid had a maximum diameter of 22.5 nm and demonstrated a clear hexagonal outline with a side length of 12.5 nm. These authors agreed that it was probably made up of 12 subunits arranged on the vertices of an icosahedron but felt the 12 subunits were in turn made up of smaller subsubunit structures. These authors also reported the presence of hollow ring-shaped structures in their preparations with an internal diameter of 4.5 nm and external diameter of 8.5 nm. We have seen similar "rings" in certain fractions of DNV, H-1, RV, Haden, and AAV types 1, 2, and 3 (see Figs. 1 and 2). Breese *et al.* (1964) reported similar ring-shaped particles with RV.

Although a number of authors had attempted visualization of the prototype parvovirus (RV) with negative staining technique (Payne *et al.*, 1964; Toolan *et al.*, 1964; Breese *et al.*, 1964), Vasquez and Brailovsky (1965) were the first to attempt a reconstruction of the viral capsid to determine the number of capsomers and the capsid symmetry.

These authors reported that RV capsids appeared angular in profile with many showing a somewhat hexagonal outline but such particles were not perfect polyhedrons; rather, they were approximate polyhedra which were deformed by concavities and protuberances. They reported seeing both five- and three-fold symmetry with triangular patterns over the surface of selected particles. Individual capsomers were approximated at 3.0×2.0 nm. Vasquez and Brailovsky pointed out that the roughly hexagonal shape of the virion indicated it may have a deltahedron structure with the most likely configuration being a pentagonal dodecahedron with pentagonal pyramids placed on each face, the total number of capsomers being 32. They rather consistently found 9–10 capsomers along the periphery of the capsid with evidence of 5, 3, 2 symmetry. They further pointed out that a protein coat might be formed with virtually no gaps with such a capsid with a triangularation number of $T = 3$ and 32 capsomers.

As has already been noted, Karasaki (1966), made a careful comparative study of H-1, HT, H-3, RV, and HB viruses. Using specific antiserum made against each virus, he was able to visualize the various viruses under the electron microscope after negative staining with PTA or uranyl acetate. The differences in size between the group of viruses composed of H-3, RV, and HB and the group composed of H-1 and H-T became clear when members of one size group were treated with antiserum and mixed with members of the other group not treated with antiserum.

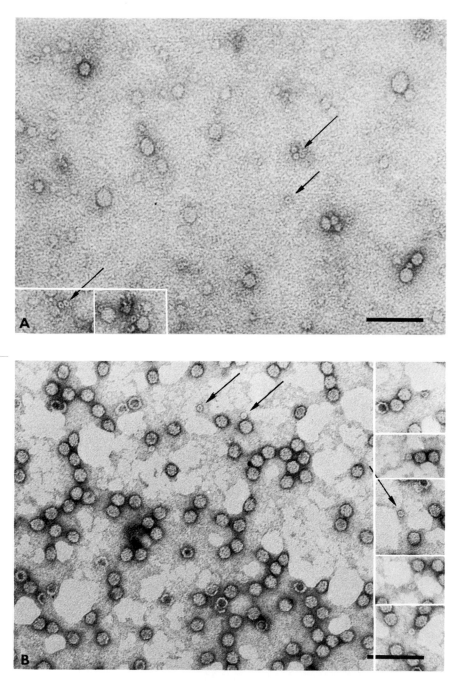

Fig. 1. Small 6-10 nm "rings" seen in (A) RV and (B) AAV-3.

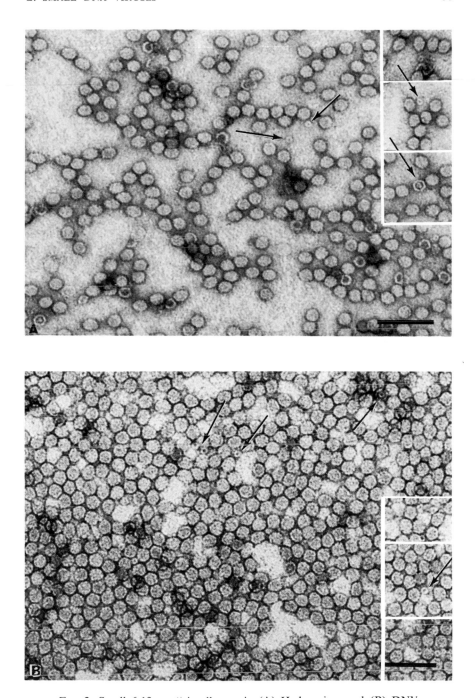

FIG. 2. Small 6-10 nm "rings" seen in (A) Haden virus and (B) DNV.

In spite of the obvious difference in size between the two groups of agents, Karasaki speculated that all of the small DNA viruses he studied have a similar general morphology with markedly angulated profiles being shown in both complete and empty particles. The particles appeared to be arranged in three- and fivefold symmetry with some particles exhibiting decagonal and some hexagonal outlines. In no case was he able to find more than 10 peripheral capsomers in preparations of any of the five viruses he examined. He constructed a model made of 32 units or capsomers with 12 placed on a fivefold rotational axis and 20 on a threefold axis. Such a model fits the formula $NC = 10 \times (n - 1)^2 + 2$. When such a model is viewed on a fivefold axis, six central capsomers and the peripheral capsomers can be observed. When such a model is observed on a sixfold axis, seven central capsomers can be observed but all of the peripheral capsomers are obscured by the overlapping of each other. The author therefore concluded that all the viruses he examined had 32 capsomers arranged as an icosahedron or pentagonal dodecahedron in agreement with the conclusions published by Vasquez and Brailovsky (1965) for RV.

The adeno-associated viruses have been studied in a number of laboratories, but so far only three groups have attempted to interpret their detailed fine structure. Archetti and Bocciarelli (1964, 1965) studied a small 25 nm particle found in tissue culture preparations of the simian adenovirus SV-11. These particles ultimately were shown to be indistinguishable from AAV-1 (I. Archetti, personal communication; Hoggan, unpublished results; Archetti et al., 1966). These authors found the particles to show a clear polyhedral profile with a surface covered by capsomers measuring about 4 nm in diameter with an average center-to-center spacing of 7.0 nm. They were able to count 15–17 structural units on the visible part of the capsid with no more than 10 capsomers being noted on the periphery of empty capsids. When they counted the number of capsomers on the border of complete particles they could see 12–14 with sometimes 6 and sometimes 5 capsomers being seen on the inner portion of the particle. The authors pointed out that if they were seeing just one-half the number of the total capsomers, the particle would be made of 20 units with one capsomer situated either on the face of an icosahedron or one at each vertex of a pentagonal decahedron. On the other hand, these authors point out that if fewer than one-half the number of capsomers are visible, the most likely number of capsomers would be 32. The authors pointed out that they could make no definite statement as to the trimeric state of the capsomers. However, if their hypothesis of 20 capsomers is correct, 20 units on the surface of a icosahedron assumes that the capsomers are equal and trimeric, therefore, not simple knobs. When examining broken down capsids, they saw that the capsomers did in fact have a complex structure.

Mayor *et al.* (1965) examined 20 nm particles from a preparation of SV15 which later turned out to be AAV-4. They reported that the 20 nm particles were definitely hexagonal and in specifically selected particles they found subunit patterns in which a central subunit was surrounded by five other subunits. They were never able to find a central capsomer surrounded by 6 capsomers. They usually counted 6 capsomers at the outer edge of selected particles and frequently saw a central group of 3 capsomers in the form of a triangle. They compared these observations with a model constructed of twelve ping-pong balls in the form of an icosahedron and concluded that the capsid was composed of twelve subunits in the form of an icosahedron as proposed for ϕX174.

A novel approach to the fine structure of AAV was taken by Smith *et al.* (1966). These workers, after adjusting their microscope to produce at least 10 Å resolution and taking a number of focal series of purified AAV, made reverse image prints. Careful examination of such prints revealed a resemblance between AAV and reovirus. They, therefore, felt the capsid of AAV was best represented by a reticulum or netlike arrangement of protein fibers similar to that suggested for reovirus by Vasquez and Tournier (1964). They further felt that the holes in the netlike structure were penetrated by stain, thus appearing dark. They acknowledged difficulty in determining the number of such "holes," but felt such "holes" were clearly demonstrated. Using a model of a rhombic triacontahedron with holes at each vertex (32), they pointed out similarities between selected particles and such a model. They thought such a model more nearly approximated the AAV structurally (see Fig. 3).

The densonucleosis virus (DNV) of *Galleria mellonella L.* has been examined in detail by negative staining by Kurstak and Cote (1969). They found the particles had an average diameter of 20 ± 1.5 nm, and that the virus exhibited icosahedral symmetry with most particles appearing to have a roughly hexagonal outline. They found two types of particles in their preparations: first, full particles whose surface was made up of regular shallow capsomers, and empty particles where only those capsomers around the periphery could be noted. In high magnification micrographs, the capsomers were found to measure between 2.0–3.5 nm with a 1.5 nm central hole. The authors pointed out that the small size of the virus made interpretations of the exact number and arrangement of the capsomers difficult. They did find a few capsids in which capsomers appeared to be arranged as equilateral triangles with 6 capsomers on the face, i.e., 3 on each side with

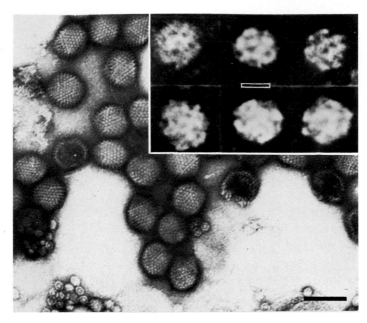

Fig. 3. Print of PTA-stained preparation of AAV-1 and adenovirus type 4 with an enlarged insert demonstrating reticulum arrangement of AAV capsid (micrograph kindly supplied by Dr. K. O. Smith).

one in the middle (see Fig. 4). Using the formula $NC = 10 \times (n - 1)^2 + 2$, where $n =$ the number of capsomers along an edge and $n = 3$, they calculated the total number of capsomers as 42. Because it has been suggested that RV and H-1 have 32 capsomers, these authors felt DNV was quite distinct from these parvovirus candidates and should be placed in their own family, Densoviridae.

C. HEMAGGLUTINATION OF VARIOUS PARVOVIRUS CANDIDATES

One physicochemical property of the parvovirus candidates which has been used to characterize them, is their varied ability to hemagglutinate different erythrocytes. This property, because of its speed and simplicity, has been used extensively for those agents that exhibit it. It will not be the object of this report to give a detailed review of all the work on the various parvovirus candidates; for this purpose the reader is referred to Toolan (1968). However, certain salient facts should be pointed out. For example, AAV types 1, 2, and 3 do not hemagglutinate either at 4° or 37°C (Hoggan, 1965, and unpublished data), but AAV-4 does agglutinate human, guinea

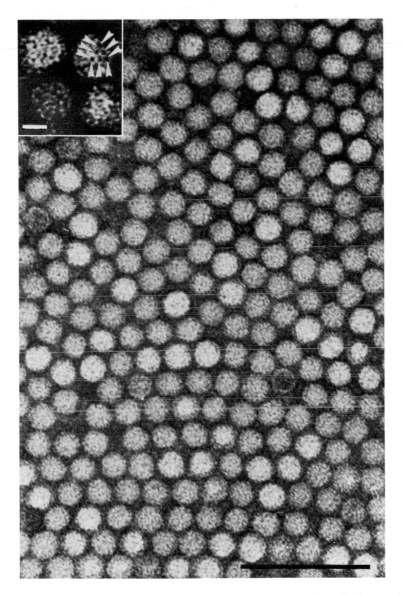

FIG. 4. Negatively stained preparation of DNV. Inset shows enlarged micrograph of a DNV particle showing a triangular facet made up of 6 capsomers (courtesy of Dr. E. Kurstak).

pig, and sheep erythrocytes at $4°$ but not at $37°C$ (Ito and Mayor, 1968). Similarly, feline panleukopenia and mink enteritis virus hemagglutinate pig erythrocytes at $4°$ but not at $37°C$ (Johnson and Cruickshank, 1966).

Haden virus was shown by Abinanti and Warfield (1961) to agglutinate guinea pig and human erythrocytes, but not rat cells. MVM can be differentiated from RV and H-1 by its ability to agglutinate mouse red blood cells (Crawford, 1966). The porcine parvovirus candidates have been shown to agglutinate mouse and rat erythrocytes (Mayr et al., 1967, 1968; Horzinek et al., 1967) and the minute virus of canines exhibits hemagglutination (Binn et al., 1968). One property which DNV shares with AAV 1, 2, and 3 is that it does not appear to agglutinate any erythrocyte tried (Hoggan, unpublished data; Bergoin, unpublished data).

D. Relative Specific Density of Various Parvovirus Candidates in CsCl

One property of the parvovirus candidates which has not only been useful in grouping them together, but also has been invaluable in purification procedures, is their relatively high specific density in CsCl. Without exception, all of the small DNA viruses band in isopycnic CsCl gradients in a range between 1.38–1.46 gm/cm^3.

However, it is imperative to point out that because different workers use different techniques, the density values have differed from report to report (Smith et al., 1966). For example, some workers use preformed gradients with shorter runs which may tend to result in lower density values. It is obvious, therefore, that in comparing density between two parvovirus agents, they should be carried out under the same conditions. It has also been found that different preparations of a given virus may show slightly different densities because of different conditions of production (i.e., different cells, different multiplicity ratios of infection, or different passage levels). These limitations must be kept in mind when comparing different results.

E. Pilot Comparative Study of Nine Parvovirus Candidates

It becomes obvious to readers and workers in the field that even though all the parvovirus candidates can be placed in the same group where the limits of each characteristic can be defined, the specific value for any given characteristic may not only vary from laboratory to laboratory, but from report to report from a given laboratory. It has already been pointed out, for example, that the average diameter for a given candidate or the relative specific density varies from report to report. It, therefore, becomes evident that when comparing the properties and looking for minor differences between different members or a potential member of the parvovirus group, one must carry out such studies under well-defined conditions and in the same tests in as much as it is possible.

We have attempted such a study and our preliminary results are summarized in Table I. This table shows the average diameter of the particles from the infectious bands obtained during a 48-hour isopycnic banding in CsCl, the density of such bands, and the ability of such virus to agglutinate various erythrocytes.

TABLE I

COMPARATIVE STUDY OF NINE PARVOVIRUS CANDIDATES

Virus	Average size (nm)	Density in CsCl (gm/cm³)	Agglutination RBC[a]		
			GP	HuO	R
AAV-1	21.8 ± 1.3	1.395	0	0	0
AAV-2	23.8 ± 2.7	1.388	0	0	0
AAV-3	21.4 ± 2.7	1.394	0	0	0
AAV-4	22.0 ± 1.8	1.445	+ +	+ +	+
RV	19.1 ± 1.2	1.400	+ + +	0	+ +
H-1	21.7 ± 1.8	1.422	+ + + +	+ + +	+
Haden	22.5 ± 1.0	1.425	+ + +	+ + + +	0
DNV	21.9 ± 2.7	1.440	0	0	0
MVM	19.3 ± 1.2	1.417	+ + +	0	+

[a] All virus preparations for hemagglutination tests were standardized to contain 16–32 complement fixing units of virus specific antigen. Complete agglutination of red blood cells at dilutions of 1:2 to 1:16 were recorded +. Similarly, 1:32 to 1:128 as + +, 1:256 to 1:1,024 as + + +, and greater than 1:1,024 as + + + +. GP, HuO, and R refer to guinea pig, human type O, and rat erythrocytes.

It should be noted that even though the range of diameter values for all of the viruses compared, overlap, the RV and MVM appear somewhat smaller and in fact could be picked out when examined as coded specimens by the microscopist.

In the particular experiments reported here, the density values of the major infectious bands of virus fell between 1.38–1.45 gm/cm³ with AAV-2 being the lightest and AAV-4 being the heaviest.

One observation which is not shown in Table I but should be noted is the repeated finding that in addition to the primary infectious band and a number of lighter minor bands, some of which contain large numbers of empty particles, a very heavy (1.44–147 gm/cm³) minor band is frequently found with all of the parvovirus candidates we have studied.

This very heavy band always contains fewer particles and less CF antigen but is antigenically indistinguishable from the major band. On the other hand, these particles have invariably been found to have a smaller diameter than those particles found in the major band. We have concluded that these particles have in fact lost or perhaps never obtained an outer layer of protein. Figures 5–7 show electron micrographs of particles from the major upper virus bands (B) and the lower, denser minor bands (A) for AAV-2, AAV-3, and Haden virus.

F. BIOCHEMICAL AND BIOPHYSICAL CHARACTERIZATION OF THE NUCLEIC ACID OF VARIOUS PARVOVIRUS CANDIDATES

By definition, members of the parvovirus group contain DNA. However, characterization of the NA from the various candidate members of the parvovirus group has been fraught with difficulty, resulting in many conflicting results from different laboratories. One problem with a number of these viruses is that they require special conditions for getting large yields (Hoggan et al., 1966b; Crawford, 1966; Toolan, 1968; Storz and Warren, 1970). They are extremely small, and because of their tight capsid it has been difficult in some cases to release the viral DNA even after sufficient quantities of viruses were obtained (Rose et al., 1966). Another problem, only recently explained, is the fact that certain candidate viruses are unique in that even though both positive and negative DNA strands are made and encapsidated, they are not found together in a single capsid. This shall be discussed at length in the light of new data (III, F, 3, pp. 65–71).

1. Early Studies which Indicated Nucleic Acid Type

Historically, early workers who studied RV and H-1 suspected that these viruses contained DNA because they all produced intranuclear inclusions in susceptible cells (Dawe et al., 1961; Rabson et al., 1961; Moore, 1962; Bernhard et al., 1963; Hampton, 1964). It was found that such inclusions were Feulgen-positive and gave staining with acridine orange, which indicated that these inclusions contained DNA (Rabson et al., 1961; Bernhard et al., 1963; Hampton, 1964). Additional evidence that the X-14 strain of

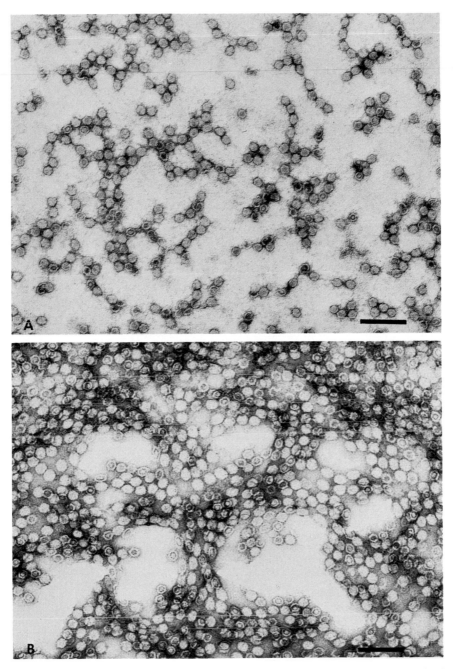

FIG. 5. Electron micrographs of (B) major AAV-2 band at a density of 1.388 gm/cm^3 and (A) heavy AAV-2 band at a density of 1.438 gm/cm^3.

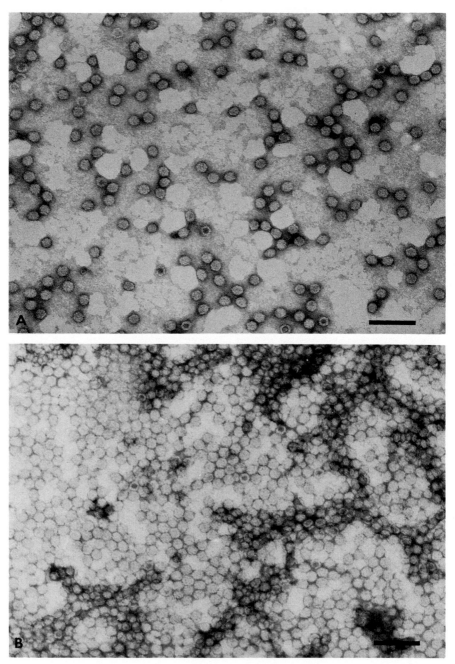

FIG. 6. Electron micrographs of (B) major AAV-3 band at a density of 1.394 gm/cm³ and (A) heavy AAV-3 band at a density of 1.475 gm/cm³.

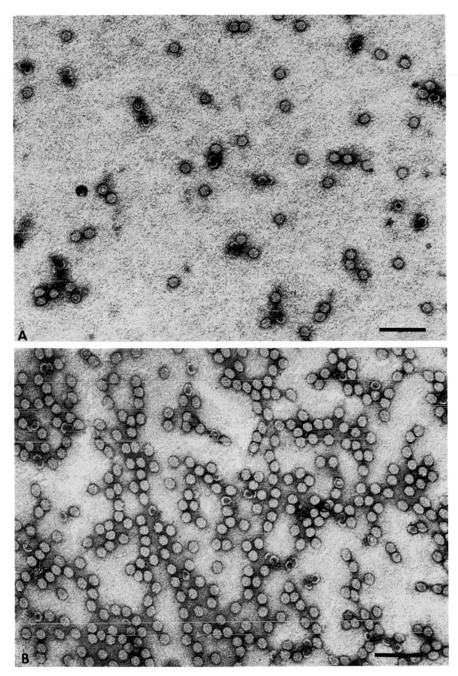

FIG. 7. Electron micrographs of (B) major Haden virus band at a density of 1.425 gm/cm³ and (A) heavy Haden virus band at a density of 1.46 gm/cm³.

RV contains DNA, was provided by Payne *et al.* (1963), who found the virus was inhibited by FUDR. That the NA was, in fact, DNA, was further substantiated by Cheong *et al.* (1965), who reported that ^{3}H-labeled thymidine was incorporated into the RV nucleic acid while ^{3}H-labeled uridine was not. Ledinko (1967) found H-1 virus was inhibited by 1-β-D-arabinofuranosylcytosine (ARAC).

2. *DNA Strandedness of Nondefective Parvovirus Candidates*

When Mayor and Jordan (1966) used acridine orange to stain X-14 infected cells, they found a yellow-green fluorescence at the edge of the nuclei, which they interpreted as double-stranded DNA being present at this particular stage of replication. May *et al.* (1967) were able to isolate small quantities of DNA from RV which they found had characteristics of double-stranded DNA. Their extracted DNA showed minimal infectivity which was reduced 25-fold by DNAase but not by RV antiserum. They showed that the DNA they studied had a sharp melting curve with a T_m of 87°C and did not react with formaldehyde by showing an optical density increase as would be expected for a single-stranded DNA molecule. They estimated the molecular weight of the DNA to be 2×10^6 daltons with a $G + C$ content of 43% when calculated from the T_m value, and 45% when calculated from buoyant density. In contrast to these results, Robinson and Hetrick (1969) extracted DNA from RV and found it did not demonstrate a distinct melting profile. They further found that either the intact RV virions or extracted DNA showed an increase in optical density when reacted with formaldehyde. They estimated RV DNA to have a molecular weight of 1.2×10^6 daltons by measuring the length of extracted DNA when examined by electron microscopy. These authors thus concluded that the DNA of RV must be single-stranded.

Crawford (1966) showed that MVM probably contained single-stranded DNA, since DNA in the intact virion reacted with dilute formaldehyde resulting in an increase in optical density and a shift of the absorption maximum to a longer wave length, as was shown for the single-stranded DNA containing bacteriophage ϕX174 (Sinsheimer, 1959). Crawford *et al.* (1969), later showed that the base composition of the DNA from MVM was indicative of a single-stranded molecule since no base pairing was found, either by base analysis after hydrolysis of ^{32}P-labeled DNA or by nearest neighbor base sequence analysis using the technique described by Josse *et al.* (1961).

Usatequi-Gomez *et al.* (1969) reported that DNA from H-1 virus was single-stranded. Their conclusion was based on similar observations, i.e., extracted H-1 DNA showed no sharp thermal melting curve and reacted with formaldehyde. They also did a base analysis by breaking the H-1 DNA into 5′-mononucleotides and found that H-1 DNA did not show base pairing and had a G + C content of 41.7%.

Salzman and Jori (1970) also found the DNA of RV to be single-stranded. Again, their data was based on the reaction of formaldehyde with RV DNA in intact virions and base analyses of extracted DNA. They found no base pairing with a G + C content of 42.5%. Again they found no sharp melting curve.

McGeoch *et al.* (1970) using nearest neighbor frequency pattern analysis on the DNA's of MVM, RV, and H-1 virus found strong evidence for single-stranded DNA in each case. There was no base pairing and the G + C contents for each were reported as 40.9% for MVM, 43.5% for RV, and 45.2% for H-1.

May and May (1970) reported three species of DNA from purified preparations of RV: a major component with a sedimentation coefficient of 27 and two minor components with sedimentation coefficients of 21.5 and 17.5 S. In CsCl, all three DNA components show relative specific densities between 1.726 and 1.729 gm/cm³ with no significant shift upon denaturization. The authors therefore concluded that RV DNA is, in fact, single-stranded and not double-stranded, as a previous report from their laboratory had indicated. They further pointed out that their earlier studies (May *et al.*, 1967) were carried out with a DNA preparation which was probably contaminated with rat cell DNA.

It now seems quite clear and it is now generally agreed that RV, MVM, and H-1 viruses all contain single-stranded DNA.

3. *Strandedness of AAV DNA*

The studies of the DNA of the various serological types of AAV proved to be more difficult to resolve.

During early studies, Atchison *et al.* (1965) using the acridine orange staining technique suggested that AAV-1 contained double-stranded DNA. Mayor and Melnick (1966) reported that the small particles they isolated from SV15 preparations stained as though they contained single-stranded DNA if care was taken to use only preparations which were free of any morphological evidence of degraded adenovirus. They further concluded that the staining seen earlier in their own and in other laboratories was probably due to contaminating double-stranded adenovirus DNA.

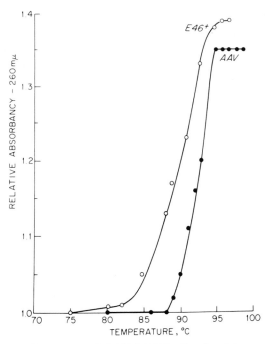

FIG. 8. Thermal melting curve of AAV-1 DNA. Reprinted by permission of *Proc. Nat. Acad. Sci. U.S.* (1966) **56**, 86–92.

Studies in our laboratory (Rose *et al.*, 1966, 1968) and a study by Parks *et al.* (1967b) on DNA extracted from the various types of AAV clearly indicated that DNA was double-stranded in such preparations. For example, Fig. 8 shows a melting curve for AAV-1 DNA, and Table II shows good base pairing when a base analysis was carried out on enzymically hydrolyzed ^{32}P-labeled DNA. It was also shown that boundary sedimenta-

TABLE II

Base Composition of AAV-1 DNA

Nucleotide	Moles/100 moles nucleotides
Adenylic acid	22.9 ± 0.1
Thymidylic acid	22.8 ± 0.1
Guanylic acid	26.9 ± 0.2
Cytidylic acid	27.3 ± 0.3

tion of alkali-denatured AAV DNA gave a molecular weight determination one-half of that of "native" DNA. These studies left little doubt that the DNA extracted from purified AAV virions was double-stranded.

As pointed out in Secion III,F,2, Crawford *et al.* (1969), showed that MVM contained single-stranded DNA. They further compared their results on MVM DNA with those obtained for the single-stranded DNA extracted from ϕX174 (Sinsheimer, 1959; Swartz *et al.*, 1962). Crawford *et al.* (1969), determined the molecular weight of MVM DNA and ϕX174 DNA by electron microscopy (Kleinschmidt and Zahn, 1959), and found them to be not more than 1.7×10^6 daltons.

In the same communication, the authors reported on their comparative studies on the physical properties of MVM, AAV-1, and ϕX174. They found that MVM, AAV-1, and ϕX174 viruses were very similar in size. They further found that these three small DNA-containing viruses had similar buoyant densities in CsCl when compared under similar conditions (1.39–1.43 gm/cm^3) and each demonstrated sedimentation coefficients between 104 and 112 $s_{20,w}$. Because of these similarities in the physical properties of these viruses, these authors argued that the AAV-1 virion could not contain more DNA than ϕX174 or MVM, which they showed to contain 1.7×10^6 daltons or less. Such a conclusion was in contradiction to the results found on the DNA extracted from the various types of AAV (Rose *et al.*, 1966, 1968; Parks *et al.*, 1967b) which was shown to be double-stranded and have a molecular weight of 3×10^6 daltons or more. Although Crawford *et al.* (1969) considered it an unlikely explanation, they proposed that the DNA inside of the AAV-1 particle is single-stranded, but that both negative and plus complementary strands are encapsidated, but in different particles in the virus population. They further suggested that these complementary strands could anneal during extraction of the DNA from the virions, giving rise to double-stranded DNA molecules.

Although such a hypothesis seemed unlikely, it would, of course, explain the apparent discrepancy between the studies of Mayor and Melnick (1966) using acridine orange, and the physical studies of the DNA extracted from the AAV particles (Rose *et al.*, 1966, 1968; Parks *et al.*, 1967b). Because of these findings, it was decided to test the hypothesis directly (Rose *et al.*, 1969). These experiments were carried out as follows: Tridium-labeled 5-bromdeoxyuridine (^{3}H-BUDR) was substituted for the thymidine in a preparation of AAV-3 to form a density label. A similar preparation of AAV-3 was labeled with ^{14}C-thymidine. Now having a heavy ^{3}H-labeled BUDR population, and a light ^{14}C-labeled population, they could be mixed and separated in a CsCl gradient. As can be seen in Fig. 9, the particles which

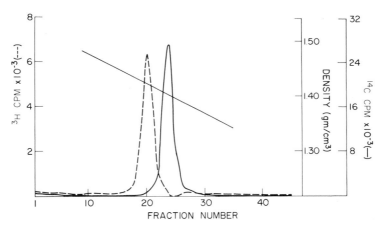

F<small>IG</small>. 9. Pattern of radioactivity in factions from CsCl equilibrium density gradient of mixed heavy and light AAV-3 particles [reprinted by permission of *Proc. Nat. Acad. Sci. U.S.* (1969) **64**, 863–869].

contained ^{3}H-BUDR substituted DNA were distinctly heavier than the ^{14}C-labeled particles and could be resolved in the gradient.

If the hypothesis of Crawford *et al.* (1969), was, in fact, true, extraction of a mixture of heavy and light particles should result in the formation of a density hybrid with about 50% of the duplexes containing both heavy and light DNA strands. It was, of course, first necessary to show that the DNA extracted separately was distinct and that the duplexes formed would either be heavy in both strands, or light in both strands. In Fig. 10A it can be seen that the DNA from heavy and light preparations could easily be distinguished when they were extracted separately before centrifugation, with no evidence of a density hybrid. The ^{3}H-labeled BUDR DNA had a buoyant density of 1.787 gm/cm^3, and the ^{14}C-labeled thymidine DNA had a density of 1.715 gm/cm^3.

On the other hand, when DNA was extracted from mixtures of heavy and light particles, as is shown in Fig. 10B, a substantial amount of hybrid DNA was formed. In the C part of Fig. 10, one can see the results when DNA was extracted separately from heavy and light particles, denatured by alkali, mixed together, and reannealed in 4X SSC at 67°C for 4 hours. Under these conditions, a sharp band of heavy and light hybrid DNA is formed which has a density similar to that density hybrid formed when the DNA was extracted from a mixture of heavy and light particles. Fig. 10D shows the results when a mixture of light and heavy DNA (which had been prepared separately) was put through the routine DNA extraction procedure. A small amount of hybrid was formed which is probably the result of poly-

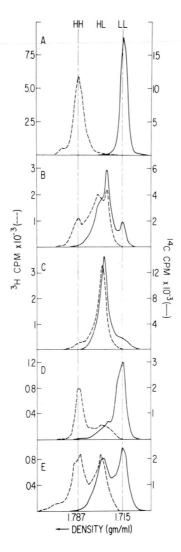

Fig. 10. Equilibrium density centrifugation of AAV DNA. (A) A mixture of heavy and light DNA. (B) DNA extracted from a mixture of heavy and light particles. (C) Sample of DNA shown in (B) which was denatured in 0.1 M NaOH and reannealed in 4X SSC at 67°C for 4 hours. (D) A mixture of heavy and light DNA which had been put through extraction procedure. (E) Sample of DNA shown in (B) after breakage in the French press. [Reproduced by permission from *Proc. Nat. Acad. Sci. U.S.* (1969) **64**, 863–869.]

nucleotide sequences which were incompletely base paired during the first extraction procedure. Figure 10E shows the density profile of AAV-DNA HL hybrid after being broken into units about one-quarter of the original molecular weight, in a French press. It should be noted that approximately one-half the total DNA has remained as hybrid, and it could thus be concluded that the hybrid was laterally formed by annealing of complementary strands.

Because of the uniqueness of the finding of positive and negative strands being encapsidated separately and annealing upon extraction, an additional control of another DNA was included. This control consisted of making heavy and light Ad-2 DNA-containing particles and submitting them to the same single and mixed extraction procedure used for AAV. Figure 11A shows the DNA density profile of Ad-2 DNA when a mixture of heavy

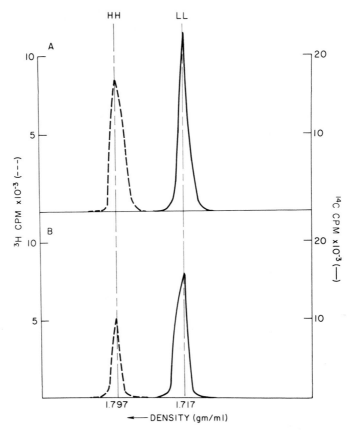

FIG. 11. Equilibrium density centrifugation of adenovirus DNA. (A) A mixture of heavy and light DNA. (B) DNA extracted from a mixture of heavy and light particles. [Reproduced by permission *Proc. Nat. Acad. Sci. U.S.* (1969) **64**, 863–869.]

and light DNA was made after extraction and then spun to equilibrium in CsCl and Fig. 11B shows the result when a mixture of heavy and light DNA containing Ad-2 particles was extracted together and then spun to equilibrium. As can be seen, no hybrid was formed. In addition to the results of Rose et al. (1969), two other papers were published in 1969 which bear on the problem of the strandedness of AAV and the hypothesis of complementary strands being encapsidated separately. In the first Mayor et al. (1969a) compared the reaction of AAV virions and AAV DNA with dilute formaldehyde and acridine orange with the reactions obtained with adenovirus 2, ϕX174, and their extracted DNA's. In the AAV virion they obtained results similar to ϕX174 indicative of single strands, while the reactions with extracted AAV DNA were indicative of double-stranded DNA's.

In a later paper, Mayor et al. (1969b) showed that AAV DNA appeared as single strands when released from the capsid at low salt concentration at temperatures below the normal T_m of AAV. These authors concluded from their data that the virion probably contained single complementary strands which annealed when released from the viral capsid.

Although AAV is the first example of single complementary strands being encapsidated separately, it may not be unique since it should be pointed out that the densonucleosis virus (DNV) has been reported to contain a double-stranded DNA with molecular weight of 4×10^6 daltons (Truffaut et al., 1967). Since this particle is about the same size as AAV (22–25 nm in diameter) and bands in CsCl at 1.40–1.43 gm/cm³, and the extracted DNA has a molecular weight between 4 and 4.5×10^6 daltons, the same objections which Crawford et al., applied to AAV can also be applied to DNV. The DNV virion should not contain that much DNA and may, in fact, contain complementary single strands. This possibility is currently being investigated in our own and other laboratories. If this is the case, all the small DNA viruses (under 30 nm) will have been shown to contain single-stranded DNA.

Of further interest along these lines is a recent paper by Robinson (1970), who has described self-annealing RNA from two myxoviruses. Newcastle disease virus (NDV) and Sendai virus were found to contain large areas of self-complementary RNA.

G. Immunological Characterization and Relatedness between Various Parvovirus Candidates

A number of serological procedures have been applied to the study of the various candidate viruses. These include serum neutralization (SN) complement fixation (CF), immunofluorescence (FA), immunodiffusion, microagglutination as observed in the electron microscope, and in those cases where the viruses have been shown to hemagglutinate, hemagglutination inhibition (HI).

1. *Comparative Studies Using SN*

According to Toolan (1968), it was noted in 1961 by Dalldorf that RV, H-1, and H-3 could be divided into two serological groups by SN. He found that H-3 and RV cross-neutralized each other while H-1 was neutralized only by H-1 serum. Later Toolan (1964) found that of two new isolates of H viruses, one (HT) was similar to H-1 by SN, but the other, HB, was not related to either H-1 or RV by SN. Using the microagglutination of virus by specific serum and examination under the electron microscope Karasaki (1966) found similar results.

A number of different virus isolates from rats which show diverse biological properties have been found by SN to be indistinguishable from RV or H-3. These include X14 virus (Payne *et al.*, 1964) and the hemorrhagic encephalopathy virus of rats (El Dadah *et al.*, 1967).

Although Crawford (1966) found that MVM did not cross-react by SN or HI with RV, J. Parker (1969, personal communication) did find that MVM, RV, and H-1 all cross-reacted by FA. We have also found that even though these viruses are distinct by SN and CF, they cross by FA (see Section III,G,2). The Haden virus was only recently suspected of being a parvovirus (Storz and Warren, 1970), and there are no reports in the literature of cross SN between this virus or other parvovirus candidates. We have prepared high titered serum against Haden virus and found no cross-reactivity by SN, CF, or FA (see Section III,G,2) with RV, H-1, MVM, DNV, or the four AAV types.

Feline panleukopenia virus and mink enteritis virus appear to be closely related to each other by SN, although not identical (Burger *et al.*, 1963; Gorham *et al.*, 1966; Johnson *et al.*, 1967).

The minute virus of canines does not cross-react by SN or HI with RV, H-1, or MVM (Binn *et al.*, 1968; Binn, 1969, personal communication).

There are no reports regarding serological studies between the porcine parvovirus candidates and other candidates, even though they are suspected as belonging to this group of viruses.

Very soon after the discovery of the viruses now known as AAV, it was noted that they had characteristics in common with the osteolytic viruses RV and H-1 (Hoggan, 1965; Atchison *et al.*, 1965). However, no serological relationships have been found between the AAV and other parvo-virus candidates.

The SN test has been valuable in studying the relationships between the various AAV types (Hoggan *et al.*, 1966b), in epidemiological studies (Blacklow *et al.*, 1967b, 1968a,b), and in showing differences between different isolates of the same type (Hoggan *et al.*, 1969). It also led to the discovery that AAV-3 field isolates from children contained small amounts of AAV-2.

Two types of serum neutralization tests have been used in our laboratory for our AAV studies. The first test used extensively in our studies on the distribution of antibodies (Blacklow *et al.*, 1968a) was a fluorescent focus reduction test. The second technique we have used was carried out as follows: Various dilutions of antiserum are reacted with 100–1000 infectious doses of the various AAV types and the reaction mixtures are inoculated onto cells preinfected with adenovirus helper. After the cells are destroyed by the adenovirus helper, the tissue culture fluids are tested for the presence or absence of AAV CF antigen.

Table III summarizes some comparative data which shows heterologous, as well as, homologous SN titer of various AAV sera. It can be seen that four serological types of AAV can be distinguished, but there is some cross-neutralization between AAV types 2 and 3. It can also be noted that little or no difference between the H and M strains of AAV-2 could be detected but the three strains of AAV-3 showed considerable differences. The AAV-

TABLE III

COMPARATIVE SERUM NEUTRALIZATION TITER OF VARIOUS AAV

Virus	Serum titer[a]						
	1H	2H	2M	3H	3K	3T	4M
AAV-1H	320	0	0	0	0	0	0
AAV-2H	0	1280	640	20	0	20	0
AAV-2M	0	640	640	20	0	10	0
AAV-3H	0	20	20	1280	640	640	0
AAV-3K	0	0	0	320	640	10	0
AAV-3T	0	20	20	320	0	1280	0
AAV-4M	0	0	0	0	0	0	40

[a] The serum neutralization titer is the reciprocal serum dilution which prevented formation of CF antigen when reacted against 100–1000 infectious doses.

3H was by far the broadest of the three strains and it could neutralize both 3K and 3T. The 3K and 3T strains also neutralized 3H but showed little or no capacity to neutralize each other.

2. *Comparative Serological Studies between Nine Parvovirus Candidates Using CF and FA*

It has generally been found that while SN is usually more sensitive, as well as more specific, than either CF or FA, cross-reactions between different related viruses are more often found with the later two techniques. Table IV is a summary of some of our results which compares the homologous and heterologous serum titers of various parvovirus candidates as measured by CF and FA.

As can be seen in Table IV, there is some cross-reactivity by both CF and FA between AAV-2 and AAV-3. One should also note that even though the RV, H-1, and MVM sera gave higher homologous titers by CF than FA, no cross-reactions between these viruses were noted by CF, but were observed in the FA tests. These later results confirm the finding of J. Parker (1969, unpublished). No other cross-reactions have been noted.

It is interesting that we have found the different AAV types are genetically related on the basis of *in vitro* DNA-RNA homology studies with relatedness values of 27–37% (Rose *et al.*, 1968) while experiments to test relatedness between AAV with H-1 and RV have been negative.

IV. Discussion and Conclusions

The members or candidate members of the group of small DNA-containing viruses named parvoviruses is rapidly increasing. Many serologically distinct isolates have been, and are continuing to be made from such widely phylogenetically divergent hosts as humans (Blacklow *et al.*, 1967b; Toolan, 1968), rodents (Kilham and Olivier, 1959; Crawford, 1966), canines (Binn *et al.*, 1968), bovines (Abinanti and Warfield, 1961), porcines (Mayr *et al.*, 1968), aves (Dutta and Pomroy, 1967), and insects (Meynadier *et al.*, 1964). It has become evident that even though these varied agents can cause many different syndromes when introduced experimentally in animals or when in their natural hosts, all apparently need rapidly dividing cells for replication. One wonders whether this property has anything to do with the antitumor effect of AAV (Kirschstein *et al.*, 1968) and H-1 virus (Toolan, 1968), or the transforming effect DNV has on mouse L cells (Kurstak *et al.*, 1969).

TABLE IV

COMPARATIVE SERUM TITERS OF PARVOVIRUS CANDIDATES WHEN TESTED BY COMPLEMENT FIXATION AND FLUORESCENT ANTIBODY

	AAV-1		AAV-2		AAV-3		AAV-4		RV		H-1		MVM		Haden		DNV	
	CF[a]	FA[b]	CF	FA	CF	FA	CF	FA	CF	FA	CF	FA	CF	FA	CF	FA	CF	FA
AAV-1	2560	160	0	0	0	0	0	0	0	0	0	0	0	—	0	0	0	0
AAV-2	0	0	2560	320	80	20	0	0	0	0	0	0	0	—	0	0	0	0
AAV-3	0	0	40	20	1280	160	0	0	0	0	0	0	0	—	0	0	0	0
AAV-4	0	0	0	0	0	0	1280	640	0	0	0	0	0	—	0	0	0	0
RV	0	0	0	—	0	—	0	—	2560	160	0	40	0	—	0	0	0	0
H-1	0	0	0	—	0	—	0	—	0	20	2560	80	0	—	0	0	0	0
MVM	0	—	0	—	0	—	0	—	0	—	0	—	1280	—	0	—	0	—
Haden	0	0	0	—	0	—	0	—	0	0	0	0	0	—	1280	40	0	—
DNV	0	—	0	—	0	—	0	—	0	—	0	—	0	—	0	—	2560	—

[a] Reciprocal serum dilution that gives a 3+ reaction with 4-8 units of antigen.
[b] Reciprocal serum dilution that gives good 4+ staining of virus-infected cells in the indirect immunofluorescence FA test.

Comparative physicochemical studies of these agents have shown that they all have a number of common characteristics regardless of their origin or specific host requirements.

It has been shown that all of the candidates are in the size range of 18–26 nm in diameter and exhibit cubic symmetry. They all lack essential lipid since all are resistant to ether and chloroform. They are resistant to high temperatures (56°C for 4 hours) and can remain viable on fomites for many years. They all show a high relative specific density in CsCl (1.38–1.45 gm/cm³), a property that has been used in purification of many of the agents.

They all appear to contain single-stranded DNA which has a molecular weight of less than 2×10^6 daltons. Some members show the unique property of encapsidating both positive and negative strands in separate virions. It is interesting to point out that it was the comparative studies of Crawford *et al.* (1969) on MVM, AAV-1, and ϕX174 that first led them to consider the possibility that AAV contained single complementary DNA strands in different virions and led to the experiments which ultimately proved such a hypothesis to be correct. Up to that time, no such case had been described in biology.

This finding alone shows the great value of such comparative studies and emphasizes the great foresight shown by Dr. Lwoff when many years ago he pointed out that many similarities exist between viruses that should allow them to be grouped together regardless of whether they originated from mammals, birds, insects, or plants, and predicted a systematic study of such groups might well lead to a better basic understanding of the viruses.

REFERENCES

Abinanti, F. R. and Warfield, M. S. (1961). *Virology* **14**, 288–289.

Amargier, A., Vago, C., and Meynadier, C. (1965). *Arch. Gesamte Virusforsch.* **15**, 659–667.

Andrewes, C. H. (1970). *Virology* **40**, 1070–1071.

Anonymous (1965). *Ann. Inst. Pasteur, Paris* **109**, 625–637.

Archetti, I. and Bocciarelli, D. S. (1964). *Proc. 3rd European Reg. Conf. Electron Microscopy, Prague*, p. 343.

Archetti, I. and Bocciarelli, D. S. (1965). *Ann. Inst. Super. Sanita* **1**, 103–106.

Archetti, I., Bereczky, E., and Bocciarelli, D. S. (1966). *Virology* **29**, 671–673.

Atchison, R. W. (1970). *Virology*, **42**, 155–162.

Atchison, R. W., Casto, B. C., and Hammon, W. McD. (1965). *Science* **194**, 754–756.

Atchison, R. W., Casto, B. C., and Hammon, W. McD. (1966). *Virology* **29**, 353–357.

Bernhard, W., Kasten, F. H., and Chang, C. (1963). *C. R. Acad. Sci. Paris* **257**, 1566–1569.

Binn, L. N., Lazar, E. C., Eddy, G. A., and Kajima, M. (1968). *Bacteriol. Proc. 69th Ann. Meeting Amer. Soc. Microbiol., New York,* p. 161.

Blacklow, N. R., Hoggan, M. D., and Rowe, W. P. (1967b). *Proc. Nat. Acad. Sci. U.S.* **58**, 1410–1415.

Blacklow, N. R., Hoggan, M. D., and Rowe, W. P. (1968a). *J. Nat. Cancer Inst.* **40**, 319–327.

Blacklow, N. R., Hoggan, M. D., Kapikian, A. Z., Austin, J. B., and Rowe, W. P. (1968b). *Amer. J. Epidemiol.* **88**, 868–878.

Blacklow, N. R., Hoggan, M. D., and McClanahan, M. S. (1970). *Proc. Soc. Exp. Biol. Med.*, **134**, 952–954.

Breese, S. S., Howatson, A. F., and Chany, C. (1964). *Virology* **24**, 598–603.

Burger, D., Gorham, J. R., and Ott, R. L. (1963). *Small Animal Clinician* **3**, 611–614.

Chandra, S. and Toolan, H. W. (1961). *J. Nat. Cancer Inst.* **27**, 1405–1459.

Chany, C. and Brailovsky, C. (1967). *Proc. Nat. Acad. Sci. U.S.* **57**, 87–94.

Cheong, L., Fogh, J., and Barcley, R. K. (1965). *Fed. Proc.* **24**, 2570.

Crawford, L. V. (1966). *Virology* **29**, 605–612.

Crawford, L. V., Follett, E. A. C., Burdon, M. G., and McGeoch, D. J. (1969). *J. Gen. Virol.* **4**, 37–46.

Dalton, A. J., Kilham, L., and Zeigel, R. F. (1963). *Virology* **20**, 391.

Dawe, C. J., Kilham, L., and Morgan, W. (1961). *J. Nat. Cancer Inst.* **27**, 221–228.

Dutta, S. K. and Pomroy, B. C. (1967). *Amer. J. Vet. Res.* **28**, 296–299.

El Dadah, A. H., Nathanson, N., Smith, K. O., Squire, R. A., Santos, G. W., and Melby, C. C. (1967). *Science* **156**, 392–394.

Gorham, J. R., Hartsough, G. R., Sato, N., and Lust, S. (1966). *Vet. Med.* **61**, 35–40.

Hall, C. E., Maclean, E. C., and Tessman, I. (1959). *J. Mol. Biol.* **1**, 192–194.

Hampton, E. G. (1964). *Cancer Res.* **24**, 1534–1543.

Hoggan, M. D. (1965). *Fed. Proc.* **24**, 248.

Hoggan, M. D. (1970). *In* "Progress in Medical Virology" (J. L. Melnick, ed.), Vol. 12, pp. 211–239, Karger, Basel.

Hoggan, M. D., Blacklow, N. R., Rafaoko, R., and Rowe, W. P. (1966a). *Bacterial Proc. 66th Ann. Meeting Amer. Soc. Microbiol., Los Angeles*, p. 109.

Hoggan, M. D., Blacklow, N. R., and Rowe, W. P. (1966b). *Proc. Nat. Acad. Sci. U.S.* **55**, 1467–1471.

Hoggan, M. D., Shatkin, A. J., Blacklow, N. R., Koczot, F. M., and Rose, J. A. (1968). *J. Virol.* **2**, 850–851.

Hoggan, M. D., Blacklow, N. R., and Rowe, W. P. (1969). *Bacterial Proc. 69th Ann. Meeting Amer. Soc. Microbiol., Miami Beach,* p. 198.

Horzinek, M., Recyko, E., and Petyoldt, K. (1967). *Arch. Gesamte Virusforsch.* **21**, 475–478.

Hosaka, Y. (1965). *Biochim. Biophys. Acta* **104**, 261–273.

Ito, M. and Mayor, H. D. (1968). *J. Immunol.* **100**, 61–68.

Johnson, R. H. (1965). *Res. Vet. Sci.* **6**, 472–481.

Johnson, R. H. and Cruickshank, J. G. (1966). *Nature (London)* **212**, 622–623.

Johnson, R. H., Margolis, G., and Kilham, L. (1967). *Nature (London)* **214**, 175–177.

Josse, J., Kaiser, A. D., and Kornberg, A. (1961). *J. Biol. Chem.* **230**, 864–875.

Karasaki, S. (1966). *J. Ultrastr. Res.* **16**, 109–122.

Kilham, L. (1961a). *Virology* **13**, 141–142.

Kilham, L. (1961b). *Proc. Soc. Exp. Biol. Med.* **106**, 825–829.

Kilham, L. and Olivier, L. J. (1959). *Virology* **7**, 428–437.

Kleinschmidt, A. and Zahn, R. K. (1959). *Z. Naturforsch.* **14b**, 770–779.

Kirschstein, R. L., Smith, K. O., and Peters, E. A. (1968). *Proc. Soc. Exp. Biol. Mod.* **128**, 670–673.

Kurstak, E. and Cote, J. R. (1969). *C. R. Acad. Sci. Paris* **268**, 616–619.

Kurstak, E., Belloncik, S., and Brailovsky, C. (1969). *C. R. Acad. Sci. Paris* **269**, 1716–1719.

Ledinko, N. (1967). *Nature (London)* **214**, 1346–1347.

Ledinko, N. and Toolan, H. W. (1968). *J. Virol.* **2**, 155–156.

Ledinko, N., Hopkins, M. S., and Toolan, H. W. (1969). *J. Gen. Virol.* **5**, 19–31.

McGeoch, D. J., Crawford, L. V., and Follett, E. A. C. (1970). *J. Gen. Virol.* **6**, 33–44.

May, P. and May, E. (1970). *J. Gen. Virol.* **6**, 437–439.

May, P., Niveleau, A., Berger, G., and Brailovsky, C. (1967). *J. Mol. Biol.* **27**, 603–614.

Mayor, H. D. and Jordan, L. E. (1966). *Exp. Mol. Pathol.* **5**, 580–589.

Mayor, H. D. and Melnick, J. L. (1966). *Nature (London)* **210**, 331–332.

Mayor, H. D., Jamison, R. M., Jordan, L. E., and Melnick, J. L. (1965). *J. Bacteriol.* **90**, 235–242.

Mayor, H. D., Ito, M., Jordan, L. E., and Melnick, J. L. (1967). *J. Nat. Cancer Inst.* **38**, 805–820.

Mayor, H. D., Jordan, L. E., and Ito, M. (1969a). *J. Virol.* **4**, 191–194.

Mayor, H. D., Torikai, K., Melnick, J. L., and Mandel, M. (1969b). *Science* **166**, 1280–1282.

Mayr, A. and Mahnel, H. (1966). *Zentrabl. Bakteriol. Parasitenk. Abt.* 1 *Orig.* **199**, 399–407.

Mayr, A., Bachman, D. A., Sheffy, B. E., and Siegl, G. (1967). *Arch. Gesamte Virusforsch.* **21**, 113–119.

Mayr, A., Bachman, P. A., Siegl, G., Mahnel, H., and Sheffy, B. E. (1968). *Arch. Gesamte Virusforsch.* **25**, 38–51.

Melnick, J. L., Mayor, H. D., Smith, K. O., and Rapp, R. (1965). *J. Bacteriol.* **90**, 271–274.

Meynadier, G., Vago, C., Plantevin, G., and Atger, P. (1964). *Rev. Zool. Agr. Appl.* **63**, 207–208.

Moore, A. E. (1962). *Virology* **18**, 182–191.

Parks, W. P., Melnick, J. L., Rongey, R., and Mayor, H. D. (1967a). *J. Virol.* **1**, 171–180.

Parks, W. P., Green, M., Pina, M., and Melnick, J. L. (1967b). *J. Virol.* **1**, 980–987.

Payne, F. E., Shellabarger, G. F., and Schmidt, R. W. (1963). *Proc. Amer. Ass. Cancer Res.* **4**, 51.

Payne, F. E., Beals, T. F., and Preston, R. E. (1964). *Virology* **23**, 109–113.

Rabson, A. S., Kilham, L., and Kirschstein, R. L. (1961). *J. Nat. Cancer Inst.* **27**, 1217–1221.

Rapp, F. (1969). In "Annual Review of Microbiology" (C. E. Clifton, S. Raffel, and M. P. Starr, eds.), Vol. 23, pp. 293–316. Annual Reviews, Inc., Palo Alto, California.

Robinson, D. M. and Hetrick, F. M. (1969). *J. Gen. Virol.* **4**, 269–281.

Robinson, W. S. (1970). *Nature (London)* **225**, 944–945.

Rose, J. A., Hoggan, M. D., and Shatkin, A. J. (1966). *Proc. Nat. Acad. Sci. U.S.* **56**, 86–92.

Rose, J. A., Hoggan, M. D., Koczot, F. M., and Shatkin, A. J. (1968). *J. Virol.* **2**, 999–1005.

Rose, J. A., Berns, K. I., Hoggan, M. D., and Koczot, F. M. (1969). *Proc. Nat. Acad. Sci. U.S.* **64**, 863–869.

Salzman, L. A. and Jori, L. A. (1970). *J. Virol.* **5**, 114–122.

Sinsheimer, R. L. (1959). *J. Mol. Biol.* **1**, 43–53.

Smith, K. O., Gehle, W. D., and Thiel, J. F. (1966). *J. Immunol.* **97**, 754–766.

Storz, J. and Warren, G. S. (1970). *Arch. Gesamte Virusforsch.*, **30**, 271–274.

Swartz, M. N., Trautner, T. A., and Kornberg, A. (1962). *J. Biol. Chem.* **237**, 1961–1967.

Tennant, R. W., Layman, K. R., and Hand, R. S., Jr. (1969). *J. Virol.* **4**, 872–878.

Toolan, H. W. (1960a). *Fed. Proc.* **19**, 208.

Toolan, H. W. (1960b). *Science* **131**, 1446–1448.

Toolan, H. W. (1964). *Proc. Amer. Ass. Cancer Res.* **5**, 64.

Toolan, H. W. (1968). *Int. Rev. Exp. Pathol.* **6**, 137–176.

Toolan, H. W., Saunders, E. L., Greene, E. L., and Fabrizio D. P. A. (1964). *Virology* **22**, 286–288.

Toolan, H. W., Daldorf, G., Barclay, M., Chandra, S., and Moore, A. E. (1960). *Proc. Nat. Acad. Sci. U.S.* **46**, 1256–1258.

Tromans, W. J. and Horne, R. W. (1961). *Virology* **15**, 1–7.

Truffaut, N., Berger, G., Niveleau, A., May, P., Bergoin, M., and Vago, C. (1967). *Arch. Gesamte Virusforsch.* **21**, 469–474.

Usatequi-Gomez, M., Toolan, H. W., Ledinko, N., Fadhil, A., and Hopkins, M. S. (1969). *Virology* **39**, 617–621.

Vago, C., Duthoit, J. L., and Delahage, F. (1966). *Arch. Gesamte Virusforsch.* **18**, 344–349.

Vasquez, C. and Brailovsky, C. (1965). *Exp. Mol. Pathol.* **4**, 130–140.

Vasquez, C. and Tournier, P. (1964). *Virology* **24**, 128–130.

Wilner, B. I. (1969). "A Classification of the Major Groups of Human and Other Animal Viruses," 4th Ed., pp. 137–141, Burgess, Minneapolis, Minnesota.

Note Added in Proof: A recent paper on the characterization of minute virus of canines (MVC) appeared in 1970. The reference is Binn, L. N., Lazar, E. C., Eddy, G. A., and Kajima, M. (1970). *Infection and Immunity* **1**, 503–508.

CHAPTER 3

The Papovavirus Group

KENNETH K. TAKEMOTO, CARL F. T. MATTERN, AND
WILLIAM T. MURAKAMI

I. Introduction

The papovaviruses are small, deoxyribonucleic acid-containing viruses consisting of the papilloma and polyoma subgroups. Certain similarities shared by members of the group, i.e., size, structure, type of nucleic acid, ability to cause tumors, etc., led to the proposal (Melnick, 1962) that they be placed in the papovavirus group. The term was derived from the first two letters of the papilloma viruses, polyoma, and vacuolating agent (now known as simian virus 40 or SV40).

Many members of the papovavirus group induce benign or malignant growths in either their natural hosts or under artificial conditions in laboratory animals. Certain papovaviruses, furthermore, can produce alterations in tissue culture cells resulting in their conversion to malignant cells in a process known as "transformation." They are of interest, therefore, as models for the study of viral carcinogenesis in defined, *in vitro* conditions.

Table I lists members of the papovavirus group. Papovaviruses are probably ubiquitous in nature, and as procedures for isolation and identification improve, others will be added to the list. A new candidate for this group is a virus which has been called frog papovavirus (Gravell, Darling-

TABLE I

PAPOVAVIRUS GROUP

Papilloma	Polyoma
Human papilloma (wart)	Polyoma (mouse)
Rabbit papilloma	SV40 (monkey)
Rabbit oral papilloma	K (mouse)
Canine papilloma	Rabbit kidney vacuolating agent (rabbit)
Bovine papilloma	
Hamster papilloma	
Equine papilloma	
Deer fibroma	

ton, and Granoff, unpublished; Granoff, 1969), which was isolated from the Lucké renal carcinoma of frogs. Viruslike particles resembling papovaviruses have been seen by electron microscopy of thin sections of specimens from cases of a human disease known as multifocal leukoencephalopathy (ZuRhein and Chou, 1965; Howatson *et al.*, 1965; Schwerdt *et al.*, 1966). The virus has not yet been isolated and propagated in cell cultures or experimental animals.

II. Biology of Papovaviruses

Two types of virus–cell interactions are known among papovaviruses: (a) productive or lytic infection resulting in synthesis of progeny virus and ending in cell death, and (b) abortive infection whereby some viral products, but little or no infectious virus are made; in this case, the infected cell survives but is permanently altered or transformed. Both kinds of interactions will be briefly considered.

A. PRODUCTIVE INFECTIONS

1. *Papilloma Viruses*

Information regarding replication of papilloma viruses is meager due to the lack of *in vitro* methods for their cultivation. There is no definitive evidence for replication of any of the papilloma viruses in tissue culture. Human wart virus was reported to induce changes in cultured cells by several investigators (Oroszlan and Rich, 1964; Morgan and Balduzzi, 1964; Noyes, 1965). Bovine papilloma virus was reported to grow on the chorio-allantoic membrane of chick embryos by Olson *et al.* (1960).

2. *Polyoma-SV*40

Unlike the papilloma viruses, most of the polyoma subgroup replicate efficiently in cell culture and have been extensively investigated over the last decade. Several detailed reviews have covered both the lytic and transforming events (Stewart, 1960; Defendi, 1966; Black, 1968; Eckhart, 1968; Eddy, 1969). Only the salient features of their replication will therefore be discussed.

Members of the polyoma subgroup have a narrow host range for productive infection and high yields of virus are generally obtained only in cells derived from a single species. Thus, efficient virus growth is obtained only in mouse cells for polyoma, primary or established African green monkey kidney cells for SV40, and rabbit kidney cells for rabbit kidney virus. K virus grows to a limited extent in mouse cell cultures. Viral replication occurs in the nucleus of infected cells. The growth cycle is long, and infectious virus is not detected until 20 hours.

Productive or abortive infection by SV40 or polyoma leads to increased synthesis of enzymes involved in DNA metabolism. Stimulation of cellular DNA synthesis is also observed in these cells. An early product of viral infection is a protein known as tumor or T antigen which appears in the nucleus 6 to 8 hours after infection, before viral DNA or protein synthesis. It has been called tumor antigen because it was first recognized as an antigen present in tumors induced by oncogenic DNA-containing viruses. Serological techniques, either complement fixation or fluorescent antibody, are used to detect their presence using serum from tumor-bearing animals. The antigen is virus specific and cross-reactions between polyoma or SV40 T antigens do not occur. The function of T antigens is unknown; they do not appear to be structural proteins of the virus particles, nor is there any evidence that they may be early enzymes induced by viral infection. The T antigen is thought to be coded for by viral genes since they are synthesized by cells undergoing productive infection and, in addition, are associated with tumors induced *in vivo* or cells transformed *in vitro* in a variety of cells from diverse species.

B. ABORTIVE INFECTION (TRANSFORMATION)

Besides its natural host, bovine papilloma virus can induce tumors in hamsters (Cheville, 1966; Robl and Olson, 1968) or mice (Boiron *et al.*, 1965). *In vitro* cell transformation by this virus has also been observed (Black *et al.*, 1963; Boiron *et al.*, 1964). There are reports of other papilloma viruses causing cell changes resembling transformation, but definitive evidence such as the synthesis of new antigens or other criteria is lacking.

Most of the large body of information regarding transformation comes from investigations with polyoma and SV40. K virus can also transform mouse cells (Takemoto and Fabisch, 1970), although it is considered to be a nononcogenic virus. SV40 has an extensive host range in its transforming capacity: the virus can transform cells from a variety of species, such as human, mouse, rat, guinea pig, hamster, and sheep. Polyoma on the other hand, transforms cells primarily from rodents—mice, hamsters, and rats.

Transformation is recognized by (1) presence of intranuclear T antigen, (2) altered cell morphology and growth characteristics, (3) loss of contact inhibition, (4) new surface or transplantation antigens, and (5) ability to cause tumors when inoculated into suitable hosts. Of special interest is the permanent integration of viral genes in the transformed cells as evidenced by the synthesis of T antigens, viral messenger RNA, and the ability to rescue infectious virus. Many workers have shown that SV40-transformed cells which do not harbor infectious virus can be induced to yield virus very readily by the technique of fusing tumor cells with permissive AGMK cells (Gerber, 1963; Watkins and Dulbecco, 1967; Koprowski *et al.*, 1967). The cell-fusing agent for rescue experiments is inactivated Sendai virus, a paramyxovirus. Polyoma virus can also be recovered from transformed cells (Fogel and Sachs, 1969; Vogt, 1970). Polyoma- or SV40-transformed cells can also be induced to produce virus by inducing agents such as ultraviolet irradiation or mitomycin C treatment (Gerber, 1964; Burns and Black, 1968; Fogel and Sachs, 1970). The evidence is thus clear that the entire genomes of SV40 and polyoma may be present in cells transformed by these viruses and the interaction of these viruses with cells bears superficial resemblances to the lysogenic bacteriophage systems.

C. ANTIGENIC PROPERTIES

The papovaviruses are antigenically distinct from each other and cross-reactions do not occur between members of the group. However, antigenic variants of polyoma (Hare and Morgan, 1963) and SV40 (Ozer et al., 1969) do exist. Human, bovine, canine, and rabbit papilloma viruses showed no cross-reacting precipitin lines by immunodiffusion tests (LeBouvier et al., 1966).

III. Chemical Composition of SV40 and Polyoma Virus

A. PROPERTIES OF PURIFIED PARTICLES

SV40 and polyoma viruses have been more extensively characterized chemically and physically than other members of the papovavirus group and will therefore be discussed in greater detail.

The physicochemical characteristics of the virus particles indicate that SV40 and polyoma virus are similar in size and chemical composition. The sedimentation coefficient ($s_{20,W}$) of polyoma virions has been determined in different laboratories and reported to be 242, 238, and 239 S (Winocour, 1963; Crawford and Crawford, 1963) and 136 S (Murakami et al., 1968) has been reported for the empty capsids. The sedimentation rate of SV40, strain 777, has been determined to be identical to that of the polyoma virion ($s_{20,W} = 240$ S; Black et al., 1964) while Koch et al. (1967) have reported a slightly lower value of 219 S for the variant, SV40-777-42A.

In comparison, the $s_{20,W}$ values of the papilloma viruses are higher for both full and empty capsids: human wart virus, 296 and 168 S; rabbit papilloma virus, 298 and 172 S; and canine papilloma virus, 300 and 165 S (Crawford and Crawford, 1963).

The molecular weight of SV40-777-42A has been calculated to be 17.3

$\times 10^6$ daltons from sedimentation velocity and diffusion measurements (Koch *et al.*, 1967). The partial specific volume was determined pycnometrically ($\bar{v} = 0.658$) and the diffusion constant ultracentrifugally ($D_{20} = 0.90 \times 10^{-7}$ cm² sec⁻¹). The molecular weight of polyoma virions and empty capsids have been estimated to be 23.6×10^6 and 15.0×10^6 daltons, respectively, by sedimentation equilibrium measurements (Fine, Bancroft and Murakami, in preparation).

Measurement of the DNA contents of the two viruses, by phosphorus analysis of the virions have yielded a value of 12.5% (Koch *et al.*, 1967; Murakami *et al.*, 1968). Kleinschmidt *et al.* (1965) reported the DNA content of rabbit papilloma virus to be 12%.

B. Viral DNA

The nucleic acids of the tumor viruses, including the DNA of the polyoma virus group, have been reviewed by Crawford (1969). Since the DNA content of the virions and the molecular weight of the DNA provides an independent means for the calculation of the particle weight, estimations of the molecular weights of the DNA contained in SV40 and polyoma virus particles will be discussed briefly.

1. *Molecular Weight*

The DNA extracted from polyoma virions has been shown to consist of three components with sedimentation coefficients of 20.3, 15.8, and 14.5 S (Weil and Vinograd, 1963). The differences in the sedimentation rates has been shown to be due to differences in the configuration of the DNA molecules (Vinograd *et al.*, 1965). The fast component represents twisted closed circular duplex molecules (component I). The intermediate form is the extended circular duplex (component II), which can be produced by the introduction of one or more single-strand scissions into component I. Double-strand scission of component I gives rise to homogeneous linear molecules which have a sedimentation rate of 14.5 S. The linear molecules extracted from virions are more heterogeneous in size and have been shown to be largely fragments of cellular DNA which have been enclosed in the polyoma virus capsid (Winocour, 1969). Estimations of the molecular weight of polyoma virus DNA by sedimentation velocity, band width measurements under conditions of equilibrium density gradient centrifugation, and electron microscopy have yielded values consistent with 3×10^6 daltons (Weil and Vinograd, 1963; Crawford, 1964; Vinograd *et al.*, 1965;

Gray *et al.*, 1967). Sedimentation constants of 21, 16, and 14 S have been reported for the three forms of SV40 DNA (Crawford and Black, 1964; Anderer *et al.*, 1967; Yoshiike, 1968a). These values indicate that the size of the SV40 DNA is probably identical to that of polyoma DNA. Anderer *et al.* (1967), however, have concluded that the molecular weight of the SV40 DNA is 2.25×10^6 daltons. This represents an average value for the molecular weight estimated in three different ways: (1) calculation, using values for the DNA content of the particles and the measured particle weight; (2) contour length measurements of DNA extracted from virions; and (3) calculation, using the sedimentation coefficient of the linear molecules. Although there is good agreement among these values, it is probable that the molecular weight has been underestimated in two of these procedures. The molecular weight was calculated to be 2.53×10^6 daltons from measurement of the lengths of linear molecules by electron microscopy. The lengths of the measured circular molecules are equivalent to a value of 2.9×10^6 daltons (Crawford, 1969). The sedimentation coefficient of the linear form (14 S) and the Eigner-Doty equation (1965), an empirical equation derived from sedimentation coefficient and molecular weight measurements of linear DNA's with molecular weights less than 4×10^6 daltons, was used to calculate a value of 2.33×10^6 daltons from the sedimentation data. It has been reported that an equation with an additive constant (Crothers and Zimm, 1965) agrees better than an exponential equation (Eigner and Doty, 1965) with the sedimentation velocities measured for low molecular weight linear DNA, where the molecular weights have been determined by light scattering (Hudson *et al.*, 1968). The Crothers and Zimm equation, as well as equations relating the sedimentation velocities of extended circular DNA (Gray *et al.*, 1967) and superhelical closed circular DNA (Hudson *et al.*, 1968) to their molecular weights, give values, of 3.1×10^6 daltons, which is in excellent agreement with 3.0×10^6 daltons, obtained by measurement of the lengths of cyclic and linear molecules of polyoma DNA (Stoeckenius; appendix to Weil and Vinograd, 1963).

The DNA of some of the papilloma viruses has been found to be considerably larger than that of polyoma or SV40: 5.3×10^6 daltons for human wart virus (Crawford, 1965), and 4.7×10^6 (Watson and Littlefield, 1960) or about 5×10^6 (Kleinschmidt *et al.*, 1965) for rabbit papilloma virus. The contour length of the DNA of rabbit papilloma virus was determined to be 2.3–2.8 μ (Kleinschmidt *et al.*, 1965) compared to 1.52 μ for polyoma virus and 1.44 μ for rabbit kidney virus (Crawford and Follett, 1967) and 1.4–1.7 μ (Anderer *et al.*, 1967) or 1.76 μ (Yoshiike, 1968a) for SV40.

2. Pseudovirions and Defective Particles

The encapsidation of mouse cellular DNA fragments within polyoma particles (pseudovirions; Michel *et al.*, 1967) has been reviewed by Winocour (1969). Although the pseudovirions have a buoyant density range which overlaps both the virion and empty capsid bands when centrifuged to equilibrium in CsCl density gradients, the bulk of the particles are distributed on the lighter side of the virion band (Winocour, 1968). Analysis of the cellular DNA fragments by sedimentation velocity and electron microscopy have shown a wide size distribution (Michel *et al.*, 1967; Winocour, 1968). Most of the cellular DNA present has a size similar to that of the virus ($\sim$14 S; 3×10^6 daltons). Some faster sedimenting forms (up to 20 S; Winocour, 1968) as well as slower sedimenting fragments (11 to 14 S; Michel *et al.*, 1967) have been observed.

Defective particles containing closed circular viral DNA with molecular weights lower than that of infectious DNA have been reported for SV40 (Yoshiike, 1968a,b) and polyoma virus (Thorne, 1968; Blackstein *et al.*, 1969). Uchida *et al.* (1966) demonstrated that serial undiluted passage of SV40 in African green monkey kidney cell cultures resulted in marked decreases in yields of infectious virus, while the yield of complement fixing viral antigen was reduced slightly. Examination of the virus by equilibrium centrifugation in CsCl revealed that the virions purified from undiluted passage lysates formed a band (O.D.$_{258}$) which was broader and skewed toward lighter densities, compared to those from dilute passages (Uchida *et al.*, 1968). The plaque-forming activity of the undiluted passage particles showed a sharp profile, distributed on the heavier side of the virion band, while the T antigen-forming activity gave a broad distribution similar to the absorbance profile, peaking at a density approximately 0.01 gm/ml lower than the plaque-forming activity. Similar analysis of dilute passage virions showed a closer correspondence between the distribution of plaque-forming activity and the absorbance profile. However, the plaque-forming activity showed a density maximum at a slightly higher density than the T antigen-forming activity or absorbance profile, suggesting the presence of small amounts of defective particles in the preparation.

The DNA contained in the particles banding at different densities has been analyzed by sedimentation velocity and electron microscopy (Yoshiike, 1968a). The DNA has been shown to have the closed circular configuration of component I with a heterogeneous size distribution, ranging from that of infectious DNA to about 15% smaller. Particles lighter than the infectious virion contain DNA of shorter lengths, depending on the density, indicating that random deletions of various lengths occurred during serial propagation of the virus.

Equilibrium centrifugation of virions produced after three undiluted serial passages of a large plaque variant of SV40, strain 777, gave rise to a split virion band (Yoshiike, 1968b). The plaque-forming and T antigen-forming activities were associated with the heavier density particle (1.34 gm/ml), while the homogeneous defective particles, lacking both activities, banded at a density of 1.33 gm/ml. The defective particles were shown to contain closed circular DNA about 12% shorter than the infectious DNA.

Size heterogeneity in the supercoiled fraction of polyoma DNA was detected by electrophoretic analysis of subfractions of 20 S DNA which had been fractionated by sucrose gradient centrifugation (Thorne, 1968). The DNA from slower sedimenting subfractions exhibited faster mobilities, when electrophoresed in agar gels, suggesting shorter lengths for the slower sedimenting molecules. Sedimentation velocity analysis has confirmed the presence of smaller molecules, which are biologically defective (Thorne *et al.*, 1968), and have yielded estimates that the defective DNA is 3 to 7% shorter than infectious DNA.

Blackstein *et al.* (1969), analyzing the effect of undiluted serial passage of different strains of polyoma virus in mouse embryo cell cultures, found that the loss of infectivity yields was related to the appearance of shorter closed circular DNA molecules. Analysis of the size distribution by alkaline sucrose gradient centrifugation revealed greater size heterogeneity than that reported by Yoshiike (1968a,b) or by Thorne (1968). Two noninfectious DNA species, which have been shown to be in the supercoiled configuration by electron microscopy, sediment more slowly than infectious DNA. Contour length measurements, after conversion to the open circular configuration by DNase treatment, yielded molecular weight estimations of 2.2×10^6 and 1.47×10^6 daltons for the two defective forms. All strains of polyoma virus tested gave rise to the two defective forms, in variable quantities, on undiluted serial passage. The defective forms represented a larger proportion of the total viral DNA as the serial passages were continued.

C. VIRAL PROTEINS

1. *SV*40

The protein of SV40 virions and empty capsids has been separated into three components (A, B, and C) by electrophoresis (Anderer *et al.*, 1967). Virus particles were dissociated in 8 *M* urea–0.1 *M* tris–0.1 *M* β-mercaptoethanol, pH 9.6, and electrophoresed on cellulose acetate membranes in

8 M urea–0.1 M tris–0.01 M β-mercaptoethanol, pH 8.6–9.4. The three components were also resolved by ion-exchange chromatography on Dowex 1-X2, using 8 M urea–β-mercaptoethanol solutions of increasing pH to develop the column at elevated temperatures (40°–50°C). When chromatography was carried out at room temperature, considerable aggregation of polypeptide chains A and B occurred.

The relative amounts of the three polypeptides were determined by densitometry of the stained electropherograms and similar results were obtained with virion and capsid preparations (A = 45%, B = 45%, C = 10%). Sedimentation velocity and diffusion measurements gave an average molecular weight of 16,350 daltons for a mixture of the three components dissolved in 4 M guanidinium chloride–0.1 M tris–0.01 M β mercaptoethanol, pH 8.0. Sedimentation equilibrium measurements by the modified Archibald method gave a value of 16,100 daltons for the mixture. Molecular weights of 16,400, 16,900, and 16,800 daltons were obtained for the purified proteins A, B, and C, respectively, by sedimentation equilibrium centrifugation (Schlumberger *et al.*, 1968). The molecular weight of the aggregate of A and B polypeptides was 35,800 daltons, consistent with a dimer.

Evidence for the internal localization of the minor C polypeptide was obtained by trypsin digestion of full and empty particles (Anderer *et al.*, 1967). The proteins of both particle types were completely digested by extended trypsin action. Partial tryptic digests showed that polypeptides A and B were more rapidly degraded than polypeptide C. This conclusion has been further substantiated by the isolation of an alkali-stable, DNA-polypeptide C complex from the pH 10.5—dissociation products of virions by sucrose gradient centrifugation (Anderer *et al.*, 1968). The molecular weight of the protein-DNA complex has been estimated to be 3.7×10^6 daltons by sedimentation velocity and diffusion measurements. The protein can be dissociated and isolated from the complex by CsCl density gradient centrifugation. It is concluded that polypeptides A and B together constitute the protein shell of the virus and that polypeptide C is an internal component, which might play a role in orienting the DNA within the virus particle (Anderer *et al.*, 1967, 1968).

The amino acid compositions of the separated polypeptides show that the C peptide contains more basic residues (about 20% of the total residues) than either polypeptides A or B. The compositions of C polypeptide isolated by ion-exchange chromatography or dissociation of the DNA complex are closely similar. The C polypeptide present in the DNA complex was shown to be unreactive with anti-SV40 *Cercopithecus* sera or with anti-T antigen hamster sera, in complement fixation tests.

2. *Polyoma Virus*

The polyoma virus protein was first analyzed by acrylamide gel electrophoresis by Thorne and Warden (1967). They concluded that the polyoma virus capsid is composed of one type of structural polypeptide. The virus protein was dissociated by suspension of virus in 0.01 M phosphate buffer, pH 7.2, containing 0.1% sodium dodecyl sulfate (SDS)–0.5 M urea–0.1% β-mercaptoethanol, and heating at 56°C for 10 minutes. Electrophoresis in acrylamide gels containing 0.1% SDS revealed one major peak with several minor peaks. The minor peaks were thought to represent, at least in part, contaminants of cellular origin. Similar results were obtained by Fine *et al.* (1968) when SDS-dissociated viral protein was electrophoresed in SDS-containing gels by a discontinuous buffer system. Polyoma virions and empty capsids were dissociated by incubation in 0.05 M sodium carbonate buffer–0.1% β-mercaptoethanol–2% SDS, pH 10.6, for 5 minutes at 80°C. A single protein band was found when the gels were stained and the proteins from full and empty particles were shown to migrate at the same rate. These results were consistent with the conclusion that only one protein component was present in the polyoma virus particles. However, comparison of the amino acid compositions of the proteins present in virion and capsid preparations suggested the possible presence of a basic polypeptide(s) in virions, in addition to the structural protein present in both virions and empty capsids. Virions and capsids were dissociated by acid-urea treatment (10 M urea–0.1 M acetic acid–0.1% β-mercaptoethanol, pH 5.3) and incubated at 80°C for 10 minutes, to investigate this possibility. Electrophoresis in acrylamide gels containing 10 M urea, pH 4.5, revealed the presence of a minor protein component in the virion preparation in addition to the slower migrating major component which was observed in both virion and empty capsid preparations. Densitometer tracing indicated that the minor component represented about 10% of the total protein present in the gels.

Subsequent investigation of SDS-dissociated proteins, using a continuous buffer system (Shapiro *et al.*, 1967), has revealed the presence of a total of five polypeptides in the virion preparations (Fine, Bancroft, and Murakami, in preparation). The proteins have been designated capsid protein, internal proteins (IP) 1, 2, 3, and 4. The molecular weights of the polypeptides have been estimated by the procedure of Shapiro *et al.* (1967) and found to be: capsid protein, 42,000; IP-1, 30,000; IP-2, 23,000; IP-3, 20,000; and IP-4, 17,000 daltons. The molecular weight of the capsid protein has been estimated to be 37,000 daltons by sedimentation equilibrium measurements in 6.5 M guanidinium chloride. The previously reported value of

50,200 (Fine *et al.*, 1968) was measured in 0.1% SDS solution and was not corrected for SDS binding.

The relative proportions of the different proteins in the gel has been estimated by measuring the distribution of radioactivity of lysine- and valine-labeled virus proteins among the protein bands. The major band (capsid protein) represents about 80% of the total radioactivity and each of the minor components 3 to 7%. Distinct amino acid compositions have been found for each of the purified proteins. The minor components are more basic than the capsid protein. The basic acidic residue ratios are: capsid protein, 0.59; IP-1, 0.77; IP-2, 0.96; IP-3, 1.6; and IP-4, 1.3).

The previous observation that SDS-dissociated polyoma virus protein gave rise to a single band when electrophoresed by a discontinuous buffer system (Fine *et al.*, 1968) has been clarified by the demonstration that the five proteins can be resolved when extracted from the band and reelectrophoresed, using the continuous buffer system. Maizel *et al.* (1968) have reported similar "stacking" of adenovirus proteins under conditions of disc electrophoresis. Proteins from the two bands obtained by acrylamide gel electrophoresis of acid-urea treated particles have been reelectrophoresed in SDS gels as described by Shapiro *et al.* (1967). The minor component has been shown to migrate as IP-1 and the major component as capsid protein in the SDS system. The absence of internal proteins 2, 3, and 4, from the acid-urea gels has been found to be due to incomplete dissociation of the virus. Sucrose gradient centrifugation of the dissociation products has revealed that although 40 to 60% of the protein is dissociated, the remainder is still complexed to DNA. The dissociated protein consists largely of capsid protein and IP-1, indicating that about 50% of the capsid protein and all of internal proteins 2, 3, and 4, are in the DNA-protein complex. The complex is present after incubation at 37°C for 12 hours but does appear to dissociate slowly with prolonged incubation (4–5 days). The urea dissociation appears to occur much more slowly in solution than in protein monolayers deposited over an aqueous hypophase containing urea (Vasquez *et al.*, 1969). It is not clear whether the basic components are associated directly with the DNA, as has been demonstrated for the C polypeptide–SV40 DNA complex, or with the residual capsid protein which remains in the complex. Attempts to isolate a protein–DNA complex, free of capsid protein, are in progress.

In contrast to SV40, where a structure unit of the capsid appears to be composed of two polypeptide chains (A and B), these results substantiate the conclusion of Thorne and Warden (1967) that the polyoma capsid is composed of a single protein component. Basic, minor components have been shown to be present in both SV40 and polyoma virus particles. Their function remains to be elucidated. It has been suggested that they might play a structural role in orienting the DNA within the virus particle (Anderer *et al.*, 1967). It is also possible that they might be involved in the regulation of the synthesis of viral products in infected cells, e.g., transcription.

No information is available as to the genetic specification of the minor protein components. The genetic capacity of SV40 and polyoma virus is limited. The DNA can specify, at most, about 2×10^5 daltons of protein. In the extreme case that all of the polypeptides found in virions are structural, nonfunctional proteins, and are specified by the viral DNA, about one-third of the SV40 and about two-thirds of the polyoma virus genome would be required for structural protein synthesis. In either case, very little residual information would be available for the synthesis of functional and/or regulatory proteins.

IV. The Structure of the Papovaviruses

A. CAPSID

1. *Number of Capsomeres*

a. All of these viruses which have been subjected to critical analysis have been shown to possess an outer protein shell or capsid consisting of 72 capsomeres. These include the rabbit and human papilloma viruses (Finch and Klug, 1965; Klug and Finch, 1968), SV40 (Klug, 1965; Anderer *et al.*, 1967), and polyoma and K viruses (Klug, 1965). Other viruses of this group include bovine (Boiron *et al.*, 1964, 1965; Tajima *et al.*, 1968; Crawford and Crawford, 1963), canine (Crawford and Crawford, 1963; Cheville and Olson, 1964), and other papilloma viruses (Rdzok *et al.*, 1966; Tajima *et al.*, 1968), rabbit kidney vacuolating (RKV) virus (Chambers *et al.*, 1966), and the virus associated with progressive multifocal leukoencephalopathy (PML) ZuRhein and Chou, 1965; Schwerdt *et al.*, 1966). Their dimensions coupled with relative size and observable features of their capsomeres strongly suggests that they will also be shown to consist of 72 capsomeres.

Reasons for the difficulty in establishing the number of capsomeres in this group of viruses have been discussed in detail elsewhere (Klug and Finch, 1968; Mattern, 1969). Basically it involves the contribution of images from both the upper and lower surfaces of most negatively stained virus particles and the fact that in the case of 72 capsomere viruses capsomeres on the upper and lower surfaces are not in register in any view of the virus. By contrast, the $T = 1$ series (12, 42, 92, 162, etc., capsomeres) and the $T = 3$ series (32, 122, 272, etc., capsomeres) of viruses seen on a two-fold axis will present capsomeres on the two surfaces in superposition permitting relatively easy identification of two five-fold axes and the distribution of capsomeres on or about the edge line joining the two five-fold axes, which is essential in establishing the capsomere number of any virus.

Unfortunately, well-oriented viruses showing a dominantly single-sided image are rarely encountered. However double-surface image analysis of these viruses has been accomplished by means of specimen tilting and comparison of such images with those determined by computer for particles viewed at various orientations (Klug and Finch, 1968). The results of these and previously mentioned investigations leave virtually no doubt that the viruses which have been so investigated contain 72 capsomeres.

b. The $T = 7$ series of icosahedral viruses exist as either dextro or levo forms (Fig. 1). The "hand" of these viruses has been established in only a few cases. Human wart virus (Klug and Finch, 1965, 1968) and SV40 (Klug, 1965; Anderer *et al.*, 1967) are dextro, whereas rabbit papilloma virus (Finch and Klug, 1965; Klug and Finch, 1968) and K virus (Klug, 1965) are levo forms. Figure 2 shows three dominantly one-sided images of K virus printed so as to present the true hand of the virus under the assumption that the dominant side is in contact with the grid (Finch and Klug, 1965). The third virus is seen very nearly on a three-fold axis and image enhancement by rotation (Markham *et al.*, 1963) confirms the selection of five-fold axes and the 721 structure of K virus.

c. Although a number of models of capsomere substructure have been proposed suggesting the presence of 60 hexameric and 12 pentameric vertex capsomeres for a total of 420 structure units (Finch and Klug, 1965; Caspar, 1966; Klug and Finch, 1968), verification of this proposal has not yet been obtained. Some difficulties encountered in this arrangement have been discussed previously (Mattern, 1969). An additional difficulty appears as the result of observations discussed elsewhere in this chapter; namely, that the polyoma capsid consists of a single protein of about 42,000 mol. wt. and SV40 capsid protein exists as two different peptide chains of about 16,600 mol. wt. in equimolar proportions (Anderer *et al.*, 1967). It is possible that polyoma capsomeres are hexamers and pentamers, the former of about 240,000 mol. wt., but it would require that the SV40 capsomeres, which are of similar size, to be dodecamers and decamers containing 6 and 5 of each of the two types of biochemical subunits, respectively. However,

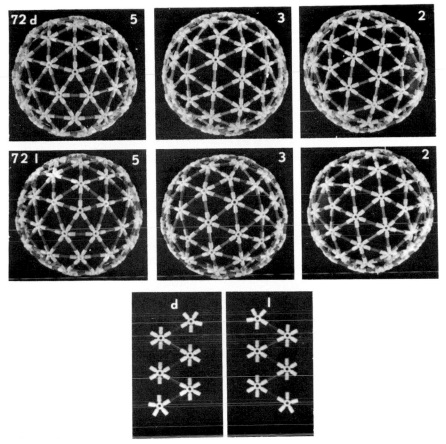

Fig. 1. Geodestix models of 72 unit structures; *upper*, dextro forms; lower, levo forms seen as indicated on five, three, and two fold rotation axes. Only one surface is shown. The lower two figures show the distribution of capsomeres along the line joining two, five fold axes for 72d and 72l structures.

it should be noted that the hexameric capsomeres of turnip yellow mosaic virus, which appear to be of comparable size are undoubtedly constructed of 6 protein subunits of 20,000 mol. wt. (Matthews and Ralph, 1966). Thus the hexameric capsomere of TYMV has a molecular weight of only one-half that of polyoma virus under the above hypothesis.

An alternative possibility for polyoma virus is that the capsomere molecular weight is also about 120,000 and that it consists of 3 subunits.*

* As discussed elsewhere (Mattern, 1969), there are in fact no symmetry requirements for the 60 capsomeres on the faces since none lie on two- or three-fold rotation axes. Only the 12 vertex capsomeres are required to have five-fold symmetry.

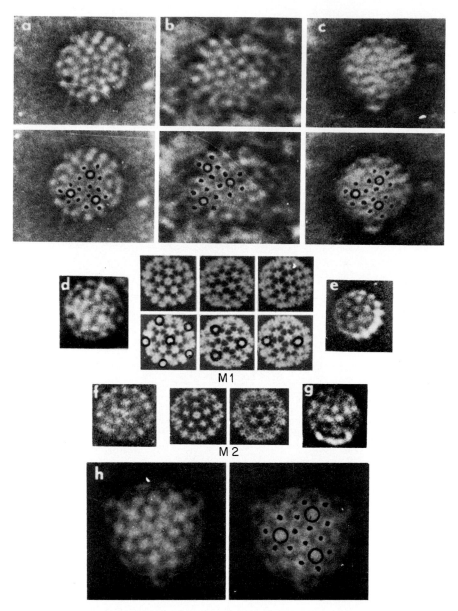

FIG. 2. Individual K virus particles (a–c) on which three, five fold axes may be identified (accompanying prints). This virus thus appears to be 721. M-1 represents a single-sided image of a 721 model. Largely single-surface images of K virus each with an identifiable five fold axis (d, e). M-2 represents double-surfaced model images on five- and three-fold axes with comparable virus particles (f and g). Fig. 2(h) presents an enlargement of particle c coupled with image enhancement by rotation.

The capsomere itself could still have a hexameric configuration. However it is difficult to visualize SV40 capsomeres consisting of 6 and 5 subunits of two types of protein in equimolar proportion since this would require at least two different kinds of pentamer.

B. DIMENSIONS

There appear to be two general size classes of the viruses of this group. The papilloma viruses in general are slightly larger than SV40, polyoma, and K viruses. Crawford and Crawford (1963) and Black *et al.* (1964) determined the ratio of diameters of papilloma viruses to polyoma virus or SV40 to be 1.2 and 1.26, respectively, by direct comparison of separate and mixed virus preparations negatively stained with neutral phosphotungstate (PTA). This was compatible with observed differences in sedimentation velocities of both full and empty particles of these two viruses. The $s_{20,W}$ values of the larger particles is about 300 (Crawford and Crawford, 1963; Watson and Littlefield, 1960), the smaller particles, about 240 (previously discussed in Section III). Since the sedimentation velocity is proportional to the square of the particle diameter, other factors being constant, these values suggest a ratio of their diameters of about 1.12.

Attempts to measure the diameter of these viruses by electron microscopy have lead to a variety of reported values. Their dimensions appear to vary with the following several factors. (1) Sectioned viruses of the smaller type (27–38 mμ) usually appear smaller than isolated negatively stained particles (38–50 mμ). The larger type is reported as 34–56 mμ sectioned and 48–58 mμ when negatively stained; this probably results from difficulty in visualizing the outer capsomere coat in sectioned viruses as well as flattening of negatively stained particles. (2) Particles containing PTA internally, whether empty or still containing DNA usually appear smaller than particles which exclude PTA, presumably due to internal stain minimizing flattening. (3) Isolated particles in arrays appear substantially smaller than unarrayed particles and particles deeply embedded in PTA appear smaller than those embedded in less PTA. (4) In sectioned cells, the center-to-center spacings of arrayed particles of both the smaller and larger type, are reported in the range of 36–42 mμ.

In Fig. 2, all the images of K virus are presented at the same total magnification. It is obvious that the upper three particles (a-c) are greatly flattened. Their diameter exceeds 50 mμ considerably and more capsomeres are readily visualized than would be expected on the basis of unflattened models. Somewhat less flattened particles are presented in Fig. 2d-g. The particle in Fig. 2e was very deeply embedded along its left edge with partial obscuring of its edge due to the thickness of the stain. Also note the diameter of the particle is quite similar to that of the model which "represents" a 39 mμ particle. The similarity of capsomere spacing in the models and the virus particles suggests that the "true" diameter of K virus is close to 40 mμ.

Similar difficulties have been encountered in our experience with polyoma virus and SV40. In Fig. 3a, SV40 particles (CsCl banded, density 1.33) were stained with uranyl acetate as previously described (Mattern *et al.*, 1967; Mattern and DeLeva, 1968). When uranyl acetate is incompletely removed by washing, particles remaining in contact with the stain are invariably flattened to a diameter of 50–60 mμ. Particles not embedded in stain appear 38–40 mμ in diameter with a densely staining core and an external ring of stain. Rarely one encounters curled supporting films which are sufficiently stable to obtain a picture. In such cases the particles not embedded in stain are always seen in side view as relatively round (Fig. 3b). Occasionally observations were made on curling films in which embedded particles were first seen as circular in "top" or normal view and subsequently with curling of the film the same particles were seen as flattened in "side" view. Finally with continued curling the same particles appear again as circular in "bottom" view. These observations suggest that the flattening is not an artifact of grid compression. The major dimension of the flattened particles in "side" view is indistinguishable from the diameter of the circular profiles. These images may be compared with the negatively stained particles in Fig. 4a and uranyl acetate stained particles in Fig. 4b, c.

As has been discussed by Kahler *et al.* (1959), an elipsoid of revolution with a minor axis (diameter) of 20 mμ and volume equal to a sphere of 40 mμ diameter would present a major axis (diameter) of about 56.5 mμ. This is compatible with the maximum degree of flattening of many virus particles under certain conditions.

From these considerations it would appear that the outer diameter of the polyoma, SV40, and K virus group of viruses is not far from 40 mμ. Koch *et al.* (1967) have estimated the size of SV40 virus as 41.1 $\pm$ 0.6 mμ. The diameter of the papilloma viruses is somewhat larger probably 45–50 mμ

C. Internal Structure

Both sectioned viruses and isolated particles stained with uranyl acetate reveal a densely staining core about 25 mμ in diameter. For reasons currently not understood, the core may appear either as a solid sphere or as

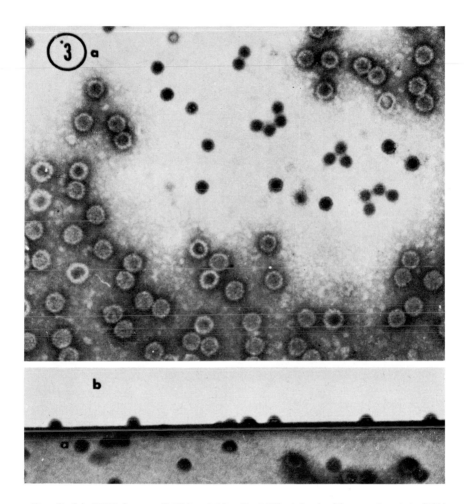

FIG. 3. (a) SV40 from a CsCl band (density 1.33) stained with uranyl acetate (UA). The darkly staining particles are 38–40 mμ in diameter; those embedded in UA are more than 50 mμ. (b) A similar preparation seen on a curling grid showing the small densely staining particle in side view to be relatively unflattened.

a shell. This undoubtedly represents the double-stranded circular DNA which represents about 12–13% of the virus mass. The lengths of the DNA molecules of the smaller viruses [polyoma, rabbit kidney vacuolating virus, and SV40 have been estimated to be 1.44–176 μ (Crawford and Follett, 1967; Anderer et al., 1967; Yoshiike, 1968a], whereas rabbit papilloma virus DNA is reported to be 2.3–2.8 μ (Kleinschmidt et al., 1965). The folding pattern of the DNA within the viral core is not known; however, as is the

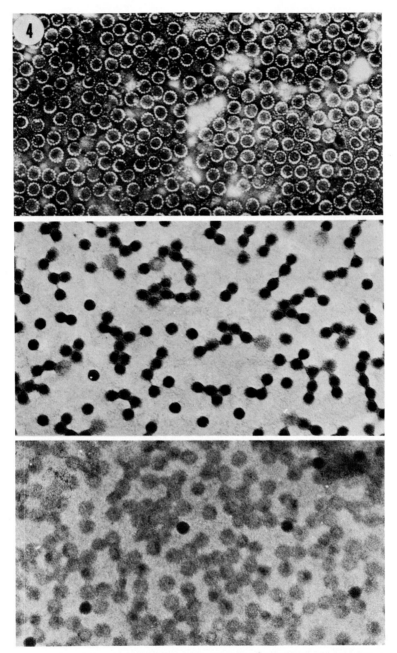

FIG. 4. SV40 particles. *Upper*, "full particles," embedded in PTA often contain internal stain and are difficult to distinguish from typical "empty" particles. *Middle*, "full" particles stained with UA. A few poorly staining and flattened "empty" particles are readily distinguished. *Lower*, "empty" particles (CsCl, density 1.29) contaminated with a few "full" particles.

case with SV40 (Anderer *et al.*, 1967) and polyoma virus (Fine *et al.*, 1968) there is very likely a basic protein (about 10% of the total virus protein) intimately associated with the nucleic acid of these viruses.

D. VARIANT STRUCTURES

Most of the members of this group of viruses have been reported to produce a number of variant, tubular, spherical, and oblate structures.

1. *Tubular Forms*

Tubular forms have been frequently reported to be associated with polyoma virus (Bernard *et al.*, 1959; Howatson and Almeida, 1960; Mattern *et al.*, 1966). Isolated polyoma filaments of three diameters (about 48, 38, and 22 mμ) and of variable diameter have been observed (Mattern *et al.*, 1967) and subsequently it was found that the larger filaments ($\geq$48 mμ) were devoid of DNA and represented flattened tubular shells (Mattern and DeLeva, 1968). An intermediate size (35–48 mμ) was found to be relatively unflattened and to contain DNA as deduced from its properties after staining with uranyl acetate. Similar DNA-containing filaments were observed *in vivo* (Mattern *et al.*, 1966) and were thought to represent a filamentous precursor of the mature icosahedral form of the small plaque strain of polyoma virus.

Kiselev and Klug (1969) have undertaken a detailed analysis of wide and narrow tubular variants of human wart and rabbit papilloma viruses which had been described earlier (Williams *et al.*, 1960; Howatson, 1962; Breedis *et al.*, 1962; Noyes, 1964; Finch and Klug, 1965). Wide tubes (45–55 mμ) appeared to consist of a hexagonal net of hexameric capsomeres, whereas narrow tubes appeared to have a novel organization consisting of two types of tessellation of pentameric units in a cylindrical form.

Other tubular variants have been observed in RKV (Chambers *et al.*, 1966; Crawford and Follett, 1967), canine papilloma virus (Cheville and Olson, 1964), SV40 (Bernard *et al.*, 1959), and PML virus (Howatson *et al.*, 1965).

2. *Spherical or Oblate Variants*

A number of spherical variants believed to consist of 32 and 12 capsomeres have been reported both in polyoma virus and SV40 preparations (Mattern *et al.*, 1967; Anderer *et al.*, 1967). The latter authors also reported a 42 capsomere particle, and, although their identification of fivefold axes is not convincing, such structures may well exist.

Somewhat elongated closed shells, some of which may represent tubules with rounded icosahedral ends, have been described in papilloma viruses (Finch and Klug, 1965b).

At present the most likely explanation for the presence of these structures is that they are naturally produced, are artifacts of purification (e.g., the empty 72 capsomere "top component" shells), or that they represent aggregation of isolated capsomeres into any of the possible icosahedral and tubular shells. In the light of recent evidence on the small size of the virus particle, the multiple capsomere shell hypothesis advanced by Mattern *et al.* (1967) seems an unlikely possibility.

REFERENCES

Anderer, F. A., Schlumberger, H. D., Koch, M. A., Frank, H., and Eggers, H. J. (1967). *Virology* **32**, 511–523.

Anderer, F. A., Koch, M. A., and Schlumberger, H. D. (1968). *Virology* **34**, 452–458.

Bernhard, W., Febvre, H. L., and Cramer, R. (1959). *C. R. Acad. Sci. Paris* **249**, 483–485.

Black, P. H. (1968). *Ann. Rev. Microbiol.* **22**, 391–426.

Black, P. H., Hartley, J. W., Rowe, W. P., and Huebner, R. J. (1963). *Nature (London)* **199**, 1016–1018.

Black, P. H., Crawford, E. M., and Crawford, L. V. (1964). *Virology* **24**, 381–387.

Blackstein, M. E., Stanners, C. P., and Farmilo, A. J. (1969). *J. Mol. Biol.* **42**, 301–313.

Boiron, M., Levy, J. P., Thomas M., Friedmann, J. C., and Bernard, J. (1964). *Nature (London)* **201**, 423–424.

Boiron, M., Thomas, M., and Chenaille, P. (1965). *Virology* **26**, 150–153.

Breedis, C., Berwick, L., and Anderson, T. F. (1962). *Virology* **17**, 84–94.

Burns, W. H. and Black, P. H. (1968). *J. Virol.* **2**, 606–609.

Caspar, D. L. D. (1966). *J. Mol. Biol.* **15**, 365–371.

Chambers, V. C., Hsia, S., and Ito, Y. (1966). *Virology* **29**, 32–43.

Cheville, N. F. (1966). *Cancer Res.* **26**, 2334–2339.

Cheville, N. F. and Olson, C. (1964). *Amer. J. Pathol.* **45**, 849–859.

Crawford, L. V. (1964). *Virology* **22**, 149–152.

Crawford, L. V. (1965). *J. Mol. Biol.* **13**, 362–372.

Crawford, L. V. (1969). *Advan. Virus Res.* **14**, 89–152.

Crawford, L. V. and Black, P. H. (1964). *Virology* **24**, 388–392.

Crawford, L. V. and Crawford, E. M. (1963). *Virology* **21**, 258–263.

Crawford, L. V. and Follett, E. A. C. (1967). *J. Gen. Virol.* **1**, 19–24.

Crothers, D. M. and Zimm, B. H. (1965). *J. Mol. Biol.* **12**, 525–536.

Defendi, V. (1966). *Progr. Exp. Tumor Res.* **8**, 125–188.

Eckhart, W. (1968). *Phys. Rev.* **48**, 513–533.

Eddy, B. E. (1969). *Monogr. Virol.* **7**, 1–114.

Eigner, J. and Doty, P. (1965). *J. Mol. Biol.* **12**, 549–580.

Finch, J. T. and Klug, A. (1965). *J. Mol. Biol.* **13**, 1–12.
Fine, R., Mass, M., and Murakami, W. T. (1968). *J. Mol. Biol.* **36**, 167–177.
Fogel, M. and Sachs, L. (1969). *Virology* **37**, 327–334.
Fogel, M. and Sachs, L. (1970). *Virology* **40**, 174–177.
Gerber, P. (1963). *Science* **140**, 889–890.
Gerber, P. (1964). *Science* **145**, 833.
Granoff, A. (1969). *Curr. Topics Microbiol.* **50**, 107–137.
Gray, H. B., Jr., Bloomfield, V. A., and Hearst, J. E. (1967). *J. Chem. Phys.* **46**, 1493–1498.
Hare, J. D. and Morgan, H. R. (1963). *Virology* **19**, 105–107.
Howatson, A. F. (1962). *Brit. Med. Bull.* **18**, 193–198.
Howatson, A. F. and Almeida, J. D. (1960). *J. Biophys. Biochem. Cytol.* **8**, 828–834.
Howatson, A. F., Nagai, M., and ZuRhein, G. M. (1965). *Can. Med. Ass. J.* **93**, 379–386.
Hudson, B., Clayton, D. A., and Vinograd, J. (1968). *Cold Spring Harbor Symp. Quant. Biol.* **33**, 435–442.
Kahler, H. Rowe, W. P., Lloyd, B. J., and Hartley, J. W. (1959). *J. Nat. Cancer Inst.* **22**, 647–657.
Kiselev, N. A. and Klug, A. (1969). *J. Mol. Biol.* **40**, 155–171.
Kleinschmidt, A. K., Kass, S. J., Williams, R. C., and Knight, C. A. (1965). *J. Mol. Biol.* **13**, 749–756.
Klug, A. (1965). *J. Mol. Biol.* **11**, 424–443.
Klug, A. and Finch, J. T. (1965). *J. Mol. Biol.* **11**, 403–423.
Klug, A. and Finch, J. T. (1968). *J. Mol. Biol.* **31**, 1–12.
Koch, M. A., Eggers, H. J., Anderer, F. A., Schlumberger, H. D., and Frank, H. (1967). *Virology* **32**, 503–510.
Koprowski, H., Jensen, F. C., and Steplewski, Z. (1967). *Proc. Nat. Acad. Sci. U. S.* **58**, 127–133.
LeBouvier, G. L., Sussman, M., and Crawford, L. V. (1966). *J. Gen. Microbiol.* **45**, 497–501.
Maizel, J. V., Jr., White, D. O., and Scharff, M. D. (1968). *Virology* **36**, 115–125.
Markham, R., Frey, S., and Hills, G. J. (1963). *Virology* **20**, 88–102.
Mattern, C. F. T. (1969). *In* "Biochemistry of Viruses" (H. B. Levy, ed.), pp. 55–100. Dekker, New York.
Mattern, C. F. T. and DeLeva, A. M. (1968). *Virology* **36**, 683–685.
Mattern, C. F. T., Takemoto, K. K., and Daniel, W. A. (1966). *Virology* **30**, 242–256.
Mattern, C. F. T., Takemoto, K. K., and DeLeva, A. M. (1967). *Virology* **32**, 378–392.
Matthews, R. E. F. and Ralph, R. K. (1966). *Advan. Virus Res.* **12**, 273–328.
Melnick, J. L. (1962). *Science* **135**, 1128–1130.
Michel, M. R. Hirt, B., and Weil, R. (1967). *Proc. Nat. Acad. Sci. U. S.* **58**, 1381–1388.
Morgan, H. R. and Balduzzi, P. C. (1964). *Proc. Nat. Acad. Sci. U. S.* **52**, 1561–1564.
Murakami, W. T., Fine, R., Harrington, M. R., and BenSassan, Z. (1968). *J. Mol. Biol.* **36**, 153–166.
Noyes, W. F. (1964). *Virology* **23**, 65–72.
Noyes, W. F. (1965). *Virology* **25**, 358–363.
Olson, C., Segre, D., and Skidmore, L. V. (1960). *Am. J. Vet. Res.* **21**, 233–242.
Oroszlan, S. and Rich, M. A. (1964). *Science* **146**, 531–533.
Ozer, H. L., Takemoto, K. K., Kirschstein, R. L., and Axelrod, D. (1969). *J. Virol.* **3**, 17–24.
Rdzok, E. J., Shipkowitz, H. L., and Richter, W. R. (1966). *Cancer Res.* **26**, 160–165.

Robl, M. G. and Olson, C. (1968). *Cancer Res.* **28**, 1596–1604.

Schlumberger, H. D., Anderer, F. A., and Koch, M. A. (1968). *Virology* **36**, 42–47.

Schwerdt, P. R., Schwerdt, C. E., Silverman, L., and Rubinstein, L. J. (1966). *Virology* **29**, 511–514.

Shapiro, A. L., Vinuela, E., and Maizel, J. V., Jr. (1967). *Biochem. Biophys. Res. Commun.* **28**, 815–820.

Stewart, S. E. (1960). *Advan. Virus Res.* **7**, 61–90.

Tajima, M., Gordon, D. E., and Olson, C. (1968). *Amer. J. Vet. Res.* **29**, 1185–94.

Takemoto, K. K. and Fabisch, P. (1970). *Virology* **40**, 135–143.

Thorne, H. V. (1968). *J. Mol. Biol.* **35**, 215–226.

Thorne, H. V. and Warden, D. (1967). *J. Gen. Virol.* **1**, 135–137.

Thorne, H. V., Evans, M. J., and Warden, D. (1968). *Nature (London)* **219**, 728–730.

Uchida, S., Watanabe, S., and Kato, M. (1966). *Virology* **28**, 135–141.

Uchida, S., Yoshiike, K., Watanabe, S., and Furuno, A. (1968). *Virology* **34**, 1–8.

Vasquez, D., Kleinschmidt, A. K., and Basilico, C. (1969). *J. Mol. Biol.* **43**, 317–325.

Vinograd, J., Lebowitz, J., Radloff, R., Watson, R., and Laipis, P. (1965). *Proc. Nat. Acad. Sci. U. S.* **53**, 1104–1111.

Vogt, M. (1970). *J. Mol. Biol.* **47**, 307–316.

Watkins, J. F. and Dulbecco, R. (1967). *Proc. Nat. Acad. Sci. U. S.* **58**, 1396–1403.

Watson, J. D. and Littlefield, J. W. (1960). *J. Mol. Biol.* **2**, 161–165.

Weil, R. and Vinograd, J. (1963). *Proc. Nat. Acad. Sci. U. S.* **50**, 730–738.

Williams, R. C., Kass, S. J., and Knight, C. A. (1960). *Virology* **12**, 48–58.

Winocour, E. (1963). *Virology* **19**, 158–168.

Winocour, E. (1968). *Virology* **34**, 571–582.

Winocour, E. (1969). *Advan. Virus Res.* **14**, 153–200.

Yoshiike, K. (1968a). *Virology* **34**, 391–401.

Yoshiike, K. (1968b). *Virology* **34**, 402–409.

ZuRhein, G. M. and Chou, S. (1965). *Science* **148**, 1477.

CHAPTER 4 *Adenoviruses**

ERLING NORRBY

I. Introduction

Adenoviruses capable of replicating in mammalian and avian cells have
been described. A comparative analysis of members of this group of viruses
might focus on a wide range of different properties. The following review
will attempt to summarize (a) structural characteristics of virion compo-

* The author's work presented in this review was supported by grants from the Swedish
Medical Research Council (projects Nos. B70-16X-548-06B and B70-16X-744-05) and
the Swedish Cancer Society (project Nos. 171-K69-02X and 171-K70-03X).

nents, (b) biological activities of virus-specific products, and (c) various aspects of virus–cell interactions.

Adenoviruses are taxonomically well defined, indicating a homogeneous phylogenetic origin. The problem of evolutionary differentiation into a variety of types with varying capacities to multiply in host cells of different species origin is of considerable interest. An understanding of factors determining whether certain cells are permissive, partially permissive, or nonpermissive for a certain serotype will be of great value as a means of providing insight into the oncogenic potential of different adenovirus types. Furthermore, knowledge of this kind will aid in an evaluation of the overall medical importance of zoonotic infections and also of the possibilities for genetic interactions between types recovered from host cells of different species origin.

A number of reviews, to a greater or lesser extent, have been concerned with comparative aspects of adenoviruses (Ginsberg, 1962; Rowe and Hartley, 1962; Brandon and McLean, 1963; Pereira et al., 1963; Sohier et al., 1965; Norrby, 1968). As time has progressed the adenovirus group has been expanded to include an ever increasing number of members. The knowledge of properties of individual members is also rapidly increasing with time. A summary of the state of knowledge as it now exists, therefore, seems necessary.

II. Definition of an Adenovirus

Adenoviruses contain doublestranded DNA with a molecular weight of 20 to 25 million daltons and with a length of 11 to 13 nm (Green et al., 1967; van der Eb et al., 1969; Doerfler and Kleinschmidt, 1970). In the virions, DNA appears associated with arginine-rich proteins (Laver et al., 1968; Prage et al., 1968). This nucleoprotein(s) is surrounded by a capsid composed of 252 capsomers forming an icosahedron with a diameter of about 80 nm. No envelope is present. Two kinds of capsomers built up of, at least, partly different polypeptides (Pettersson and Höglund, 1969) can be distinguished. These are nonvertex capsomers (hexons) and vertex capsomers (penton bases) (Valentine and Pereira, 1965; Norrby, 1966). The latter carry projections called fibers. The complex of a penton base and a fiber is referred to as a penton. This terminology of capsid components was proposed in 1966 by Ginsberg et al.

Hexons, penton bases, and also some fibers carry antigenic specificities, which are shared between adenoviruses or mammalian host cell origin (see below). These antigenic specificities have not been detected in components of avian adenoviruses. This fact occasionally was used as an argument for not including the latter types in this group. However, considering the structural similarity of avian and mammalian adenoviruses, and in the light of criteria adopted for establishing other virus groups, a separate grouping of the two seems not to be warranted.

Virus maturation occurs within nuclei of cells, although the translation of virus-specific messenger RNA into proteins occurs in the cytoplasm (Velicer and Ginsberg, 1968; Pina and Green, 1969). After having been synthesized the proteins are rapidly transported into the nucleus.

The description above is based mainly on data obtained in studies of human adenoviruses. It remains to be seen whether it is true in all details for other groups of adenoviruses. However, it seems likely that this will be the case and that adenoviruses will remain a relatively homogeneous group of viruses, readily distinguishable from other virus groups.

III. Summary of Identified Adenoviruses

Table I represents an attempt to list all types of adenoviruses which have hitherto (March, 1970) been described. The number of serotypes is of the order of 78, when specificities demonstrated by neutralization tests are considered. Criteria to be used for differentiation of adenoviruses into serotypes have not as yet been unequivocally established. As was discussed in a recent review (Norrby 1969a), a choice has to be made between specificities determined in neutralization and hemagglutination-inhibition tests, i.e., the immunological uniqueness of hexons and fibers, respectively. It was proposed that a separation into serotypes should be based on an individual behavior (more than eightfold homotypic preference of antiserum) in neutralization tests. Isolates distinguishable only in hemagglutination-inhibition tests should be considered as subtypes (strains) of the serotype identified by results of neutralization tests.

Some figures in Table I deserve special comments. The number of serotypes of human adenoviruses is usually given as 31. However, by the criteria given above serotype 29 should be regarded as a subtype of type 15 (cf. Rosen et al., 1962; Stevens et al., 1967). Thus 30 serotypes remain, to which should be added 2 types described in a very recent publication (Blacklow et al., 1969). The group of simian adenoviruses should be divided into two

TABLE I

ADENOVIRUSES OF DIFFERENT HOST CELL SPECIES ORIGIN

Species of "natural" host cells	Number of serotypes[a]	Key references[b]
Human	32	Stevens et al., 1967; Blacklow et al., 1969
Simian		
ape	7	Hillis and Goodman, 1969
monkey	16	Rapoza, 1967; Kim et al., 1967
Bovine		
cattle	7 or more	Rondhuis, 1968; Matsumoto et al., 1969, 1970; Bartha, 1969; Guenov et al., 1969
sheep	?	McFerran et al., 1969
Porcine	4	Clarke et al., 1967; Bibrack, 1969
Canine	2	Swango et al., 1969
Murine	2	Hartley and Rowe, 1960; Reeves et al., 1967
Avian		
chicken	8	Kawamura et al., 1964; Burke et al., 1968
goose	1 (?)	Csontos, 1967

[a] Concerning criteria for identification see text.
[b] References were selected not to be the original descriptions of isolations, but rather to represent the most recent publication containing available knowledge.

(Hillis and Goodman, 1969). The 7 serotypes isolated from chimpanzees display characteristics which clearly relate them to certain human serotypes (see below). Characteristics of certain chimpanzee serotypes might even justify their association with the human adenovirus subgroup III, were they not primarily isolated from nonhuman sources. The number of simian-monkey serotypes was given as 18 by Rapoza (1967), but 6 of these were pairwise closely related in neutralization tests. Therefore these pairs should be reduced to individual serotypes, although concerning one pair (SV33 and SV38) somewhat contradictory results were obtained (Hull et al., 1965). Findings by Kim et al. (1967) suggest the occurrence of one more simian-monkey serotype, giving a total number of 16. Seven different serotypes of bovine adenoviruses have been distinctly identified and it seems likely that additional serotypes may occur (Bartha, 1969; Wilcox, 1969; Guenov *et al.*, 1969). Sheep adenoviruses were identified only recently. No comparative serological analysis of the 8 isolates made by McFerran *et al.* (1969) has as yet been undertaken. They may therefore be identical with, for example, previously known bovine serotypes. The number of established porcine serotypes is 4, but isolates hitherto not completely characterized have been described (Horzinek and Uberschär, 1966). The problem whether the canine adenoviruses, infectious canine hepatitis (ICH), and laryngotracheitis (ICL) viruses, are immunologically sufficiently different to allow their differentiation into two serotypes, has not been completely settled. Accumulated evidence, however, suggests that they should be described as two distinct canine adenoviruses (Swango *et al.*, 1969; Marusyk *et al.*, 1970). The number of serologically distinct avian adenoviruses also has been a subject for discussion. However, data by Kawamura *et al.* (1964) demonstrated 8 different serotypes among which GAL (Gallus adenolike) and CELO (chicken embryo lethal orphan) viruses were represented. Data by Clemmer (1964) and by Burke *et al.* (1968) demonstrated that some different isolates previously believed to be serologically distinct should be grouped together as single serotypes.

IV. Structural Characteristics of Virions and Virion Components of Adenoviruses

The fine structure of the core components of adenoviruses, i.e., DNA plus some basic (arginine-rich) proteins, as revealed by the negative contrast technique, was recently described (Laver *et al.*, 1968; Stasny *et al.*, 1968). It appeared as a fluffy spherical structure. No distinct characteristics of symmetry of isolated core structures were demonstrated. This internal component(s) is surrounded by an icosahedral capsid to form the complete virion. The structural complexity of this capsid has been extensively studied during the last 5 years. Most of these studies have concerned human adenoviruses. The current state of knowledge has been recently reviewed (Norrby, 1969a).

The characteristic icosahedral shape of the capsid was first shown for human adenovirus type 5 (Horne *et al.*, 1959) and later also for a number of other serotypes of human adenoviruses as well as representatives of different groups of animal adenoviruses (cf. Schlesinger, 1969). The capsid is composed of 252 capsomers. Nonvertex capsomers (hexons) and vertex capsomers (penton bases) are of a different nature (Valentine and Pereira, 1965; Norrby, 1966). The morphological appearance of nonvertex capsomers is somewhat dependent upon conditions of negative contrasting. Occasionally they were described as spherical, but more recent data suggest a tubular or otherwise hollow structure. In one report it was described that GAL virus differs structurally from other adenoviruses (MacPherson et al., 1961). Both the diameter of virions and of individual capsomers appeared somewhat larger. Additional reports (Davies and Englert, 1961; Watson et al., 1963) did not suggest any significant size difference from previously studied adenoviruses. Shadowed particles had a diameter of 78 to 90 nm (Atanasiu and Lepine, 1960) and in sections the diameter was estimated to be 86 nm (Sharpless et al., 1961). Finally, GAL virus was found to pass through filters with a diameter of 100 nm (Burke et al., 1968). Thus the weight of available evidence indicates the dimensions of GAL virus not to be different from that of other adenoviruses. Further information to settle this question would be appreciated. Of interest in this connection is the fact that structural characteristics of the serologically related avian adenovirus (CELO) (Kawamura and Tsubahara, 1963) do not seem to differ from those of nonavian adenoviruses (Dutta and Pomeroy, 1963; Petek et al., 1963; Clemmer, 1964).

Vertex capsomers seem to differ from other capsomers by having a wedgelike appearance (Wadell et al., 1969; Laver et al., 1969; Pettersson and Höglund, 1969). Furthermore they carry a projection, not only when forming a part of the capsid, but frequently also when occurring in an isolated form. Exceptions to this rule were found in studies of soluble components of human adenoviruses types 12 (Norrby and Ankers, 1969) and 31 (Ahmad-Zadeh et al., 1968). In preparations of these serotypes the major fraction of all soluble vertex capsomers did not occur associated with fibers, i.e., forming penton structures, but in a freelying form.

Vertex projections (fibers) vary in length. Among human adenoviruses five different lengths of fibers (10–11, 12–13, 17–19, 23–28, and 28–31 nm) have been encountered (cf. Norrby, 1969a). Studies on three simian-monkey adenoviruses (SA-7, SV15, and SV20) have revealed fiber lengths of 28 to 31 nm (Norrby, unpublished). A virion of SV20 is illustrated in Fig. 1. This figure also includes a picture of a virion of CELO virus. This avian adenovirus also carries projections. These are 14 nm long (Norrby, unpublished). Fibers have also been identified to be associated with virions of the two dog adenoviruses. The length of fibers were 25 to 27 nm and 35 to 37 nm in ICH and ICL virions, respectively (Marusyk et al., 1970).

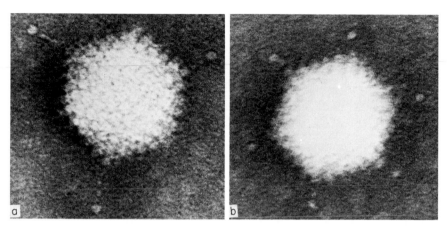

FIG. 1. Ultrastructure of virions of simian-monkey virus SV20 (a) and of CELO virus (b). Negative contrasting with sodium tungstosilicate. Magnification, × 570,000.

V. Hemagglutinating Activity of Adenoviruses; Association with Virus Products

The hemagglutinating activity of adenoviruses can be subjected to a comparative analysis from two different points of view. First, the capacity of a certain serotype to agglutinate red cells of different species origin might be considered. Second, a comparison of components—nonsoluble and soluble—carrying hemagglutinating activity may be of interest.

Table II represents a summary of hemagglutinating characteristics, which have been demonstrated for adenoviruses of different species origin. Some features deserve special comment. Generally, rat cells have been most successfully employed. This kind of cell was also described to be agglutinated by some viruses specific for avian cells. The serological specificity of this activity remains to be demonstrated. Most adenoviruses which do not agglutinate rat cells are capable of agglutinating monkey red cells.

A separation of human adenoviruses on the basis of their hemagglutinating properties yields subgroups comprising members which also share other biological properties. Exceptions to this rule are serotypes 20, 25, and 28 (poor monkey cell agglutinius; subgroup I) and 4 (partial agglutination of rat cells; subgroup III), which are biologically related to members of subgroup II (Wigand, 1970) and I (Norrby and Wadell, 1967), respectively. Recently it was proposed that on the basis of their hemagglutinating properties, types 20, 25, and 28 should in fact be placed in subgroup II instead

TABLE

HEMAGGLUTINATING ACTIVITY

"Natural" host cells	Subgroup designation	Serotype	Agglutinability of red cells		
			Rat	Monkey	Human
Human	Ia	3, 7, 14, 21	—	+(37)	—
	Ib	11, 16	—	+(4,37)	—
	Ic	20, 25, 28	—	+(37)	—
	IIa	8, 9, 10, 19	+(4,37)	—	+(4,37)
	IIb	13, 26, 27	+(4,37)	—	N.t.[f]
	IIc	15, 17, 22, 23, 24, 29, 30	+(4,37)	—	+$_p$(4,37)[d]
	IIIa,b,c+IIId	1, 2, 4, 5, 6+12, 18, 31	+$_p$(4,37)	—	+$_p$(4,37)
Simian-ape	I	One type	—	+(4,37)	—
	II	Four types	+$_p$(4,37)	—	—
	III	Two types	—	—	—
Simian-monkey	I	SV36	—	+(4,37)	N.t.[f]
	II	SV15, SV17, SV23, SV27, SV31, SV32, SV37, SV39 + one more type	+(4,37)	+(4)	+$_p$(4)
	IIIa	SV1, SV20, SV33, SV34, SV38	+$_p$(4,37)	+(4)	N.t.
	IIIb[e]+IIIc	SV11, SV25, SV30+SA7	+$_p$(4,37)	—	N.t.
Bovine	—	1, 2	+	N.t.	—
	—	3	—	N.t.	N.t.
Porcine	—	1	+(4,37)	+(4)	+(4,37)
	—	2, 3	—	—	—
Canine	—	ICH, ICL	—(4,37)	—	+(4,37)
Murine	—	—	—	—	—
Avian	—	1 (CELO)	+(4,37)	N.t.	N.t.
	—	2 (GAL-1)	—	N.t.	N.t.

[a] p, Partial agglutination. Figures within parenthesis denote temperature of agglutination.
[c] Only type 8 gives agglutination at 37°C.
[e] Subgroup IIIb separated out here.

II

OF ADENOVIRUSES

from different species[a]		Comments	References[b]
Guinea pig	Others		
—	—	Elute at 4°C, with the possible exception of type 21	*Rosen, 1960*
—	—	Agglutinate wide range of monkey red cells	Simon, 1962
—	—	Poor agglutinins. Related to subgroup II members	Schmidt et al., 1965; Bauer and Wigand, 1967;
+(4,37)[c]	—		Dreizin and Zolotarskaya, 1967
—	—	Types 9, 13, 15 and 23 are capable of agglutinating monkey cells	Henry et al., 1968
—	—		Wadell, 1969
—	—	Type 4 related to subgroup I Types 12, 18, 31 are poor agglutinins	Wigand, 1970
—	—	No agglutination of chimpanzee red cells	*Hillis and Goodman, 1969*
—	—		
—	—		
—	—		Tyrrell et al., 1960
+(4)	—		*Rapoza, 1967*
			Kim et al., 1967
+(4)	—	Higher titer with rat cell at 4° than at 37°C	
—	—	Data on SA-7 (Norrby, unpublished)	
—	Mouse (type 2 only)	Temperature of agglutination not described. Serotypes 4 to 6 not	*Klein, 1962*
N.t.	Mouse	studied	Kern et al., 1969
+(4,37)	Mouse (4)		
—	—	Serotype 4 not studied	*Clarke et al., 1967*
+	—		*Kunishige and Hirato, 1960;*
			Espmark and Salenstedt, 1961
—	—		Hartley and Rowe, 1960
N.t.	—	Avian cells are not agglutinated	*Clemmer, 1964*
N.t.	—	Other serotypes have not been studied	Burke et al., 1968

[b] References are given in a chronological order for each group of viruses. Key references are in italics.

[d] Only type 15 was studied.

[f] N.t., not tested.

of I (Hierholzer and Dowdle, 1970). Furthermore serotypes 12, 18, and 31 form a separate group within subgroup III, based on a number of distinct biological characteristics.

One member of subgroup II, type 9, was found to be eluted from human 0 erythrocytes by *Cholera vibrion* filtrate (Norrby *et al.*, 1967). Further studies employing preparations of purified enzymes, have suggested that this effect is due to neuraminidase (Norrby, unpublished). In addition to certain members of subgroup II, which were previously described to give a complete agglutination of human 0 erythrocytes (Rosen, 1960), other members of the same subgroup, as well as members of subgroup III, were found to be capable of agglutinating these cells (Wadell, 1969; Norrby and Ankerst, 1969). However, in these cases, only a partial agglutination was obtained. The phenomenon of partial agglutination has been found to be due to a competition between incomplete and complete hemagglutinins (see below) for receptors on red cells (Wadell, 1969).

The highly oncogenic simian adenovirus SA-7 was described by Rapoza (1967) to lack hemagglutinating activity. However studies in the author's laboratory (Norrby, unpublished) have demonstrated that this serotype can give a partial agglutination of rat erythrocytes and that this activity is specifically inhibited by homotypic antiserum.

Two incomplete and three complete soluble hemagglutinins (HA's) have

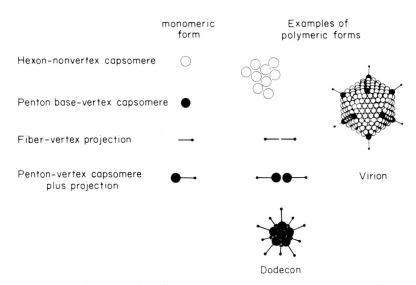

FIG. 2. Schematic description of monomeric and some experimentally identified polymeric forms of adenovirus capsid components. (Reproduced by due permission from Norrby, 1969a.)

been demonstrated in studies of human adenoviruses. The incomplete HA's were isolated fibers and pentons and the complete HA's dodecons (composed of 12 pentons plus some extra structure), dimers of pentons, and dimers of fibers (Fig. 2; cf. Norrby, 1969a). These soluble complete HA's occur in various combinations in preparations of different types of human adenoviruses. All possible combinations have been encountered except the simultaneous occurrence of dodecons and dimers of pentons.

Few data are available concerning the occurrence of nonsoluble and different soluble HA's among adenoviruses of extrahuman host cell specificity. In preparations of dog adenoviruses (ICH and ICL) it was found that, in contrast to the situation with most human adenoviruses, the major fraction of all HA activity resided in virions (Marusyk *et al.*, 1970). Soluble HA of the penton dimer type was also demonstrated. Similarly, preparations of some simian adenoviruses (SA-7, SV15) display HA activity carried by virions and empty capsids and by penton dimers, although in these cases, the major part of the activity was carried by the soluble HA (Norrby, unpublished).

VI. Comparison of Immunological Characteristics of Adenovirus-Specific Proteins

A comparative analysis of immunological characteristics of virus products may concern both nonstructural and structural proteins. The major part of all studies of nonstructural proteins has concerned tumorlike and T antigens, which appear as early proteins during the lytic multiplication of adenoviruses and occur in tumor cells, respectively. At present no data are available suggesting that tumorlike and T antigens of individual serotypes can be differentiated. They will, therefore, be referred to by the common term T antigen.

T antigen specificities, as determined by immunofluorescence and complement fixation, have been used for subgrouping of human adenoviruses (Huebner, 1967; Gilden *et al.*, 1968a; McAllister *et al.*, 1969). Four different subgroups were distinguished and correlation between this classification and oncogenic potentiality *in vivo* was demonstrated. Low level cross-reactivity between representatives of different subgroups were seen by use of the immunofluorescence technique (Riggs *et al.*, 1968). However, the significance of these reactions may be somewhat doubtful (Rapp, 1969). T antigens have also been identified in cells infected with simian-monkey type (Hull *et al.*, 1965; Gilden *et al.*, 1967, 1968b), bovine, canine (Kern *et al.*, 1969), and avian adenoviruses (Potter and Oxford, 1969; Anderson *et al.*, 1969). Simian-monkey serotypes could be separated into three different subgroups, not including nononcogenic serotypes. No cross-reactivity has been demonstrated between T antigens induced by human, simian-monkey, bovine, canine, and avian oncogenic adenoviruses except in one study of the first two groups (Riggs and Lennette, 1967). This observation remains to be confirmed. It is obvious that T antigens include a range of different virus products. In order to make a critical comparative immunological analysis, a separation of these different products has to be made. Some attempts to purify T antigens have been described (Gilead and Ginsberg, 1968a,b; Tockstein *et al.*, 1968; Jainchill *et al.*, 1969; Riggs *et al.*, 1969; Potter *et al.*, 1970), but frequent problems have been encountered due to the lability of the products concerned.

Noncapsid structural components have been identified in adenovirus virions (Russell and Knight, 1967). However, only in one study was a preliminary comparative immunological analysis undertaken (Prage *et al.*, 1968). A homotypic preference was found in gel diffusion tests, but more recent studies indicate the occurrence of group specificities (Prage, personal communication).

Capsid components as studied in human adenoviruses carry a wide range of antigenic specificities (cf. Norrby, 1969a). Group-specific CF antigen which, as generally measured, represents a component of hexons (Valentine and Pereira, 1965; Norrby, 1966), has been found to be shared not only between human adenoviruses, but also to occur in extrahuman adenoviruses. However this antigenic component was not found in avian adenoviruses (Rowe and Hartley, 1962; Kawamura and Tsubahara, 1963; Clemmer, 1964).

In order to analyze the qualitative and quantitative immunological relationships between hexons of adenoviruses of different host cell origin, cross-absorption experiments were performed in the author's laboratory. Purified hexons and antisera against this component prepared from four different human adenoviruses representing different subgroups (types 3, 15, 6, and 12), two simian adenoviruses (SA-7 and SV20), and two dog adenoviruses (ICH and ICL) were used. In addition crude cell harvests of CELO virus and an antiserum against this virus were included. Prior to performance of absorption experiments, titers of sera against eight units of the different antigens were determined. Results obtained are summarized in Table III. In previous studies of relationships between hexon CF antigens, one-way cross-reactions were often encountered (Rowe and Hartley, 1962). In the present study no

TABLE III

COMPLEMENT FIXATION (CF) ANTIBODY TITERS OF SERA AGAINST PURIFIED HEXONS OF
SELECTED HUMAN (WITHOUT PREFIX) AND EXTRAHUMAN ADENOVIRUSES IN TESTS WITH
HOMOTYPIC AND HETEROTYPIC HEXONS

Serum against hexons of type[a]	CF antibody activity in tests with hexons of types[b]								
	3	15	6	12	SA-7	SV20	ICH	ICL	CELO
3	*640*	320	80	320	40	80	160	80	<20
15	640	*1280*	320	1280	80	320	320	640	<40
6	2560	1280	*2560*	2560	320	640	640	640	<80
12	2560	1280	1280	*2560*	160	640	640	640	<80
SA-7	640	320	320	640	*640*	640	320	640	<20
SV20	1280	640	640	1280	320	*1280*	640	640	<40
ICH[c]	320	160	160	160	80	320	*640*	640	<20
ICL	320	320	320	160	80	320	1280	*1280*	<40
CELO	<10	<10	<10	<10	<10	<10	<10	<10	80

[a] Hexons to be used for immunization and absorption were prepared from naturally occurring soluble components by anion-exchange chromatography, in some cases in combination with zonal centrifugation (Norrby and Wadell 1969, and unpublished; Marusyk et al., 1970). The preparations were controlled by biological tests not to contain pentons, fibers or free vertex capsomers.

[b] The technical procedure for CF testing was previously described (Norrby 1969b). Different serum dilutions were tested against 8 units of CF antigen. Antigen titers were determined with an antiserum against virions of type 3 except for type 3 hexons and crude CELO virus material in which tests anti-adenovirus type 9 virion and homotypic antiserum were used, respectively.

[c] The antisera against hexons of ICH gave a complete lysis of red cells only after incubation with ICH or ICL hexons. In tests with hexons of adenoviruses of different host cell species partial lysis and prozone phenomena were seen.

one-sided preference of this kind was found except to some extent regarding the cross-reactions between human adenoviruses and ICH, as previously noticed (Carmichael and Barnes, 1962). Cross-reactions were recorded between all hexons except those present in CELO virus material, as should be expected. There was some variation between titers obtained with one and the same serum against different hexons. The difference varied between 2- to 16-fold. SA-7 hexons seemed to contain only a limited spectrum of antigen specificities as demonstrated by the relatively low titers obtained with all heterotypic sera and also by the small variation in serum titers recorded with anti-SA-7 hexon serum in tests with different antigens. No preference for sera to give high titers with hexons of serotypes of the same species origin was seen except to some extent with the two dog adenoviruses. These findings suggest the absence of any hexon antigenic component shared between adenoviruses of a certain host cell species origin. Further evidence for this was found in cross-absorption experiments (Table IV). Thus, for example, absorption of an antiserum against type 3 hexons with SV20 hexons or a serum against SA-7 hexons with type 3 hexons, removed all antibody activity against some members of both the same and also of a different host cell species origin. However, absorption of an antiserum against ICH hexons with hexons of type 3 removed all antibodies except those reacting with both ICH and ICL virus hexons. This might have been anticipated since ICH and ICL viruses have immunologically closely related hexons as indicated by the occurrence of mutual cross-reactions in neutralization tests (Swango *et al.*, 1969). The species-specific antigen shared between these two types might therefore be more comparable to type (or

TABLE IV

The Effect of Absorption with Adenovirus Hexons of Different Host Cell Species Origin on the CF Activity of Antihexon Sera Against Hexons of Certain Types of Adenoviruses

Serum against hexons of type[a]	Absorbed with hexons of type[a,b]	CF antibody activity in tests with hexons of types[a]							
		3	15	6	12	SA-7	SV20	ICH	ICL
3	—	320	320	80	320	40	80	160	80
	SV20	320	20	<10	40	<10	<10	<10	<10
6	—	2560	1280	2560	2560	320	640	320	640
	SA-7	<40	80	1280	80	<40	<40	40	<40
SA-7	—	640	320	320	640	640	640	320	640
	Ad 3	40	40	40	80	640	640	80	<40
SV20	—	1280	640	640	1280	320	1280	640	640
	ICL	320	160	80	320	40	1280	<40	<40
ICH	—	320	160	160	160	80	320	1280	1280
	Ad 3	<20	<20	<20	<20	<20	<20	640	320

[a] See corresponding footnotes of Table III.

[b] The technique used for absorption was previously described (Norrby, 1969b; Norrby and Wadell, 1969).

possibly subgroup) specificities as demonstrated in hexons of human adenoviruses (Norrby, 1969b; Norrby and Wadell, 1969) rather than a species
group specificity.

Cross-reactions between vertex capsomers of adenoviruses of different
species origin were studied by use of the penton hemagglutination-enhance-
ment (HE) test. Antisera against pentons or penton oligomers were tested
against purified pentons. Serum "profiles" (cf. Wadell and Norrby, 1969)
recorded in these tests have been summarized in Fig. 3. It was found that
vertex capsomers of adenoviruses of different species origin share one or

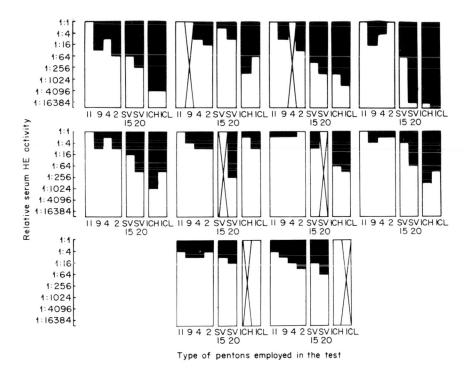

FIG. 3. Hemagglutinating enhancement (HE) activity of sera against monomers or
polymers of pentons of selected serotypes of adenoviruses of different host cell origin
in tests with different heterotypic pentons (denoted along bottom part of columns).
Crossed columns mark the serotype origin of antibodies examined in individual sets of
HE tests. *Top row*: sera against human adenovirus types 3 and 9 dodecons, types 4 and 6
pentons. *Middle row*: sera against vertex capsomers of human adenovirus type 12, monkey
adenovirus types SV15, and SV20 pentons and whole virus material of type SA-7. *Bottom
row*: sera against pentons of dog adenoviruses ICH and ICL. The HE titer of sera in
tests with different pentons is expressed in a relative fashion. The unfilled part of a column
gives the ratio of serum HE activity with a specific penton over maximal titer obtained
with any heterotypic penton.

more antigenic specificities. However, in contrast to the immunological relationships between hexons there was a preference for sera to react with pentons of members of the same host cell species rather than those of other species. Thus is some cases, e.g., antisera against type 6 pentons, cross-reactions were demonstrable only with sera containing very high titers of HE antibodies. One exception to this general rule was that antisera against human type 9 dodecons gave an equally good enhancement of some pentons of both human and simian origin. Furthermore antisera against pentons of SV15, SV20, and SA-7 displayed a higher HE activity in tests with pentons of human serotypes than of heterologous simian types. SV15 pentons generally were more readily demonstrated than SV20 pentons. SA-7 pentons could be obtained only in low titers and were notoriously difficult to clearly demonstrate. They have therefore not been included in Fig. 3. Anti-CELO virus serum did not display any penton HE activity against any of the different pentons studied.

A possible occurrence of common antigen(s) in fibers of adenoviruses of different host cell species origin was analyzed by the use of HE tests in a way analogous to that described above for penton incomplete HA. Purified fiber incomplete HA's of members of subgroups II and III of human adenoviruses and of monkey (SV15, SV20) and dog (ICH, ICL) adeno-viruses, and antisera against these components were employed (Fig. 4). Fibers of SV20 virus were found to behave identically with fibers of members of subgroup III of human adenoviruses (except type 4). In addition low-grade one- or two-way crossings were found in certain combinations be-tween all other fibers studied. Antisera against fibers of ICH and ICL viruses (not included in Fig. 4) gave a slight aggregation of SV20 fibers, but not of fibers from any other serotype of a different host cell species origin.

The only previous publications dealing with possible immunological relationships between vertex capsomers and fibers of adenoviruses of dif-ferent host cell species origin concerned human and simian-ape types. Cross-reactions were found in HI tests between fibers of human adenovirus type 16 and the simian-ape type of subgroup I (identical with C-1 virus originally described by Rowe et al., 1958) and of human adenovirus types 2 and 5 and two members of simian-ape subgroup II (Hillis et al., 1969). In the presence of antisera against members of human subgroups III, a considerable enhancement of the hemagglutinating activity of members of simian-ape subgroup II was obtained, indicating additional immunological relationships between fiber (or possibly penton) components of these adeno-viruses. It can finally be mentioned that Rowe et al. in their early publica-

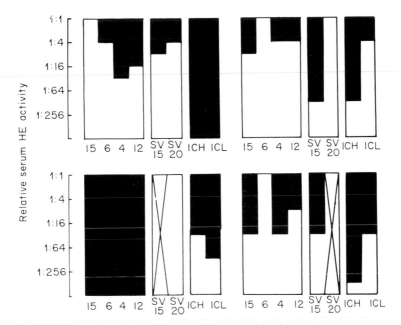

FIG. 4. Hemagglutination enhancement (HE) activity of sera against fibers of selected serotypes in tests with different heterotypic fibers. Upper part of staple diagram: sera against fibers of human adenoviruses types 9 (left) and 2 (right). Lower part of diagram: sera against fibers of monkey adenoviruses types SV15 and SV20. Concerning the arrangement of the diagram, see Fig. 3.

tion (1958) demonstrated that the simian-ape member of subgroup I cross-reacted in neutralization tests with human serotype 14. These data taken together with those on the characteristic host cell specificity of simian-ape adenoviruses (see below) emphasize the close relationship between this group and human adenoviruses.

VII. Some Aspects on Adenovirus-Cell Interactions; Cytopathology, Lytic and Nonlytic Multiplication

Adenovirus-cell interactions can be discussed from a number of different points of view. First, the host cell range, i.e., the permissive or nonpermissive character of cells of different species origin, for certain types of adenoviruses may be analyzed. Second, the sequence of events in permissive and nonpermissive cells may be examined on a morphological (cytopathology) or on a biochemical level. Third, a comparative analysis of different characteristics of the state of nonpermissiveness and the tendency of certain cells to become transformed or otherwise affected by the virus infection might the profitably carried out. The present state of knowledge with regard to these problems was summarized in the excellent review by Schlesinger (1969). A more penetrating analysis of adenovirus-cell interactions is outside the scope of the present comparative review. The presentation therefore will be limited to some comments on (a) cytopathology in permissive cells, (b) host cell range, and (c) oncogenic activity of adenoviruses of different species origin.

A. Cytopathology in Permissive Cells

A number of studies have been undertaken to determine the cytopathology of human adenoviruses. In spite of this it is difficult to give a simple schematic presentation of characteristic patterns which have been encountered. This is partly due to the conflicting nature of some of the data presented and also, on some points, due to the scarcity of information. To some extent, different members of the same HA subgroup give similar intranuclear changes. Serotypes 3 and 7, subgroup I (plus the aberrant member of subgroup III, type 4), give a cytopathology different from that of members of subgroup III, types 1, 2, and 5 (Boyer et al., 1957, 1959; Barski and Cornfeit, 1958). The exact position of type 4 with regard to its cytopathology is somewhat unclear since not only has it been described as belonging to subgroup I, but also to subgroups II (Wigand and Bauer, 1964; Loytved and Wigand, 1969) and III (Rafajko, 1965). Members of subgroup II (including the three aberrant members of subgroup I, types 20, 25, and 28 (Wigand, 1970) display a kind of cytopathology different from that of members of subgroups I and III (Loytved and Wigand, 1969). However, one member of subgroup II, type 8, gives a cytopathology to some extent resembling that of both subgroup I and III members. Concerning types 12 and 18, representatives of subgroup IIId, information on cytological changes seems to be available only on the ultrastructural level. Some morphological changes, e.g., structural characteristics of paracrystals, clearly differentiate them from other members of human adenoviruses (Martinez-Palomo et al., 1967; Weber and Liao, 1969). Morphological features (lattice spacing, etc.) of paracrystals, when present (types 2, 4, 5, 6, 12, and 18), are different for members representing different subgroups (cf. Weber and Liao, 1969).

Information concerning cytopathology of different nonhuman adenoviruses as compared to that of members of different subgroups of human adenoviruses have been tabulated in Table V. The relative uncertainty of this compilation of data should be emphasized. Varying cytological techniques were employed to carry out the more or less complete studies which have been published. Cytopathology classification of some simian-monkey adenoviruses is partly based on unpublished data mentioned by Rapoza (1967). The only adenovirus outside the group of human adenoviruses which has been demonstrated to cause the formation of paracrystalline structures is ICH virus (Givan et al., 1967; Matsui and Bernhard, 1967). These structures are different from those appearing during multiplication of certain human serotypes.

B. HOST CELL RANGE

The host cell range of adenoviruses is generally described as "relatively species specific." However, a number of serotypes have been found to induce a cytopathic degeneration in cells from certain species other than those of the "natural" host. In some cases this degeneration may have been caused by "toxic" activity of the virus preparation or by an incomplete multiplication of the virus, but in other cases a complete, cytolytic virus multiplication was demonstrated. Different cell lines derived from a certain species have occasionally been found to display a varying capacity to support the multiplication of adenoviruses of a different host cell species origin. One more factor complicating the analysis of the host cell range is the fact that in certain systems adenoviruses can replicate in nonpermissive cells by assistance of a helper virus (cf. Rapp, 1969).

Certain human adenoviruses besides replicating in human cells, also were found capable of multiplying in primary cultures of established cells lines of simian-ape (Schell et al., 1968), simian-monkey (Beardmore et al., 1965; Jerkofsky and Rapp, 1968; Schell et al., 1968), bovine (Warren and Cutchins, 1957; Liao and Weber, 1969), pig (Guerin and Guerin, 1957; Hitchcock et al., 1960), rabbit (Rowe et al., 1955; Kelly and Pereira, 1957, Ankudas and Khoobyarian, 1962), rat (Rowe et al., 1955), and hamster origin (Takahashi et al., 1966; Cooper et al., 1967; Shimojo and Yamashita, 1968). Multiplication of human adenoviruses in vivo in rabbits (Pereira and Kelly, 1957), pigs (Betts et al., 1962), and dogs (Sinha et al., 1960) has been described. Conversely nonhuman adenoviruses in some cases were found to give a multiplication in vitro in human cells, e.g., simian-monkey serotypes (Hammon et al., 1963) and a mouse serotype (Sharon and Pollard, 1964).

TABLE V

CYTOPATHOLOGY OF ADENOVIRUSES IN PERMISSIVE CELLS

"Natural" host cells	Hemagglutinating subgroup	Serotype studied	Nuclear changes similar to members of Rosen's subgroup				References
			Ia,b	II	IIIa,b	IIId	
Human	Ia	3, 7	+	—	—	—	Boyer et al., 1957, 1959; Barski and Cornfeit, 1958
	Ic	20, 25, 28	—	+	—	—	
	IIa	8	+	—	+	—	Loytved and Wigand, 1969
	IIa to c	9, 10, 13, 15, 17, 19, 22, 23, 24, 26, 27	—	+	—	—	
	IIIa,b	1, 2, 5	—	—	+	—	Boyer et al., 1957, 1959; Barski and Cornfeit, 1958
	IIIc	4	+	—	—	—	
	IIId	12, 18	—	—	—	+	Martinez-Palomo et al., 1967; Weber and Liao, 1969
Simian-ape	I	C-2	+	Not studied	Not studied	Not studied	Hillis et al., 1969
Simian-monkey	II	SV15, SV17, SV27	+	—	+	—	Prier and LeBeau, 1958; Tyrrell et al., 1960; Fong et al., 1968
	III + V	SA-7	—	—	+	—	Malherbe et al., 1963; Rapoza, 1967
Bovine	—	1, 2, 3	—	—	+	—	Klein et al., 1959, 1960
	—	4, 5, 6	+	—	—	—	Bartha, 1969
Porcine	—	1, 2, 3	+	—	—	—	Chandler et al., 1965; Clarke et al., 1967
Canine	—	ICH	+	—	+	—	Tajima et al., 1961; Garg et al., 1967
	—	ICL	—	—	+	—	Yamamoto 1969a,b
Murine	—		—	—	+	—	Starr et al., 1965
Avian	—	GAL	—	—	+	—	Atanasiu and Lepine, 1960; Sharpless et al., 1961
	—	CELO	—	—	+	—	Petek et al., 1963

Simian-ape serotypes represent an interesting group of adenoviruses which multiply equally well in both simian and human cells (Rowe *et al.*, 1958; Hillis and Goodman, 1969; Hillis *et al.*, 1968, 1969), whereas simian adenoviruses of the monkey type in most cases are restricted to simian cells (Hammon *et al.*, 1963, Hull, 1968). Infection of cells from a species other than the "natural" one, in the majority of all cases, leads to an abortive cycle of virus replication. This abortive cycle may include a varying number of the steps of the productive cycle of multiplication. In primary monkey kidney cells certain human adenoviruses cause the appearance of T antigen, virus-specific DNA, but no structural components. In contrast, type 12 replication in hamster cells leads to production of T antigen, but no DNA. In the latter system, transformation may occur. Further information on virus-cell interactions in nonpermissive systems and their significance with respect to oncogenesis may be found in reviews by Schlesinger (1969) and Rapp (1969).

C. Oncogenic Activity of Adenoviruses of Different Species Origin

The oncogenic activity of adenoviruses is expressed by transformation of rodent cells *in vivo* or *in vitro*. Serotypes proved to be capable of displaying an activity of this kind have been summarized in Table VI. Oncogenic subgroups of human adenoviruses are identified by the occurrence of cross-reacting T antigens and cross-hybridizable virus-specific RNA in tumor cells. Members of subgroups C and D so far have only been shown to transform cells *in vitro*. The composition of the oncogenic subgroups compares with those of HA subgroups of Rosen. The only differences are that serotype 4 which is a subgroup III member, belongs to oncogenic subgroup B (Huebner, 1966), and that serotypes 20, 25, and 28 (subgroup I members) share a T antigen with members of oncogenic subgroup D (McAllister *et al.*, 1969). Members of different oncogenic subgroups may share tumor-specific transplantation antigen(s) (Sjögren *et al.*, 1967).

Human adenoviruses are frequently classified as highly, moderately, or weakly oncogenic. Attempts have been made to correlate the guanine/cytosine (GC) content of virus DNA as compared to that of cellular DNA with oncogenic potential of certain serotypes (cf. Schlesinger, 1969). As regards human adenoviruses it was demonstrated that the lower the GC content, i.e., the more the composition of virus DNA resembled cellular DNA, the higher was the oncogenic capacity. However, recent studies demonstrated that SA-7, which is the adenovirus carrying the highest onco-

TABLE VI

Oncogenic or Transforming Capacity of Adenoviruses[a]

"Natural" host cells	Oncogenic subgroup designation	Antigenic type	References[b] for transformation in vivo	in vitro
Human	A	12	Trentin et al., 1962	Reed, 1967; Freeman et al., 1967a
		18	Huebner et al., 1962	
		31	Pereira et al., 1965	
	B	7	Girardi et al., 1964	
		3	Huebner et al., 1965	Freeman et al., 1967b; van der Nordaa, 1968
		14, 16, 21		
		4		van der Nordaa (personal communication)
	C	1, 2		Freeman et al., 1967c; Gilden et al., 1968a
				van der Nordaa, 1968
				McAllister et al., 1969
Simian (monkey)	D	9, 10, 13, 15, 17, 19, 26	Hull et al., 1965	Altstein et al., 1967 (SA-7)
	I	SV1, SV11, SV25, SV35, SV34, SV38		
	II	SV20, SV23	Gilden et al., 1968b	Riggs and Lennette 1967 (SA-7, SV20)
	III	SA-7		
Bovine	—	3	Darbyshire, 1966	—
			Gilden et al., 1967	
Canine	—	ICH	Sarma et al., 1967	—
	—	ICL	Dulac et al., 1970	
Avian	—	CELO	Sarma et al., 1965	Anderson et al., 1969

[a] Modified from Schlesinger, 1969.

[b] References are to the first observations of tumor induction or in cases of groups of serotypes to the most comprehensive publication available.

genic potential, has a DNA with a high GC content (Pina and Green, 1968).

Concerning the capacity to transform cells, it should be pointed out that variations may occur between the *in vivo* and *in vitro* conditions. Members of the human oncogenic subgroups C and D are not capable of inducing neoplastic transformation *in vivo*, whereas this is readily demonstrated with members of subgroup A. In spite of this, members of these different subgroups were found to be approximately equal in efficiency of inducing transformation of rat cells *in vitro* (McAllister *et al.*, 1969).

VIII. Concluding Remarks

Comparative virology is of particular interest in cases when experiences gained from studies of a certain kind of structural or biological property can be used to predict other characteristics of a type, subgroup, or group of viruses. The biological activity originally employed for subgrouping of human adenoviruses was their hemagglutinating activity (Rosen, 1960). As the focus of interest came to be shifted to the oncogenic properties of adenoviruses other schemes of subdivision based on immunological cross-reactions between T antigens were proposed (Huebner, 1967; Gilden *et al.*, 1968a; McAllister *et al.*, 1969). A comparison of different subgroups obtained after separations based on these two different biological properties reveal considerable similarities. It seems therefore that the original subgrouping proposed by Rosen, supplemented with some additional subdivisions, may in itself be satisfactory and that other schemes with a different terminology are superfluous. Thus Rosen's subgroup IIId and Ia, Ib, and IIIc will represent Huebner's oncogenic subgroups A and B, etc. The usefulness of Rosen's subgrouping is furthermore evident from a comparison of clinical-epidemiological behavior, cytopathology, and structural characteristics of different serotypes.

The possibilities for a comparative analysis of members or groups of adenoviruses of a different host cell species origin are restricted by the lack of information with regard to certain essential biological characteristics. A comparison of information on hemagglutinating activity (Table II), cytopathology (Table V), and oncogenic or transforming activity (Table VI) only allows some different comments. The subgrouping of simian-ape adenoviruses was recently proposed (Hillis and Goodman, 1969) and it remains to be seen whether a correlation with other properties will be found. Somewhat more is known about simian-monkey adenoviruses. It is apparent that most oncogenic members belong to Rapoza's hemagglutinating subgroup III (Gilden *et al.*, 1968b). The only exceptions are SV1, which may be a special case (Gilden *et al.*, 1968b), and SV23, which is a member of subgroup II. No particular features of correlation between cytopathology in permissive cells and oncogenicity can be deduced. It is perhaps more appropriate to attempt finding correlations between relative host cell range and oncogenicity *in vivo* and *in vitro* and between capacity to induce transplantation antigen(s) and oncogenicity *in vivo*.

A comparison of biological properties of different adenoviruses allows some speculation on the evolutionary differentiation of this group of viruses. In this connection, it is of interest to note that although no immunological cross-reactions have been found between adenoviruses of avian and mammalian cell origin their general architecture, including the presence of vertex projections as described above, is the same. It would be of interest to know whether vertex and nonvertex capsomers of avian adenovirus virions display different characteristics similar to those found in studies of human adenoviruses. As regards mammalian adenoviruses, immunological cross-reactions are found not only between hexons, but also as described above between vertex capsomers and fibers of some types of different host cell species origin. It should be noticed that possibilities for cross-reactions between fibers may be limited by their physical characteristics, i.e., that only few antigenic sites are available in fibers of the smallest dimensions (Norrby, 1969a). The degree of immunological relationship seems to be lower between vertex capsomers than between hexons and even less so between fibers than between hexons. The intimate cross-reactions between fibers of some simian-ape and human types (Hillis and Goodman, 1969) may represent a special case. One might postulate that the degree of immunological relationship between different kinds of structural components in an inverse way reflects the degree of their evolutionary specialization. Thus an axis of evolutionary differentiation might be given as hexons → vertex capsomers → fibers, which was already suggested by Kilkson (1966). It should be noticed, however, that an equally important "axis" of evolutionary differentiation can be visualized within hexon components (Norrby *et al.*, 1969), which, in one way, is expressed by the overall type specificity of the surface of adenovirus virions. With reference to the postulated evolutionary differentiation of hexon structure units into penton structure units, results of attempts to characterize the proteins of these components are awaited with interest. A possible occurrence of identical polypeptide region(s) could serve as an indicator of a common evolutionary origin.

Very few data have been published on the genetic relatedness between human and nonhuman adenoviruses as determined by analysis of the degree of DNA-DNA homology in hybridization experiments. In one study (Pina and Green, 1968) it was found that the simian-monkey serotypes SV15 and SA-7 displayed 9 to 23% DNA homology with human adenovirus serotypes 2, 7, and 12. A corresponding degree of homology is also found between some different human serotypes. Further studies of this kind would be appreciated and in particular it would be of interest to find out whether any region of DNA homology exists between avian and mammalian adenoviruses.

As a final topic of relevance to comparative virology of adenoviruses, the possible importance of zoonotic infections may be discussed. This topic can be divided into two. One of these concerns the situation in which an adenovirus of one host cell species origin infects cells from a different species, in which a transformation may occur as a result of a nonpermissiveness of these cells. *In vitro* transformation of human amnion cells by an avian adenovirus was mentioned in one report (Mancini *et al.*, 1969). A possible demonstration of further systems of this kind to be compared with systems involving rodent cells would be of interest. It is very difficult to evaluate to what extent an *in vivo* situation of this kind may be of practical importance, but it certainly should be kept in mind. The second subtopic of possible zoonotic infections refers to a situation when a certain adenovirus serotype may have a relatively broad host cell specificity, i.e., that it can cause cytolytic infections in cells of different species origin. Again it is difficult to evaluate the medical significance of situations of this kind. It has been described that neutralizing antibodies against bovine (Klein *et al.*, 1959) and some simian-monkey adenoviruses (Aulisio *et al.*, 1964) can be demonstrated in human sera. In the latter study a much higher frequency of sera-positives was found among a population in a relatively closer geographical contact with monkeys. Recently it was described that certain human sera contain antibodies capable of neutralizing the dog adenovirus ICH (Smith *et al.*, 1970). It has furthermore been found that some serotypes of human adenoviruses can multiply in rabbits (Pereira and Kelly, 1957), dogs (Sinha *et al.*, 1960), and pigs (Betts *et al.*, 1962). Pneumonic lesions were observed in some pigs. Further studies of this kind are warranted. Of particular interest from the point of view of human infections may be members of the simian-ape group. Judging from their host cell range and close biological relationship to human denoviruses (Rowe *et al.*, 1958; Hillis and Goodman, 1969) they should be considered as potential human pathogens.

ACKNOWLEDGEMENTS

Some of the author's results presented in this review were obtained as products of collaborative efforts with Dr. Göran Wadell and Mr. Raymond Marusyk. Mrs. Halyna Marusyk gave a valuable assistance in the electron microscopy examinations.

REFERENCES

Ahmad-Zadeh, C., Herzberg, M., Paccaud, M. F., and Regamey, R. H. (1968). *Pathol. Microbiol.* **32**, 254–266.

Altstein, A. D., Sarycheva, O. F., and Dodonova, N. N. (1967). *Science* **158**, 1455–1457.

Anderson, J., Yates, V. J., Jasty, V., and Mancini, L. O. (1969). *J. Nat. Cancer. Inst.* **42**, 1–7.

Ankudas, M. M. and Khoobyarian, N. (1962). *J. Bacteriol.* **84**, 1287–1291.

Atanasiu, P. and Lepine, P. (1960). *Ann. Inst. Pasteur* (Paris) **98**, 915–919.

Aulisio, L. G., Wong, D. C., and Morris, J. A. (1964). *Proc. Soc. Exp. Biol. Med.* **117**, 6–11.

Bartha, A. (1969). *Acta Vet. Hung.* **19**, 319–321.

Barski, G., and Cornfeit, F. (1958). *Ann. Inst. Pasteur* (Paris) **94**, 724–731.

Bauer, W., and Wigand, R. (1967). *Arch. Gesamte Virusforsch.* **21**, 11–24.

Beardmore, W. G., Harlick, M. J., Seratini, A. and McLean, I. W., Jr. (1965). *J. Immunol.* **45**, 422–435.

Betts, A. O., Jennings, A. R., Lamont, P. A., and Page, Z. (1962). *Nature* (*London*) **193**, 45–46.

Bibrack, B. (1969). *Zentralbl. Veterinaermed.* **16**, 327–334.

Blacklow, N. R., Hoggan, M. D., Austin, J. B., and Rowe, W. P. (1969). *Amer. J. Epidemiol.* **90**, 501–505.

Boyer, G. S., Leuchtenberger, C., and Ginsberg, H. S. (1957). *J. Exp. Med.* **105**, 195–216.

Boyer, G. S., Denny, F. W., Jr., and Ginsberg, H. S. (1959). *J. Exp. Med.* **110**, 827–844.

Brandon, F. B. and McLean, I. W., Jr. (1963). *Advan. Virus Res.* **9**, 157–193.

Burke, C. N., Luginbuhl, R. E., and Williams, L. F. (1968). *Avian Dis.* **12**, 483–505.

Carmichael, L. E. and Barnes, F. D. (1962). *Proc. Soc. Exp. Biol. Med.* **110**, 756–758.

Chandler, R. L., Haig, D. A., and Smith, K. (1965). *J. Royal Microscop. Soc.* **84**, 133–142.

Clarke, M. C., Sharp, H. R. A., and Derbyshire, J. B. (1967). *Arch. Gesemte Virusforsch.* **21**, 91–97.

Clemmer, D. I. (1964). *J. Infec. Dis.* **114**, 386–400.

Cooper, J. E. K., Stich, H. F., and Yohn, D. S. (1967). *Virology* **33**, 533–541.

Csontos, L. (1967). *Acta Vet. Hung.* **17**, 217–219.

Darbyshire, J. H. (1966). *Nature* (*London*) **211**, 102.

Davies, M. C. and Englert, M. E. (1961). *Virology* **13**, 143–144.

Doerfler, W. and Kleinschmidt, A. K. (1970). *J. Mol. Biol.* **50**, 579–593.

Dreizin, R. S. and Zolotarskaya, E. E. (1967). *Vopr. Virusol.* **5**, 577–581.

Dulac, G. C., Swango, L. J., and Burnstein, T. (1970). *Can. J. Microbiol.* **16**, 391–394.

Dutta, S. K. and Pomeroy, B. S. (1963). *Proc. Soc. Exp. Biol. Med.* **114**, 539–451.

Espmark, Å. and Salenstedt, C. R. (1961). *Arch. Gesamte Virusforsch.* **11**, 64–71.

Fong, C. K. Y., Bensch, K. G., and Hsiung, G. D. (1968). *Virology* **35**, 297–310.

Freeman, A. E., Black, P. H., Wolford, R., and Huebner, R. J. (1967a). *J. Virol.* **1**, 362–367.

Freeman, A. E., Vanderpool, E. A., Black, P. H., Turner, H. C., and Huebner, R. J. (1967b). *Nature (London)* **216**, 171–173.

Freeman, A. E., Black, P. H., Vanderpool, E. A., and Henry, P. H. (1967c). *Proc. Natl. Acad. Sci. U.S.* **58**, 1205–1212.

Garg, S. P., Moulton, J. E., and Sekhri, K. K. (1967). *Amer. J. Vet. Res.* **28**, 725–730.

Gilden, R. V., Kern, J., Beddow, T. G., and Huebner, R. J. (1967). *Virology* **31**, 727–729.

Gilden, R. V., Kern, J., Freeman, A. E., Martin, C. E., McAllister, R. C. Turner, H. C., and Huebner, R. J. (1968a). *Nature (London)* **219**, 517–518.

Gilden, R. V., Kern, J., Heberling, R. L., and Huebner, R. J. (1968b). *Appl. Microbiol.* **16**, 1015–1018.

Gilead, Z. and Ginsberg, H. S. (1968a). *J. Virol.* **2**, 7–14.

Gilead, Z. and Ginsberg, H. S. (1968b). *J. Virol.* **2**, 15–20.

Ginsberg, H. S. (1962). *Virology* **18**, 312–319.

Ginsberg, H. S., Pereira, H. G., Valentine, R. C., and Wilcox, W. C. (1966). *Virology* **28**, 782–783.

Girardi, A. J., Hilleman, M. R., and Zwickey, R. E. (1964). *Proc. Soc. Exp. Biol. Med.* **115**, 1141–1150.

Givan, K. F., Turnbull, C., and Jezequel, A. M. (1967). *J. Histochem. Cytochem.* **15**, 688–694.

Green, M., Pina, M., and Kimes, R. (1967). *Proc. Nat. Acad. Sci. U.S.* **53**, 1302–1309.

Guenov, I., Schopov, I., Schlabinkov, Z., Sartmadskiev, K., Fjodorov W., and Simeenov, S. (1969). *Zentralbl. Veterinaermed.* **16**, 335–340.

Guerin, L. F. and Guerin, M. M. (1957). *Proc. Soc. Exp. Biol. Med.* **96**, 322–323.

Hammon, W. McD., Yohn, D. S., Casto, B. C., and Atchison, R. W. (1963). *J. Nat. Cancer Inst.* **31**, 329–345.

Hartley, J. W. and Rowe, W. P. (1960). *Virology* **11**, 645–647.

Henry, M., Sohier, R., and Armynot du Chatelet, M. F. (1968). *C. R. Acad. Sci. (Paris)* **267**, 1903–1904.

Hierholzer, J. C. and Dowdle, W. R. (1970). *Proc. Soc. Exp. Biol. Med.* **134**, 482–488.

Hillis, W. D. and Goodman, R. (1969). *J. Immunol.* **103**, 1089–1095.

Hillis, W. D., Holmes, A. W., and Davison, V. (1968). *Proc. Soc. Exp. Biol. Med.* **129**, 366–369.

Hillis, W. D., Garner, A. C., and Hillis, A. I. (1969). *Amer. J. Epidemiol.* **90**, 344–353.

Hitchcock, G., Tyrrell, D. A. J., and Bynoe, M. L. (1960). *J. Hyg.* **58**, 277–282.

Horne, R. W., Brenner, S., Waterson, A. P., and Wildy, P. (1959). *J. Mol. Biol.* **1**, 84–86.

Horzinek, M. and Uberschär, S. (1966). *Arch. Gesamte Virusforsch.* **18**, 406–421.

Huebner, R. J. (1967). *In* "Perspectives in Virology" (M. Pollard, ed.), p. 147–166. Academic Press, New York.

Huebner, R. J., Rowe, W. P., and Lane, W. T. (1962). *Proc. Nat. Acad. Sci. U.S.* **48**, 2051–2058.

Huebner, R. J., Casey, M. J., Chanock, R. M., and Schell, K. (1965). *Proc. Nat. Acad. Sci. U.S.* **54**, 381–388.

Hull, R. N. (1968). *Virol. Monogr.* **2**, 1–66.

Hull, R. N., Johnson, I. S., Culbertson, C. G., Reimer, C. B., and Wright, H. P. (1965). *Science* **150**, 1044–1046.

Jainchill, J. L., Candler, E. L., and Anderson, N. G. (1969). *Proc. Soc. Exp. Biol. Med.* **130**, 770–775.

Jerkofsky, M. and Rapp, F. (1968). *J. Virol.* **2**, 670–677.

Kawamura, H. and Tsubahara, H. (1963). *Nat. Inst. Animal Health Quart.* **3**, 77–82.

Kawamura, H., Shimizu, F., and Tsubahara, H. (1964). *Nat. Inst. Animal Health Quart.* **4**, 183–193.

Kelly, B. and Pereira, H. G. (1957). *Brit. J. Exp. Pathol.* **38**, 396–400.

Kern, J., Gilden, R. V., and Huebner, R. J. (1969). *Appl. Microbiol.* **17**, 914–919.

Kilkson, R. (1966). *J. Theoret. Biol.* **12**, 435–438.

Kim, C. S., Sveltenfuss, E. A., and Kalter, S. S. (1967). *J. Infec. Dis.* **117**, 292–300.

Klein, M. (1962). *Ann. N.Y. Acad. Sci.* **101**, 493–497.

Klein, M., Ealey, E., and Zellat, J. (1959). *Proc. Soc. Exp. Biol. Med.* **102**, 1–4.

Klein, M., Zellat, J., and Michaelson, T. V. (1960). *Proc. Soc. Exp. Biol. Med.* **105**, 340–342.

Kunishige, S. and Hirato, K. (1960). *Jap. J. Vet. Res.* **8**, 271–275.

Laver, W. G., Pereira, H. G., Russel, W. C., and Valentine, R. C. (1968). *J. Mol. Biol.* **37**, 379–386.

Laver, W. G., Wrigley, N. G., and Pereira, H. G. (1969). *Virology* **39**, 599–605.

Liao, S. K. and Weber, J. (1969). *Can. J. Microbiol.* **15**, 847–850.

Loytved, D. and Wigand, R. (1969). *Z. Med. Mikrobiol. Immunol.* **154**, 277–286.

McAllister, R. M., Nicolson, M. O., Reed, G., Kern, J., Gilden, R. V., and Huebner, R. J. (1969). *J. Nat. Canc. Inst.* **43**, 917–923.

McFerran, J. B., Nelson, R., McCracken, J. M., and Ross, J. G. (1969). *Nature (London)* **221**, 194–195.

Macpherson, I., Wildy, P., Stoker, M., and Horne, R. W. (1961). *Virology* **13**, 146–149.

Malherbe, H., Harnin, R., and Ulrich, M. (1963). *S. Afr. Med. J.* **37**, 407–411.

Mancini, L. O., Yates, V. J., Jasty, V., and Anderson, J. (1969). *Nature (London)* **222**, 190–191.

Martinez-Palomo, A., LeBuis, J., and Bernhard, W. (1967). *J. Virol.* **1**, 817–829.

Marusyk, R., Norrby, E., and Lundqvist, U. (1970). *J. Virol.* **5**, 507–512.

Matsui, K. and Bernhard, W. (1967). *Ann. Inst. Pasteur* (Paris) **112**, 773–780.

Matsumoto, M., Inaba, Y., Tanaka, Y., Sato, K., Ito, H., and Omori, T. (1969). *Jap. J. Microbiol.* **13**, 131–132.

Matsumoto, M., Inaba, Y., Tanaka, Y., Sato, K., Ito, H., and Omori, T. (1970). *Jap. J. Microbiol.* **14**, 430–431.

Norrby, E. (1966). *Virology* **28**, 236–248.

Norrby, E. (1968). *Curr. Topics Microbiol. Immunol.* **43**, 1–43.

Norrby, E. (1969a). *J. Gen. Virol.* **5**, 221–236.

Norrby, E. (1969b). *Virology* **37**, 565–576.

Norrby, E. and Ankerst, J. (1969). *J. Gen. Virol.* **5**, 183–194.

Norrby, E. and Skaaret, P. (1968). *Virology* **36**, 201–211.

Norrby, E. and Wadell, G. (1967). *Virology* **31**, 592–600.

Norrby, E. and Wadell, G. (1969). *J. Virol.* **4**, 663–670.

Norrby, E., Nyberg, B., Skaaret, P., and Lengyel, A. (1967). *J. Virol.* **1**, 1101–1108.

Norrby, E., Marusyk, H., and Hammarskjöld, M.-L. (1969). *Virology* **38**, 477–482.

Pereira, H. G. and Kelly, B. (1957). *Nature (London)* **180**, 615–616.

Pereira, H. G., Huebner, R. J., Ginsberg, H. S., and van der Veen, J. (1963). *Virology* **20**, 613–620.

Pereira, M. S., Pereira, H. G., and Clarke, S. K. R. (1965). *Lancet* i, 21–23.

Petek, M., Felluga, B., and Zoletto, R. (1963). *Avian Dis.* **7**, 38–49.

Pettersson, U. and Höglund, S. (1969). *Virology* **39**, 90–106.

Pina, M. and Green, M. (1968). *Virology* **36**, 321–323.
Pina, M. and Green, M. (1969). *Virology* **38**, 573–586.
Potter, C. W. and Oxford, J. S. (1969). *J. Gen. Virol.* **4**, 287–289.
Potter, C. W., Oxford, J. S., and McLaughlin, B. (1970). *J. Gen. Virol.* **6**, 105–116.
Prage, L., Pettersson, U., and Philipson, L. (1968). *Virology* **36**, 508–511.
Prier, J. E. and LeBeau, R. W. (1958). *Amer. J. Pathol.* **34**, 789–795.
Rafajko, R. R. (1965). *Proc. Soc. Exp. Biol. Med.* **119**, 975–982.
Rapoza, N. P. (1967). *Amer. J. Epidemiol.* **86**, 736–744.
Rapp, F. (1969). *Ann. Rev. Microbiol.* **23**, 293–316.
Reed, S. E. (1967). *J. Gen. Virol.* **1**, 405–412.
Reeves, W. C., Scrivani, R. P., Pugh, W. E., and Rowe, W. P. (1967). *Proc. Soc. Exp. Biol. Med.* **124**, 1173–1175.
Riggs, J. L. and Lennette, E. H. (1967). *Proc. Soc. Exp. Biol. Med.* **126**, 802–806.
Riggs, J. L., Takemori, N., and Lennette, E. H. (1968). *J. Immunol.* **100**, 348–354.
Riggs, J. L., Teitz, Y., Cremer, N. E., and Lennette, E. H. (1969). *Proc. Soc. Exp. Biol. Med.* **132**, 527–532.
Rondhuis, P. R. (1968). *Arch. Gesamte Virusforsch.* **25**, 235–236.
Rosen, L. (1960). *Amer. J. Hyg.* **71**, 120–128.
Rosen, L., Hovis, J. F., and Bell, J. A. (1962). *Proc. Soc. Exp. Biol. Med.* **110**, 710–713.
Rowe, W. P. and Hartley, J. W. (1962). *Ann. N.Y. Acad. Sci.* **101**, 466–474.
Rowe, W. P., Huebner, R. J., Hartley, J. W., Ward, T. G., and Parrott, R. H. (1955). *Amer. J. Hyg.* **61**, 197–218.
Rowe, W. P., Hartley, J. W., and Huebner, R. J. (1958). *Proc. Soc. Exp. Biol. Med.* **97**, 465–470.
Russell, W. C. and Knight, B. E. (1967). *J. Gen. Virol.* **1**, 523–528.
Sarma, P. S., Huebner, R. J., and Lane, W. T. (1965). *Science* **149**, 1108.
Sarma, P. S., Wass, W., Huebner, R. J., Igel, H., Lane, W. T., and Turner, H. C. (1967). *Nature* (*London*) **215**, 293–294.
Schell, K., Young, J., and Rhim, J. S. (1968). *Proc. Soc. Exp. Biol. Med.* **129**, 320–325.
Schlesinger, R. W. (1969). *Advan. Virus Res.* **14**, 1–61.
Schmidt, N. J., King, C. J., and Lennette, E. H. (1965). *Proc. Soc. Exp. Biol. Med.* **118**, 208–211.
Sharon, J. and Pollard, M. (1964). *Nature* (*London*) **202**, 1139–1140.
Sharpless, G. R., Levine, S., Davies, M. C., and Englert, M. F. (1961). *Virology* **13**, 315–322.
Shimojo, H. and Yamashita, T. (1968). *Virology* **36**, 422–433.
Simon, M. (1962). *Acta Microbiol. Acad. Sci. Hung.* **9**, 45–54.
Sinha, S. K., Fleming, L. W., and Sokoles, S. (1960). *J. Amer. Vet. Ass.* **136**, 481–483.
Sjögren, H. O., Minowada, J., and Ankerst, J. (1967). *J. Exp. Med.* **125**, 689–701.
Smith, K. O., Gehle, W. D., and Kniker, W. T. (1970). *J. Immunol.* **105**, 1036–1039.
Sohier, R., Chardonnet, Y., and Prunieras, M. (1965). *Progr. Med. Virol.* **7**, 257–325.
Starr, T. J., Kajima, M., and Holmes, A. W. (1965). *Texas Rep. Biol. Med.* **23**, 607–620.
Stasny, J. T., Neurath, A. R., and Rubin, B. A. (1968). *J. Virol.* **2**, 1429–1442.
Stevens, D. A., Schaeffer, M., Fox, J. P., Brandt, C. D., and Romano, M. (1967). *Amer. J. Epidemiol.* **86**, 617–633.
Swango, L. J., Eddy, G. A., and Binn, L. N. (1969). *Amer. J. Vet. Res.* **30**, 1381–1387.
Tajima, M., Motohashi, T., and Samejima, T. (1961). *Amer. J. Vet. Res.* **22**, 236–249.
Takahashi, M., Ueda, S., and Ogino, T. (1966). *Virology* **30**, 742–743.

Tockstein, G., Polasa, H., Pina, M., and Green, M. (1968). *Virology* **36**, 377–386.

Trentin, J. J., Yabe, Y., and Taylor, G. (1962). *Science* **137**, 835–841.

Tyrrell, D. A. J., Buckland, F. E., Lancaster, M. C., and Valentine, R. C. (1960). *Brit. J. Exp. Pathol.* **41**, 610–616.

Valentine, R. C. and Pereira, H. G. (1965). *J. Mol. Biol.* **13**, 13–20.

van der Eb, A. J., van Kesteren, L. W., and van Bruggen, E. F. J. (1969). *Biochim. Biophys. Acta* **182**, 530–541.

van der Noordaa, J. (1968). *J. Gen. Verol.* **3**, 303–304.

Velicer, L. F., and Ginsberg, H. S. (1968). *Proc. Nat. Acad. Sci. U.S.* **61**, 1264–1271.

Wadell, G. (1969). *Proc. Soc. Exp. Biol. Med.* **132**, 413–421.

Wadell, G. and Norrby, E. (1969). *J. Virol.* **4**, 671–680.

Wadell, G., Norrby, E., and Skaaret, P. (1969). *Arch. Gesamte Virusforsch.* **26**, 33–52.

Warren, J. and Cutchins, E. C. (1957). *Virology* **4**, 297–304.

Watson, D. H., MacPherson, I. A., and Davies, M. C. (1963). *Virology* **19**, 418–419.

Weber, J. and Liao, S. K. (1969). *Can. J. Microbiol.* **15**, 841–845.

Wigand, R. (1970). *J. Gen. Virol.* **6**, 325–328.

Wigand, R., and Bauer, H. (1964). *Arch. Gesamte Virusforsch.* **14**, 674–682.

Wilcox, C. F. (1969). *Aust. Vet. J.* **45**, 265–270.

Yamamoto, T. (1969a). *J. Gen. Virol.* **4**, 397–401.

Yamamoto, T. (1969b). *Microbios* **4**, 371–384.

CHAPTER 5

Herpesviruses: Current Information on the Composition and Structure

BERNARD ROIZMAN AND PATRICIA G. SPEAR

I. Introduction

The herpesviruses are formally defined as large enveloped virions with an icosahedral capsid consisting of 162 capsomeres and arranged around a DNA core (Fig. 1). The word herpes is derived from ερπειν meaning to creep and recurs in medical texts dating back at least 25 centuries. A virus isolated from herpes simplex, a disease of man which subsequently gave the virus its name, was first experimentally transmitted more than half a century ago (Grüter, 1924). Nevertheless, despite the suspicion that nearly

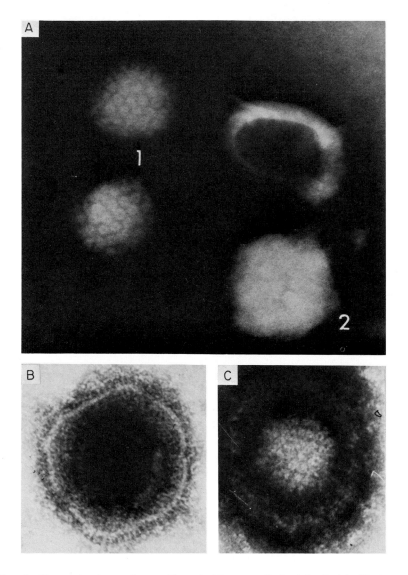

Fig. 1. Herpes simplex nucleocapsids and virions negatively stained with silicotungstic acid. (A) Nucleocapsids (1) and intact virion (2) which is impervious to negative stain. (B) and (C) Enveloped nucleocapsids into which the stain has permeated revealing some details of the structure of the envelope.

every species in the animal kingdom harbors at least one and possibly more herpesviruses, the family Herpesviridae is not blessed with many known members. There are several reasons for this: herpesviruses are complex viruses, cantankerous, and at times difficult to grow. Recent findings of herpesviruses associated with certain cancers of man and animals should greatly spur the discovery of new members of this group.

The objective of this paper is to delineate briefly the available information concerning the structure and morphogenesis of the herpesvirus virion.

II. Chemical Composition of the Herpesvirion

A. PURIFICATION

Chemical analyses of viruses require preparations of reasonable purity with respect to the specific component being analyzed. Preparations of virus relatively free from contamination with host DNA have been reported. The procedure for purification with respect to host DNA is based on the observation that enveloped virions migrate into the cytoplasm which is free of host DNA as long as the nuclei remain intact. Provided the cytoplasm is extracted with care to prevent nuclear breakage and then treated with DNAse to digest DNA adhering to the surface of the virions (Smith, 1963), nearly every purification procedure employed subsequently yields satisfactory results.

Virus preparations free from contamination with host proteins or glycoproteins have not been reported. Precipitation of virions with corresponding antibody (Shimono et al., 1969) does not appear to be a useful procedure. No published procedure to date has successfully separated virions from amino acid-labeled debris of uninfected cells. Some of the problems encountered so far may be summarized as follows:

(1) Fractionation procedures involving single steps, as for example centrifugation in 10–50% w/w sucrose density gradients prepared in isotonic saline, yield three discrete bands containing varying amounts of empty, full, and enveloped nucleocapsids. However, the bands are visibly contaminated with fragments of cell membranes and other host debris.

(2) Multistep fractionation procedures have generally been unsuccessful. The virus does not withstand well several changes in osmotic pressure. It is particularly labile in CsCl solutions (Spring and Roizman, 1967) and we suspect, with some cause, in high salt solutions in general. Multistep frac-

tionation procedures are more deleterious to naked nucleocapsids than to enveloped nucleocapsids.

Very recently (January, 1971) it was found that two purification procedures yield relatively pure preparations of enveloped and naked nucleocapsids, respectively. The first procedure involves the centrifugation of post-mitochondrial supernatant from the infected cell cytoplasm on Dextran-10 gradients (average mw-10,000 daltons; Pharmacia Fine Chemicals, Piscataway, New Jersey) instead of the sucrose gradients commonly used to prepare enveloped virus. Microsomal vesicles, which heavily contaminate enveloped virus preparations from sucrose gradients, sediment much more slowly than the virus in Dextran, so that virus free of cellular membranes may be obtained. The purity of enveloped nucleocapsids prepared in this fashion was established by electron microscopy and by the very large increase in the ratio of viral to host proteins in the final product (Spear and Roizman, unpublished observation).

The second purification procedure which has been reported to yield relatively pure nucleocapsid preparations involves treatment of the cytoplasmic extract with 1% w/v of sodium deoxycholate prior to centrifugation in a sucrose density gradient (Olshevsky and Becker, 1970). It appears that a nonionic detergent Nonidet P-40 (O'Brien, 1964) can also be used. The purpose of the detergent treatment is to solubilize viral envelopes and debris containing lipids. It should be pointed out, however, that deoxycholate and Nonidet P-40 may have deleterious effects on other viral structures; naked nucleocapsids have no apparent biological activity (see discussion in Roizman et al., 1969) and the fact that the treated nucleocapsids appear unaltered in the electron microscope is no assurance of structural integrity.

B. DNA

1. Composition and Structure

The base compositions of the DNA's of the herpesviruses are shown in Table I. The following should be pointed out. (1) There is agreement that the DNA of herpesviruses is double-stranded (Russell, 1962; Darlington and Randall, 1963). (2) The molecular weights of the DNA of several herpesviruses (equine abortion, herpes simplex, and Marek's disease) are of the order of 10^8 daltons (Soehner et al., 1965; Becker et al., 1968; Lee et al., 1971). Minor differences however, have been noted. Thus, comparative studies have shown that Marek's disease virus DNA is 1.2×10^8 daltons compared to 1.1×10^8 daltons for the DNA of herpes simplex subtype 1 (Lee et al., 1971), and that the DNA of the MP mutant of subtype 1 is

TABLE I

BASE COMPOSITION OF THE DNA OF HERPESVIRUSES

Virus	Authors	Method	C + G Mole %
Pseudorabies	Ben-Porat and Kaplan, 1962	Fractionation of ^{32}P labeled deoxyribonucleotides	74
	Russell and Crawford, 1964	Buoyant density in CsCl	74
	Kaplan and Ben-Porat, 1964	Buoyant density in CsCl	73
	Plummer et al., 1969	Buoyant density in CsCl	71.2
Infectious bovine Rhinotracheitis	Russell and Crawford, 1964	Buoyant density in CsCl	71
	Plummer et al., 1969	Buoyant density in CsCl	71.2
Monkey lymphoma virus	Goodheart, 1970	Buoyant density in CsCl	70.2
Herpes simplex subtype 1	Ben-Porat and Kaplan, 1962	Fractionation of ^{32}P-labeled deoxyribonucleotides	74
	Russell and Crawford, 1963	T_m	68
		Buoyant density in CsCl	68
		Spectrophotometrically (260/280 at pH 3)	67
	Roizman et al., 1969	Buoyant density in CsCl	67
	Russell and Crawford, 1964	Buoyant density in CsCl	68
	Lando et al., 1965	Fractionation of ^{32}P-labeled deoxyribonucleotides	65 ± 2.1
	Goodheart et al., 1968	Buoyant density in CsCl	68.3
Herpes simplex subtype 2	Goodheart et al., 1968	Buoyant density in CsCl	70.4

TABLE I (*Continued*)

Virus	Authors	Method	C + G Mole %
Herpes saimiri virus	Goodheart, 1970	Buoyant density in CsCl	68.6
Bovine mammilitis virus	Martin et al., 1966	Buoyant density in CsCl	64
Canine herpes virus	Aurelian, 1969	Buoyant density in CsCl	61
	Plummer et al., 1969	Buoyant density in CsCl	31.4
Equine herpes virus	Darlington and Randall, 1963	Fractionation of deoxyribonucleotides	56
	Russell and Crawford, 1964	Buoyant density in CsCl	55
	Soehner et al., 1965	Buoyant density in CsCl and T_m	57
	Plummer et al., 1969	Buoyant density in CsCl	58–59
Human Cytomegalovirus	Crawford and Lee, 1964	Buoyant density in CsCl	58
	Plummer et al., 1969	Buoyant density in CsCl	58.6
Mouse Cytomegalovirus	Plummer et al., 1969	Buoyant density in CsCl	59.1
Vervet Cytomegalovirus	Plummer et al., 1969	Buoyant density in CsCl	50.6
Burkitt lymphoma virus	Weinberg and Becker, 1969	Cocentrifugation with herpes simplex	67–68
	Goodheart, 1969	Buoyant density in CsCl	60–61
	Wagner et al., 1970	Buoyant density in CsCl	56–58
Cat herpes virus	Plummer et al., 1969	Buoyant density in CsCl	48
Marek's disease herpesvirus	Lee et al., 1971	Buoyant density in CsCl	46
Chicken herpes virus	Plummer et al., 1969	Buoyant density in CsCl	45
Lucke frog virus	Wagner et al., 1970	Buoyant density in CsCl	44–46

some 5% smaller than that of the wild type (Kieff, Bachenheimer, and Roizman, manuscript in preparation). (3) The DNA is linear and not cross-linked (Kieff, Bachenheimer, and Roizman, manuscript in preparation). (4) Unusual bases have not been reported. (5) the DNA does not appear to be methylated (Low *et al.*, 1969). (6) There is agreement that the guanine (G) plus cytosine (C) content of the DNA of herpesviruses varies considerably (Table 1).

Recently controversy has arisen concerning the base composition of canine herpesvirus, a newcomer to the family. Plummer *et al.* (1969) reported that the DNA of canine herpesvirus has a guanine+cytosine content roughly similar to that of mammalian cells, i.e., 44 G+C moles %. On the other hand, Aurelian (1969) reported that the DNA of her canine herpesvirus has a buoyant density approximately equal to that of herpes simplex DNA and therefore contains 68 G+C moles %. A priori it would not be expected that the DNA's of two immunologically related viruses such as herpes simplex and canine herpesvirus (Aurelian, 1968) would differ by 24 G+C moles %. The discrepancy in the results may well be due to the techniques used in two studies. Both workers analyzed DNA from infected cells and from partially purified virus. However, Aurelian's results are based on the analysis by isopycnic centrifugation in a preparative ultracentrifuge of only the DNA synthesized after infection and labeled with radioactive thymidine, whereas Goodheart's determinations of buoyant density were done in an analytical ultracentrifuge which makes apparent all the DNA present in the preparation. The analytical ultracentrifuge can give very precise buoyant density determinations, but it may not be possible to detect small amounts of viral DNA in the presence of a large excess of cellular DNA. Even the partially purified virus may be contaminated with enough cellular DNA to completely obscure the presence of any viral DNA. However, when one analyzes, as Aurelian did, only the DNA synthesized after infection, the probability of detecting small amounts of viral DNA is greatly enhanced.

2. *The Synthesis of Viral DNA*

The synthesis of viral DNA has been recently reviewed in detail (Kaplan 1969; Roizman, 1969) and very little new information has come forth since. The available data may be summarized briefly as follows:

(1) It has been known for many years that, in cells infected with herpes simplex virus, the amount of DNA in the nucleus increases after infection and moreover, that the intranuclear inclusion typical of herpes-infected

cells is Feulgen positive. Recent studies summarized elsewhere (Roizman, 1969) show that thymidine is incorporated into viral DNA in the nucleus. The site of thymidine incorporation into DNA corresponds topologically to the inclusion body and appears in the electron microscope to be electron translucent (Schwartz, unpublished data reported by Roizman, 1969).

(2) Viral DNA synthesis is preceded by aggregation and displacement of host chromosomes to the periphery of the nucleus and by inhibition of host DNA synthesis. There is some question as to whether the inhibitor of DNA synthesis is a structural component of the virion (Newton, 1968) or a product made after infection (Ben-Porat and Kaplan, 1965). The mechnanism by which DNA synthesis is inhibited is unknown.

(3) The kinetics of synthesis of viral DNA appear to vary somewhat. In HEp-2 cells infected with herpes simplex the bulk of viral DNA is made between 3 and 6 hours postinfection (Roizman, 1969). In cells infected with equine abortion virus DNA synthesis extends almost to the end of the reproductive cycle. The onset of synthesis of viral DNA requires protein synthesis (Roizman and Roane, 1964). There is some question as to whether continued synthesis of viral DNA requires concomitant protein synthesis (Kaplan et al., 1967). In a series of studies reported between 1963 and 1967 Kaplan and Ben-Porat reported that (a) pseudorabies virus DNA replicates in a semiconservative fashion, (b) less than one-half of the DNA not integrated into virions and presumed available to function as templates is actually replicating, and (c) the withdrawal of viral DNA from the DNA pool into virions is slow and inefficient (Kaplan and Ben-Porat, 1963, 1966; Ben-Porat and Kaplan, 1963; Kaplan et al., 1967).

(4) Several enzymes involved in DNA synthesis and differing with respect to activity, physical properties, and immunologic reactivity from corresponding enzymes of uninfected cells have been reported in cells infected with herpes simplex and pseudorabies virus. Among these enzymes are thymidine kinase (Klemperer et al., 1967; Kit et al., 1967), DNA nucleotidyltransferase (Keir, 1968; Stevens and Jackson, 1967; Shedden et al., 1966), and deoxyribonuclease (Keir, 1968; Morrison and Keir, 1967).

(5) Several papers, notably those of Kamiya et al. (1964, 1965) and Zemla et al. (1967), reported that in herpesvirus-infected cells the synthesis of DNA acts as a regulatory event. The papers have been reviewed in detail elsewhere (Roizman, 1969).

3. Transcription of Viral DNA

There is as yet very little information on the transcription of viral DNA. The available data may be summarized as follows:

(1) Lando and Ryhiner (1969) reported the preparation of infectious herpes simplex virus DNA. If it can be shown that the infectious DNA is free of protein contamination, the data would argue very strongly that the virus utilizes host DNA-dependent RNA polymerase for the initial transcription of viral DNA.

(2) Virus specific RNA is made in the nucleus. This conclusion is based on the finding that uridine-^{3}H incorporated into RNA hybridizable with viral DNA first appears in the particulate fraction of the nuclei of infected cells. Virus specific RNA appears in the cytoplasm only after a lag of 10 to 25 minutes. The identity of nuclear and cytoplasmic virus-specific RNA was established by hybridization competition tests (Wagner and Roizman, 1969b; Roizman et al., 1970b).

(3) Viral mRNA defined as virus-specific RNA associated with poly-ribosomes is readily demonstrable between 4 and 8 hours postinfection. There is evidence that at least one class of viral mRNA is made as a high molecular weight precursor which is then cleaved in the nucleus in a manner as yet undetermined (Wagner and Roizman, 1969b; Roizman et al., 1970b).

(4) Approximately 20 to 30% of the cytoplasmic virus-specific RNA is associated with membranes (Roizman et al., 1970b). It is not yet entirely clear whether the proteins made on bound polyribosomes are different from those made on free polyribosomes.

4. The Information Content of Herpesvirus DNA

One objective of the studies of the multiplication of herpesviruses is a complete description of the structure, functions, amount, and time of synthesis of all the products specified by the virus in the infected cells. At present the number of products and their functions are uncertain. According to current accounting practices, herpesvirus DNA carries information sufficient to specify the sequence of 140,000 amino acids. Assuming that the genes for the structural proteins are not duplicated, only 10 to 20% of the total genetic information relates to structural protein synthesis. Herpesviruses are not as complex as some of the large DNA T phages and, moreover, the amount of information carried by the virus seems astronomical by comparison with the information content of papova- and adenoviruses, which also multiply in the nucleus, or with that of myxoviruses and arboviruses, which also have an envelope. What then is the need for so much genetic information?

The answer is unknown. A point that is worth considering is that all of the genetic information carried by the virus may not be expressed in cell cultures. The argument is based on the fact that most laboratory strains were recovered at one time or another from sick individuals. Prior to infection of cells *in vitro* the information content of the virus was shaped and molded for many millenia for better survival in the complex multicellular organism it normally infects. Unlike the small DNA and RNA viruses, herpesviruses have established, in the course of evolution, a unique relationship with the host they usually infect (Roizman, 1966). The main feature of this relationship is that following primary infection, and in spite of the appearance of antibody, herpesviruses survive asymptomatically in some specific tissue for the life-span of the host. Perhaps even more extraordinary, there are subtle but reproducible biochemical and biophysical differences among viruses isolated from recurrent infections occurring in different parts of the body (Dowdle *et al.*, 1967; Ejercito *et al.*, 1968; Terni and Roizman, 1970; Roizman *et al.*, 1970a); these observations suggest that the site of survival is determined by the virus. The capacity to coexist is not an indication that the virus is incapable of inflicting injury; herpesviruses frequently cause death or very severe illness in species other than their natural hosts. Thus herpes simplex virus infection of man is usually inapparent, infrequently serious, and rarely fatal; in the rabbit the virus causes severe damage to the central nervous system. Pseudorabies is a mild disease of pigs resembling herpes simplex in man; it is fatal in sheep and cattle. Virus B causes recurrent eruptions in old-world monkeys reminiscent of recurrent herpes infections of man. As several unfortunate virologists have involuntarily demonstrated, virus B infection of man is almost invariably fatal. In the light of these unique features of their natural history, it seems reasonable to postulate that herpesviruses express their genetic potentialities more fully and effectively in the hosts with which they coexist than in ones they destroy. Alas it is difficult to carry out meaningful biochemical experiments in experimental animals infected with a virus native to them. The cell culture is best suited for this purpose, but it is not the native habitat of herpesviruses in the evolutionary sense. For this reason, it seems unlikely that the entire information content of herpesviruses will ever be determined from studies of the infection in cultures of dispersed undifferentiated cells.

C. Structural Proteins

1. *General Considerations*

All of the available data are based on acrylamide gel electropherograms of solubilized proteins from sucrose density gradient bands containing varying amounts of partially purified empty, filled, and enveloped nucleocapsids labeled with amino acids or with glucosamine. These data suffer from several serious shortcomings as follows:

a. Limitations Imposed by Materials Available for Analysis. As pointed out in Section II,A, pure virus free of host proteins and membranes is not available. Consequently the studies on viral proteins are confined to analyses of partially purified virus obtained from cells labeled with radioactive amino acids or with glucosamine late in infection, i.e., at the time when host protein synthesis is greatly diminished. The data obtained from such studies are useful in that they permit an estimation of the number of structural proteins of the virus. They cannot be used for the estimation of the relative *amounts* of structural components for two reasons. First, preparations containing partially purified virions and nucleocapsids are almost invariably contaminated with membranes. Glycoproteins contained in fractions rich in enveloped virions are also present in purified membranes suggesting that cellular membranes bind structural glycoproteins of the virus (Spear *et al.*, 1970). The contribution of viral proteins in the membranes contaminating virus preparations cannot be estimated with any degree of accuracy. Second, estimates of relative amounts of structural proteins in the virion necessarily assume that all structural proteins are uniformly labeled. Since the time course of synthesis of structural proteins is variable (Spear and Roizman, 1968, and unpublished studies) it cannot be assumed that no structural components were made before the addition of radioactive label to the medium.

b. Limitations Imposed by Analytical Techniques. The electropherograms available to date (Spear and Roizman, 1968; Shimono *et al.*, 1969; Olshevsky and Becker, 1970) were obtained on material solubilized according to one of several variations of the procedure described by Maizel (1966) and Maizel *et al.* (1968) and employing sodium dodecyl sulfate (SDS), β-mercaptoethanol, urea, and in some instances, glacial acetic acid. The solubilized material is then electrophoresed in acrylamide gels containing 0.1% SDS. This procedure appears to separate and retain in solution poliovirus peptides. Questions have arisen concerning the usefulness of this procedure in the separation of adenovirus polypeptides. With respect to characterization of herpesvirus proteins the usefulness of the procedure is not clear. In one study (Olshevsky and Becker, 1970) material was solubilized by standing at room temperature for 10 hours in 0.1 M phosphate buffer, pH 7.2, containing 0.5% SDS, 0.01% β-mercaptoethanol, and 0.5 M urea. In this study naked empty nucleocapsids yielded one major protein band (no. II, 110,000 daltons), naked full nucleocapsids yielded one major (no. II) and several minor bands (nos. V, VI, VII, and VIII corresponding to 58,000, 51,000, 37,000, and 30,000 daltons, respectively) whereas enveloped nucleocapsids yielded bands I-IX (no. I, 140,000 daltons, no. III, 100,000 daltons, no. IV, 85,000 daltons). With the exception of the minor bands obtained from material rich in naked nucleocapsids the electropherograms are esthetically pleasing and in general a numerically plausible case has been made by Becker (personal communication) in support of the hypothesis that each band contains one protein and that the sum of the proteins accounts for all of the structural components of the virus. In our hands a variation of the procedure used by Olshevsky and Becker (1970) involving (a) lipid extraction with acidified dimethylformamide prior to solubilization, (b) the use of a discontinuous buffer system (0.01 M phosphate buffer in the sample layered on top of the gel), and (c) electrophoresis through a 20 cm, 5.7% acrylamide gel cross-linked with ethylene diacrylate yielded the profile shown in Fig. 2. By this technique band III of Olshevsky and Becker (1970) resolves into 3 bands, a slow moving band (S_2) containing nonglycosylated, acid-extractable proteins, and two bands (G_1 and G_2) containing glycosylated proteins. However, questions arise whether the various bands each contain only one protein and whether the proteins in each band consist of

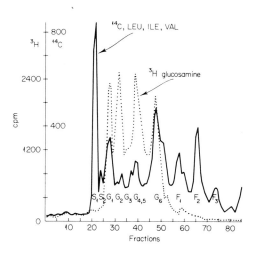

FIG. 2. Electropherograms of herpes simplex virus (strain MP) proteins and glycoproteins labeled between 4 and 18 hours after infection of HEp-2 cells. The enveloped virions were purified by a rate centrifugation of cytoplasmic extract on a 10–50% w/w sucrose gradient in 0.15 M NaCl, 0.01 tris, pH 7.2, at 20,000 rpm for 45 minutes in a Spinco SW25.3 rotor followed by isopycnic flotation through a discontinuous sucrose gradient as described previously (Spear et al., 1970; Keller et al., 1970). The direction of electrophoresis is from left to right.

single polypeptide chains rather than aggregates. At the time of writing this chapter (*i*) The band designated as I (Olshevsky and Becker, 1970) or *a* (Spear and Roizman, 1968) appears highly variable in size and may be an artifact of incomplete solubilization. This band is almost totally absent from material solubilized in 8 *M* urea, 0.1% SDS, 0.5% dithiothreitol, 0.05 *M* tris buffer, pH 8.6, and dialyzed against 0.5 *M* iodoacetamide prior to electrophoresis. (*ii*) The electrophoretic mobilities of the proteins in bands II and III are not appreciably altered by 8 *M* urea and 0.1% dithiothreitol in solutions buffered at pH 8.6 or by 5 *M* guanidine and 0.1% dithiothreitol also in solutions buffered at pH 8.6. However, the amounts of material recovered in bands II and III (bands S_1, S_2, G_1, G_2) are considerably reduced if virus preparations are solubilized by heating at 60°C for 1 hour in 1% SDS, 0.1% β-mercaptoethanol, 0.5 *M* urea, and 10% glacial acetic acid. The material solubilized in this fashion yields electropherograms showing numerous overlapping bands and a relatively high background. (*iii*) The band designated as $G_{4,5}$ (Fig. 2) is wider than either G_2 or G_6. Protein extracted from the left and right halves of band $G_{4,5}$ formed two distinct bands on reelectrophoresis.

It seems worthwhile to point out that there are a number of obvious pitfalls inherent in the analyses of proteins and glycoproteins by the techniques described above. Thus (1) 0.1% SDS, β-mercaptoethanol, and relatively small concentrations of urea do not dissociate effectively protein aggregates as evidenced from the fact that bands with protein aggregates of molecular weights of 140,000 daltons and greater are occasionally detected in polyacrylamide gels. (2) If left unmodified, reduced sulfhydryl groups may easily become oxidized by atmospheric oxygen with the formation of disulfide bonds somewhat at random, resulting in numerous poorly defined, spurious bands in acrylamide gels, and (3) it is not clear whether acetic acid acts as a potent denaturing agent exposing more disulfide bonds to SDS and β-mercaptoethanol or whether the acid degrades preferentially certain proteins.

2. *The Synthesis of Viral Proteins*

In this discussion we will refer to the proteins contained in partially purified virus preparations as viral even though there is no definitive evidence that they are specified by the virus. The available information concerning the synthesis of the proteins is summarized below.

a. Site of Protein Synthesis. As far as we can tell, cytoplasmic protein synthesis accounts for all protein synthesis in infected cells (Sydiskis and Roizman, 1966; Fujiwara and Kaplan, 1967).

b. Viral mRNA and Polyribosomes. Within a short time after infection host polyribosomes begin to disaggregate (Sydiskis and Roizman, 1967) and the spectrum of proteins and glycoproteins made in the infected cell changes drastically. Beginning with 4–6 hours postinfection the spectrum of proteins made in the infected cell cytoplasm resembles in part the spectrum of proteins present in partially purified virus preparations (Spear and Roizman, 1968 and unpublished data).

Coincidental with the disappearance of host polyribosomes and the change in the spectrum of proteins, there appears in the cytoplasm new, more homogeneous, and more rapidly sedimenting polyribosomes than those present in the uninfected cells (Sydiskis and Roizman, 1966, 1968). The evidence that the polyribosomes are "viral" is based on the finding that RNA associated with these ribosomes and presumably functioning as messenger hybridizes with viral DNA even in the presence of a very large excess of host RNA (Wagner and Roizman, 1969a,b). It is of interest to note that approximately 10–20% of the polyribosomes are bound to membranes. It is relevant to point out that the actual mechanism by which the virus inhibits host protein synthesis is uncertain. It is noteworthy that host RNA continues to be synthesized at a reduced rate. However the RNA that is made is processed aberrantly, rapidly transported to the cytoplasm, and does not appear to function in protein synthesis.

c. Nutritional Requirement. The nutritional requirements of herpesviruses do not appear to be unique. However, three interesting observations have been reported:

(*i*) HEp-2 cells infected with herpes simplex virus and maintained in mixture 199 (Morgan *et al.*, 1950) produce more infectious virus than those maintained in Eagle's (1959) medium containing 1 to 4 times the recommended concentration of amino acids and vitamins. Calf serum is not required (Roizman and Spear, 1968).

(*ii*) Some herpesviruses require specific differentiated cells for their multiplication. Thus the virus associated with Marek's disease of fowl produces infectious progeny in the cells lining the feather follicle but not in the cells of the viscera or in cells grown in culture. Nazerian and Witter (1970) reported that the cells producing infectious Marek's herpesvirus contained cytoplasmic inclusions in which the envelope of the virion appeared to undergo extensive modification in shape and appearance. These observations are exceedingly significant in that they suggest that the host phenotype at the time of infection is a major determinant of virus multiplication. Cytomegaloviruses may also require differentiated cells for multiplication.

(*iii*) It has been reported that the multiplication of herpes simplex virus in cultures of human embryonic fibroblasts, chick embryo fibroblasts, and of monkey kidney cells is not affected by absence of arginine (Jeney *et al.*, 1967). However, herpes simplex virus does not multiply in the absence of arginine in cells in continuous cultivation (Tankersley, 1964; Roizman *et al.*, 1965, 1967; Becker *et al.*, 1967; Jeney *et al.*, 1967; Inglis, 1968). The differences between primary and continuous cell cultures with respect to the capacity to support virus multiplication in arginine-free medium may reflect the size of the arginine pool (Piez and Eagle, 1958; Gönczöl *et al.*, 1967) or the presence of mycoplasma which degrade arginine (Rouse *et al.*, 1963). (Parenthetically, the failure to isolate mycoplasma on suitable media is not very reassuring since obligate intracellular forms may be present.) Arginine starvation of cells in continuous cultivation does not prevent adsorption, penetration or uncoating of herpes simplex virus (Inglis, 1968), the reproductive events occurring during the first 4 hours after infection (Roizman *et al.*, 1967), and viral DNA synthesis (Becker *et al.*, 1967). There is general agreement that the effects of arginine starvation are more pronounced than the effects of deprivation of other amino acids essential for animal cells. There is less agreement on the significance of the effects of arginine starvation. Subak-Sharpe and his colleagues (Subak-Sharpe and Hay, 1965; Subak-Sharpe *et al.*, 1966; Hay *et al.*, 1966, 1967; Subak-Sharpe, 1968) presented data in support of the hypothesis that herpesviruses, which have DNA of a high $G+C$ content and might code for proteins rich in arginine, direct the synthesis of arginine-specific tRNA. Extensive studies by Morris *et al.* (1970) failed to uncover arginine-specific tRNA in HEp-2 cells infected with herpes simplex virus, but the data do not exclude the possibility that herpesviruses specify a tRNA specific for arginine derivatives. It is conceivable that the requirement for arginine reflects the amino acid composition of the structural proteins of the virus. Findings in support of this hypothesis emerged from recent analyses of viral proteins made in infected cells incubated in media with and without arginine (Spring *et al.*, 1969). The data show that the probability that a viral protein is completed in arginine-deprived cells is inversely proportional to the molecular weight and arginine content. In the absence of arginine several relatively large proteins rich in arginine are made in smaller amounts than arginine-poor proteins of the same size.

(*iv*) Henle and Henle (1968) reported that the production of viral products in Burkitt lymphoma cells is enhanced by arginine deprivation. This observation has been confirmed and extended by Weinberg and Becker (1969). Experiments involving the deprivation of various amino acids have established that the effects of arginine deprivation in inducing the multiplication of the virus are unique. This observation is of considerable importance for induction of virus multiplication and for synchronization of virus multiplication for maximal yields. Again, for optimal virus yields, arginine must be restored to the medium once virus multiplication begins. The mechanism by which arginine deprivation induces the virus to multiply is uncertain. The effect of arginine deprivation may be to depress host macromolecular synthesis—a necessary prerequisite for the synthesis of viral macromolecules in herpes infected cells (Aurelian and Roizman, 1965).

d. Time Course of Protein Synthesis. In herpes simplex virus-infected cells the synthesis of protein, as measured from the rate of incorporation of amino acids and the amount of polyribosomes present in the cytoplasm, reaches peak levels between 4 and 6 hours postinfection and thereafter slowly declines. However, there is substantial protein synthesis as late as 12 hours postinfection and, moreover, analysis of proteins made at different times after infection suggests that the synthesis of structural proteins may be asynchronous (Spear and Roizman, 1968).

e. The Transport of Viral Proteins from the Cytoplasm into the Nucleus. As pointed out elsewhere (Roizman *et al.*, 1969) the nucleocapsids and probably the inner envelope are assembled in the nucleus whereas the outer envelope is acquired in transit from the nucleus. Viral proteins therefore must migrate from the cytoplasm into the nucleus. The transport of proteins from the cytoplasm into the nucleus has been measured in independent studies by Olshevsky *et al.* (1967), Spear and Roizman (1968), and Ben-Porat *et al.* (1969). These studies indicate that the rate of migration of the proteins is relatively slow. In all three studies the measurements were done by separation of nuclei from cytoplasm and the major criticism of the technique is that proteins may have leaked from the nuclei into the cytoplasm. In defense of the techniques used in these studies two points should be made. First, there is relatively little leakage of RNA from nuclei obtained by treating cells with Nonidet P-40, and only slightly more from nuclei obtained from cells broken with a Dounce homogenizer (Wagner and Roizman, 1969a). Second, autoradiography on whole cells pulse labeled with amino acids and then chased with unlabeled amino acids is an alternative technique that has been used to measure the rate of migration of proteins into the cytoplasm. However, this techniques does not readily differentiate between proteins actually in the nucleus and those accumulating at the nuclear membrane. At the moment thin section autoradiography appears to be the most promising technique. This, however, has not been done.

There is evidence that the transport from the cytoplasm to the nucleus is selective. In earlier studies (Spear and Roizman, 1968) it was shown that at least three proteins did not accumulate in appreciable amounts in the nuclei of infected cells even after a 3-hour chase. More recently it has been found that, when Nonidet P-40 is used to fractionate infected cells, all the glycoproteins made after infection appear in the Nonidet P-40 extractable, cytoplasmic fraction (Spear and Roizman, 1970). It would appear from these studies that glycoproteins become bound to cytoplasmic membranes and possibly to the nuclear membranes but are not part of the complex macromolecular structures in the interior of the nuclei. It seems pertinent to note that nuclei of infected cells obtained with the aid of Nonidet P-40 contain numerous nucleocapsids and on analysis by acrylamide gel electrophoresis appear to contain all proteins associated with the virion except the glycoproteins.

3. *Glycoproteins*

Recent studies from two laboratories (Spear *et al.*, 1970; Olshevsky and Becker, 1970) have shown that several proteins detected by acrylamide gel electrophoresis of whole herpes simplex virus-infected cells, purified smooth membranes, and partially purified virions are glycosylated. The available information concerning the glycoproteins may be summarized as follows:

a. Source of Genetic Information for the Glycoproteins. There is indirect but entirely consistent evidence that the glycoproteins are specified by the virus and not by the host. The evidence is based on two observations. First the electropherograms of the glycoproteins made 4 hours postinfection and thereafter, i.e., after host polyribosomes disaggregate and host protein synthesis ceases (Sydiskis and Roizman, 1966, 1967), are substantially different from the glycoproteins made before infection (Spear *et al.*, 1970). The second observation is that the glycoproteins specified by different strains of herpes simplex virus differ in quantitative and qualitative characteristics (Keller *et al.*, 1970). Extensive analyses cited in part in Section II,C,1b showed that partially purified virions (MP strain) contain at least 6 glycosylated proteins which in acrylamide gels (Fig. 2) form four major and one minor bands. The glycoproteins associated with other strains of herpes simplex virus are also arranged in 4-5 bands, but the distribution of the glycosylated proteins in the bands differs somewhat (Fig. 3). Perhaps more convincing are the distribution of glycoproteins in purified membranes (Fig. 4). The glycoproteins extracted from membranes of cells infected with the F, mP, or with VR-3 strains of herpes simplex virus form two major

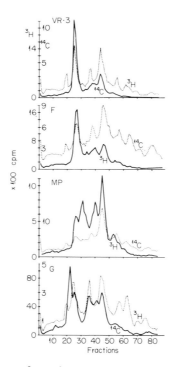

FIG. 3. Electropherograms of proteins prepared from partially purified herpes simplex virions of strains differing with respect to surface properties and with respect to their effects on the interaction of infected cells among themselves. The virions are prepared from HEp-2 cells labeled simultaneously with glucosamine and with a mixture of amino acids (leucine, isoleucine, and valine) between 4 and 22 hours postinfection with the indicated viral strain. The procedure for the partial purification of the virus was described by Keller *et al.* (1970). The solubilization and electrophoresis in acrylamide gels was described by Spear and Roizman (1968). The distribution of the incorporated glucosamine is indicated by solid lines; the distribution of incorporated amino acids is indicated by the broken lines. Data from Keller *et al.* (1970). Note the difference in the electrophoretic profile of the proteins in the MP virion from those shown in Fig. 2. The difference is due to the solubilization procedures employed in these studies.

bands migrating with proteins 100,000 and 50,000 daltons, respectively, and two or possibly three bands migrating with proteins of intermediate size. The glycoproteins extracted from membranes of cells infected with the MP strain form bands with identical electrophoretic mobility. However, there is only one predominant glycoprotein which migrates with proteins 50,000 daltons in molecular weight. The predominant glycoprotein of membranes of cells infected with the G strain of herpes simplex also migrates with proteins 50,000 daltons in molecular weight. However, in place of one of

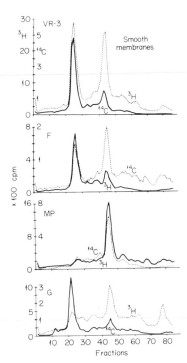

FIG. 4. Electropherograms of proteins prepared from purified smooth membranes of HEp-2 cells infected with different strains of herpes simplex virus. The smooth membranes and virions (Fig. 3) were prepared from the same batch of infected cells. The procedure for purification of the smooth membranes and of the solubilization and electrophoresis of viral proteins was described by Spear *et al.* (1970). The distribution of the incorporated glucosamine is indicated by the solid lines. The distribution of the incorporated amino acids is indicated by the broken line. Data from Keller *et al.* (1970).

the glycoproteins migrating with proteins between 50,000 and 100,000 daltons, electropherograms showed a small band of glycoprotein migrating more slowly than proteins 100,000 daltons in molecular weight.

It should be pointed out that the discussion in the preceding section, concerning the number and properties of the proteins in the herpesvirion, is applicable to glycoproteins as well, in that we do not really know the exact number of glycosylated polypeptides made in herpesvirus infected cells. On the other hand, since the glycoproteins from cells infected with different strains of virus were solubilized at the same time and sometimes simultaneously in artificial mixture, there is little doubt that the glycoproteins specified by different strains have different properties, consistent with the hypothesis that they are specified by the virus (Keller *et al.*, 1970).

b. Evidence that Glycoproteins are Structural Components of the Virus.
Since preparations of enveloped virus free from contamination with membranes are not available, the evidence that glycoproteins are structural
components of the virion is indirect and based on two observations. First,
the glycoproteins present in smooth membranes of infected cells differ in
relative composition from those present in partially purified virions prepared
from the same batch of infected cells. These data indicate that the membrane
contaminants present in the partially purified enveloped nucleocapsid preparation do not account for all the glycoproteins and that at least some
glycoproteins must be present in the virion. The nonionic detergent Nonidet
P-40 solubilizes the envelope of the virion and the glycoproteins. This
observation indicates that the glycoproteins are structural components of
the envelope and not of the nucleocapsid. Since the envelope is derived
from a modified cellular membrane, and since the glycoproteins in smooth
membranes and in the virion appear to differ in relative amounts, it necessarily follows that the binding of glycoproteins may be ordered and not
random.

c. Site of Glycosylation. Numerous indications point to the membranes
as the most likely site of glycosylation of viral proteins. The data are based
on several observations as follows: (1) Following a 5-minute pulse, only
trace amounts of glucosamine are associated with nascent peptides and,
moreover, the amount of precipitated radioactive glucosamine in nascent
peptides is a very small fraction of total acid precipitable and ethanol-
insoluble glucosamine present in the cell (Spear and Roizman, 1970). It
seems pertinent to point out that glucosamine is not actively metabolized
by infected cells; following intervals of labeling as long as 20 hours more
than 90% of the radioactive label is recovered in the form of glucosamine
and galactosamine (Keller, unpublished studies 1970). The addition of
puromycin causes cessation in the incorporation of amino acids into proteins
and the disaggregation of polyribosomes. However, the incorporation of
glucosamine into an acid-precipitable, alcohol-insoluble fraction continues
albeit at a slightly reduced rate (Spear *et al.*, 1970). In a pulse-chase experiment the pool of nonmembrane-bound glycoproteins was very small compared to the amount of membrane-bound glycoproteins detected after a
1-hour pulse; furthermore the transfer of glucosamine label from free
glycoproteins to membrane-bound glycoproteins could not be detected after
incubation in nonradioactive medium (Spear and Roizman, 1970). (2)
Acrylamide gel electrophoresis of the proteins in purified smooth membranes
prepared immediately after a 1-hour pulse with radioactive amino acids and

glucosamine and again after a chase period of several hours reveals the following: (a) Nonglycosylated proteins appear in the membranes after a pulse and subsequently disappear after the chase with the concomitant appearance of more slowly migrating glycoproteins. (b) The glycoproteins synthesized during the 1-hour pulse and detected in membranes form broad bands in the gel, indicating heterogeneity in size, with the glucosamine peak always trailing the amino acid peak. After the chase period the major glycoproteins synthesized during the pulse form narrow bands in the gels with complete coincidence of glucosamine and amino acid label (Spear and Roizman, 1970).

d. *The Time Course of Synthesis and Insertion of Glycoproteins into Smooth Membranes.* To determine whether the rates of synthesis of the various membrane glycoproteins change during the course of infection, cells were pulse labeled with radioactive glucosamine and amino acids from 5–6, 7–8, and 12–13 hours postinfection. Smooth membranes were then purified from the cells immediately after the pulse or after further incubation in non-radioactive medium. Analysis by acrylamide gel electrophoresis of the proteins synthesized during each time interval and ultimately incorporated into the membranes reveals that the membrane proteins are synthesized at the same rate relative to one another throughout the course of infection. Furthermore there seems to be no detectable differences in the patterns of binding of the proteins to membranes throughout infection. Since the proteins are probably first bound to the membranes and then glycosylated (see preceding section), information concerning the rate of insertion and glycosylation of the proteins will require further experiments to identify each glycoprotein with its nonglycosylated counterpart.

e. *The Source of Genetic Information for the Glycosylation of Viral Proteins.* In one series of experiments it has been found that the proteins in the minor glycoprotein bands are more extensively glycosylated in African green monkey kidney (VERO) cells than in HEp-2 cells. The data are consistent with the hypothesis that the extent of glycosylation may be determined in part by the host (Keller *et al.*, 1970). The source for the genetic information for the transferases involved in the glycosylation is unknown. It is conceivable that both viral and host enzymes may be involved.

D. LIPIDS

The presence of lipids is deduced primarily from the loss of biological activity following exposure of the virus to lipid solvents (Roizman and

Roane, 1963) and to lipases (Spring and Roizman, 1968). Lipids thus appear as essential constituents for the infectivity of both nucleocapsids extracted from nuclei and virions extracted from the cytoplasm of infected cells; however, the nature of the essential lipid is unknown. There have been few studies of lipid in infected cells (Falke, 1967). It seems pertinent to mention here, however, that the source of genetic information for the synthesis of the lipid contained in the herpesvirion has not been unequivocably established. Traces of new glycolipids in infected cells are readily apparent, but to date have not been analyzed in detail (Kieff, unpublished data). Virus grown in cells prelabeled with choline or in the presence of labeled choline became labeled and retained the label on isopycnic and rate centrifugations. Alas, virus was labeled not as well but with equal tenacity by artificially mixing unlabeled virus with labeled debris of uninfected cells (Roizman, unpublished studies).

III. The Architecture of the Herpesvirion

A. SOURCES OF INFORMATION, DEFINITIONS, AND GENERAL DESCRIPTION OF THE VIRION

Information concerning the structure of herpesviruses is derived from three sources: (1) thin sections of material embedded in plastic and stained with heavy metal salts, (2) dried preparations of particles permeated with phosphotungstic or silicotungstic acid (negative stain techniques), and (3) chemical and biological studies of isolated components. Correlations among these techniques are difficult enough at best, but nearly impossible without adequate terminology. For our purposes the terminology proposed by Lwoff *et al.* (1962), Casper *et al.* (1962), and Lwoff and Tournier (1966) is a useful starting point. There is general agreement that the herpesvirion consists of a core, a capsid, and an envelope. The core is defined as a centrally located body containing DNA and, probably, protein with particular affinity for DNA. The core is enclosed in a capsid. The structural subunit of most capsids of animal viruses consists of proteins, which may be associated in clusters or capsomeres. The envelopes of animal viruses, when present, consist of lipids, glycolipids, and glycoproteins probably arranged in an orderly structure. The capsid and core form the nucleocapsid. The herpesvirion is defined as an enveloped nucleocapsid.

The literature on the structure of herpesviruses is extensive; to avoid the morass of contradictions and terms that make sense only to their authors, it seems desirable to take an argumentative rather than a descriptive approach in the discussion of the fine structure of the virion. The available data suggest that the structure of the virion consists of (1) a core 25–30 nm in diameter containing viral DNA and probably protein, (2) an inner capsid 8–10 nm thick, (3) a middle capsid 15 nm thick, (4) an outer capsid 12.5 nm thick and consisting of 162 capsomeres, (5) an inner envelope 10 nm thick, probably containing structural components with low affinity for electron-opaque heavy metal salts, and last (6) an outer envelope approximately 20 nm thick. The data supporting this hypothesis concerning the structure of the herpesvirion and some of the properties of the architectural components follow.

B. THE ARCHITECTURAL COMPONENTS

1. *The Number of Architectural Components*

The presence of at least six architectural components in the virion is discerned from thin sections of infected cells. A similar number of architectural components may be deduced from biophysical data summarized in Table II. It is generally agreed that nucleocapsids approximately 100 nm in diameter either dispersed or in crystalline arrays are a common feature of the nuclei of infected cells, whereas enveloped nucleocapsids or virions 150–170 nm in diameter are sequestered in structures delineated by membranes and projecting into the nucleus or, more frequently, into the cytoplasm. The electron photomicrographs reported by Felluga (1963), Carmichael et al. (1965), Rabin et al. (1968), and by Siegert and Falke (1966) show intranuclear particles 100 nm in diameter consisting of a semitranslucent center surrounded by three shells: electron opaque, translucent, and opaque, in that order. In electron photomicrographs published by others (Morgan et al. 1953, 1954, 1958, 1959; Stoker et al., 1958; Falke et al., 1959; Morgan and Rose, 1960; Epstein, 1962a,b; Watson et al., 1964; Lunger, 1964; Lunger et al., 1965; McGavran and Smith, 1965; Becker et al., 1965; Shipkey et al., 1967; Nii et al., 1968; Spring et al., 1968; Schwartz and Roizman, 1969a,b) the semitranslucent center and the first electron-opaque shell fuse into an electron-opaque body. The mature virion as seen in thin sections contains two additional shells, the inner being electron translucent whereas the outer is electron opaque. Epstein (1962a) and Zambernard and Vatter (1966) demonstrated by means of nuclease digestion that the electron-opaque center of the virion contains DNA.

TABLE II

SUMMARY OF THE PROPERTIES OF THE VIRION AND OF THE ARCHITECTURAL SUBUNITS OF HERPES SIMPLEX VIRUS[a]

Designation	How obtained	Size (nm)		Properties
Core	Isopycnic centrifugation in CsCl solutions of	25	$p = 1.41$	Smallest particle containing viral DNA
Core + inner capsid	formaldehyde-stabilized preparations	40–45	$p = 1.31$	
Core + inner + middle capsid	Degradation of unstabilized nucleocapsids at low pH or in high salt solutions	75	$p = 1.31$	
Core + inner + middle + outer capsid = nucleocapsid	Extract of nuclei or whole cells centrifuged in sucrose, etc.	100	$p = 1.31$	Not infectious; contains antigen common to many herpesviruses
Nucleocapsid + inner envelope	Nuclei of infected cells	120		Unstable in CsCl; hydrodynamically smaller than virion; infectious, sensitive to lipases
Nucleocapsid + inner + outer envelope = virion	Cytoplasm of infected cells Extracellular fluid	150–160	$p = 1.27$	Relatively stable in CsCl; infectious, sensitive to lipases

[a] Roizman et al. (1969).

2. *The Inner and Outer Envelopes*

It is generally inferred that the envelope (henceforth designated as the outer envelope) of many herpesviruses is derived from the inner lamella of the nuclear membrane (Morgan *et al.*, 1954; Falke *et al.*, 1959; Siegert and Falke, 1966; Felluga, 1963; Nii *et al.*, 1968; Schwartz and Roizman, 1969b). However, the envelope of other herpesviruses may be derived from cytoplasmic membranes (Epstein, 1962b; Siminoff and Menefee, 1966; Stackpole, 1969; Schwartz and Roizman, 1969b). The intact outer envelope is impervious to phosphotungstic acid. The stain readily penetrates virions modified by drying or by exposure to antibody reactive with the envelope (Watson, 1968). The thickness of the outer envelope is difficult to estimate since the envelope tends to collapse on drying and under centrifugal stress. We estimate it to be 20 nm thick (Spring *et al.*, 1968). It appears to consist of two layers each showing a repeating unit structure. Occasionally spikes seem to project from the surface of the outer envelope; perhaps these spikes are visualized only when excess negative stain fills the space between the repeating unit. We have the impression that the envelopes of virions freshly extracted from infected cells form a tight sheath around the nucleocapsids and do not collapse as readily as those of particles stored for several days.

The evidence of an inner envelope is based on two sets of observations. First, in order to be enveloped, the nucleocapsid comes in opposition to the nuclear membrane where it is enfolded. Electron photomicrographs of thin sections show that nucleocapsids maintain a fixed distance from the membrane and, moreover, the space between the membrane and the nucleocapsids is filled by a substance which seems to adhere to the particle rather than to the membrane. This substance adhering to the nucleocapsid is not a property of all intranuclear nucleocapsids. This conclusion is based on the observation that the minimal distance between adjacent nucleocapsids in crystalline arrays in the center of the nucleus is less than the minimal distance between adjacent nucleocapsids at the nuclear membrane.

The second set of observations deals with physical and biological differences between infectious nuclear and cytoplasmic viruses prepared from 8-hour infected cells i.e., before the nuclei become very fragile and difficult to handle, and are as follows: (1) Infectious nuclear virus is very unstable in CsCl solution in that only 0.01% of infectivity is recovered after centrifugation (Spring and Roizman, 1968). (2) Infectious nuclear virus sediments more slowly than cytoplasmic virus in sucrose density gradients (Spring and Roizman, 1968; Roizman *et al.*, 1969). (3) Infectious cytoplasmic and nuclear viruses can be differentiated in neutralization tests and in tests employing protamine sulfate as an inhibitor of adsorption to cells (Roizman *et al.*, 1969). (4) The infectivity of both nuclear and cytoplasmic virus, however, is inactivated by phospholipase C with simultaneous release of choline. The treatment with lipases does not affect the structure of the nucleocapsid as seen with negative staining.

It should be pointed out that the electron photomicrographs of thin sections of cells infected with some herpesviruses, but most notably with Marek's disease virus (Nazerian and Witter, 1970) and the virus associated with Lucké adenocarcinoma (Stackpole, 1969), suggest that the envelope of extracellular virus is different from that of the virus which has just budded into the cytoplasm. The differences vary. The envelope of herpes simplex virions released into the extracellular fluid frequently is somewhat larger and "looser" than that of enveloped cytoplasmic herpesvirions. Some electron photomicrographs show a fuzzy fringe surrounding the envelope of herpesvirions released into the extracellular fluid. Perhaps the most dramatic modification of the envelope occurs in permissive feather-follicle cells infected with Marek's virus. In these cells the enveloped nucleocapsids emerging from the nucleus are sequestered into cytoplasmic inclusions in which the envelope becomes greatly enlarged (Nazerian and Witter, 1970).

The evidence that lipids and glycoproteins are contained in the envelopes is based on two lines of evidence. First, lipid solvents and lipases inactivate infectivity of the virus (Roizman and Roane, 1963; Spring and Roizman, 1968). Since the intact envelope is impermeable to molecules the size of phosphotungstic acid (Watson, 1968), it seems likely that the action of lipases is on the surface of the virion and not on the interior of the virion. The second line of evidence stems from the fact that Nonidet P-40 strips the envelope releasing the glycoprotein in soluble form. The nucleocapsid does not appear to be greatly affected by this treatment [see Section II,C,3b].

The distribution of the lipids and glycoproteins in the two envelopes is uncertain, but the weight of the evidence favors the hypothesis that the glycoproteins are contained in the outer envelope only. This conclusion is based on the fact that nuclei prepared with Nonidet P-40 lack glycosylated proteins but contain the same number of nonglycosylated structural proteins as nuclei prepared by Dounce homogenization. Nonidet P-40 solubilizes the outer nuclear and cytoplasmic membranes but apparently does not affect the contents of the nucleus any more than Dounce homogenization does. It could be argued that the glycoprotein constituents of the inner envelopes surrounding the nucleocapsids in the nucleus leak out, but at the moment this seems improbable. The leakage of RNA characteristic of nuclear RNA from nuclei obtained with Nonidet P-40 is minimal (Wagner and Roizman, 1969a). Moreover, as pointed out earlier, the nonglycosylated structural proteins are retained in the nucleus.

To summarize, the outer envelope is defined morphologically, chemically, and operationally. The inner envelope is defined operationally and to a lesser extent morphologically. Biochemical information concerning its structure and composition is lacking. The inner and outer envelope in thin sections account for the two outer shells—one electron opaque and one electron translucent. Some of the postulated properties of the inner envelope are that (1) it is electron translucent, (2) it is readily permeable to phosphotungstic and silicotungstic acids, (3) it contains structural components which are unstable and tend to disperse in solutions of high ionic strength and, last, (4) it confers upon naked nucleocapsids affinity for the nuclear membrane thus promoting envelopment.

3. The Outer and Middle Capsid

The most extensive studies of the outer capsid were those of Wildy et al. (1960). They reported that the staining of herpes simplex virus with phosphotungstic acid at low pH resulted in a better definition of the capsid structure at the expense of the envelope. Their data show that the surface of the nucleocapsid consists of 162 capsomeres arranged to form an icosahedron showing a 5:3:2 axial symmetry. The capsomeres on the surface of the nucleocapsids were described as prisms with hexagonal cross sections approximately 9.5×12.5 nm; since they fill partially with phosphotungstic acid it has been inferred that the distal end is hollow. The hole is estimated at 4.0 nm in diameter running down the middle along the long axis.

The evidence for the existence of a middle capsid situated underneath the outer capsid comes from several sources. Wildy et al. (1960) described the existence of a "core" approximately 77.5 nm in diameter. According to their calculations a particle of this dimension could be obtained by stripping from the 105 nm nucleocapsid a shell of capsomeres 12.5 nm thick. This particle was seen only as a part of the disintegrating virion and never by itself. A particle of similar size was seen by Spring et al. (1968) in herpes simplex virus partially disaggregated by centrifugation to equilibrium in CsCl. The nucleocapsids seen in the centrifuged preparation ranged from a small minority lacking a few capsomeres of the outer capsid to numerous particles with only a few capsomeres still projecting from an inner structure impermeable to negative stain. In size and appearance the structure was similar to that seen by Wildy et al. (1960).

The size of the middle capsid may be deduced from published photomicrographs of negatively stained preparations and thin sections. The photomicrographs reported by Stackpole and Mizell (1968) and by Toplin and Schidlovsky (1966) show that phosphotungstic acid penetrates the inner structure of herpesvirus nucleocapsids extracted from frog adenocarcinoma and Burkitt lymphoma cells, respectively, much more readily than that of the herpes simplex virion. The phosphotungstic acid delineates an inner shell which, in the photomicrographs by Stackpole and Mizell, has an outer diameter of 75–77 nm and an inner diameter of 45 nm. The outer dimension corresponds to the internal structure seen in partially disintegrated particles by Wildy *et al.* (1960) and Spring *et al.* (1968). In thickness and size, the shell corresponds to the concentric electron translucent shell seen in thin sections of nucleocapsids. The properties and structural components of the internal capsid are uncertain. It does not have affinity for electron-opaque salts of heavy metals and hence particles arrested in that stage of development could not be differentiated from particles consisting of cores and inner capsids.

4. *The Core and Inner Capsid*

Epstein (1962a) and Zambernard and Vatter (1966) demonstrated by means of nuclease digestion the presence of DNA in the electron opaque or semiopaque center of the nucleocapsid of herpes simplex virus and of the virus found in adenocarcinoma of the frog, respectively. Evidence concerning the dimension of the core and the existence of an inner capsid surrounding the core have come from several different sources.

Spring *et al.* (1968) described two small particles in preparations of herpes simplex stabilized with formaldehyde and centrifuged to equilibrium in solutions of CsCl. One particle banded between 1.37 and 1.45 gm/cm^3 and contained viral DNA. Electron photomicrographs of preparations stained with silicotungstic acid show a small particle approximately 25 nm in diameter; it was impervious to the negative stain and revealed no morphological features. The particle was absent from unstabilized preparations centrifuged to equilibrium.

The second particle described by Spring *et al.* (1968) banded in CsCl density gradients at a density of 1.352 gm/cm^3. Electron photomicrographs show a body 25 nm in diameter surrounded by beadlike projections which increased the diameter of the particle to about 45 nm. This particle was seen in fresh, unfixed lysates of infected cells. Particles of similar size were seen by others. Photomicrographs of negatively stained preparations of herpes type virus from Burkitt lymphoma cells by Toplin and Schidlovsky (1966) show a centrally located body 25–30 nm in diameter surrounded by a shell 0.8–1.0 nm in thickness. Particles of 30–40 nm, according to Siegert and Falke (1966), and 35–45 nm, according to Rabin *et al.* (1968), were reported in the nuclei of cells infected with herpes simplex. The particles described by these authors consist of an electron-opaque polygonal-shaped ring with a semielectron-translucent center. Similar particles somewhat larger in size (50–55 nm) and containing nucleoprotein were reported in preparations of bovine herpes virus by Bocciarelli *et al.* (1966). Particles of similar size were also reported by Stackpole and Mizell (1968) in cell lysates and in nuclei of cells from adenocarcinoma of the frog and by Epstein *et al.* (1968) in cells infected with the herpesvirus present in fowl afflicted with Marek's disease. In thin sections of nuclei, the particle appeared as a semitranslucent body surrounded by an electron-opaque shell. In negatively stained preparations the particle appeared impervious to PTA and poorly defined.

In light of the fact that the small 45–55 nm particles have been seen by a number of authors working with different herpesviruses and cells it seems rather unlikely that they are artifacts of adventitious contaminants. This conclusion is supported by the studies of Sydiskis (1969) who showed that particles containing DNA and banding in positions similar to the 25 nm and 45 nm particles of Spring *et al.* (1968) are precursors of complete herpesvirions.

The association of the 25 nm particle with viral DNA (Spring *et al.*, 1968) suggests that it is the core. In size it corresponds to the semitranslucent center seen in thin sections by Siegert and Falke (1966), Rabin *et al.* (1968), and by Toplin and Schidlovsky (1966) and delineated with phosphotungstic acid in photomicrographs by Toplin and Schidlovsky. The electron-opaque shell surrounding the semitranslucent body in the photomicrographs reported by Siegert and Falke (1966) and by Rabin *et al.* (1968) correspond in size to the beadlike projections around the 25 nm particle reported by Spring *et al.* (1968). If the 25 nm particle is the core, it seems likely that the beadlike projections seen in negatively stained preparations and the inner electron opaque shell seen in thin sections represent the inner capsid.

IV. Conclusions

In this paper we have shown the following:

(1) The structural components of the herpesvirion are DNA, proteins, glycoproteins, and lipids. Of these structural components, the DNA has

been most thoroughly defined and has been found to be quite variable in composition from one herpesvirus to another. The proteins, glycoproteins, and lipids have not been thoroughly and systematically analyzed largely because of the difficulties in obtaining purified virions and subviral components.

(2) The architectural components of the herpesvirion are a core, a multilayered capsid, and a complex envelope. We can operationally differentiate between an inner and an outer envelope. Both confer ability to infect cells upon the nucleocapsid, but the size and stability of the particles differ. Biophysical and electron microscopic data suggest the existence of three components in the capsid, i.e., an inner, middle, and outer capsid layers. However, cores without capsids or with one or two capsid layers have not been adequately studied and an operational definition is still lacking.

(3) At present we have but a vague idea as to the composition of the architectural components of the virion. It seems likely that the lipids and glycoproteins are structural components of the envelopes. The glycoproteins of the herpes simplex virion form six bands in acrylamide gels indicating that the envelopes contain at least six glycoproteins. If each of the remaining architectural components is made of different structural components it would follow that the virion contains at least six glycoproteins and four nonglycosylated proteins. This however, is a very low estimate, since it is based on the assumption that the hexagonal and pentagonal capsomeres in the outer capsid are made of the same proteins and that the inner and middle capsid layers are similarly constructed.

In this paper we made every effort to make apparent the difficulties and problems inherent in the studies of the structure of the herpesvirion. We hope that it is also its redeeming feature.

ACKNOWLEDGEMENTS

We would like to express our thanks to the editors for their patience, to Mrs. Norma Coleman for the secretarial assistance, and to United States Public Health Service (No. CA 08494), the American Cancer Society (No. E314E), and Whitehall Foundation for the grants which enabled us to work at the bench—in the interim between writing reviews.

REFERENCES

Aurelian, L. (1968). *Proc. Soc. Exp. Biol. Med.* **127**, 485–488.
Aurelian, L. (1969). *J. Virol.* **4**, 197–202.
Aurelian, L. and Roizman, B. (1965). *J. Mol. Biol.* **11**, 539–548.
Becker, P., Melnick, J. L., and Mayor, H. D. (1965). *Zxp. Mol. Pathol.* **4**, 11–23.

Becker, Y., Olshevsky, U., and Levitt, J. (1967). *J. Gen. Virol.* **1**, 471–478.

Becker, Y., Dym, H., and Sarov, I. (1968). *Virology* **36**, 184–192.

Ben-Porat, T. and Kaplan, A. S. (1962). *Virology* **16**, 261–266.

Ben-Porat, T. and Kaplan, A. S. (1963). *Virology* **20**, 310–317.

Ben-Porat, T. and Kaplan, A. S., (1965). *Virology* **25**, 22–29.

Ben-Porat, T., Shimono, H., and Kaplan, A. S. (1969). *Virology* **37**, 56–61.

Bocciarelli, D. S., Orfei, Z., Mondino, G., and Persechino, A. (1966). *Virology* **30**, 58–61.

Carmichael, L. E., Strandberg, J. D., and Barnes, F. D. (1965). *Proc. Soc. Exp. Biol. Med.* **120**, 644–650.

Caspar, D. L. D., Dulbecco, R., Klug, A., Lwoff, A., Stoker, M. G. P., Tournier, P., and Wildy, P. (1962). *Cold Spring Harbor Symp. Quant. Biol.* **27**, 49.

Crawford, L. V. and Lee, A. J. (1964). *Virology* **23**, 105–107.

Darlington, R. W. and Randall, C. C. (1963). *Virology* **19**, 322–327.

Dowdle, W. R., Nahmias, A. J., Harwell, R. W., and Pauls, F. P. (1967). *J. Immunol.* **99**, 974–980.

Eagle, H. (1959). *Science* **130**, 432–437.

Ejercito, P. M., Kieff, E. D., and Roizman, B. (1968). *J. Gen. Virol.* **3**, 357–364.

Epstein, M. A. (1962a). *J. Exp. Med.* **115**, 1–11.

Epstein, M. A. (1962b). *J. Cell. Biol.* **12**, 589–597.

Epstein, M. A., Achong, B. G., Churchill, A. E., and Biggs, P. M. (1968). *J. Nat. Cancer Inst.* **41**, 805–820.

Falke, D. (1967). *Z. Naturforsch.* **22**, 1360–1362.

Falke, D., Siegert, R., and Vogell, W. (1959). *Arch. Gesamte Virusforsch.* **9**, 484–496.

Felluga, B. (1963). *Ann. Sclavo* **5**, 412–424.

Fujiwara, S. and Kaplan, A. S. (1967). *Virology* **32**, 60–68.

Gönczöl, E., Jeney, E., and Váczi, L. (1967). *Acta Microbiol. Acad. Sci. Hung.* **14**, 39–43.

Goodheart, C. R. (1969). *Bacteriol. Proc.*, pp. 153–154.

Goodheart, C. R. (1970). *J. Amer. Med. Ass.* **211**, 91–96.

Goodheart, C. R., Plummer, G., and Waner, J. L. (1968). *Virology* **35**, 473–475.

Grüter, W. (1924). *Muench. Med. Wochschr.* **71**, 1058–1060.

Hay, J., Koteles, G. J., Keir, H. M., and Subak-Sharpe, H. (1966). *Nature (London)* **210**, 387–390.

Hay, J., Subak-Sharpe, H., and Shepherd, W. M. (1967). *Biochem. J.* **103**, 69.

Henle, W. and Henle, G. (1968). *J. Virol.* **2**, 182–191.

Inglis, V. B. M. (1968). *J. Gen. Virol.* **3**, 9–18.

Jeney, E., Gönczöl, E., and Váczi, L. (1967). *Acta Microbiol. Acad. Sci. Hung.* **14**, 31–37.

Kamiya, T., Ben-Porat, T., and Kaplan, A. S. (1964). *Biochem. Biophys. Res. Commun.* **16**, 410–415.

Kamiya, T., Ben-Porat, T., and Kaplan, A. S. (1965). *Virology* **26**, 577–589.

Kaplan, A. S. (1969). "Herpes Simplex and Pseudorabies." Springer-Verlag, New York.

Kaplan, A. S., and Ben-Porat, T. (1963). *Virology* **19**, 205–214.

Kaplan, A. S., and Ben-Porat, T. (1964). *Virology* **23**, 90–95.

Kaplan, A. S. and Ben-Porat, T. (1966). *Symp. Int. Congr. Microbiol. Moscow*, pp. 463–482.

Kaplan, A. S., Ben-Porat, T., and Coto, C. (1967). *In* "Molecular Biology of Viruses" (J. Colter, ed.), pp. 527–545. Academic Press, New York.

Keir, H. M. (1968). *Symp. Soc. Gen. Microbiol.*, pp. 67–99.

Keller, J. M., Spear, P. G., and Roizman, B. (1970). *Proc. Nat. Acad. Sci. U.S.* **65**, 865–871.
Kit, S., Dubbs, D. R., and Anken, M. (1967). *J. Virol.* **1**, 238–240.
Klemperer, H. G., Haynes, G. R., Shedden, W. I. H., and Watson, D. H. (1967). *Virology* **31**, 120–128.
Lando, D. and Ryhiner, M. L. (1969). *C. R. Acad. Sci. Paris* **269**, 527–530.
Lando, D., De Rudder, J., and Garilhe, De Michel P. (1965). *Bull. Soc. Chim. Biol.* **47**, 1033–1043.
Lee, L. F., Kieff, E. D., Bachenheimer, S. L., Roizman, B., Spear, P. G., Burmester, B. R., and Nazerian, K. (1971). *J. Virol.* **7**, 289–294.
Low, M., Hay, J., and Keir, H. M. (1969). *J. Mol. Biol.* **46**, 205–207.
Lunger, P. D. (1964). *Virology* **24**, 138–145.
Lunger, P. D., Darlington, R. W., and Granoff, A. (1965). *Ann. N.Y. Acad. Sci.* **126**, 289–314.
Lwoff, A. and Tournier, P. (1966). *Ann. Rev. Microbiol.* **20**, 45–74.
Lwoff, A., Horne, R., and Tournier, P. (1962). *Cold Spring Harbor Symp. Quant. Biol.* **27**, 51–55.
McGavran, M. H. and Smith, M. C. (1965). *Exp. Mol. Pathol.* **4**, 140.
Maizel, J. V., Jr. (1966). *Science* **151**, 988–990.
Maizel, J. V., Jr., White, D. O., and Scharff, M. D. (1968). *Virology* **36**, 126–136.
Martin, W. B., Hay, D., Crawford, L. V., LeBouvier, G. L., and Crawford, E. M. (1966). *J. Gen. Microbiol.* **45**, 325–332.
Morgan, C., and Rose, H. M. (1960). *4th Int. Conf. Electron Microscopy*, Berlin, 1958, Vol. 2, pp. 590–602. Springer-Verlag, Berlin.
Morgan, C., Ellison, S. A., Rose, H. M., and Moore, D. H. (1953). *Proc. Soc. Exp. Biol. Med.* **82**, 454–457.
Morgan, C., Ellison, S. A., Rose, H. M., and Moore, D. H. (1954). *J. Exp. Med.* **100**, 195–202.
Morgan, C., Jones, E. P., Holden, M., and Rose, H. M. (1958). *Virology* **5**, 568–571.
Morgan, C., Rose, H. M., Holden, M., and Jones, E. P. (1959). *J. Exp. Med.* **110**, 643–656.
Morgan, J. F., Morton, H. J., and Parker, R. C. (1950). *Proc. Soc. Exp. Biol. Med.* **73**, 1–8.
Morris, V. L., Wagner, E. K., and Roizman, B. (1970). *J. Mol. Biol.,* **52**, 247–263.
Morrison, J. M. and Keir, H. M. (1967). *Biochem. J.* **103**, 70–71.
Nazerian, K. and Witter, R. L. (1970). *J. Virol.,* **5**, 388–397.
Newton, A. A. (1968). *In* "International Virology I" (J. L. Melnick, ed.), p. 65. S. Karger Barel, New York.
Nii, S., Morgan, C., and Rose, H. M. (1968). *J. Virol.* **2**, 517–536.
O'Brien, B. R. A. (1964). *J. Cell Biol.* **20**, 521.
Olshevsky, U. and Becker, Y. (1970). *Virology,* **40**, 948–960.
Olshevsky, U., Levitt, J., and Becker, Y. (1967). *Virology* **33**, 323–334.
Piez, K. A. and Eagle, H. (1958). *J. Biol. Chem.* **231**, 533–545.
Plummer, G., Goodheart, C. R., Henson, D., and Bowling, C. P. (1969). *Virology* **39**, 134–137.
Rabin, E. R., Jenson, A. B., Phillips, C. A., and Melnick, J. L. (1968). *Exp. Mol. Pathol.* **8**, 34–48.
Roizman, B. (1966). *In* "Perspectives of Virology IV" (M. Pollard, ed.), pp. 283–304. Harper Row, New York.
Roizman, B. (1969). *Curr. Topics Microbiol. Immunol.* **49**, 1–79.

Roizman, B. and Roane, P. R., Jr. (1963). *Virology* **19**, 198–204.

Roizman, B. and Roane, P. R., Jr. (1964). *Virology* **22**, 262–269.

Roizman, B. and Spear, P. G. (1968). *J. Virol.* **2**, 83–84.

Roizman, B., Borman, G. S., and Kamali-Rousta, M. (1965). *Nature (London)* **206**, 1374–1375.

Roizman, B., Spring, S. B., and Roane, P. R., Jr. (1967). *J. Virol.* **1**, 181–182.

Roizman, B., Spring, S. B., and Schwartz, J. (1969). *Fed. Proc.* **28**, 1890–1898.

Roizman, B., Keller, J. M., Spear, P. G., Terni, M., Nahmias, A. J., and Dowdle, W. R. (1970a). *Nature (London)*, **227**, 1253–1254.

Roizman, B., Bachenheimer, S. L., Wagner, E. K., and Savage, T. (1970b). *Cold Spring Harbor Symp. Quant. Biol.* **35**, 753-771.

Rouse, H. C., Bonifas, V. H., and Schlesinger, R. W. (1963). *Virology* **20**, 357–365.

Russell, W. C. (1962). *Virology* **16**, 355–357.

Russell, W. C. and Crawford, L. V. (1963). *Virology* **21**, 353–361.

Russell, W. C. and Crawford, L. V. (1964). *Virology* **22**, 288–292.

Schwartz, J. and Roizman, B. (1969a). *Virology* **38**, 42–49.

Schwartz, J. and Roizman, B. (1969b). *J. Virol.* **4**, 879–889.

Shedden, W. I. H., Subak-Sharpe, H., Watson, D. H., and Wildy, P. (1966). *Virology* **30**, 154–157.

Shimono, H., Ben-Porat, T., and Kaplan, A. S. (1969). *Virology* **37**, 49–55.

Shipkey, F. H., Erlandson, R. A., Bailry, R. B., Babcock, V. I., and Southan, C. M. (1967). *Exp. Mol. Pathol.* **6**, 39–67.

Siegert, R. and Falke, D. (1966). *Arch. Gesamte Virusforsch.* **19**, 230–249.

Siminoff, P., and Menefee, M. G. (1966). *Exp. Cell Res.* **44**, 241–255.

Smith, K. O. (1963). *J. Bacteriol.* **86**, 999–1009.

Soehner, R. L., Gentry, C. A., and Randall, C. C. (1965). *Virology* **26**, 394–405.

Spear, P. G. and Roizman, B. (1968). *Virology* **35**, 545–555.

Spear, P. G. and Roizman, B. (1970). *Proc. Nat. Acad. Sci. U.S.* **66**, 730–737.

Spear, P. G., Keller, J. M., and Roizman, B. (1970). *J. Virol.* **5**, 123–131.

Spring, S. B. and Roizman, B. (1967). *J. Virol.* **1**, 294–301.

Spring, S. B. and Roizman, B. (1968). *J. Virol.* **2**, 979-985.

Spring, S. B., Roizman, B., and Schwartz, J. (1968). *J. Virol.* **2**, 384–392.

Spring, S. B., Roizman, B., and Spear, P. G. (1969). *Virology* **38**, 710–712.

Stackpole, C. (1969). *J. Virol.* **4**, 75–93.

Stackpole, C. and Mizell, M. (1968). *Virology* **36**, 63–72.

Stevens, J. G. and Jackson, N. L. (1967). *Virology* **32**, 654–661.

Stoker, M. G. P., Smith, K. M., and Ross, R. W. (1958). *J. Gen. Microbiol.* **19**, 244–249.

Subak-Sharpe, H. (1968). *In* "Molecular Biology of Viruses" (L. V. Crawford and M. G. P. Stoker, eds.), pp. 47–66. Cambridge Univ. Press, Cambridge.

Subak-Sharpe, H. and Hay, J. (1965). *J. Mol. Biol.* **12**, 924–928.

Subak-Sharpe, H., Shepherd, W. M., and Hay, J. (1966). *Cold Spring Harbor Symp. Quant. Biol.* **31**, 583–594.

Sydiskis, R. J. (1969). *J. Virol.* **4**, 283–291.

Sydiskis, R. J. and Roizman, B. (1966). *Science* **153**, 76–78.

Sydiskis, R. J. and Roizman, B. (1967). *Virology* **32**, 678–686.

Sydiskis, R. J. and Roizman, B. (1968). *Virology* **34**, 562–565.

Tankersley, R. W., Jr. (1964). *J. Bacteriol.* **87**, 609–613.

Terni, M. and Roizman, B. (1970). *J. Infec. Dis.* **121**, 212–216.

Toplin, I. and Schidlovsky, G. (1966). *Science* **152**, 1084–1085.

Wagner, E. K. and Roizman, B. (1969a). *J. Virol.* **4**, 36–46.

Wagner, E. K. and Roizman, B. (1969b). *Proc. Nat. Acad. Sci. U.S.* **64**, 626–633.

Wagner, E. K., Roizman, B., Savage, T., Spear, P. G., Mizell, M., Durr, F. E., and Sypowicz, D. (1970). *Virology*, in press.

Watson, D. H. (1968). *Symp. Soc. Gen. Microbiol.* **18**, 207–229.

Watson, D. H., Wildy, P., and Russell, W. C. (1964). *Virology* **24**, 523–538.

Weinberg, A. and Becker, Y. (1969). *Virology* **39**, 312–321.

Wildy, P., Russell, W. C., and Horne, R. W. (1960). *Virology* **12**, 204–222.

Zambernard, J. and Vatter, A. E. (1966). *Virology* **28**, 318–324.

Zemla, J., Coto, C., and Kaplan, A. S. (1967). *Virology* **31**, 736–738.

MAX BERGOIN* AND SAMUEL DALES†

* Station De Recherches Cytopathologiques INRA-CNRS, 30, Saint Christol-les-Ales, France. Work of Max Bergoin on *Amsacta* virus was carried out at the Boyce Thompson Institute for Plant Research, Inc., Yonkers, New York under the sponsorship of a USPHS Grant No. AI-08836.

† Department of Cytobiology, The Public Health Research Institute of the City of New York, Inc., New York, New York. This work was supported by a USPHS Grant No. AI-07477.

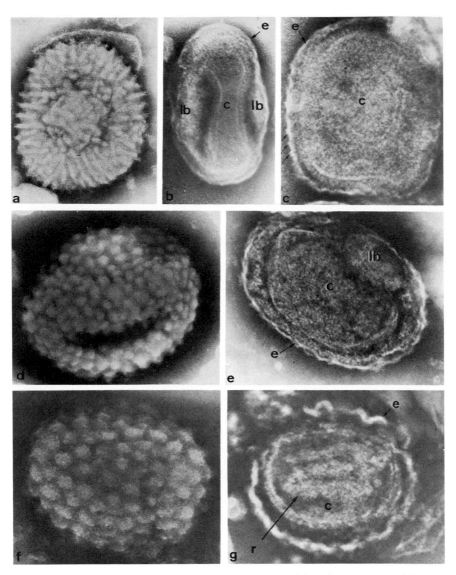

FIG. 1. Comparative morphology of mature virus particles of vertebrate and inverte-brate poxviruses as observed in negatively stained whole mount preparations. In this and subsequent figures the letters denote the following: c, core; cc, core coat; r, ropelike or cable structure within the core; lb, lateral body; e, envelope. (a) Vaccinia virus: the surface is covered by ridges which may be rodlets or tubules. × 160,000. (From Dales, 1963.) (b) Vaccinia virus viewed in its narrow aspect: the internal structure comprises a central biconcave core and two lateral bodies. × 155,000. (From Dales and Mosbach, 1968.) (c) Vaccinia virus viewed in its broad rectangular aspect. The folds of the envelope (arrows) are probably related to the surface ridges of the intact virion shown in (a).

I. Introduction

An interesting development in the field of insect pathology has been the identification in recent years of infectious agents possessing the morphology of poxviruses. The original isolation of the infectious principle was made from the grub of the common cockchaffer *Melolontha melolontha* L. by Hurpin and Vago (1963) and Vago (1963). Subsequently, similar disease agents were recognized in natural populations of several orders of insects collected in Europe, Africa, North and South America, Australia, and India, indicating a global distribution of these pathogens. The recently published morphological and biochemical information has established more firmly the similarities between the insect agents and poxviruses infecting vertebrate hosts. In compliance with the objective of the editors to emphasize the comparative aspects of virology, the first part of this review is devoted to the brief description of the architecture, chemical constitution, and development of poxviruses parasitizing both insect and vertebrate phyla. In the latter part of this article we consider exclusively and in greater detail some recently published and as yet unpublished information about the biology of the insect agents.

II. Comparative Observations on the Structure and Composition of Poxviruses

A. STRUCTURE OF MATURE FORMS AND SUBVIRAL COMPONENTS

The basic construction of poxviruses as exemplified by vaccinia, *Melolontha*, and *Amsacta* pox is illustrated in Figs. 1 and 2. In each of these agents three major components are evident including a lipoprotein envelope at the surface, lateral bodies, and a centrally positioned platelike core. These components can be identified both in whole mounts and in thin sections of

---

Short projections emanating from the core coat are visible in the lower left, subjacent to the outer envelopes. × 200,000. (From Dales, 1963.) (d) *Melolontha* poxvirus: the surface is covered by globular evaginations giving the particle a mulberry appearance. × 150,000. (From Bergoin *et al.*, 1971.) (e) *Melolontha* poxvirus with a disrupted envelope. Note the kidney-shaped core and a single lateral body. × 130,000. (From Bergoin *et al.*, 1971.) (f) *Amsacta* poxvirus: the individual surface globules are larger than those of *Melolontha* poxvirus. × 175,000. (From Granados and Roberts, 1970.) (g) A stain-penetrated damaged particle of *Amsacta* poxvirus. The cable or ropelike coiled structure is evident inside the core. × 140,000. (From Granados and Roberts, 1970.)

mature virus particles (Peters, 1956; Easterbrook, 1966; Westwood *et al.*, 1964; Dales and Siminovitch, 1961; Dales, 1962, 1963). The surface of envelopes of particles viewed by negative contrast appears to be convoluted into rodlets or tubular structures (Fig. 1). In *Amsacta* and *Melolontha* pox, the convolutions are highly prominent endowing the particles with a mul-

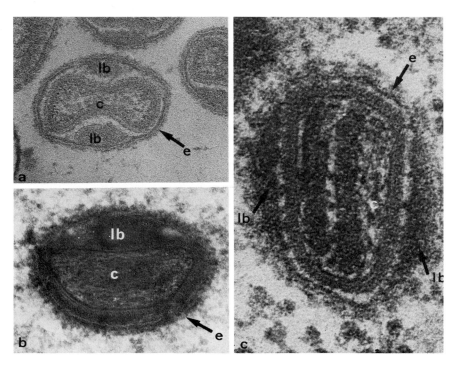

FIG. 2. Thin sections illustrating the comparative morphology of mature forms of vertebrate and invertebrate poxviruses. (a) Vaccinia virus displayed in its narrow aspect. The biconcave core and two lateral bodies are clearly visible. × 175,000. (From Pogo and Dales, 1969a.) (b) *Melolontha* poxvirus: the kidney-shaped core contains a filamentous dense substance. A single lateral body is positioned at the hilus of the core. × 122,000. (From Bergoin *et al.*, 1969a.) (c) *Amsacta* poxvirus: the core, rectangular in profile and containing a ropelike, dense structure, is flanked by two lateral bodies. × 208,000. (From Granados and Roberts, 1970.)

berrylike appearance (Fig. 1d and f). When vaccinia virus is prepared for electron microscopy by freeze-etching procedures, obviating artifacts which may result from chemical fixation or dehydration, the convolutions of the envelope are not evident. However, the entire surface of the virus has distributed on it double parallel rows of globular units (Medzon and Bauer,

1970), forming ridges that presumably correspond to the rodlets as evident in negatively stained preparations (Fig. 1a). The chemical nature or function of the globules on the envelope is unknown at present.

The lateral bodies are amorphous, proteinaceous masses that occupy the space between the envelope and central core. In agents with a core that in its narrow aspect is constricted bilaterally at the center, there are two lateral bodies attached to the biconcave plate, as in the case of vaccinia (Figs. 1b and 2a). In *Amsacta* pox the core is not constricted and the lateral bodies are represented by two discrete masses on either side of the cylindrical core (Fig. 2c; see also Fig. 13). However, in *Melolontha* pox only a unilateral constriction exists, providing space for a single lateral body within the hilus of the kidney-shaped core (Figs. 1e and 2b).

The core consists of a thick, proteinaceous coat enclosing the DNA genome. On its outer surface the core of vaccinia bears closely packed projections which emanate from the subjacent layer (Fig. 1c; cf. Fig. 3g). In *Amsacta* and *Melolontha* pox the core has on its surface fewer projections, each one larger than those of vaccinia (Figs. 3b and d) (Granados and Robers, 1970; Bergoin *et al.*, 1971). At high resolution these projections in *Amsacta* pox are observed to be flexible filaments that are tightly coiled into short, tubular elements (Fig. 3d). It has not been established chemically whether these tubules in *Amsacta* cores and projections on the cores of vaccinia are homologous.

In all poxviruses that have been examined by means of thin sections the space within the core is filled with tightly coiled dense threads, each 20–25 Å in width and presumed to be the DNA genome. In some mature particles of both vertebrate and invertebrate poxviruses a folded cablelike lucent structure is also evident inside the core (Figs. 1g and 2c; cf. Fig. 7). To date, neither the chemical nature nor function of this structure, originally identified by Peters and Muller (1963), has been elucidated. It is conceivable that the cable represents an enzyme–DNA complex which is organized or assembled in mature progeny during the inert stage of the replicative cycle.

B. CHEMICAL AND ANTIGENIC COMPOSITION OF MATURE VIRUS

Since the original work of Hoagland, Smadel, Rivers, and their colleagues (Hoagland *et al.*, 1940a,b; 1941a,b; Smadel *et al.*, 1940a,b), detailed chemical and serological analyses of poxviruses from vertebrates have been reported repeatedly (Zwartouw, 1964; Joklik, 1962, 1966; Woodson, 1968).

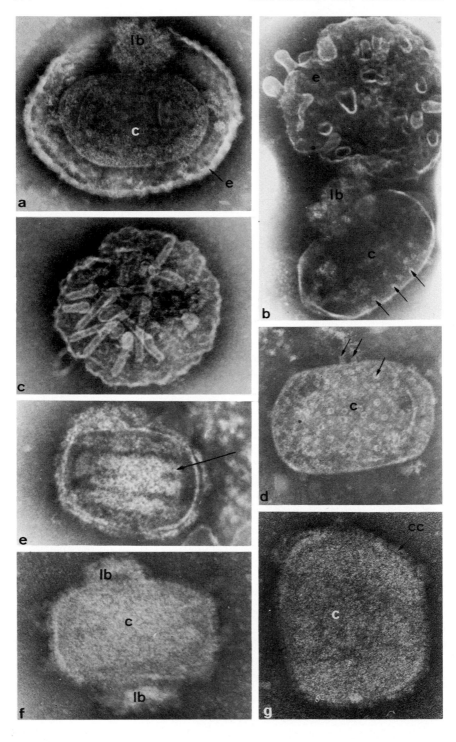

Most recent data based on purified material show that poxviruses have a weight of 5.5×10^{-15} gm per particle, containing approximately 5% DNA, 90% protein, 2% lipid, and 2% phospholipid (Joklik, 1966), of which lecithin is a major component (Dales and Mosbach, 1968), and traces of riboflavin and copper. Although most of the carbohydrate contained in the virus can be accounted for by the deoxyribose moiety of DNA, small quantities may also occur as glycoproteins in the lipoprotein envelope. The lipids are known to be a component of the envelopes since they are extractable by appropriate solvents and can be removed during stripping of the envelope by means of detergent and 2-mercaptoethanol (Dales and Mosbach, 1968) according to Easterbrook's (1966) procedure (Fig. 3f and g).

The double-stranded DNA genome possesses the highest molecular weight of any animal virus nucleic acids reported to date. Physical and chemical determinations have indicated a molecular weight of $150–200 \times 10^6$ daltons (Gafford and Randall, 1967; Joklik, 1962). The DNA genome has also been freed from purified virus particles and spread on the surface of protein films in preparation for electron microscopy. Such dispersed DNA molecules have been shown to be linear. In the case of fowl pox they possess a contour length of about 100 μ equivalent to a molecular weight of 192×10^6 (Hyde et al., 1967) and in the case of vaccinia the contour length measures 87 μ, equivalent to a molecular weight of 167×10^6 daltons (Sarov and Becker, 1967). Although the nucleic acid species of the genome of at least three insect poxviruses is known to be DNA (Bergoin et al., 1969b; Götz

---

Fig. 3. Whole mounts of degraded particles illustrating the components of vertebrate and invertebrate poxviruses. (a)–(c) Appearance of *Melolontha* poxvirus following prolonged contact with a reducing, alkaline solution employed for releasing virions from the proteinaceous inclusions. (a) Note a single lateral body lying in the vicinity of the ruptured envelope. × 112,000. (From Bergoin et al., 1971.) (b) A lateral body and the viral core have been separated from the envelope. Tubular elements are evident on the surface of the core (arrows). Globular evaginations of the envelope emanate from the surface. × 140,000. (From Bergoin et al., 1971.) (c) The separated envelope is folded into tubular structures. × 145,000. (From Bergoin et al., 1971.) (d) and (e) Appearance of *Amsacta* poxvirus particles following treatment with alkaline reducing solution. (d) The surface of the core is covered with coiled elements in the form of fine tubules (arrows). × 135,000. (From R. R. Granados, unpublished.) (e) A coiled ropelike component occupies the interior of the core. × 140,000. (From Granados and Roberts, 1970.) (f) Vaccinia virus stripped of its envelope by Easterbrook's (1966) method. Note the attached lateral bodies. × 185,000. (From Pogo and Dales, 1969a.) (g) Vaccinia virus particle stripped of its envelope and subjected to controlled digestion with trypsin. The lateral bodies have been removed leaving the isolated core. × 185,000. (From Pogo and Dales, 1969a.)

et al., 1969; Roberts and Bergoin, manuscript in preparation), at the time of writing molecular weight determinations have not been reported.

It is evident, from the high molecular weight of the DNA in poxviruses, that the genome may contain information sufficient to code for several hundred proteins. By analogy with information about tailed bacteriophages such as T4 (Edgar *et al.*, 1964), only some of these proteins are expected to be present in the virus itself, since the others would serve a catalytic function during the replicative sequence. Complete degradation of mature virus particles to individual polypeptides followed by electrophoresis in acrylamide gels indicates that at least 17 to 20 polypeptides are present in the virus (Holowczak and Joklik, 1967). Immunological investigations revealed that in extracts prepared from vaccinia or rabbit pox-infected rabbit skin, chorioallantoic membranes or HeLa cells there are 17 to 20 antigenic components which react with hyperimmune sera. However, only about one-half of these antigens are those of the mature virus particles (Westwood *et al.*, 1965). Among the antigens present in the virus, some occur as the internal component and others, of high molecular weight and possessing great avidity for the virus-neutralizing antibody, are most probably surface antigens (Woodroofe and Fenner, 1962, 1965).

The serological relationships between various agents of the vertebrate pox group have been studied by means of a variety of immunological procedures including complement fixation, gel diffusion, hemagglutination-inhibition, pox or plaque neutralization, and staining with fluorescein-coupled antibody. Close antigenic relationships were found to exist between vaccinia, cowpox, rabbit pox, and ectromelia, on the one hand, and myxoma and fibroma agents, on the other. However, no cross-antigenicity was found between either of the above subgroups or between them and fowl or monkey pox. However, when antisera were prepared against an alkaline extract containing the so-called NP internal antigen of vaccinia or rabbit pox, such antisera reacted with the NP antigen of all the vertebrate poxviruses tested (Woodroofe and Fenner, 1962), indicating the presence of a common NP antigen in all of them. It remains to be determined whether an identical NP antigen can be demonstrated in poxviruses of invertebrates.

During the past few years several investigators have demonstrated that highly purified particles of vertebrate poxviruses contain in the core four enzymatic activities, including a DNA-dependent RNA polymerase or transcriptase (Kates and McAuslan, 1967a; Munyon *et al.*, 1967), a nucleotide phosphohydrolase (Gold and Dales, 1968; Munyon *et al.*, 1968), and an alkaline and neutral DNase (Pogo and Dales, 1969a). These enzymatic activities are synthesized as late virus functions but are presumed to play a role during the early stages of the replicative process. Our recent experiments (Pogo *et al.*, 1971) have demonstrated that three of the four above enzyme activities, including the nucleotide phosphohydrolase and two DNases occur in the *Amsacta* pox. The transcriptase activity could not be detected, presumably because the enzyme was inactivated during the stringent purification procedures. The identification of similar enzymatic activities in poxviruses infecting both invertebrate and vertebrate hosts can now be employed as a biochemical diagnostic criterion to supplement the morphological similarities used heretofore, in identifying insect agents as bona fide poxviruses.

III. The Replicative Cycle

A. Penetration

The cycle of poxvirus biogenesis, as exemplified by vaccinia, commences with the uptake and penetration of inoculum particles in preparation for intracellular uncoating (Dales and Siminovitch, 1961; Dales, 1963; Joklik, 1964). During the sequence of penetration the virus cores (Fig. 4a), with their complement of DNA and enzymatic activities, are introduced into the cytoplasmic matrix of the host cell. Penetration of the inoculum occurring *in vivo* can be simulated by a stepwise dissection of purified virus particles in the test tube, employing Easterbrook's (1966) procedure, as illustrated in Fig. 3f and g. After stripping of the lipoprotein envelope (Fig. 3f), the lateral bodies can be digested specifically with trypsin (Fig. 3g). Coincident with removal of the lateral bodies the neutral and acid DNase activities in the virus cores are enhanced strikingly (Pogo and Dales, 1969a), providing evidence suggesting that the lateral bodies contain a specific inhibitor of the DNases. In this connection it may be significant that during penetration *in vivo*, the lateral bodies become separated from the core (Fig. 4a) at the time that inoculum particles escape from the phagocytic vacuoles into the cytoplasmic matrix (Dales, 1963). Perhaps dissociation of the core from the lateral bodies is the natural *in vivo* mechanism for activating the DNases. Presence of intracytoplasmic cores of *Amsacta* pox, morphologically like those of vaccinia, in insect tissue cells that have been asynchronously infected (see Figs. 15 and 16) (Granados and Roberts, 1970), implies that this insect agent undergoes the same sequence during penetration as that described for vaccinia (Dales, 1963).

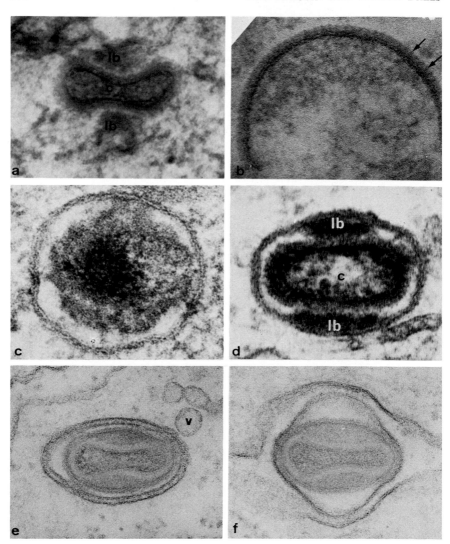

Fig. 4. Stages in the replicative cycle of vaccinia. (a) Peripheral cytoplasm of a cell sampled 20 minutes after inoculation. At this early stage the inoculum particle has penetrated into the host cell and is undergoing intracellular dismemberment. The envelope is apparently lysed and the two lateral bodies have become separated from the biconcave core. × 155,000. (From Dales, 1963.) (b) Early stage in the morphogenesis of vaccinia, observed 3–3.5 hours after inoculation. A partially assembled envelope consists of a unit membrane coated externally by a layer of closely packed, dense spicules (arrows). × 182,000. (From Dales and Mosbach, 1968.) (c) A maturing vaccinia observed in a cell 4 hours after inoculation. The trilaminar lipoprotein envelope surrounds the particle. The internal material consists of a dense fibrous nucleoid (DNA) surrounded by a less dense homogeneous substance thought to be material of the presumptive lateral bodies

B. INCEPTION OF VIRUS-SPECIFIED SYNTHESIS

During the second stage of the replicative cycle molecules of "early" messenger RNA (mRNA) are transcribed from the DNA template enclosed inside the core by the transcriptase and are released into the cytoplasm of recently infected host cells (Kates and McAuslan, 1967b; Kates and Beeson, 1970). One of the RNA's synthesized in this manner may be mRNA from which the uncoating factor is translated. Synthesis of the uncoating factor is required for rupturing the core so as to uncoat (i.e., release) (Joklik, 1964) the DNA genome contained within. Kates and Beeson (1970) have suggested that prior to uncoating another core enzyme present in vaccinia and *Amsacta* pox, the nucleotide phosphohydrolase may play a role in the transfer of nascent mRNA out of the cores.

Following uncoating of the DNA, the genome is reduplicated in a synchronous wave of synthesis, about 2 to 4 hours in duration (Joklik, 1966). Synthesis of the DNA occurs within discrete cytoplasmic factories into the mass of which are integrated the proteins and other macromolecules associated with the virus-regulated syntheses (Cairns, 1960; Magee *et al.*, 1960; Loh and Riggs, 1961).

C. MORPHOGENESIS

During multiplication of vaccinia, virus-specified envelopes appear in the viroplasmic matrix material, commencing at 3 to 3.5 hours following inoculation. First evidence of membrane formation is the appearance of short, arched segments consisting of a unit membrane and spicule coat (Fig. 4b) (Dales and Siminovitch, 1961; Dales and Mosbach, 1968). During development of these membranes the surface area is presumably increased while the original curvature is maintained, so as to enclose completely each immature virus particle in a spherical envelope (Fig. 4c). In the morphogenesis of *Melolontha* and *Amsacta* pox (illustrated in Figs. 17, 18, and 21)

---

and core. × 150,000. (From Dales and Mosbach, 1968.) (d) A maturing or mature particle lying in the cytoplasmic matrix of a cell sampled 7 hours after inoculation. The inner core and the lateral bodies are clearly differentiated within the viral envelope. × 162,000. (From Dales and Mosbach, 1968.) (e) Mature progeny particle enclosed in a cisterna apparently derived from vesicles (v) of the endoplasmic reticulum which become fused with one another. × 120,000. (From Dales, 1963.) (f) A mature extracellular particle lying between two adjacent "L" cells. The virus is enclosed by a single membrane originating from the inner membrane of the cisterna. × 120,000. (From Dales, 1963.)

(Granados and Roberts, 1970; Bergoin *et al.*, 1969a), a similar process in the formation of envelopes probably occurs. Judging by the mode in which these envelopes are developed, it appears that they are unique structures, discontinuous from any pre-existing membranes of host cells. In contrast, the envelopes of other animal and plant agents, such as influenza and rhabdoviruses, are acquired by budding at the surface, whereby segments of the host's endoplasmic reticulum or plasma membrane are utilized (see discussion in Dales, 1965). The envelopes of poxviruses meet several of the criteria for cellular membranes, including possession of a trilaminar unit membrane and a composition of lipid and protein, also perhaps including glycoprotein.

Experiments using inhibitors of DNA replication and transcription show that synthesis of proteins destined for assembly into vaccinia-specific membranes belongs in the category of early functions and commences about 2.5 hours after inoculation (Rosenkrantz *et al.*, 1966; Dales and Mosbach, 1968; Nagayama *et al.*, 1970). The assembly of immature virus particles, which commences at 3 to 3.5 hours is a continuous, exponential process of 10 or more hours duration. In this process of assembly the particles are constituted from an expanding intracytoplasmic pool of virus protein.

According to a reconstructed sequence of morphopoiesis all poxviruses appear to undergo differentiation from the spherical immature stage into mature particles by alterations occurring inside the virus envelopes (Fig. 4c and d) (Dales and Mosbach, 1968). An early stage of differentiation is evident when discrete nucleoids consisting of bundles of dense, 20–25 Å DNA filaments, appear inside the envelopes (Fig. 4c). In some transitional forms of developing progeny evidence of a condensation into lateral bodies and cores (Fig. 4d) has been obtained. Once they have been developed the maturing or mature forms of vaccinia virus migrate away from the region of the viroplasmic matrix and may undergo further processing in preparation for release from the host.

Regarding the maturation process, biochemical experiments show that the four virus-contained enzymatic activities present in the core are synthesized as late functions (McAuslan and Kates, 1967; Kates *et al.*, 1968; Pogo and Dales, 1969b). Some of them may be the late factors required for differentiation from the immature to mature forms of poxviruses (Nagayama *et al.*, 1970). Thus transcription into mRNA's associated with the early functions including the process of uncoating, replication of DNA, and formation of virus-specific envelopes and other structures of immature particles, may occur from parental genomes but transcription into mRNA's related to the late functions, including maturation factors, is regulated from

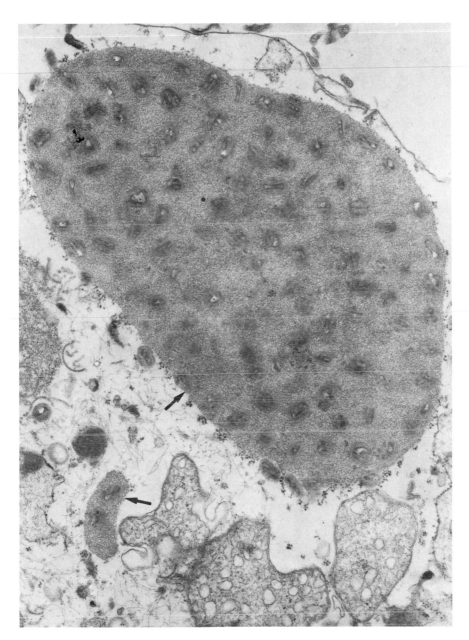

FIG. 5. "A"-type inclusion (Downie body, arrow) lodged in the cytoplasm of a HeLa cell infected with the CP-58 strain of cowpox virus. Such inclusions consist of a protein-aceous matrix in which a large number of mature virions are occluded. × 33,000. (From Ichihashi and Matsumoto, 1968.)

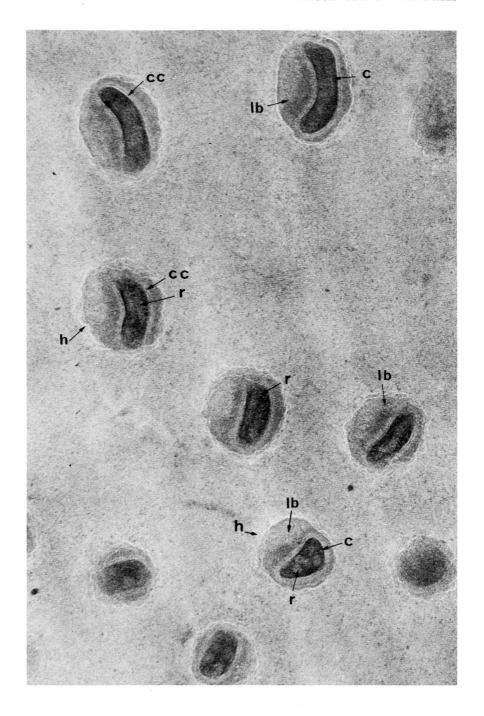

progeny templates. It remains to be determined whether the same kind of regulation of early and late functions prevails in the synthesis of insect poxviruses.

D. DISSEMINATION OF PROGENY

Two patterns appear to have evolved among the poxviruses for the dissemination of progeny particles between cells and organisms. In one process, as exemplified by vaccinia, particles of progeny become associated with smooth cisternae, preferentially in the golgi region of the cell. Electron microscopy of thinly sectioned infected cells has provided evidence suggesting that several smooth cisternae become attached to the virus surface and then coalesce to form a double membrane sac investing either single or several particles (Fig. 4e). From a reconstructed sequence it appears that the virus and its enveloping sac are moved to the cell surface in preparation for virus release. Once the outer membrane of the cisterna comes into contact with the plasma membrane, fusion occurs, so as to provide a channel for the release of the progeny virus to the cell exterior (Fig. 4f). In some cases particles are not wrapped in smooth cisternae but nevertheless are moved to the cell surface, sometimes into microvilli (Dales and Siminovitch, 1961). Release of "naked" progeny may occur from these surfaces. Similar ways of virus dissemination have been observed in *Melolontha* pox (see Section IV).

There is a second and completely different mechanism for dissemination of poxviruses which involves the integration of progeny within specific inclusions. In the case of *Melolontha* and *Amsacta* poxviruses the particles become occluded in proteinaceous, crystalline masses (see Fig. 19) (Bergoin *et al.*, 1969a; Granados and Roberts, 1970). Presumably this type of encapsulation ensures maximum protection from fluctuations in the extra-organismic environment. Among vertebrate agents including cowpox, ectromelia, fowl pox, and myxoma these inclusions, termed either "A"-type or Marchal bodies (Kato and Cutting, 1959) are formed in the cytoplasm relatively late in the infectious cycle. The A-type inclusions consist of a homogenous proteinaceous matrix (Fig. 5) unlike that of the crystalline

---

FIG. 6. "A"-type inclusion of *Melolontha* poxvirus showing a group of occluded virions in their longitudinal and transverse aspect, embedded within crystalline matrix. Structural components of the virions are clearly identifiable. Surrounding the virions is a halo (h) of material with a low density. × 56,000. (From Bergoin *et al.*, 1971.)

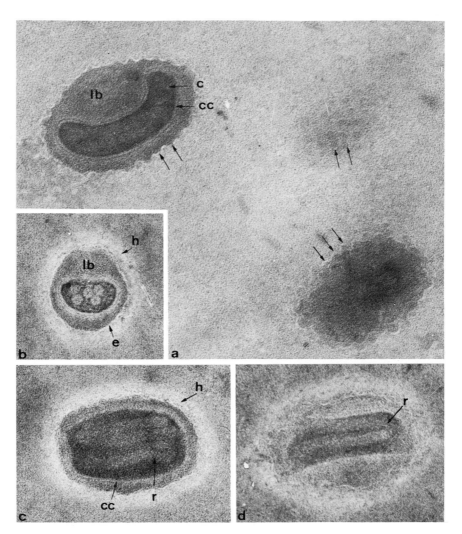

Fig. 7. The structure of occluded particles of *Melolontha* poxvirus. A halo (h) of low electron opacity surrounds the particles. (a) The crenated outline of the virions indicates that the section passed through the globular surface units (arrows). The three coats of the core consist of a denser central layer within two layers of lesser density. × 162,000. (From Bergoin *et al.*, 1971.) (b) In this transverse section there are five profiles of the cable structure within the core. × 117,000. (From Bergoin *et al.*, 1971.) (c) The virion sectioned parallel with its broad aspect possesses a rectangular outline to which the shape of the core conforms. The core is filled with four turns of the cable component. × 187,000. (From Bergoin *et al.*, 1971.) (d) Another example illustrating the ropelike coiled filament within the core. × 135,000. (From Bergoin *et al.*, 1968a.)

material of insect poxviruses. Some variants of cowpox and ectromelia can induce formation of an A-type inclusion in which progeny particles fail to be occluded while other variants induce inclusions containing large numbers of virus progeny (Morita, 1960; Kato *et al.*, 1963; Ichihashi and Matsumoto, 1968).

IV. Detailed Comparisons of Insect Poxviruses

The fourth section of this review deals more specifically with *Melolontha* and *Amsacta* pox, two of the agents that to date have been investigated most extensively.

A. STRUCTURE OF MATURE FORMS

1. *Melolontha Poxvirus*

Details about the architecture of the mature particles of *Melolontha* pox have been derived from both sectioned material and whole mounts (Vago, 1963; Vago and Croissant, 1964; Bergoin *et al.*, 1968a; Bergoin *et al.*, 1971). Thinly sectioned virions lodged within the crystalline matrix of inclusions, or spherules, are illustrated in Figs. 6 and 7. It is apparent that virus particles sectioned parallel to their long axis are 4000 Å in length and possess an oval profile. Those cut perpendicular to the long axis are about 2500 Å wide and have a circular profile. The dense core occupies an eccentric position within the envelope. In its longitudinal, narrow aspect it is kidney-shaped, due to an invagination on one side and an evagination on the opposite side. The core coat is 150 Å in thickness and is apparently constituted from three layers, of which the middle one is the most dense. On its surface the core coat bears short, cylindrical structures that are clearly revealed in whole mount preparations (Fig. 3b). Inside the core resides a folded cable or ropelike structure of low density approximately 200 Å in diameter (Figs. 6 and 7). In particles cut perpendicular to the long axis (Fig. 7b) four or five cross sections of the cable are evident. In particles sliced in a plane parallel to the long axis of the virus, so as to reveal either the broad (Fig. 7c) or narrow (Fig. 7d) aspect of the core, the cable appears as a continuous coiled structure, folded into three or four segments. The hilus between the core and the envelope is occupied by the mass of the lateral body, as evident in both sections (Figs. 6 and 7) and whole mounts (Fig. 1e). Convolutions at the

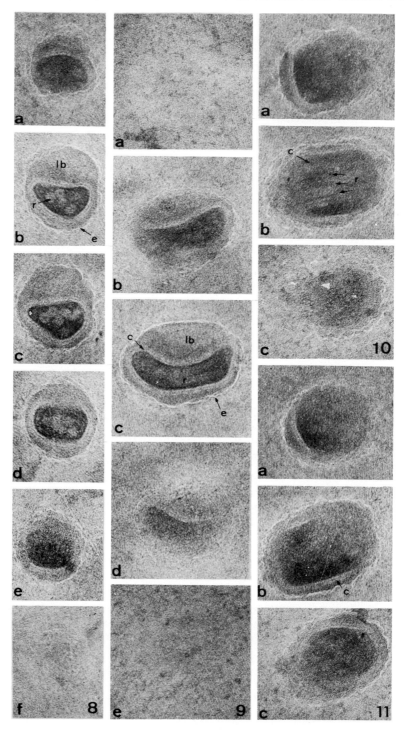

surface may be seen in both the whole mounts of intact virus (Fig. 1d) and in thin sections especially in cuts grazing the surface (Fig. 7a). Where virions become occluded within the inclusion a narrow rim of low density and variable width separates the region adjacent to the virus envelope surface from the enclosing crystalline protein (Fig. 7b,c, and d). Such halos might occur as an artifact arising from a shrinkage during processing or are caused by a disorganization of the crystalline lattice in the vicinity of occluded virions.

A three-dimensional reconstruction of virus structure can be made from serial sections, as exemplified in Figs. 8–11. When cut perpendicularly to its long axis the virus has a quasi-circular profile (Fig. 8) and at the center of the particle the cable structure is clearly evident (Fig. 8c). In cuts made parallel to the long axis and perpendicular to the broad side (Fig. 9) the virion has an oval outline. In the example illustrated in Fig. 9c the section traverses almost exactly the vertical plane of longitudinal symmetry. When the virus is sectioned parallel to its long axis and perpendicular to the narrow aspect the profile evident is rectangular (Figs. 10 and 11). The coat of the core is closely apposed on all sides to the envelope. From the results with serial sections it is possible to reconstruct diagrammatically the three-dimensional structure of this agent, as illustrated in Fig. 12a–c.

When virions are freed from the proteinaceous inclusions by agents such as 2-mercaptoethanol or sodium thioglycolate (Bergoin et al., 1967), many damaged particles are obtained in suspension. The inadvertent rupture of the envelope permits one to observe within the membrane the lateral body (Fig. 3a). In damaged, but incompletely dismembered particles, one can observe both a deformation of the folds of the isolated membrane, which appear as fingerlike evaginations (Fig. 3b and c), and the separation of cores (Fig. 3b and d). These images should be compared with those of subcomponents of vaccinia virus obtained after controlled degradation using the procedure of Easterbrook (1966) (Fig. 3f and g).

FIGS. 8–11. Occluded virions of *Melolontha* poxvirus serially sectioned in their transverse, longitudinal, and horizontal aspects.

FIG. 8. Transversely sectioned: in (c) the section passes approximately through the center of the particle. × 62,000. (From Bergoin *et al.*, 1971.)

FIG. 9. Longitudinally cut: in (c) the section passes almost exactly through the center. × 62,000. (From Bergoin *et al.*, 1971.)

FIGS. 10 and 11. Horizontally sectioned: four turns of the cable structure are evident in 10 (b) (arrows). × 62,000. (From Bergoin *et al.*, 1971.)

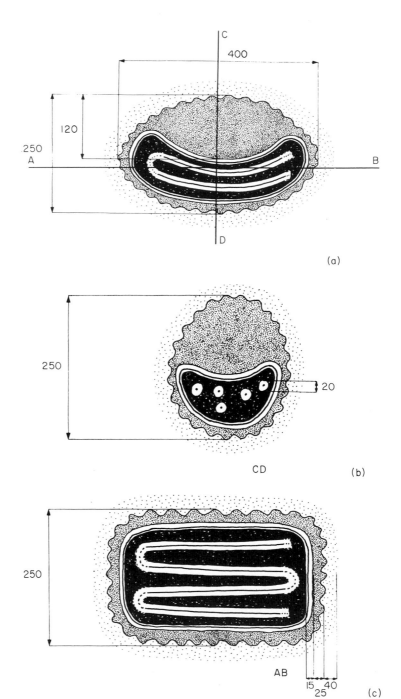

(a)

(b)

CD

(c)

AB

2. *Amsacta Poxvirus*

The structure of the *Amsacta* agent conforms generally to that of other poxviruses. The virions are, however, smaller than those of *Melolontha* pox, measuring approximately 3500 × 2500 Å (Granados and Roberts, 1970). In whole mount, the surface possesses the characteristic beaded appearance (Fig. 1f), which is also discernable in thin sections (Fig. 14). Each convolution is approximately 400 Å in diameter (Fig. 1f) making it larger than that present at the surface of *Melolontha* poxvirus (Fig. 1d). The viral core, unlike the kidney-shaped structure of *Melolontha* virus, is symmetrical in both its rectangular broad aspect (Fig. 13) and in longitudinal circular cross section (Fig. 14). The coat of the core consists of a continuous inner smooth layer 75 Å wide and an outer layer of discontinuous projections, each 110 Å long and 60 Å wide (Fig. 15). Fibrous dense elements are loosely packed within the core (see Fig. 14). In some particles the fibrous component is aggregated into dense rods three to five in number, each approximately 300 Å in diameter (Figs. 1g and 2c; cf. also Fig. 14). The homology between the rods and the ropelike elements observed in *Melolontha* pox (Fig. 7) is not certain.

3. *Other Insect Poxviruses*

As mentioned in the introduction other poxvirus types have been identified in several insect hosts. Their morphological characteristics are summarized and compared in Table I. Generally the differences between the structure of *Melolontha* poxvirus and those found in other Coleoptera (Vago *et al.*, 1968a,b, 1969; Goodwin and Filshie, 1969) are only minor. All the agents of this subgroup are relatively large with average dimensions of 4000 × 2500 Å. Each has the characteristic single lateral body, a unilateral depression of the core, and shows evidence of the cable structure within the core. Agents infecting the Lepidoptera are generally smaller, having an average size of 3500 × 2500 Å (Weiser and Vago, 1966; Meynadier *et al.*, 1968) and possess a symmetrical core structure. A poxvirus found to infect the grasshopper *Melanoplus sanguinipes* (Orthoptera) resembles closely agents of the Lepidoptera (Henry *et al.*, 1969). Two agents associated with Dipteran species are the smallest of the three major insect poxvirus groups averaging only 3200 × 2300 Å in size (Götz *et al.*, 1969; Weiser, 1969). In both their size

---

Fig. 12. Diagrammatic representation of *Melolontha* poxvirus in longitudinal (a), transverse (b), and horizontal (c) plane of symmetry. Numbers are the average dimensions in nanometers. (From Bergoin *et al.*, 1971.)

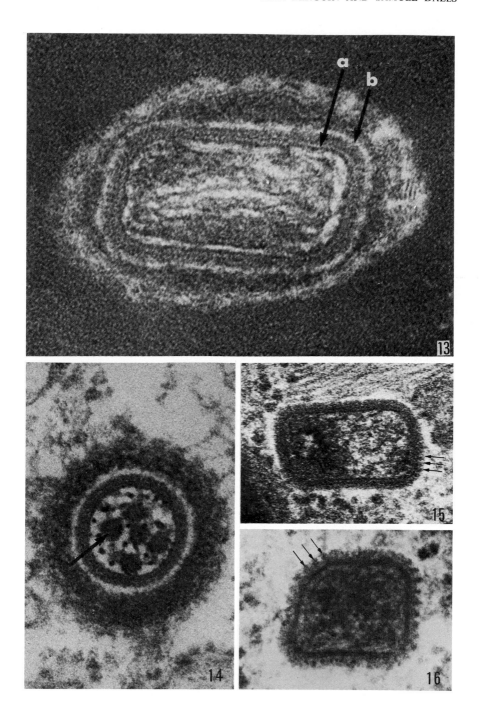

and construction they resemble most closely the poxviruses of vertebrates. They are cuboidal in outline and like many vertebrate poxviruses possess a biconcave core and two well developed lateral bodies. The material inside the core consists of dense threads. None of the organized cable structures have been described in the Dipteran viruses.

B. MORPHOGENESIS AND DISSEMINATION

The replicative cycle of vertebrate poxviruses has been studied in a precise time sequence using synchronously infected cell cultures (see Section III). Unlike the situation with vertebrate agents, synchronous *in vitro* systems have as yet not been devised for studying the replicative cycle of insect poxviruses. It is well appreciated by workers in this field (Amargier *et al.*, 1964; Bergoin *et al.*, 1969a,b; Henry *et al.*, 1969), that following inoculation insect viruses fail to initiate a simultaneous infection of the susceptible target cells, thus the infectious process in different cells within a tissue becomes highly asynchronous and the sequence of morphopoiesis has to be deduced using the vertebrate poxviruses as a model.

1. *Melolontha Poxvirus*

Development of the *Melolontha* agent can be observed in adipose tissue cells by means of thin sections (Bergoin *et al.*, 1969a). Immature forms occur in association with dense amorphous cytoplasmic masses, termed viroplasmic inclusions, that are lodged in the vicinity of the nucleus. Complete and incomplete spherical envelopes of immature progeny virus are evident

---

FIG. 13. Horizontal section of occluded *Amsacta* poxvirus. The rectangular core is clearly differentiated and delimited by a coat consisting of inner (a) (arrow) and outer (b) (arrow) layers. A fibrous material occurs within the core. × 250,000. (From Granados and Robers, 1970.)

FIG. 14. Transverse section of intracellular *Amsacta* poxvirus. The core contains five larger, dense masses, presumably the component described in Fig. 2(c). × 187,000. (From Granados and Roberts, 1970.)

FIG. 15. A core of *Amsacta* poxvirus lying free in the cytoplasm of an adipose cell. The outer layer of the core coat is formed of subunits (arrows). × 150,000. (From Granados and Roberts, 1970.)

FIG. 16. A core of vaccinia lying in the cytoplasmic matrix of a L cell shown for comparison with Fig. 15. The core is covered with regularly spaced external projections (arrows). × 160,000. (From Dales and Kajioka, 1964.)

TABLE I

MAIN CHARACTERISTICS OF POXVIRUSES OF INSECTS

Host	Size (nm)	Shape	Internal structure (vertical sections)	Nucleic acid	References
Coleoptera					
Melolontha melolontha L. (Scarabaeidae)	450×250	Oval-shaped, mulberry-like surface	Unilaterally concave core containing a filamentous rod	DNA	Vago, 1963; Vago and Croissant, 1964; Bergoin *et al.*, 1968a
Demodena boranensis BRUCH (Scarabaeidae)	420×230	Oval-shaped, mulberry-like surface	Unilaterally concave core containing a filamentous rod		Vago *et al.*, 1968b
Othnonius batesi (Scarabaeidae)	470×265	Oval-shaped, mulberry-like surface	Unilaterally concave core containing a filamentous rod		Goodwin and Filshie, 1969
Figulus sublaevis (Lucanidae)	370×250	Oval-shaped, mulberry-like surface	Unilaterally concave core containing a filamentous rod		Vago *et al.*, 1968a
Dermolepida albohirtum					Goodwin and Filshie, 1969
Phylloperha horticola L. (Rutelidae)	400×240	Oval-shaped, mulberry-like surface	Unilaterally concave core containing a filamentous rod		Vago *et al.*, 1969

TABLE I (*continued*)

Host	Size (nm)	Shape	Internal structure (vertical sections)	Nucleic acid	References
Lepidoptera					
Operophtera brumata HB (Geometridae)	370 × 250	Oval-shaped			Weiser and Vago, 1966
Acrobasia zelleri RAG. (Pyralididae)		Oval-shaped			Weiser and Vago, 1966
Amsacta moorei (Artiidae) (original host) adapted on *Estigmene acrea* (Arctiidae)	350 × 250	Oval-shaped, mulberry-like surface	Rectangular core containing rodlike structures	DNA	Granados and Roberts, 1970
Oreopsyche angustella H.S. (Bsychidae)	360 × 260	Oval-shaped, mulberry-like surface	Rectangular or slightly biconcave core		Meynadier *et al.*, 1968
Orthoptera					
Melanoplus sanguinipes F. (Acrididae)	320 × 250	Oval-shaped	Rectangular or dumbbell-shaped core		Henry *et al.*, 1969
Diptera					
Chironomus luridus (Chironomidae)	320 × 230 × 110	Cuboidal	Dumbbell-shaped core	DNA	Götz *et al.*, 1969
Camptochironomus tentans (Chironomidae)	200–230 × 270–300 × 130–150	Cushion-shaped	Dumbbell-shaped core		Weiser, 1969

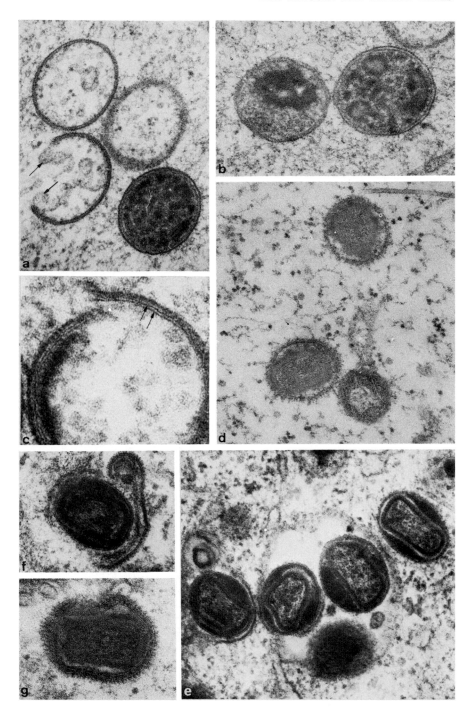

at the periphery of such inclusions. The envelopes appear to possess two dense layers, each approximately 75 Å wide separated by a less dense region about 100 Å in width (see Fig. 17c). The inner layer conforms to the trilamellar or unit membrane structure while the dense outer layer shows poorer definition. Some immature forms, lacking the dense internal matrix normally present within the spherical envelopes, may be only virus membrane shells (Fig. 17a and b). It is not clear at present whether such shells represent aberrant particles or are an intermediate stage in morphogenesis. Development of the immature particles by a process of budding from the viroplasmic masses is also evident (Fig. 18a). Intermediate stages in the process of conversion from immature spherical forms into mature virions are infrequent in cell sections. It is assumed that spherical particles become compressed laterally and differentiate internally into the core and lateral body components evident in mature virus (Fig. 17d and e). Mature particles are usually dispersed away from the site of the viroplasmic foci where early morphopoiesis is evident.

The virus progeny may be disseminated in three ways. Some mature intracytoplasmic particles become wrapped in membranes apparently originating from the host's endoplasmic reticulum and may be carried to the cell surface and released with their additional membrane into the intercellular spaces (Fig. 17f and g). Cell-to-cell transmission of the infection within a tissue might be effected by progeny released in this manner. Virus freed in this way accounts for only a minor fraction of the total progeny. The majority of the progeny become integrated within large, dense proteinaceous crystalline spherules. The spherules, generally one or two per cell, develop in the cytoplasm away from the viroplasmic foci. It is deduced from a reconstructed sequence of events that the progeny virus becomes concentrated in the vicinity of developing spherules which grow in volume

FIG. 17. Reconstructed maturation sequence of *Melolontha* poxvirus. (a) Complete and incomplete spherical forms of immature virus. Arrows indicate sheets of unit membrane in continuity with the inner membrane of the viral envelope. × 48,000. (b) Dense, immature, spherical particles possibly undergoing differentiation. The particle on the left possesses a nucleoid. × 51,000. (c) Higher resolution showing the structure of the envelope consisting of a unit membrane (arrows) and an outer dense substance reminiscent of spicules illustrated in Fig. 4(b). × 200,000. (d) Two maturing particles with a differentiated inner core lack the dense substance of the lateral bodies. The envelope bears a fuzzy external coat. × 41,000. (e) Maturing or mature particles lying in the cytoplasm possess well-differentiated lateral bodies and cores. × 50,000. (f) Mature particle in the process of envelopment in a cisterna originating from the endoplasmic reticulum. × 78,000. (g) Mature particle completely enveloped by a cisterna. Note the spiny coat on the outer membrane leaflet. × 75,000. (From Bergoin et al., 1969a.)

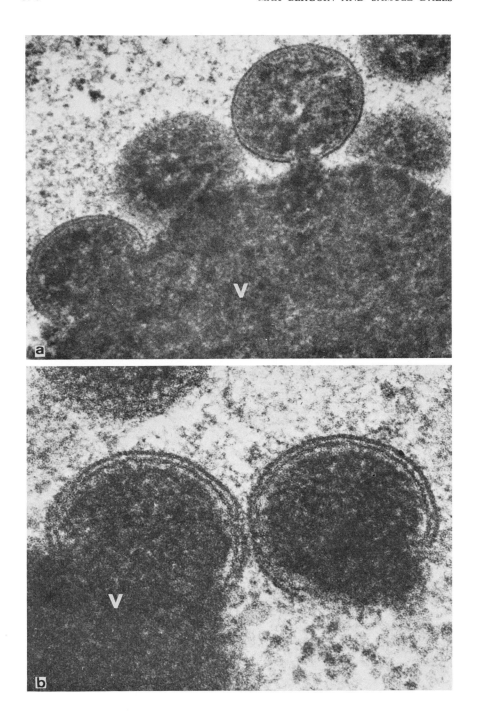

by accretion of protein and a simultaneous integration of progeny. Usually the occluded virus particles are well separated from one another and have their long axes disposed in a radial orientation with respect to the center of inclusion (Fig. 19). The crystalline inclusions sometimes grow to a large size attaining diameters of 20 to 24 μ and occupying a large fraction of the total volume within the cytoplasm of the host cell. These inclusions are liberated into the soil following the death and decomposition of infected larvae thereby providing a reservoir of the agent for the spread of the infection to animals feeding on the contaminated substrate.

Release of the occluded virus occurs in the milieu of the gut lumen where a high pH and other factors favor the rapid dissolution of the crystalline matrix. The liberated virions may pass from the lumen through the gut epithelium and invade the body cavity.

A third mechanism for spreading virus progeny from the hemocytes of *Melolontha* has been described. In hemocytes integration of virus into spherules is uncommon. Instead progeny appear to be disseminated by a process of exocytosis (Devauchelle *et al.*, 1970), whereby individual particles migrate to the surface (Fig. 20a and b). Individual particles may be enveloped by the modified cell membrane (Fig. 20c) and released by budding. Virus particles released into the hemolymph are enclosed within the wrapping membrane which possesses a layer of spicules on its outer surface (Fig. 20d and e).

2. *Amsacta Poxvirus*

Development of this agent has been studied in adipose cells of *Estigmene acrea* larvae by Granados and Roberts (1970). A reconstructed sequence indicates that as with all poxviruses, viroplasmic foci are established in the cytoplasm. Immature stages may be assembled at the periphery of foci where spherical membranes develop (Fig. 18b) in a manner similar to that described for the *Melolontha* agent (Fig. 21a). Envelopes of immature particles appear to consist of two layers both having the structure of a unit membrane. The inner layer is approximately 60 Å and the outer 75 Å in width. A region of low density about 75 Å wide separates the two dense layers. Crescent-shaped segments of membranes are also evident in association with viroplasmic foci.

FIG. 18. Development of spherical membranes in (a) *Melolontha* and (b) *Amsacta* poxvirus at the periphery of viroplasm foci (V). (a) × 87,000; (b) × 162,000. (From Bergoin *et al.*, 1969a; Granados and Roberts, 1970.)

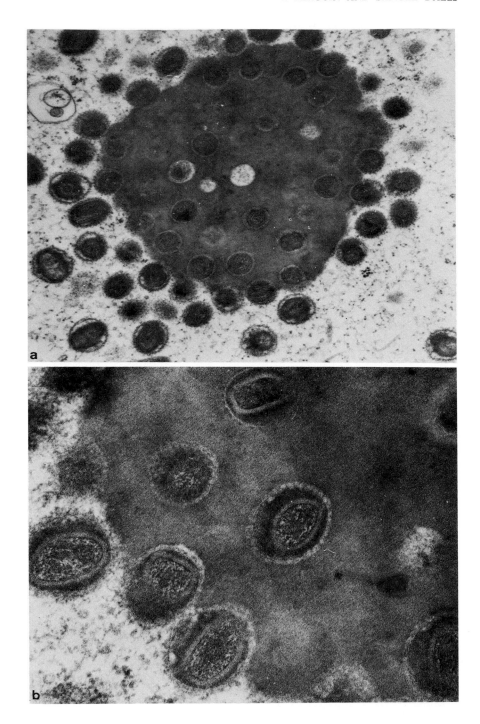

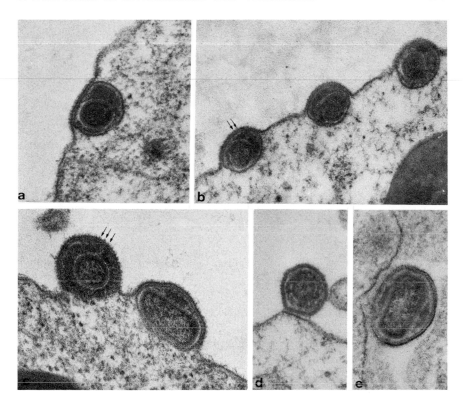

FIG. 20. Stages in the process of release of *Melolontha* poxvirus from hemocytes. (a) Where the virus is in close contact with the surface the cell membrane bears an external, spiny coat. × 45,000. (b) and (c) Evagination at the cell surface where a modified cell membrane bearing spicules (arrows) partly envelopes the virions. (b) × 41,000; (c) × 60,000. (d) A particle apparently free from the host cell. × 36,000. (e) An extracellular particle apparently surrounded by a supplementary membrane. × 60,000. (From Devauchelle *et al.*, 1970.)

Selected images illustrating differentiation from immature into mature virus particles are shown in Fig. 21. The first stage in morphogenesis may be the assembly of the coat of the virus core, as suggested in Fig. 21b–d. The space between the core and the external envelope is occupied by material of the lateral bodies. At this stage of development the virus envelope

FIG. 19. Developing spherules or "A"-type inclusions of *Melolontha* poxvirus. (a) Particles are in the process of being occluded within a proteinaceous matrix. Note the large number of maturing or mature forms of the virus in the vicinity of the inclusion. × 22,000. (From Bergoin *et al.*, 1968b). (b) High resolution of a selected area from another inclusion. The virions are embedded in the crystalline matrix with their long axes in a preferentially radial orientation. × 68,000. (From Bergoin *et al.*, 1969a.)

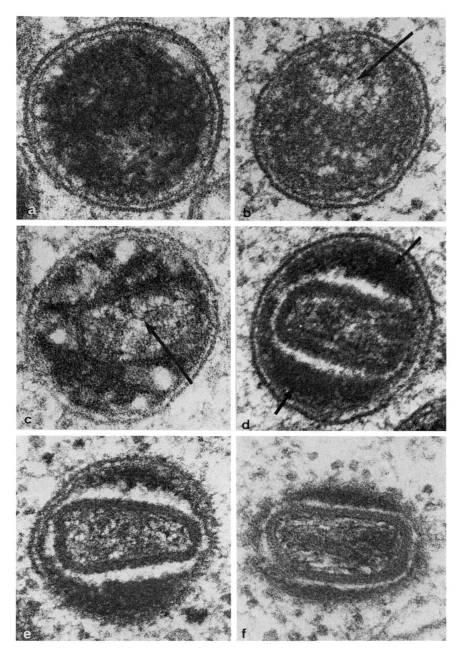

FIG. 21. Reconstructed sequence of maturation in *Amsacta* poxvirus. (a) A spherical immature form surrounded by two electron-dense membranes. × 153,000. (b) An eccentric dense region area (arrow). × 153,000. (c) Outline of the coat of the core is evident

becomes less distinctive (Fig. 21e). During the terminal stages of maturation the virus surface becomes convoluted and has a beaded appearance. The core contains a dense filamentous component presumably DNA (Fig. 21f). Intracytoplasmic mature virus occurs as both occluded forms and free particles many of which are concentrated in the vicinity of the oblate, spherical inclusions measuring usually 4 × 2.5 μ. Individual cells may contain several inclusions. Virions are absent from the peripheral region of those inclusions that are considered to be fully developed.

3. *Other Insect Poxviruses*

Apart from the *Amsacta* and *Melolontha* agents, immature forms of poxviruses have also been described by Goodwin and Filshie (1969) in tissues of the black soil scarab *Othnonius batesi* and by Henry *et al.* (1969) in the grasshopper *Melanoplus sanguinipes*. In both species of insects developmental forms of the virus occur in cells of the fat body. Differentiation from immature to mature particles appears to follow a similar pattern to that observed with poxviruses already described. The mature progeny become occluded within crystalline proteinaceous inclusions. Henry *et al.* (1969) have published evidence showing that the *Melanoplus* poxvirus may also be assembled in both the nucleus and cytoplasm of adipose cells.

C. CHEMICAL COMPOSITION

By comparison with the information available concerning the vertebrate agents, only scanty data have been published about insect poxviruses. Not surprisingly chemical and histochemical analyses show that DNA is the nucleic acid of the genome in *Melolontha* pox (Bergoin *et al.*, 1969b), *Amsacta* pox (Roberts and Bergoin, to be published), and *Chironomus luridus* pox (Götz *et al.*, 1969). Analysis of purified virions of *Amsacta* pox by Roberts and Bergoin (to be published) reveals a DNA content of at least 1.5% which is less than the amount of DNA reported for the vertebrate poxviruses (Joklik, 1962; Gafford and Randall, 1967). As mentioned previously, three out of the four enzymatic activities that have been found in

---

(arrow). × 153,000. (d) A differentiated core and two lateral bodies (arrows) are evident within the envelope. × 153,000. (e) A maturing particle is compressed laterally. At this stage the two outer membranes become less distinctive. × 153,000. (f) Mature virus possessing a central core and lateral bodies. × 153,000. (From Granados and Roberts, 1970.)

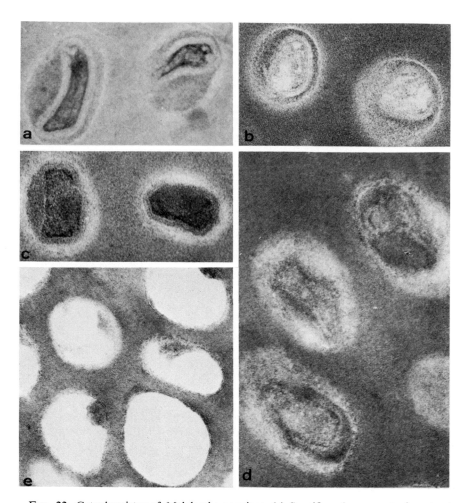

Fig. 22. Cytochemistry of *Melolontha* poxvirus. (a) Specific enhancement of contrast for DNA according to the method of Bernhard (1969) applied to sections of virus within inclusions embedded in water-miscible methacrylate. Note the highly dense material (DNA) within the viral core except in the zone occupied by the cable structure. × 79,000. (b) Section incubated with a 0.1% DNase solution for 1 hour. The dense DNA-rich area within the core has apparently been removed by hydrolysis. × 59,000. (c) Treatment as in (b) but without DNase. The densely stained area within the core has remained. × 58,000. (d) Section incubated with 0.01% pepsin for 30 minutes. The proteinaceous material surrounding the viral core has been partially digested. × 67,000. (e) After incubation of sections with 0.1% pepsin for 30 minutes the dense components of virions are completely removed leaving a lucent area in the crystalline matrix of the inclusion. × 41,000. (From Bergoin *et al.*, 1971.)

the cores of vaccinia and other vertebrate poxviruses are also present in *Amsacta* pox (Pogo *et al.*, 1971).

The chemical composition of some structural components of *Melolontha* pox has been investigated using electron microscopy in combination with cytochemistry. A specific staining method for DNA devised by Bernhard (1969) when applied to thin sections by Bergoin *et al.* (1971), reveals the presence of this nucleic acid within the core (Fig. 22a). When tested by Bernhard's method the cable component in the core does not appear to contain appreciable quantities of DNA. Following floatation of sections on a 0.1% solution of DNase for 30 to 120 minutes some dense material is extracted from the DNA-containing region of the core (Fig. 22b), while in the absence of DNase a strongly positive reaction is retained (Fig. 22c). A 0.01% solution of pepsin partially removes the dense substance surrounding the core (Fig. 22d). At a higher (0.1%) concentration of pepsin the whole sectioned virus particles are hydrolyzed (Fig. 22e).

To date we have no definitive information regarding the antigenic homology between any members of the vertebrate and invertebrate poxviruses, although preliminary comparisons by immunodiffusion between vaccinia and either *Melolontha* or *Amsacta* pox have failed to demonstrate the presence of common antigens.

Since it is now possible to obtain the insect agents in a highly purified state, using A-type inclusions as starting material, we anticipate that in the near future detailed and accurate analyses of the DNA, protein, lipid, and carbohydrate composition will be forthcoming. Hopefully, when suitable *in vitro* insect cultures, like those now used with vertebrate agents are developed, rapid quantitative experimentation on the biology of invertebrate poxviruses will become possible.

ACKNOWLEDGMENTS

The authors wish to thank Drs. Robert R. Granados and Donald W. Roberts of the Boyce Thompson Institute for Plant Research, Inc. and Dr. Yasuo Ichihashi of The Public Health Research Institute of the City of New York, Inc. for illustrative material.

REFERENCES

Amargier, A., Vago, C., and Meynadier, G. (1964). *Mikroskopie* **19**, 309–315.
Bergoin, M., Scalla, R., Duthoit, J. L., and Vago, C. (1967). *In* "Insect Pathology and Microbial Control" (P. A. Van der Laan, ed.), pp. 63–68. North Holland Publ. Co., Amsterdam.
Bergoin, M., Devauchelle, G., Duthoit, J. L., and Vago, C. (1968a). *C. R. Acad. Sci. (Paris)* **D266**, 2126–2128.

Bergoin, M., Devauchelle, G., and Vago, C. (1968b). *C. R. Acad. Sci. (Paris)* **D267**, 382–385.

Bergoin, M., Devauchelle, G., and Vago, C. (1969a). *Arch. Gesamte Virusforsch.* **28**, 285–302.

Bergoin, M., Veyrunes, J. C., and Vago, C. (1969b). *C. R. Acad. Sci. (Paris)* **D269**, 1464–1466.

Bergoin, M., Devauchelle, G., and Vago, C. (1971). *Virology* (in press).

Bernhard, W. (1969). *J. Ultrastr. Res.* **27**, 250–265.

Cairns, H. J. F. (1960). *Virology* **11**, 603–623.

Dales, S. (1962). *J. Cell Biol.* **13**, 303–322.

Dales, S. (1963). *J. Cell Biol.* **18**, 51–72.

Dales, S. (1965). *Amer. J. Med.* **38**, 699–715.

Dales, S. and Kajioka, R. (1964). *Virology* **24**, 278–294.

Dales, S. and Mosbach, E. (1968). *Virology* **35**, 564–583.

Dales, S. and Siminovitch, L. (1961). *J. Biophys. Biochem. Cytol.* **10**, 475–502.

Devauchelle, G., Bergoin, M., and Vago, C. (1970). *C. R. Acad. Sci. (Paris)* **D271**, 1138–1140.

Easterbrook, K. B. (1966). *J. Ultrastr. Res.* **14**, 484–496.

Edgar, R. S., Denhardt, G. H., and Epstein, R. H. (1964). *Genetics* **49**, 635–648.

Gafford, L. G. and Randall, C. C. (1967). *J. Mol. Biol.* **26**, 303–310.

Götz, P., Huger, A. M., and Krieg, A. (1969). *Naturwissenschaften* **56**, 145.

Gold, P. and Dales, S. (1968). *Proc. Nat. Acad. Sci. U. S.* **60**, 845–852.

Goodwin, R. H. and Filshie, B. K. (1969). *J. Invertebr. Pathol.* **13**, 317–329.

Granados, R. R. and Roberts, D. W. (1970). *Virology* **40**, 230–243.

Henry, J. H., Nelson, B. P., and Jutila, J. W. (1969). *J. Virol.* **3**, 605–610.

Hoagland, C. L., Lavin, G. I., Smadel, J. E., and Rivers, T. M. (1940a). *J. Exp. Med.* **72**, 139–147.

Hoagland, C. L., Smadel, J. E., and Rivers, T. M. (1940b). *J. Exp. Med.* **71**, 737–750.

Hoagland, C. L., Smadel, J. E., Ward, S. M., and Rivers, T. M. (1941a). *J. Exp. Med.* **74**, 133–144.

Hoagland, C. L., Ward, S. M., Smadel, J. E., and Rivers, T. M. (1941b). *J. Exp. Med.* **74**, 69–80.

Holowczak, J. A. and Joklik, W. K. (1967). *Virology* **33**, 717–725.

Hurpin, B. and Vago, C. (1963). *Rev. Pathol. Vegetale Entomol. Agr. Fr.* **42**, 115–117.

Hyde, J. M., Gafford, L. G., and Randall, C. C. (1967). *Virology* **33**, 112–120.

Ichihashi, Y. and Matsumoto, S. (1966). *Virology* **29**, 264–275.

Ichihashi, Y. and Matsumoto, S. (1968). *Virology* **36**, 262–270.

Joklik, W. K. (1962). *Virology* **18**, 9–18.

Joklik, W. K. (1964). *J. Mol. Biol.* **8**, 263–276.

Joklik, W. K. (1966). *Bacteriol. Rev.* **30**, 33–66.

Kates, J. and Beeson, J. (1970). *J. Mol. Biol.* **50**, 1–18.

Kates, J., Dahl, R., and Mielke, M. (1968). *J. Virol.* **2**, 894–900.

Kates, J. R. and McAuslan, B. R. (1967a). *Proc. Nat. Acad. Sci. U. S.* **57**, 314–320.

Kates, J. R. and McAuslan, B. R. (1967b). *Proc. Nat. Acad. Sci. U. S.* **58**, 134–141.

Kato, S. and Cutting, W. (1959). *Stanford Med. Bull.* **7**, 34.

Kato, S., Hara, J., Ogawa, M., Miyamoto, H., and Kamahora, J. (1963). *Biken's J.* **6**, 233.

Loh, P. C. and Riggs, J. L. (1961). *J. Exp. Med.* **114**, 149–160.

McAuslan, B. R. and Kates, J. R. (1967). *Virology* **33**, 709–716.

Magee, W. D., Sheek, M. R., and Burrows, M. J. (1960). *Virology* **11**, 296–299.

Medzon, E. L. and Bauer, H. (1970). *Virology* **40**, 860–867.

Meynadier, G., Fosset, J., Vago, C., Duthoit, J. L., and Bres, N. (1968). *Ann. Epiphyties* **19**, 703–705.

Morita, K. (1960). *Biken's J.* **3**, 213.

Munyon, W., Paoletti, E., and Grace, Jr., J. T. (1967). *Proc. Nat. Acad. Sci. U. S.* **58**, 2280–2288.

Munyon, W., Paoletti, E., Ospina, J., and Grace, J. T., Jr. (1968). *J. Virol.* **2**, 167–172.

Nagayama, A., Pogo, B. G. T., and Dales, S. (1970). *Virology* **40**, 1043–1051.

Peters, D. (1956). *Nature (London)* **178**, 1453.

Peters, D., and Müller, G. (1963). *Virology* **21**, 266–269.

Pogo, B. G. T. and Dales, S. (1969a). *Proc. Nat. Acad. Sci. U. S.* **63**, 820–827.

Pogo, B. G. T. and Dales, S. (1969b). *Proc. Nat. Acad. Sci. U. S.* **63**, 1297–1303.

Pogo, B. G. T., Dales, S., Bergoin, M., and Roberts, D. W. (1971). *Virology* (in press).

Roberts, D. W. and Bergoin, M., manuscript in preparation.

Rosenkranz, H. S., Rose, H. M., Morgan, C., and Hsu, K. C. (1966). *Virology* **28**, 510–519.

Sarov, I. and Becker, Y. (1967). *Virology* **33**, 369–375.

Smadel, J. E. and Hoagland, C. L. (1942). *Bacteriol. Rev.* **6**, 79–110.

Smadel, J. E., Hoagland, C. L., and Shedlovsky, T. (1943). *J. Exp. Med.* **77**, 165–171.

Smadel, J. E. and Shedlovsky, T. (1942). *J. Gen. Microbiol.* **34**, 115–123.

Vago, C. (1963). *J. Insect Pathol.* **5**, 275–276.

Vago, C. and Croissant, D. (1964). *Entomophaga* **9**, 207–210.

Vago, C., Amargier, A., Hurpin, B., Meynadier, G., and Duthoit, J. L. (1968a). *Entomophaga* **13**, 373–375.

Vago, C., Monsarrat, P., Duthoit, J. L., Amargier, A., Meynadier, G., and Van Waerebeke, D. (1968b). *C. R. Acad. Sci. (Paris)* **D266**, 1621–1623.

Vago, C., Robert, P., Amargier, A., and Duthoit, J. L. (1969). *Mikroskopie* **25**, 378–386.

Weiser, J. (1969). *Acta Virol.* **13**, 549–553.

Weiser, J. and Vago, C. (1966). *J. Invertebr. Pathol.* **8**, 314–319.

Westwood, J. C. N., Harris, W. J., Zwartouw, H. T., Titmuss, D. H. J., and Appleyard, G. (1964). *J. Gen. Microbiol.* **34**, 67–78.

Westwood, J. C. N., Zwartouw, H. T., Appleyard, G., and Titmuss, D. H. J. (1965). *J. Gen. Microbiol.* **38**, 47–53.

Woodroofe, G. M. and Fenner, F. (1962). *Virology* **16**, 334–341.

Woodroofe, G. M. and Fenner, F. (1965). *Aust. J. Exp. Biol. Med. Sci.* **43**, 123–142.

Woodson, B. (1968). *Bacteriol. Rev.* **32**, 127–137.

Zwartouw, H. T. (1964). *J. Gen. Microbiol.* **34**, 115–123.

CHAPTER 7 *A Comparative Study of the Structure and Biological Properties of Bacteriophages*

D. E. BRADLEY

I. Introduction

A. ELECTRON MICROSCOPY OF BACTERIOPHAGES

It was recognized with the advent of the electron microscope that the morphology of a virus was one of its most important distinguishing characteristics. However, the earlier specimen preparation techniques, of which shadow-casting was the most important, provided only an outline of the virion. When applied to bacteriophages, shadowing was nevertheless sufficient to indicate that they existed in a number of different forms. Electron micrographs such as shown in Fig. 1 indicated that the T-even phages of Demerec and Fano (1945) consisted of a head and a thick straight tail. It was found that this tail contracted to a different shape on adsorption to the host cell wall (Kellenberger and Arber, 1955). Other phages such as T5 had long flexible tails (Fig. 2) which did not contract.

One of the most important pieces of work using shadow-casting was the study of the *Escherichia coli* cell wall receptor molecules for phage T5 by Weidel and Kellenberger (1955) and their attachment to, and effect upon, the virion. In Fig. 3 the receptors are shown as small round objects attached to the tip of the phage tail: virions which had not been mixed with receptors had no such appendages. This study revealed several important facts, the first being that phages with noncontractile tails used them for adsorption. The second observation was that the attachment of the receptor to the virion often caused the loss of its nucleic acid (Fig. 3, arrowed) suggesting that the energy for injection lay in the virion itself.

Shadowing was largely responsible for the visualization of the structure of a bacteriophage which had no obvious tail, namely ϕX174 (Hall *et al.*, 1959).

The similar phage ϕR is shown in Fig. 4; it can be seen to consist of a small round object with a regular arrangement of knobs on the surface. Thus, even at this comparatively early stage in the electron microscopy of bacteriophages, it was evident that a number of completely different forms existed. However, it required the development of a much more efficient specimen preparation process before the fine details of phage structure could be resolved and the significance of differences between various morphological forms assessed.

The negative staining process, which consists of embedding a particle in a solid electron-dense matrix, was developed using viruses as test specimens (Hall, 1955; Huxley, 1956; Brenner and Horne, 1959) and later used in various systematic studies of the morphology of bacteriophages specific to various host genera (Bradley and Kay, 1960; Bradley, 1963a,b, 1965a,b). This method produced what amounted to a revolution in structural electron microscopy. Its simplicity and its ability to resolve details as small as 1.5 nm permitted the rapid study of phages at high resolution, and it became clear that morphology was a criterion which could be used in the classification of all viruses.

B. BACTERIOPHAGE NUCLEIC ACIDS

It had, of course, been known for many years that phages contained nucleic acid as their genetic material but it was not realized that the type might vary from one morphological form to another until a detailed study of ϕX174 by Sinsheimer (1959) showed that this bacteriophage contained single-stranded deoxyribonucleic acid (1-DNA) instead of the double-stranded DNA (2-DNA) found in the tailed phages. The discovery of phages containing 1-RNA (Loeb and Zinder, 1961) emphasized the variability of nucleic acid type and its value in comparative studies.

As a fundamental character for classification its importance is obvious and with the development of a simple fluorescent staining procedure (Mayor and Hill, 1961; Bradley, 1966a), nucleic acid type and strandedness became a vital but easily determined label for a virus.

An ingenious procedure for unravelling the strand of nucleic acid and studying it directly in the electron microscope due to Kleinschmidt et al. (1962) permitted the measurement of the length of the molecule within a virus. It was also possible to find out whether or not the molecule was in the form of a closed ring, as was the case with the 1-DNA of phage ϕX174 (Kleinschmidt et al., 1963). Thus, in addition to comparing the type and strandedness of phage nucleic acids, one could also compare the lengths. Whether by accident or design, structural studies of bacteriophages were providing the necessary basic data for their classification into meaningful groups.

C. BIOLOGICAL PROPERTIES

Since the structure and nucleic acid of phages showed so much variation, there seemed every reason to believe that other properties, such as effects on the host cell, would show differences of a similar degree. Many of the biological properties of phages had, of course, been studied before the

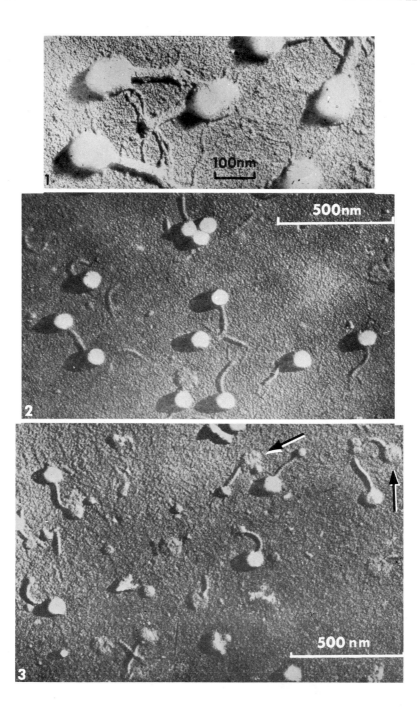

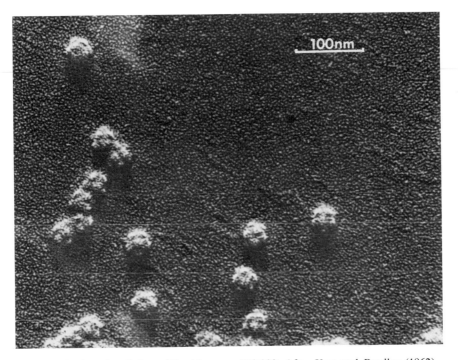

FIG. 4. Shadowed coliphage ØR virions, × 170,000. After Kay and Bradley (1962).

electron microscope was perfected. Indeed, Twort (1915) and d'Herelle (1917), the discoverers of phages, isolated viruses for staphylococci and dysentery bacteria suggesting at the outset that phages had a defined host range. The activity of a particular isolate is usually restricted to a single host species and sometimes to one or two strains within that species. Phages were also found to differ in other biological properties. Some were able to attach their genomes to the host chromosome without damaging the host cell, replicating in synchrony with it until some external influence such as radiation caused the formation and release of vegetative phage particles. Such phages were called temperate as opposed to virulent ones which did not possess this ability and were only able to multiply vegetatively.

FIG. 1. Shadowed T2 coliphage virions with extended sheaths, × 100,000. By permission of the Royal Microscopical Society.

FIG. 2. Shadowed T5 coliphage virions, × 60,000. After Weidel and Kellenberger (1955), with permission of *Biochim. Biophys. Acta.*

FIG. 3. Shadowed coliphage T5 virions mixed with receptor molecules, × 60,000. After Weidel and Kellenberger (1955), with permission of *Biochim. Biophys. Acta.*

It was also found that differences existed in the manner in which phages multiplied within their host and the mechanisms by which they caused lysis. Some intracellular virions were seen in the nucleoplasm and some in the cytoplasm. Some phages caused lysis by the induction of lysozyme synthesis, the enzyme actually dissolving the cell wall. Others caused host cells to become spheroplasts, presumably by inhibiting the synthesis of cell wall material; lysis followed, probably caused by osmotic effects.

So far we have discussed obvious differences between phages: appearance, type of genetic material, and some biological properties. These indicate gross variations from one phage isolate to another. At the other end of the comparative scale where two isolates look the same and act against the same host in the same way, one can attempt to separate them on a serological basis. The rate of inactivation of a phage by an antiserum made from a different isolate gives information about small differences in their antigenic structure. Having considered all the various properties mentioned one is bound to ask the question: at what point can one say that two given phage isolates are different? While this cannot be answered with confidence, one can at least say that the characteristics mentioned, when taken together, provide a phage with individuality and allow it to be distinguished from other phages and viruses, so providing the basis for some form of classification.

II. Morphology of Bacteriophages

A. Basic Morphological Types

As has been mentioned, the electron microscope shows that bacteriophages exist in a number of fundamentally different forms. Some are tadpole-shaped, with a head and a tail, some have a spherical appearance and no tails, and some are in the form of long thin filaments. These forms fall into obvious groups as suggested by Bradley (1965b). The illustration in Fig. 5 shows five basic morphological types and it is important to note that any known bacteriophage can be allocated to a group; there is no overlap, the differences between each group being distinct and clear-cut.

Group A is composed of phages with contractile tails like the T-even group (Fig. 1). A great number of isolates have been found showing considerable variation in head shape, size, etc. This variation is also present in group B, which is likewise very common. It contains phages with long flexible tails like T5 (Fig. 2) and also those very short tails like T3 (Fig. 6).

Group C, the tailless phages with large apical subunits like ϕX174 (Fig. 4), has two forms, the second being a marine phage, larger than ϕX174 and containing lipid. There is only very slight morphological variation (the number of protein subunits) in group D, the tailless phages with no large apical subunits (Fig. 7); the members all contain 1-RNA. Group E, the filamentous phages are unique and of great interest; they have various interesting biological properties which will be described below. A preparation of one of these phages consists of a tangled mass of filaments (Fig. 8).

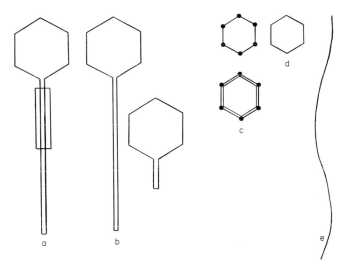

FIG. 5. Basic morphological types of bacteriophages.

B. SYMMETRY IN PHAGE STRUCTURE

It was predicted by Crick and Watson (1956) and Crick (1957) that viruses would obey the laws of solid geometry in their construction and Horne and Wildy (1961) described the application of the principles of symmetry to viruses in general. The two basic forms of symmetry, cubic and helical, are found in regular solids and rod-shaped structures, respectively; thus, the tailed phages have heads with cubic and tails with helical symmetry. A regular solid or two-dimensional object has axes of symmetry about which it can be rotated to give a series of identical appearances. For example, a square appears identical in any of four different positions when rotated about its center and therefore has fourfold symmetry, the axis of rotation (perpendicular to the plane of the square) being the fourfold

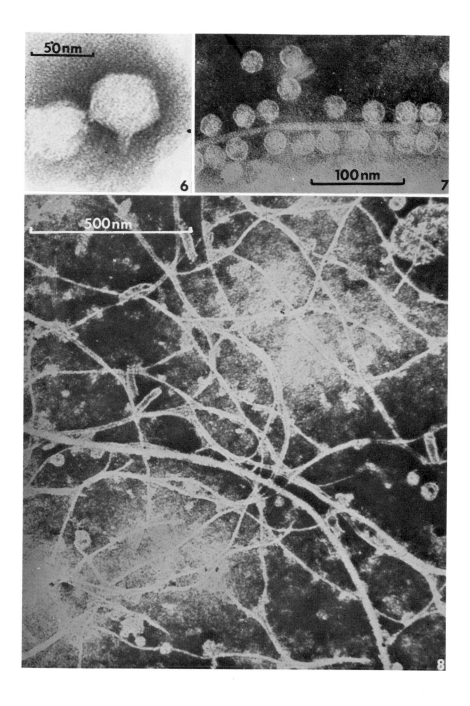

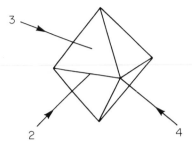

FIG. 9. Diagram of an octrahedron showing symmetry axes.

symmetry. Perhaps the commonest regular solid found among the bacterio-phages is the octahedron, shown with its symmetry axes in Fig. 9. Smaller phages (groups C and D) favor the icosahedron with twenty faces (Fig. 10). Irregular solids, such as the elongated head of the T-even phages, are often complex derivatives of the icosahedron.

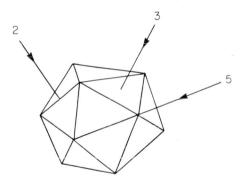

FIG. 10. Diagram of an icosahedron showing symmetry axes.

In rod-shaped bodies, helical symmetry may reveal itself in two different forms either as a stacked disc arrangement or a coil. In both cases, an end-on view presents an axis of radial or rotational symmetry like that of a regular solid and a side view shows the symmetry perpendicular to the long axis. A combination of the two forms provides the three-dimensional helical

FIG. 6. Negatively stained coliphage T3, × 333,000. With permission of *Bacteriol. Rev.*

FIG. 7. RNA coliphage ZIK/1 negatively stained, × 250,000. With permission of *J. Ultrastruct. Res.*

FIG. 8. Filamentous coliphage ZJ/2 negatively stained, × 80,000. With permission of *J. Ultrastruct. Res.*

symmetry. The concept of symmetry is important in comparative studies of virus fine structure.

The electron microscope has shown that a virus particle is made up of a number of regularly arranged subunits, obeying the rules of symmetry, surrounding a coiled nucleic acid molecule. To clarify descriptions, a special terminology was suggested by Lwoff *et al.* (1959). The virus particle as a whole, the *virion*, is made up of a protein coat called the *capsid*, which consists of a number of morphological subunits or *capsomeres*, surrounding the nucleic acid *core*.

C. THE STRUCTURE OF BACTERIOPHAGES WITHIN EACH MORPHOLOGICAL GROUP

It has been inferred above (Section II,A) that while the basic form of each morphological group is distinct and clearly defined, the members within a group may show a great deal of structural variation. A comparative study of such variations, where they exist, is important in considering the relationship of one phage to another. The following descriptions are given with this in mind.

1. *Group A*

The best known structural type of contractile phage is represented by the T-even series (T2, T4, T6) which are morphologically identical. Since the detailed structure of these isolates is well known, only the major features are given here for comparison with other contractile phages. As shown in Figs. 1 and 11, the head has an elongated shape; despite the advances in electron microscopy, its actual geometrical form is still uncertain. Williams and Fraser (1953) suggested it was a hexagonal prism with a pyramid at each end (a bipyramidal hexagonal prism), Bradley (1965c) thought it was a bipyramidal hexagonal antiprism, formed by rotating the two pyramids of the first model by 30° with respect to one another, while Moody (1965) and Boy de la Tour and Kellenberger (1965) favored a prolate icosahedron (an icosahedron with an extended middle section). The tail has a thick sheath surrounded by a network of fibers attached to a baseplate at the distal end and a thin disc or collar adjacent to the head (Figs. 5a, 11). Under conditions favorable for adsorption to the host cell wall, the fibers become detached from the collar and extend outward and downward from the baseplate. On adsorption, the sheath contracts to about half its length (Fig. 12) by rearranging its morphological subunits (Moody, 1967a,b). It

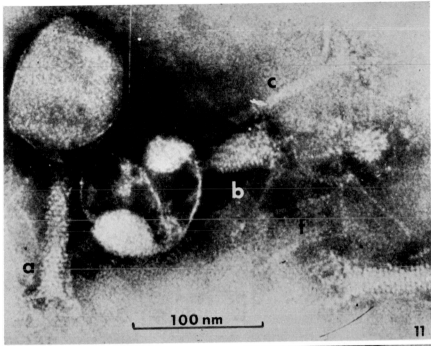

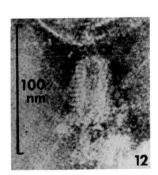

FIG. 11. Negatively stained coliphage T2 showing structural components: a, sheath; b, contracted sheath; c, core of tail; f, tail fibers. × 333,000. From *Proc. Intern. Conf. Electron Microscopy, Philadelphia*, 1962.

FIG. 12. Contracted sheath of T-even coliphage, × 333,000. With permission of *Bacteriol. Rev.*

FIG. 13. *Pseudomonas aeruginosa* phage PB-1 showing octahedral head, × 333,000. With permission of *J. Gen. Virol.*

is, of course, this contraction which distinguishes group A phages from others. The baseplate consists of a disc with six prongs and has been studied in detail by Bradley (1961) and Anderson and Stephens (1964).

Contractile phages with symmetrical heads are much more common than those with asymmetrical heads and are found associated with a wide range of bacterial genera. They have a variety of sizes and structural variations in the tail assembly (Fig. 16). The head sizes and tail lengths of a selection of these phages is given in Table I and it can be seen that there is no clear-cut division as far as size is concerned. The following brief descriptions indicate the degree of morphological variation within the group.

One of the commonest forms is typified by the *Pseudomonas aeruginosa* phage PB-1 (Bradley and Robertson, 1968). This has an octahedral head (Figs. 13, 16e); its tail has four fibers at the tip (Fig. 14), reflecting the fourfold radial symmetry of the tail, which is attached to a fourfold symmetry axis of the head (at an apex). In Fig. 14, the fibers are extended from the sheath, but more often they appear in a retracted position (Fig. 15). The contracted sheath is shown in Fig. 13, and the dimensions given in Table I. There are numerous similar isolates specific for most species of the Enterobacteriaceae.

In contrast, the *Serratia marcescens* phage SMP (Bradley, 1965a) has a massive head (see Table I) and a long tail with a bar-shaped structure at the tip, shown as a diagram in Fig. 16c. The head is probably octahedral.

While it is difficult to be sure of the geometrical form of a phage head except in cases where particularly good negative staining is achieved, the great majority of contractile phages with regular heads seem to be octahedra. In one case, however, an icosahedral shape seems likely, this being the *Bacillus subtilis* phage SP50 (Eiserling and Boy de la Tour, 1965; Figs. 16h, 17, and 18). It can be seen that the edges of the capsid are not absolutely straight but appear slightly bent, which would be typical of an icosahedron. More significant, however, is the fact that the number of prongs on the baseplate is five. This would reflect a fivefold radial symmetry in the tail, which is attached to an apex of the head, and since the rules of symmetry must be obeyed, this should correspond to a fivefold symmetry axis (Fig. 10), which would belong to an icosahedron. It is clear from these descriptions and Fig. 16, that the regular-headed contractile phages form a heterogeneous group.

2. *Group B*

Bacteriophages with noncontractile tails are more common and show as much structural variation as contractile ones. It can be seen from Fig. 19

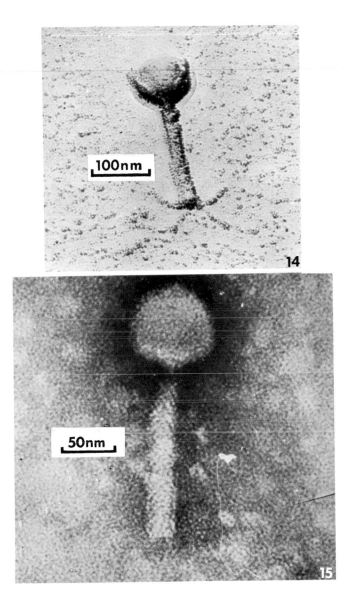

FIG. 14. Shadowed *Pseudomonas aeruginosa* phage PB-1 showing four tail fibers, × 150,000. With permission of *J. Gen. Virol.*

FIG. 15. *Pseudomonas aeruginosa* phage PB-1 with extended sheath, × 300,000. With permission of *J. Gen. Virol.*

TABLE I

COMPARISON OF HEAD AND TAIL DIMENSIONS OF CONTRACTILE PHAGES WITH SYMMETRICAL HEADS

Phage number	Host species	Head size (nm)	Tail length (nm)	Reference
E1	*Escherichia coli*	75	110	Bradley (1963a)
SM2	*Serratia marcescens*	135	235	Bradley (1965a)
5T, 78	*Proteus hauseri*	60	90	Prozesky *et al.* (1965)
13vir, 67B	*Proteus hauseri*	60	100	Prozesky *et al.* (1965)
A11/79	*Alcaligenes faecalis*	65	130	Maré *et al.* (1966)
A6	*Alcaligenes faecalis*	90	110	Maré *et al.* (1966)
PB1	*Pseudomonas aeruginosa*	75	150	Bradley and Robertson (1968)
12S	*Pseudomonas syringae*	80	100	Bradley (1963b)
SP50	*Bacillus subtilis*	95	220	Eiserling and Boy de la Tour (1965)
206, 222a, 315, 514	*Lactobacillus*	70	140	Deklerk *et al.* (1965)
200, 356	*Lactobacillus*	80	125	Deklerk *et al.* (1965)
MX1	*Myxococcus xanthus*	75	100	Burchard and Dworkin (1966)
—	*Chondrococcus columnaris*	60	100	Kingsbury and Ordal (1966)
—	*Saprospira grandis*	115	135	Lewin *et al.* (1964)

that variations exist in head shape, tail length, and the nature of terminal appendages. Perhaps the commonest form is typified by the well-known phage lambda (λ) (Figs. 19b, 20). Eiserling and Boy de la Tour (1965) showed that it consisted of a 60 mn head with a 160 nm long striated tail having a fine fiber at the tip. One cannot be certain of the head shape; it might be an icosahedron or an octahedron. The well-known coliphage T5 is not unlike lambda: the large 90 nm head is probably octahedral (Fig. 21) and the tail (200 nm long) has a cluster of fine fibers at the tip, correspond-

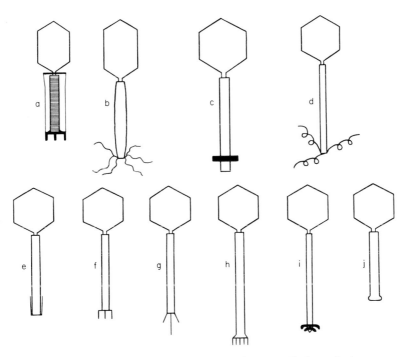

FIG. 16. Different morphological forms of contractile bacteriophages.

ing to the type shown in Fig. 19c. Many isolates of this form have knobs at the tip of the tail (Bradley, 1966b) such as the *Pseudomonas aeruginosa* phage Pc (Fig. 19d). The *Flavobacterium* phage 384 (Bradley, 1965b) has a more complex assembly which looks like two discs seen from the side in negative stained preparations (Fig. 19e).

It will be noted that the noncontractile phages described so far all have regularly shaped heads, presumably either octahedra or icosahedra. Those with irregular oblong heads are also common. The *Pseudomonas aeruginosa* phage in Figs. 19g and 22 (Bradley, 1966b) has a barlike structure at the tail tip. This is a fairly large virion with a head measuring 100 × 70 nm and the tail 175 nm.

Another particularly large type is specific for *Staphylococcus aureus* (Figs. 19h, 23); it has a tail no less than 300 nm long with a star-shaped disc at

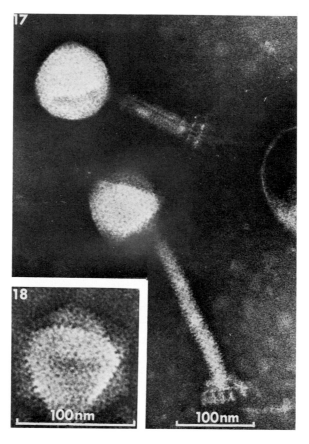

FIG. 17. *Bacillus subtilis* phage SP 50, × 200,000. After Eiserling and Boy de la Tour, with permission of Karger, Basel.

FIG. 18. *Bacillus subtilis* phage SP 50 head showing capsomeres, × 300,000. After Eiserling and Boy de la Tour, with permission of Karger, Basel.

the tip. A form with a similar head shape but a shorter tail (170 nm) is commonly associated with species of *Caulobacter* (Schmidt and Stanier, 1965), *Streptomyces* (Coyette and Calberg-Bacq, 1967), and *Clostridium* (Betz, 1968). There is no obvious terminal appendage (Fig. 19i).

In addition to phages with comparatively long tails many isolates have been described with extremely short ones. The best known are the coliphages T3 and T7 which are identical (Fig. 6). They have a regular head, probably an octahedron (Bradley, 1963a) and a short wedge-shaped tail only about 12.5 nm long (Fig. 19j). This form is particularly common for

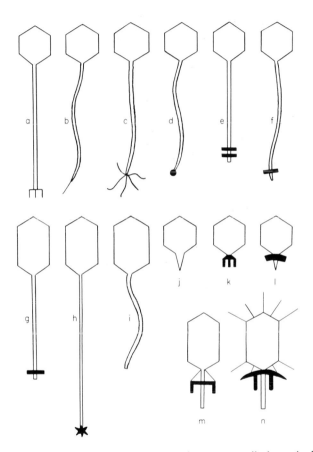

FIG. 19. Different morphological forms of noncontractile bacteriophages.

species of *Pseudomonas* and *Brucella* and is the only kind known for *Hydrogenomonas facilis* (Pootjes *et al.*, 1966). As with long-tailed phages, the structure of the tail assembly can vary. For example, Fig. 19k shows the *Salmonella* phage P22 (Anderson, 1960) which has a six-pronged disc shown in profile on the drawing. Some *Serratia* phages (Bradley, 1965a) have a thick collar (Fig. 19) but still a regular head.

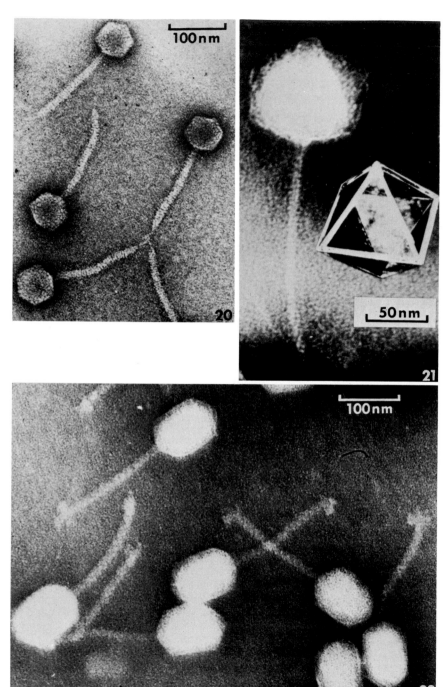

There are two short-tailed forms with oblong heads specific for *Bacillus subtilis* (Fig. 19m,n), the most important (Anderson *et al.*, 1966) being shown in Fig. 24. Perhaps its most interesting feature is the fibers protruding from the head; this is unique.

A glance at Fig. 19 suggests that this group could be divided into four: regular heads and long tails, irregular heads and long tails, regular heads and short tails, and irregular heads and short tails.

3. *Group C*

The tailless coliphages with large capsomeres are all identical in appearance and consist of a group of about twenty isolates most of which grow on *Escherichia coli* and/or *Salmonella typhimurium*. Their structure is well documented (Hall *et al.*, 1959; Tromans and Horne, 1961; Bradley, 1961; Kay and Bradley, 1962) and is shown as a shadowed preparation in Fig. 4. A negatively stained virion is included (Fig. 25) and from the drawing (Fig. 5) and micrographs this morphological type can be seen to consist of an icosahedron some 27 nm in diameter with apical subunits about 3 nm in size. Edgell *et al.* (1969) removed the apical capsomeres from the virion. They suggested a model consisting of 20 capsomeres with 12 complex capsomeres at the apices of an icosahedron.

It is possible that a lipid-containing marine bacteriophage which is parasitic for a marine pseudomonad (Espejo and Canelo, 1968) should be included in this group. It consists of a 60 nm polyedron with a more or less regular hexagonal outline. At each apex there is a small knob very similar to those of ϕX174. There appears to be no tail. The taxonomic position of this phage will be further discussed below.

4. *Group D*

The smooth tailless phages which contain 1-RNA are not restricted to *Escherichia coli* but isolates have also been found for *Caulobacter* (Schmidt and Stanier, 1965) and *Pseudomonas aeruginosa* (Feary *et al.*, 1964; Bradley, 1966b). They have a more or less similar appearance in the electron micro-

---

FIG. 20. Coliphage lambda, × 150,000. After Eiserling and Boy de la Tour, with permission of Karger, Basel.

FIG. 21. Coliphage T5, × 333,000. With permission of *Bacteriol. Rev.*

FIG. 22. *Pseudomonas aeruginosa* phage PB-2, × 275,000. With permission of *J. Gen. Microbiol.*

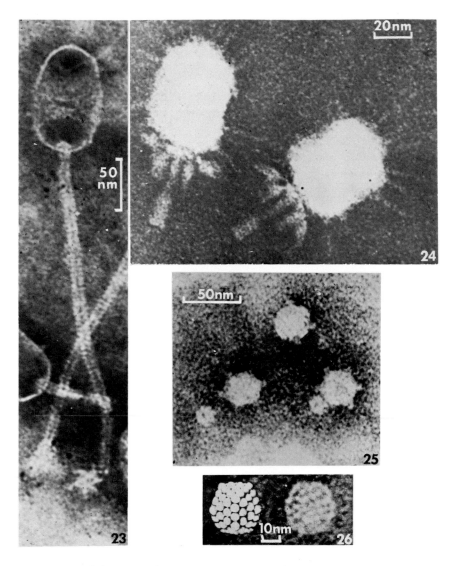

FIG. 23. Staphylococcus phage 594n, × 333,000. With permission of *J. Ultrastruct. Res.*

FIG. 24. *Bacillus subtilis* phage φ29, × 700,000. After Anderson, Hickman, and Reilly (1966), with permission of *J. Bacteriol.*

FIG. 25. φX174 type phage α3, × 333,000. With permission of *J. Gen. Virol.*

FIG. 26. RNA phage ZIK/1 with model of an icosahedron, × 500,000. With permission of *J. Gen. Microbiol.*

scope and appear to consist of icosahedra. Detailed studies have been carried out on the coliphages (Loeb and Zinder, 1961) and there seem to be two distinct morphological forms. In the case of phage ZIK/1, Bradley (1964) found that the viral capsid consisted of 92 capsomeres which were visible only under exceptional negative staining conditions (Fig. 26). More usually the virion had a smooth appearance (Fig. 7). Vasquez et al. (1966) studied the isolate R17 and concluded that it was an icosahedron with 32 capsomeres. This was based on electron micrographs of empty virions (Fig. 27) in which the capsomeres are clearly visible.

Overby et al. (1966a) quote the base ratios of the RNA of ZIK/1 as being different to R 17 so that one might expect a difference in capsid structure. Overby et al. (1966b) compared the coat proteins of MS2, an isolate with similar base ratios to R17 and Qβ (Watanabe, 1964) which has identical base ratios to ZIK/1. They found chemical differences and slightly different molecular weight. The two apparently distinct groups of RNA plages will be compared more fully below.

5. Group E

Bacteriophages in the form of long thin filaments have been found for *Escherichia coli* (Marvin and Hoffman-Berling, 1963; Zinder et al., 1963; Bradley, 1964), *Pseudomonas aeruginosa* (Takeya and Amako, 1966; Minamishima et al., 1968), *Vibrio parahaemolyticus* (Nakanishi et al., 1966), and *Salmonella typhimurium* (Meynell and Lawn, 1968). While appearing similar in the electron microscope, the isolates have different lengths, the coliphages being about 850 nm long and the *Pseudomonas* and *Salmonella* phages 1300 nm. Their diameters are similar: 5.5 nm. Figure 28 shows two coliphage virions which appear to be smooth, and even at higher magnification no structure can be discerned. Marvin (1966) suggests from X-ray evidence that the capsid is in the form of a supercoiled α-helix with a repeat of about 3.2 nm. The 1-DNA core lies in an uncoiled or loosely coiled condition up the center, since it is in the form of a closed ring only slightly more than twice the length of the virion (Marvin and Schaller, 1966; Bradley and Dewar, 1967). A tranverse section through the virion appears as a circle with a central black dot suggesting a simple tubular form (Fig. 29). Rossamando and Zinder (1968) have made a detailed study of the structure of the phage by alkali-induced disassembly into DNA and protein, and have found two distinct proteins, one being at an end of the filament. The filamentous phages have been compared in detail by Marvin and Hohn (1969).

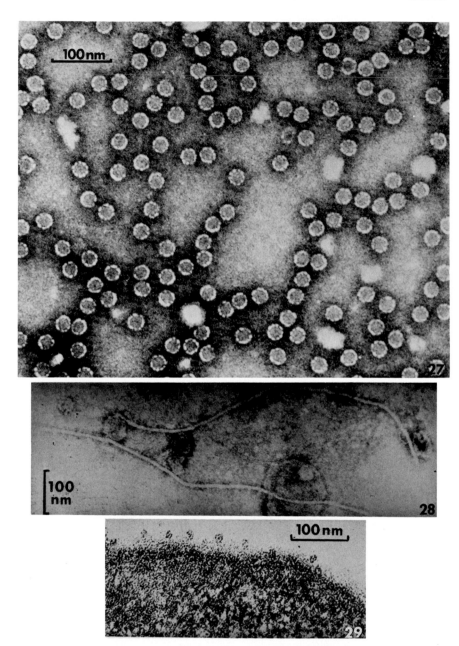

FIG. 27. RNA phage R17, empty virions showing capsomeres, × 150,000. After Vasquez *et al.* (1966), with permission of *J. Bacteriol.*

FIG. 28. Coliphage ZJ/2. With permission of *J. Gen. Microbiol.*

FIG. 29. Sections through coliphage ZJ/2, × 150,000. With permission of *J. Gen. Virol.*

III. Bacteriophage Nucleic Acids

All the characters of a virus are, of course, determined by the coding in its genome and as has been mentioned, different phages contain different types of nucleic acid. The following discussion relates nucleic acid type and strandedness to morphological type.

A. PHAGES CONTAINING DOUBLE-STRANDED DNA

As far as is known all tailed phages (groups A and B) contain 2-DNA, but, with the exception the lipid-containing phage (Section II,C,3 above) other morphological types do not. While the base ratios of many phage nucleic acids have been determined perhaps the most relevant information from a comparative point of view is the length of the nucleic acid strand. This has been measured in the electron microscope by spreading (Klein-schmidt *et al.*, 1962, 1963) in a relatively small number of cases, but the figures show, as one might expect, a relationship with head size: the larger the head the longer the molecule. In addition one can compare the volume of the nucleic acid molecule (assuming 2-DNA is a 2.0 nm diameter cylinder) with that of the phage head and so determine the amount of head volume taken up by nucleic acid. Since there is likely to be a certain amount of error in electron microscope measurements, particularly of head dimensions, such calculations as shown in Table II, are only approximate. The present author calculates that the small short-tailed *Bacillus subtilis* phage ϕ29 (Anderson *et al.*, 1966) has 82% of its head occupied by 2-DNA. Anderson *et al.* (1966) appear to underestimate the head size since their DNA volume is 25% greater than that of the head. The results in Table II are based on figures obtained by various authors and show a significant measure of agreement. It can be seen that assuming a phage head to be either an icosahedron or an octahedron makes relatively little difference in the result. The percentage of head volume occupied by DNA varies between 35.5% for an octahedral T5 to 82% for oblong ϕ29 (assumed to be a bipyramidal hexagonal prism).

Thus one can say with some confidence that phage nucleic acid occupies around 50% of the head volume and is therefore in the form of a loose coil. Such information could be useful in studying the way in which nucleic acid is coiled within viral capsids. Apart from the length, one can see whether or not the molecule of nucleic acid is in the form of a closed loop by the spreading technique. With T2 (Fig. 30) both ends of the 49 μm length are visible.

TABLE II

Comparison of the Head and Nucleic Acid Volumes of Some Bacteriophages

Phage	Nucleic acid type	Morphological type	Head shape (probable)	Side length (nm)	Head volume (nm³)	Nucleic acid length (μm)	Nucleic acid volume (nm³)	Percentage head occupied	Reference for nucleic acid length
T-even	2-DNA	Contractile	Oblong	81 × 125	3.58×10^5	55	1.72×10^5	48	Abelson and Thomas (1966)
PB1	2-DNA	Contractile	Octahedron	75	1.96×10^5	25	7.9×10^4	40	Bradley and Robertson (1968)
Lambda	2-DNA	Noncontractile	Octahedron	60	1×10^5	17.2	5.4×10^4	54	Abelson and Thomas (1966)
T5	2-DNA	Noncontractile {	Octahedron	75	3.4×10^5	38.8	1.2×10^5	35.5	Abelson and Thomas (1966)
			Icosahedron	48	2.4×10^5	38.8	1.2×10^5	50	Abelson and Thomas (1966)
T7	2-DNA	Noncontractile (short)	Octahedron	55	7.8×10^4	12.5	3.9×10^4	50	Abelson and Thomas (1966)
φ29	2-DNA	Noncontractile (short)	Oblong	31.6 × 43	2.59×10^4	5.7	1.8×10^4	82	Anderson et al. (1966)
φX174	1-DNA	Tailless, large capsomeres	Icosahedron	15	7.35×10^3	1.765	2.7×10^3	37	Section III, B
R17	1-RNA	Tailless, smooth	Icosahedron	12	3.7×10^3	1.0	1.47×10^3	40	Franklin and Granboulan (1966)

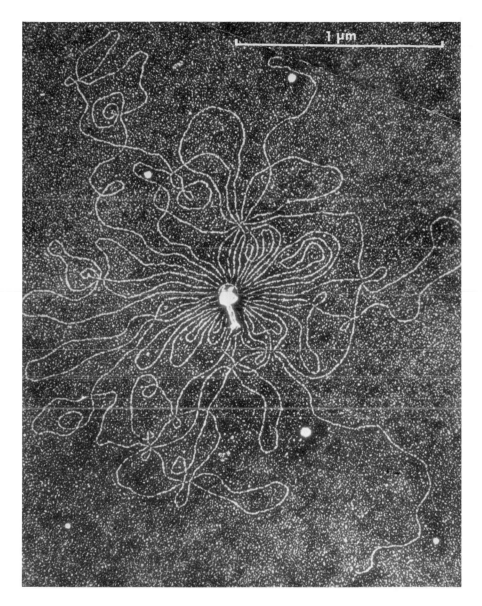

FIG. 30. DNA of coliphage T2. After Kleinschmidt *et al.* (1962), with permission of *Biochim. Biophys. Acta.*

B. Phages Containing Single-Stranded DNA

Phages containing 1-DNA are represented in morphological groups C and E. They are tailless with large capsomeres and filamentous, respectively. The forms of the molecules of the ϕX174 type phages (group C) and the filamentous phages are similar. Both are closed loops (Fig. 31 for ϕX174), demonstrated for the filamentous phage fd by Marvin and Schaller (1966). The molecular weight of ϕX174 DNA is 1.7×10^6 daltons (Sinsheimer, 1959), but that of the filamentous coliphages is 2×10^6 daltons. They therefore contain about 5500 and 6600 nucleotides, respectively. The larger filamentous *Salmonella* and *Pseudomonas* phages have longer molecules but these have not been measured. ϕX174 DNA (replicative form) measured in the electron microscope is 1.88 μm long (Chandler *et al.*, 1964) compared to 2.15 μm for the filamentous coliphage fd. The way in which the 1-DNA of the filamentous phage fits into the virion has been discussed above (Section II,C,5). Table II shows that the 1-DNA of the ϕX174 phages occupies about 37% of the volume available within the capsid, assuming that the diameter of 1-DNA is about 1.4 nm (Das Gupta *et al.*, 1966).

C. Phages Containing Single-Stranded RNA

As has been mentioned above (Section II,4), there seem to be two slightly different structural forms of RNA coliphages, the R17/MS2 and Qβ groups. Their 1-RNA has been extensively studied by various means. That of R17 is about 1 μm long (Granboulan and Franklin, 1966, 1968) and, assuming a strand diameter of 1.4 nm occupies 40% of the capsid volume (Table II). This is about the same as for the 1-DNA ϕX174 types and is somewhat lower than the figure obtained for 2-DNA phages; it might reflect a slightly looser packing of a single-stranded nucleic acid molecule.

The RNA of the coliphages, which is in fact the same as messenger RNA, has been well studied for MS2 and Qβ, and some characteristics have been compared by Overby *et al.* (1966a). There is a slight difference in molecular weights (1×10^6 and 0.9×10^6 daltons, respectively) and a higher adenine/uracil ratio for MS2 (0.95) than for Qβ (0.75). These authors provide the base ratios for several other RNA coliphages, two of which [ZJ/1 and ZIK/1, Bishop and Bradley (1965)] were similar to Qβ, the remainder, including R17, being similar to MS2.

Sinha *et al.* (1965) compared the RNA of the serologically related phages M12 and R17 and concluded that there were several differences in the nucleotide sequences in a small part of the total polynucleotide chain and that the nucleic acids were not identical like the protein coats. The taxonomic significance of these observations will be discussed below.

IV. The Infective Cycle of Bacteriophages

It has already been mentioned (Section I,C) that there is a great deal of variation in the infective processes of different phages and these will now be considered in greater detail.

A. TEMPERATE AND VIRULENT INFECTIVE PROCESSES

Nature has devised two ways in which a bacteriophage can be proliferated and these are exemplified by the virulent and temperate infective processes. In the first, four distinct stages are apparent: adsorption to the host cell, injection of nucleic acid, intracellular multiplication of virions, and the release of phage progeny by lysis. With temperate phages it is slightly more complicated: adsorption, injection, incorporation of the phage genome with the host chromosome, synchronous division of the phage genome with the host chromosome, intracellular multiplication, and lysis. The latter two stages may be spontaneous or brought about by external influences such as metabolic inhibitors. Also it is important to note that a temperate phage can behave in a virulent fashion toward a sensitive host and conversely a potentially lysogenizable host might possibly be found for an apparently strictly virulent phage. While at first sight a comparison of these infective cycles may suggest a fundamental difference with possible taxonomic significance, it must be stressed that a strictly virulent phage may merely have lost its ability to combine with the host chromosome, or more likely, that lysogenizable strains of bacteria have become extinct. One must thus link these two infective cycles since temperance does not exclude virulence but is combined with it.

B. ADSORPTION AND INJECTION

The majority of bacteriophages adsorb directly to the cell surface. With both contractile and noncontractile types this is done tail-first. A detailed study of T-even phage adsorption has been described by Simon and An-

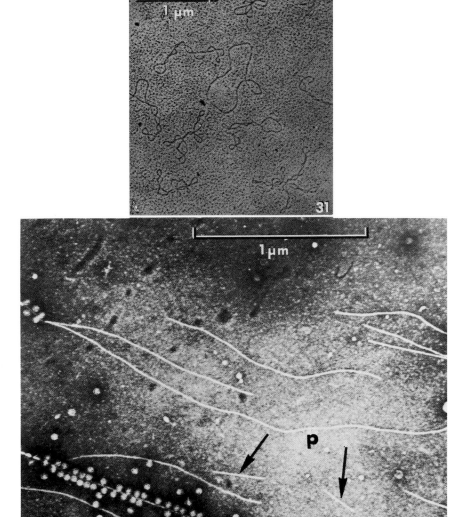

FIG. 31. φX174 replicative form DNA, × 25,000. After Chandler *et al.* [*Science* **143**, 47–49 (1964)], with permission of the Editors.

FIG. 32. Coliphage f1 adsorbed to F-pili labeled with RNA phage. p, filamentous phage virion. Arrows mark fragments, × 47,000. After Caro and Schnös (1966), with permission of *Proc. Natl. Acad. Sci. U.S.*

derson (1967). Several phages have been found to attach to bacterial flagella (Meynell, 1961; Frankel and Joys, 1966): phage PBS1, which has a contractile sheath and helical tail fibers, adsorbs by wrapping them around the flagella of *Bacillus subtilis* (Raimondo *et al.*, 1968). The group B phage T5 was shown to attach to isolated cell wall receptor molecules by the tail tip (Fig. 1,A; Weidel and Kellenberger, 1955). With the ϕX174 types (group C), which are univalent (Kay, 1962), adsorption to the cell wall (Stouthamer *et al.*, 1963) is by a specific site on the capsid. The RNA and filamentous

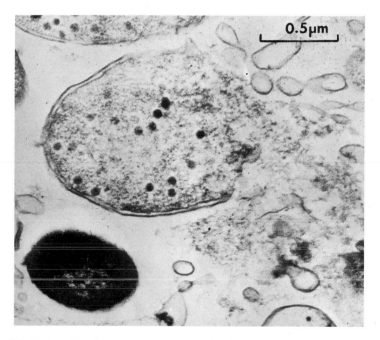

FIG. 33. Lysing *Pseudomonas aeruginosa* cell infected with phage PB-1, × 40,000. With permission of *J. Gen. Virol.*

coliphages both adsorb to F-pili, which are thin filaments extending from the cell surface and are involved in the genetic transfer of the F (fertility) factor (Brinton, 1965; Crawford and Gesteland, 1964; Caro and Schnös, 1966). As can be seen from Fig. 32, the RNA phages attach to the sides of the filament and the filamentous types to the tip.

With the tailed phages, the injection stage of the infective cycle is achieved by the passage of the nucleic acid molecule down the tail which has penetrated the cell wall, and into the cytoplasm. The work of Lanni (1965) on phage T5 indicates that injection is a complex process rather than a simple one-step transfer of the DNA molecule. In the cases of the ϕX174, RNA, and filamentous phages, comparatively little is known about nucleic acid transfer. With RNA and filamentous phages the nucleic acid probably passes down the inside of the F-pilus for a short distance, then the pilus retracts into the cell so initiating injection.

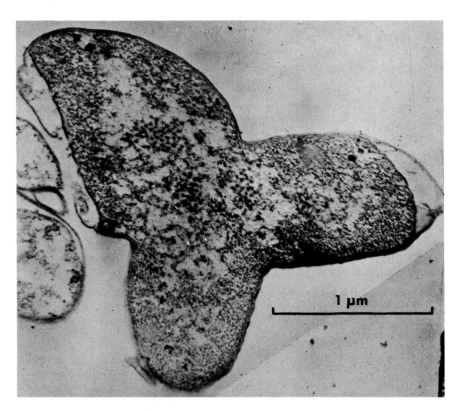

FIG. 34. *Escherichia coli* cell infected with phage α3 showing spheroplastlike form, × 41,000. With permission of *J. Gen. Virol.*

C. INTRACELLULAR MULTIPLICATION AND LYSIS

In comparing electron micrographs of phage-infected cells, one of the most obvious differences is the site of the intracellular particles. With contractile types this can be within the nucleoplasm as with T2 (Kellenberger *et al.*, 1959) or randomly distributed through the cytoplasm as with the *Pseudomonas* phage PB1 (Bradley and Robertson, 1968). Noncontractile phages usually appear in the cytoplasm and are not obviously associated

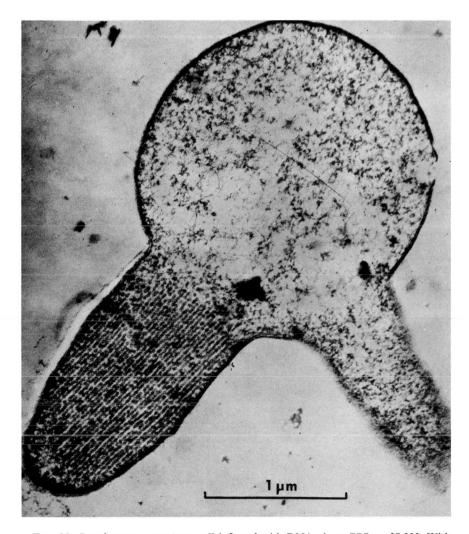

Fig. 35. *Pseudomonas aeruginosa* cell infected with RNA phage PP7, × 37,000. With permission of *J. Gen. Microbiol.*

with the nucleoplasm. The 1-DNA ϕX174 types are found in the nucleoplasm (Bradley *et al.*, 1969) but the RNA phages appear to be strictly cytoplasmic (Franklin and Granboulan, 1966; Bradley, 1966b). The filamentous phages are not obviously visible in infected cells, but Bradley and Dewar (1967) suggest that they are assembled just adjacent to the plasma membrane and Schwartz and Zinder (1967) indicate that they are associated with plasma membrane synthesis.

Another phage, the tailless lipid-containing marine isolate PM2 is assembled immediately adjacent to or attached to the plasma membrane (Cota-Robles *et al.*, 1968).

Lysis of the host cell is manifested by a number of markedly different effects. Contractile phages cause a general breakdown of the whole cell envelope, for example, T2 (Cota-Robles, 1964) and the *Pseudomonas* phage PB1 (Bradley and Robertson, 1968) shown in Fig. 33. The ϕX174 types

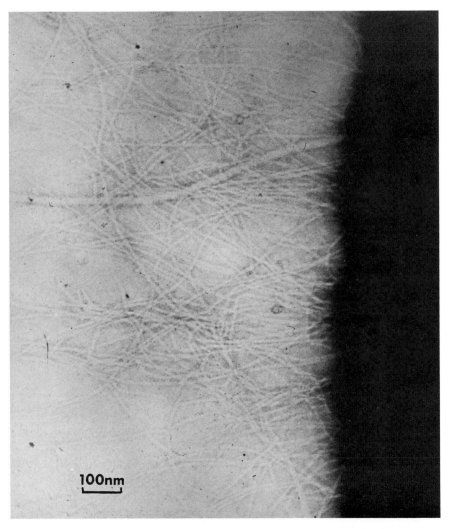

100nm

FIG. 36. Part of *Escherichia coli* cell infected with filamentous phage ZJ/2 showing extruding virions, × 100,000. With permission of J. *Gen. Virol.*

usually cause the formation of spheroplasts, illustrated in Fig. 34 (Bradley *et al.*, 1969); this is followed by a two-step breakdown of the cell envelope, first the cell wall, then the plasma membrane. Similar spheroplasts are formed by *Pseudomonas* RNA phages (Bradley, 1966b) but the intracellular virions are located in the arms of the spheroplast (Fig. 35) and not in the bulge as with the ϕX174 types. The cell wall and plasma membrane break down simultaneously on the spheroplast bulge. RNA coliphages do not form spheroplasts.

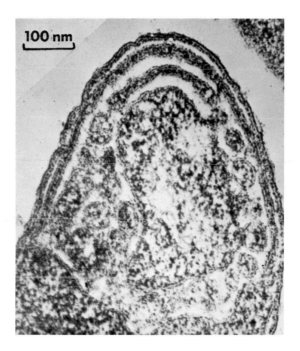

Fɪɢ. 37. Polar region of *Escherichia coli* cell after ZJ/2 phage infection, × 100,000. With permission of *J. Gen. Virol.*

The most interesting mechanism of phage release is exhibited by all the filamentous isolates regardless of host species. Here, the cell continues to divide and virions are "extruded" through the cell wall to produce a tangled mass shown in Fig. 36 (Hofschneider and Preuss, 1963; Pratt *et al.*, 1966; Bradley and Dewar, 1967). With the isolate ZJ/2, cell membrane invaginations are formed during phage release (Fig. 37) but with f1, only conditional lethal mutants produce this effect (Schwartz and Zinder, 1967). The principle features of different infective processes are summarized in Table III.

TABLE III

PRINCIPLE DIFFERENCES IN THE INFECTIVE PROCESSES OF SOME BACTERIOPHAGES

Phage	Host species	Morphological type	Adsorption	Site of intracellular phage	Mechanism of lysis
T-even	*Escherichia coli*	Contractile	Cell wall	Nucleoplasm	General cell wall and plasma membrane breakdown
PB 1	*Pseudomonas aeruginosa*	Contractile	Cell wall	Randomly distributed	General cell wall and plasma membrane breakdown
PBS 1	*Bacillus subtilis*	Contractile	Flagella	—	—
T 5	*Escherichia coli*	Noncontractile	Cell wall	Cytoplasm	General cell wall and plasma membrane breakdown
φX174	*Escherichia coli*	Tailless, large capsomeres	Cell wall	Nucleoplasm	Formation of unstable short-lived spheroplasts
α 3	*Escherichia coli*	Tailless, large capsomeres	Cell wall	Nucleoplasm	Spheroplasts and breakdown of cell wall then plasma membrane
PM 2	*Pseudomonas* (marine)	Tailless, large capsomeres, lipid-containing	Cell wall	Adjacent to plasma membrane	Plasma membrane then cell wall breakdown
R 17, Qβ, etc. (RNA)	*Escherichia coli*	Tailless, smooth	F-pili, sides	Cytoplasm	General cell wall and plasma membrane breakdown
PP 7 (RNA)	*Pseudomonas aeruginosa*	Tailless, smooth	Polar pili	Cytoplasm	Spheroplasts then rupture of bulge
fd, M13	*Escherichia coli*	Filamentous	F-pili, tip	Invisible	Extrusion, no lysis

V. Host Ranges of Bacteriophages

Bacteriophages have been found for a large number of bacterial species and for the closely related blue-green algae. In general the tailed phages are widely distributed and neither morphological group favors a particular organism.

However, the ϕX174 types are restricted to *Escherichia coli* and *Salmonella typhimurium*; only the lipid-containing PM2, which is much larger, infects cells outside the Enterobacteriaceae. RNA phages, on the other hand, are distributed among three genera: *Escherichia*, *Pseudomonas*, and *Caulobacter*. The filamentous forms again are found associated with three genera: *Escherichia*, *Pseudomonas*, and *Vibrio*. However, it is to be noted that these genera are gram-negative and not too far apart taxonomically.

The specificity of a phage isolate may often be restricted to a single species or even a single strain of a species, but an identical morphological type may well infect a different species or genus. This is particularly prevalent amongst the Enterobacteriaceae where a single phage isolate can have a host range extending over several genera. Because of this variability in specificity, host range is not considered reliable as a taxonomic criterion. Some doubt may even be cast on the restriction of bacteriophages to bacteria. The so-called cyanophages, which are parasites of blue-green algae (Safferman and Morris, 1963) closely resemble the short noncontractile tailed coliphage T3, and have an infective cycle basically similar to many phages (Smith *et al.*, 1966).

VI. Bacteriophages as Antigens

In general, bacteriophages are good antigens, producing antisera with high inactivation constants. It can be said that the inactivation of one phage isolate by antiserum produced by another shows that the phage capsids contain similar antigens and that the two isolates are one and the same. This holds true provided the inactivation rates for the homologous and heterologous phages are the same or nearly so. Such an observation indicates a close serological relationship; conversely if the heterologous phage is not inactivated at all the two isolates are unrelated. However, if the rate of inactivation of the heterologous phage by the homologous phage antiserum is slow, one must ask what sort of relationship this implies. The simplest way to answer this is to consider the mathematical calculations involved and to cite an example.

The inactivation constant K for a given antiserum is determined by incubating the phage under test with diluted antiserum and titrating samples for surviving plaque-forming units at intervals of time. K is calculated from the formula

$$K = \frac{2.3\,D}{t} \cdot \log \cdot \frac{100}{x}$$

where K is expressed in min^{-1}, $D =$ antiserum dilution (100 for 1:100), and $x =$ percent survivors at time t (minutes). The 100 refers to 100% survivors at zero time. Because K for a homologous phage varies for each batch of antiserum, it is only possible to obtain a universal comparison between heterologous and homologous phages for different antiserum batches by calculating the ratio k/K where k is the inactivation constant for the heterologous phage and K that of the homologous phage at the same antiserum dilution. With the homologous phage this will be 1 $(k/K = 1)$, a fraction for related phages and zero for unrelated ones. The question is, what is the smallest value of k/K indicating a significant serological relationship? It has been suggested (Fry and Waites, 1969) that a value of 0.25 or less indicates a difference in the antigenic structures of the two phages. This could be due to different amino acid sequences in the antigens or different spatial arrangements of the same antigens.

Morphological group C (the ϕX174 type phages) provides a useful example. It comprises a number of isolates (see Bradley, 1967) which form three distinct host range groups (designated A, B, and C for convenience), the principle members being given in Table IV. Representatives of each

TABLE IV

HOST RANGES OF THE ϕX174 BACTERIOPHAGES ON STRAINS OF *Escherichia coli*

Group	Phage	C2	C1[a]	B	H[b]	C+/L[c]	K12 (F−)	K12 (F+)	K12 (Hfr)
A	α3	+	+	+	−	−	−	−	−
	ZD13	+	+	+	−	−	−	−	−
B	ϕX174	+	−	−	+	−	−	−	−
	ϕR	+	−	−	+	−	−	−	−
	S13	+	−	−	+	−	−	−	−
C	St/1	−	−	−	−	+	+	+	+

[a] A mutant of *Escherichia coli* C2 resistant to ϕX174.
[b] An F− strain supplied by D. Kay, Oxford.
[c] An F+ strain used for isolating St/1.

group were tested for serological relatedness with one another (D. E. Brad-ley, unpublished), examples being given in Table V. Clearly $\alpha3$ and ZD/13 are closely related serologically and are members of the same host range group. ϕX174 and ϕR (host range group B) are also related; it is to be noted that anti-ϕX174 gives a much higher ratio that anti-ϕR indicating that in obtaining a serological comparison between two phages antisera from both should be used unless a high ratio (say 0.5) is obtained with the first

TABLE V

VALUES OF THE RATIO k/K FOR ANTISERA OF THE ϕX174 BACTERIOPHAGES

Phage	Host range group	Anti-$\alpha3$	Anti-ϕR	Anti-ϕX174[a]	Anti-St/1
$\alpha3$	A	1	0	—	0.084
ZD13	A	0.654	0	—	0.139
ϕX174	B	0	0.23	1	0
ϕR	B	0	1	0.76	0
S13	B	0	0.233	0.785	0
St/1	C	0.046	0	0	1

[a] Results kindly supplied by J. F. Bleichrodt and E. R. Berends-van-Abkoude.

combination tested. Phage $\alpha3$ is unrelated to group B members, but shows a very slight relationship with St/1. Conversely, anti-St/1 shows a slight relationship with $\alpha3$ and ZD/13. Bearing in mind the remarks of Fry and Waites (1969) these results are interpreted most logically as follows. Re-ferring to Table IV, members *within* host range groups A and B are sero-logically related but not chemically identical. Members of group A are not related to members of group B and members of group B are not related to members of group C. Groups A and C are slightly related but have signifi-cantly different antigenic structures. However, it should be borne in mind that serological unrelatedness can indicate either a gross difference in the capsid such as the presence or absence of a tail, or equally well a chemical difference which may be undetectable morphologically. Thus serology can provide a very fine or a very gross comparison between two phages. In a taxonomic sense, serological unrelatedness is therefore a somewhat incon-clusive comparison. However, if two phages are serologically related, even slightly, one can say that they are very similar save for a degree of difference in their antigenic structures which could have been produced by natural mutations from a common ancestor.

VII. The Taxonomy of Bacteriophages

If one wishes to classify bacteriophages, or any other group of entities, one must first select taxonomic criteria at various levels. It can be seen from the foregoing that there are certain obvious choices: nucleic acid type and strandedness, morphology, host range, and serological relationships. As has been said, all the genetic information for a bacteriophage is contained in its nucleic acid molecule and since this varies in type and strandedness from one phage to another this characteristic obviously constitutes a major criterion. It has been shown that the type of nucleic acid is reflected in morphology, the tailed phages having 2-DNA and the simpler forms 1-DNA or 1-RNA. These criteria are thus of nearly equal importance. Host range, whether at the level of the bacterial genus or species has been considered as unreliable. Serological relationships should, in the context of the discussion above, be used at a very low level of classification equivalent, perhaps to bacterial strains. Whatever criteria are used and whatever system is adopted, a level of classification is eventually reached when one has to say that two phage isolates are the same. The problem is to define this point. If they are serologically related to any degree, then they are very similar indeed, but phages with chemically identical capsids can still have variations in the nucleotide sequences of their genomes as has been described for the RNA phages M12 and R17 (Section III,C). However, these differences were small and it seems reasonable to postulate that serological relatedness signifies identity. But serological unrelatedness does not necessarily signify non-identity other than a small chemical difference in the capsid. For example, are the serologically unrelated but structurally identical ϕX174 phages different at the "species" level? The question is answered by considering the RNA coliphages. The R17/MS2 types have a capsid consisting of 32 morphological subunits whereas ZIK/1, which is probably similar to Qβ, has 92; the two groups are serologically unrelated. This constitutes a real difference in the capsid but not in the basic form of the viruses which is icosahedral. Thus, a very reasonable approach is to say that a difference in the number of capsomeres on two structurally similar viruses constitutes a significant dissimilarity. MS2 and Qβ have a 10% difference in the molecular weights of their nucleic acids and also different base ratios (Section III,C), which are reflected in their capsids. These criteria can thus constitute the lowest level of taxonomic distinction in bacteriophages and one must therefore consider the ϕX174 phages, serologically related or otherwise, as a single group or "species." The other characters which have been mentioned must be used at higher levels. At the present time the trend is to use a combination of morphology and nucleic acid type at something equivalent to a generic level. One should perhaps make a clear distinction between morphology or shape and molecular structure, which has been considered with respect to R17 and Qβ.

The lipid-containing phage, which is similar in shape but larger than the ϕX174 types must be considered here since it is included in morphological group C. However, it has just been stated that this generic level is defined by both nucleic acid type and morphology. PM2 is the only tailless phage with 2-DNA so that on this basis it must be placed in a separate morphological group. Its capsid, which contains lipid to the extent that it is visible in the electron microscope, is markedly different from any other phage and has two distinct layers. By definition, therefore, the morphological groups should be rearranged as follows: group A: tails contractile; group B: tails noncontractile; group C: tailless, double-layered capsid, apical capsomeres; group D: tailless, single-layered capsid, apical capsomeres; group E: tailless, no apical capsomeres; group F: filamentous. The morphological groups now have a single nucleic acid type each, whereas before the tailless phages with large capsomeres had viruses with either 1-DNA or 2-DNA. While it may seem logical to have higher divisions based on nucleic acid type and strandedness, it has been decided by the International Committee on the Nomenclature of Viruses that viruses should be classified as a whole. In this case each morphological group would be included with other viruses with the same nucleic acid type at a level of classification above the genus.

VIII. The Comparison of Bacteriophages with Bacteriocins and Related Entities

A bacteriocin is defined as a highly specific bactericidal particle produced by a bacterium and generally active against other strains of the same species. Some bacteriocins have a wider spectrum of activity including genera other than that of the producer species. A bacteriocin is able to kill a cell, but not to multiply within it. It would appear therefore that bacteriocins are outside the scope of this review, but in fact they have a great deal in common with bacteriophages and some types are identical to phages save that they lack the ability to multiply intracellularly in an infected cell. The structure and function of bacteriocins has been reviewed elsewhere (Reeves, 1965; Bradley, 1967) and it appears that they fall into two distinct classes. One type consists of phagelike particles or components such as the head or tail.

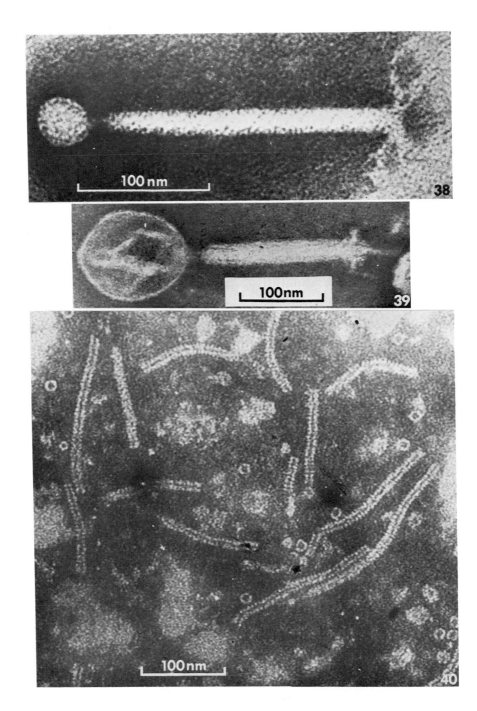

The other type is much smaller, invisible in the electron microscope, not sedimentable in the ultracentrifuge, and probably a component of the cell wall of the producer bacterium. For simplicity these are referred to as large and small bacteriocins. A useful distinguishing property is heat stability: small bacteriocins are thermostable (withstand boiling for several minutes) whereas large ones are thermolabile and generally inactivated at about 80°C or below.

There appear to be three different forms of large bacteriocin. The first, exemplified by colicin 15, an *Escherichia coli* bacteriocin, is a complete contractile phage particle identical in appearance to the bacteriophage shown in Fig. 15 (Mennigmann, 1965). It will be noted that the head is full, and in fact contains DNA, yet the particle is unable to infect and multiply in a cell, but is only able to kill it. The so-called *Bacillus* killer particles (Fig. 38) fall into this category and have been particularly well studied (Ionesco *et al.*, 1964; Seaman *et al.*, 1964; Okamoto *et al.*, 1968). Here it was found that the DNA in the head is of bacterial origin and is not injected into a sensitive cell despite the fact that the particle has a contractile tail.

The killing effect is produced by the protein alone, causing the cessation of DNA, RNA, and protein synthesis within 5 minutes. This is similar to the effect that all other bacteriocins have. A second form consists of a complete phagelike particle but with an empty head. An example is shown in Fig. 39. This is a monocin (bacteriocin for *Listeria monocytogenes*) and can be seen to have a contractile tail. The third form is the phage component which usually consists of the headless tail of a contractile phage like that shown in Fig. 15. However, Takeya *et al.* (1967) described a pyocin (*Pseudomonas aeruginosa* bacteriocin) which was like a headless noncontractile phage (Fig. 40).

Clearly bacteriocins are a heterogeneous collection of entities, the large ones obviously being related to bacteriophages in one way or another but the small ones seem to be of a fundamentally different nature. This difference has been demonstrated by Goebel *et al.* (1955) who showed that colicin K was the O-somatic antigen of the producer strain of *Escherichia coli*.

FIG. 38. *Bacillus subtilis* killer particle, × 333,000. With permission of *J. Gen. Microbiol.*

FIG. 39. Phage-like *Listeria monocytogenes* bacteriocin, × 220,000. With permission of *J. Gen. Microbiol.*

FIG. 40. Pyocin 28, × 220,000. After Takeya *et al.* (1967), with permission of *Virology.*

This is located in the bacterial cell wall and clearly unrelated to any bacterio-phage or component which would be found in the cytoplasm or nucleo-plasm. While bacteriocins are considered as a single group of bactericidal particles at present, it is likely that this conception will have to be changed. It also seems probable that the large bacteriocins can be divided into the three types mentioned above. As will be discussed below it seems reasonable to suppose that the DNA-containing bacteriocins represent a step in the evolution of bacteriophages, whereas the other two forms without DNA are a devolutionary step from a fully formed phage.

IX. Extrachromosomal Elements

Episomes or plasmids are lengths of DNA existing either in the cytoplasm or attached to the chromosome of a bacterium: they replicate in synchrony with the bacterial chromosome, and are thus perpetuated as long as the parent strain exists. Both temperate phages and bacteriocins exist as epi-somes but there are, of course, many others such as the F factors and the drug-resistant factors. While it is outside the scope of this work to compare them in detail, they do represent an important physiological link between bacteriocins and must therefore be mentioned.

X. The Origin and Evolution of Bacteriophages

The comparative study of bacteriophages and bacteriocins is important not only for taxonomic purposes but also in considering how they evolved. The following entities are likely to be involved in an evolutionary spectrum terminating in a functioning bacteriophage: extrachromosomal elements of the F factor type, large bacteriocins, and bacteriophages. In suggesting any sequence one must provide a link between each member of the chain, and also a basic hypothesis from which to reason. Lwoff (1953) suggested that phages originated from a series of mutations which caused part of the bacterial chromosome to split off and attain a degree of autonomy. In order to exist outside the cell, it coded for itself a protein coat and some form of injection system to permit its entry into another cell. The basis of this, which is that phages originated from the host chromosome, provides a starting point for the following discussion.

By definition, an extrachromosomal element or plasmid is a portion of bacterial chromosome which has attained autonomy within the cell. The F factor is the best known example. The first step in achieving autonomy could be represented by the F factor in Hfr bacteria when it is still attached to the chromosome. Hfr bacteria attach themselves to cells lacking the F factor and a genetic transfer takes places in which the episome passes into the recipient and joins its chromosome. Thus a form of infection has taken place, but the F factor DNA has only achieved autonomy for that period of time taken to transfer itself from one cell to the other. In order to achieve transfer, the F factor codes for the formation of thin filaments on the donor cell surface (F-pili). It is not entirely certain how these are involved but transfer cannot be effected without them. They could be tubes down which the F factor DNA passes in which case they represent a primitive form of temporary capsid. At all events, F factor transfer represents a type of infection analogous with that of a bacteriophage in that the F-pilus attaches to receptors in the recipient cell wall before infection. The next step would consist of the splitting of the F factor from the chromosome as is the case in F$^+$ bacteria where the episome is present in the cytoplasm. Apart from this, genetic transfer takes place in the same sort of way. It is important to stress that the F factor is not being labeled as the actual ancestor of any particular phage; it is being presented as indicating the existence of a group of similar entities at the beginning of an evolutionary spectrum leading to a bacteriophage.

In the next step, this length of DNA, now located in the cytoplasm, must code for itself a capsid so that it can exist outside the cell. The F-pilus could be the first stage in the evolution of such a protein shell. A more sophisticated version could be the filamentous phage which has a structural resemblance to the F-pilus. The filamentous phage is a near-perfect parasite in that it does not kill its host cell which continues to produce it over a long period. It thus has an advantage over other phages which cause lysis and death, and could well be the result of a single evolutionary chain. In this case the more complex phages would have to evolve separately. Going back to an episomelike entity one looks for something which has coded a tailed capsid for itself. First there are the *Bacillus* killers, bacteriocins which actually contain a segment of the host cell chromosome. The episome responsible was thus merely a piece of bacterial genome without the ability to enter the recipient cell. It will be remembered that the DNA of killers is not injected; the capsid is not efficient enough to effect genetic transfer. An episome which has this ability is the temperate phage. However, these behave in a virulent fashion toward sensitive indicators and in fact represent the end of the chain. An intermediate step would be a temperate phage which could infect or lysogenize a recipient but never behave in a virulent fashion. It seems highly probable that this exists. A hypothetical strictly virulent phage (it is impossible to prove that a phage is strictly virulent) may have lost the ability to lysogenize and thus be a devolutionary step. Those bacteriocins which are merely phage components are also probably devolutionary since it is unlikely that an episome would code for an injection system before arranging for itself to be surrounded by a capsid.

In summary it is most reasonable to suggest that phages evolved in several chains starting from different episomes. The end products could well be represented by the different morphological forms. It seems certain that the RNA phages are the result of a completely different series of evolutionary steps in which the messenger RNA of the host cell, carrying the information originally on the DNA episome, became enveloped in a capsid and achieved autonomy. The chain about which we have the most evidence is that which starts with an episome consisting of a piece of host chromosome carrying no additional information, this becoming encased in protein and forming a killer particle. The killer particle codes for an injection mechanism and becomes a strictly temperate phage. Finally mutations occur to the temperate phage genome allowing it to redirect the recipient's metabolism and multiply intracellularly causing death and lysis thus forming a virulent phage.

ACKNOWLEDGMENTS

I am most grateful to Miss M. McCulloch for preparing the prints for this chapter. This paper was prepared with the assistance of a Visiting Scientist Grant from the Medical Research Council of Canada.

REFERENCES

Abelson, J. and Thomas, C. A. (1966). *J. Mol. Biol.* **18**, 262–291.
Anderson, D. L., Hickman, D. D., and Reilly, B. E. (1966). *J. Bacteriol.* **91**, 2081–2089.
Anderson, T. F. (1960). *Proc. European Reg. Conf. Electron Microscopy, Delft,* 1960, **2**, 1008–1011.
Anderson, T. F. and Stephens, R. (1964). *Virology* **23**, 113–117.
Betz, J. V. (1968). *Virology* **36**, 9–19.
Bishop, D. H. L. and Bradley, D. E. (1965). *Biochem. J.* **95**, 82–93.
Boy de la Tour, E. and Kellenberger, E. (1965). *Virology,* **27**, 222–225.
Bradley, D. E. (1961). *Virology* **15**, 203–205.
Bradley, D. E. (1963a). *J. Gen. Microbiol.* **31**, 435–445.
Bradley, D. E. (1963b). *J. Ultrastruct. Res.* **8**, 552–565.
Bradley, D. E. (1964). *J. Gen. Microbiol.* **35**, 471–472.
Bradley, D. E. (1965a). *J. Gen. Microbiol.* **41**, 233–241.

Bradley, D. E. (1965b). *J. Roy. Microsc. Soc.* **84**, 257–316.

Bradley, D. E. (1965c). *J. Gen. Microbiol.* **38**, 395–408.

Bradley, D. E. (1966a). *J. Gen. Microbiol.* **44**, 383–391.

Bradley, D. E. (1966b). *J. Gen. Microbiol.* **45**, 83–96.

Bradley, D. E. (1967). *Bacteriol. Rev.* **31**, 230–314.

Bradley, D. E. and Dewar, C. A. (1967). *J. Gen. Virol.* **1**, 179–188.

Bradley, D. E. and Kay, D. (1960). *J. Gen. Microbiol.* **23**, 553–563.

Bradley, D. E. and Robertson, D. (1968). *J. Gen. Virol.* **3**, 247–254.

Bradley, D. E., Dewar, C. A., and Robertson, D. (1969). *J. Gen. Virol.* **5**, 113–121.

Brenner, S. and Horne, R. W. (1959). *Biochim. Biophys. Acta* **34**, 103–110.

Brinton, C. C. (1965). *Trans. N. Y. Acad. Sci.* **27**, 1003–1054.

Burchard, R. P. and Dworkin, M. (1966). *J. Bacteriol.* **91**, 1305–1313.

Caro, L.G. and Schnös, M. (1966). *Proc. Nat. Acad. Sci. U. S.* **56**, 126–132.

Chandler, B., Hayashi, M., Hayashi, M. N., and Spiegelman, S. (1964). *Science* **143**, 47–49.

Cota-Robles, E. H. (1964). *J. Ultrastruct. Res.* **11**, 112–122.

Cota-Robles, E. H., Espejo, T. R., and Haywood, P. W. (1968). *J. Virol.* **2**, 56–58.

Coyette, J. and Calberg-Bacq, C. M. (1967). *J. Gen. Virol.* **1**, 13–18.

Crawford, E. M. and Gesteland, R. F. (1964). *Virology* **22**, 165–167.

Crick, F. H. C. (1957). *In* "The Nature of Viruses" (G. E. V. Wolstenholme and E. C. P. Millar, eds.), pp. 5–18. Churchill, London.

Crick, F. H. C. and Watson, J. D. (1956). *Nature (London)* **177**, 473–475.

Das Gupta, N. N., Sarkar, M., and Misra, D. N. (1966). *J. Mol. Biol.* **15**, 619–623.

Deklerk, H. C., Coetzee, J. N., and Fourie, J. T. (1965). *J. Gen. Microbiol.* **38**, 35–38.

Demerec, M. and Fano, U. (1945). *Genetics* **30**, 119–136.

Edgell, M. H., Hutchison, C. A., and Sinsheimer, R. L. (1969). *J. Mol. Biol.* **42**, 547–557.

Eiserling, F. A. and Boy de la Tour, E. (1965). *Pathol. Microbiol.* **28**, 175–180.

Espejo, R. T. and Canelo, E. S. (1968). *Virology* **34**, 738-747.

Feary, T. W., Fisher, E., and Fisher, T. N. (1964). *J. Bacteriol.* **87**, 196–208.

Frankel, R. W. and Joys, T. M. (1966). *J. Bacteriol.* **92**, 388–389.

Franklin, R. M. and Granboulan, N. (1966). *J. Bacteriol.* **91**, 834–848.

Fry, B. A. and Waites, W. M. (1969). *J. Gen. Virol.* **4**, 47–53.

Goebel, W. F., Barry, G. T., Jesaitis, M. A., and Miller, E. M. (1955). *Nature (London)* **176**, 700–701.

Granboulan, N. and Franklin, R. M. (1966). *J. Mol. Biol.* **22**, 173–177.

Granboulan, N. and Franklin, R. M. (1968). *J. Virol.* **2**, 129–148.

Hall, C. E. (1955). *J. Biophys. Biochem. Cytol.* **1**, 1–12.

Hall, C. E., Maclean, E. C., and Tessman, I. (1959). *J. Mol. Biol.* **1**, 192–194.

d'Herelle, F. (1917). *C. R. Acad. Sci.* **165**, 373–375.

Hofschneider, P. H. and Preuss, A. (1963). *J. Mol. Biol.* **7**, 450–451.

Horne, R. W. and Wildy, P. (1961). *Virology* **15**, 348–373.

Huxley, H. E. (1956). *Proc. European Reg. Conf. Electron Microscopy, Stockholm, 1956*, pp. 260–261.

Ionesco, H., Ryter, A. and Schaeffer, P. (1964). *Ann. Inst. Pasteur* **107**, 764–776.

Kay, D. (1962). *J. Gen. Microbiol.* **27**, 201–207.

Kay, D. and Bradley, D. E. (1962). *J. Gen. Microbiol.* **27**, 195–200.

Kellenberger, E. and Arber, W. (1955). *Z. Naturforsch.* **b10**, 698–699.

Kellenberger, E., Séchaud, J., and Ryter, A. (1959). *Virology* **8**, 478–498.

Kingsbury, D. T. and Ordal, E. J. (1966). *J. Bacteriol.* **91**, 1327–1332.

Kleinschmidt, A. K., Lang, D., Jackerts, D., and Zahn, R. K. (1962). *Biochim. Biophys. Acta* **61**, 857–864.

Kleinschmidt, A. K., Burton, A., and Sinsheimer, R. L. (1963). *Science* **142**, 961–963.

Lanni, Y. T. (1965). *Proc. Nat. Acad. Sci. U. S.* **53**, 969–973.

Lewin, R. A., Crothers, D. M., Correll, D. L., and Reimann, B. E. (1964). *Can. J. Microbiol.* **10**, 75–85.

Loeb, T. and Zinder, N. D. (1961). *Proc. Nat. Acad. Sci. U. S.* **47**, 282–289.

Lwoff, A. (1953). *Bacteriol. Rev.* **17**, 269–337.

Lwoff, A., Anderson, T. F., and Jacob, F. (1959). *Ann. Inst. Pasteur* **97**, 281–289.

Maré, I. J., Deklerk, H. C., and Prozesky, O. W. (1966). *J. Gen. Microbiol.* **44**, 23–26.

Marvin, D. A. (1966). *J. Mol. Biol.* **15**, 8–17.

Marvin, D. A. and Hoffmann-Berling, H. (1963). *Nature (London)* **197**, 517–518.

Marvin, D. A. and Hohn, B. (1969). *Bacteriol. Rev.* **33**, 172–209.

Marvin, D. A. and Schaller, H. (1966). *J. Mol. Biol.* **15**, 1–7.

Mayor, H. D. and Hill, N. O. (1961). *Virology* **14**, 264–266.

Menningmann, H. D. (1965). *J. Gen. Microbiol.* **41**, 151–154.

Meynell, E. W. (1961). *Nature (London)* **190**, 564.

Meynell, G. G. and Lawn, A. M. (1968). *Nature (London)* **217**, 1184–1186.

Minamishima, K. T., Takeya, Y., Ohnishi, Y., and Amako, K. (1968). *J. Virol.* **2**, 208–213.

Moody, M. F. (1965). *Virology* **26**, 567–576.

Moody, M. F. (1967a). *J. Mol. Biol.* **25**, 167–200.

Moody, M. F. (1967b). *J. Mol. Biol.* **25**, 201–208.

Nakanishi, Y., Iida, Y., Maeshima, K., Teramoto, T., Hosaka, Y., and Ozaki, M. (1966). *Biken J.* **9**, 149–157.

Okamoto, K., Mudd, J. A., Mangan, J., Huang, W. M., Subbiah, T. V., and Marmur, J. (1968). *J. Mol. Biol.* **34**, 413–428.

Overby, L. R., Barlow, G. H., Doi, R. H., Jacob, M., and Spiegelman, S. (1966a). *J. Bacteriol.* **92**, 739–745.

Overby, L. R., Barlow, G. H., Doi, R. H., Jacob, M., and Spiegelman, S. (1966b). *J. Bacteriol.* **91**, 442–448.

Pootjes, C. F., Mayhew, R. B., and Korant, B. D. (1966). *J. Bacteriol.* **92**, 1787–1791.

Pratt, D., Tzagoloff, H., and Erdahl, W. S. (1966). *Virology* **30**, 397–410.

Prozesky, O. W., Deklerk, H. C., and Coetzee, J. N. (1965). *J. Gen. Microbiol.* **41**, 29–36.

Raimondo, L. M., Lundh, N. P., and Martinez, R. J. (1968). *J. Virol.* **2**, 256–264.

Reeves, P. (1965). *Bacteriol. Rev.* **29**, 24–45.

Rossamondo, E. F. and Zinder, N. D. (1968). *J. Mol. Biol.* **36**, 387–399.

Safferman, R. S. and Morris, M. (1963). *Science* **140**, 679–680.

Schmidt, J. M. and Stanier, R. Y. (1965). *J. Gen. Microbiol.* **39**, 95–107.

Schwartz, F. M. and Zinder, N. D. (1967). *Virology* **34**, 352–355.

Seaman, E., Tarmy, E., and Marmur, J. (1964). *Biochemistry* **3**, 607–613.

Simon, L. D. and Anderson, T. F. (1967). *Virology* **32**, 298–305.

Sinha, N. K., Enger, M. D., and Kaesberg, P. (1965). *J. Mol. Biol.* **12**, 299–301.

Sinsheimer, R. L. (1959). *J. Mol. Biol.* **1**, 43–53.

Smith, K. M., Brown, R. M., Goldstein, D. A., and Walne, P. L. (1966). *Virology* **28**, 580–591.

Stouthamer, A. H., Daems, W. T., and Eigner, J. (1963). *Virology* **20**, 246–250.

Takeya, K. and Amako, K. (1966). *Virology* **28**, 163–164.

Takeya, K., Minamishima, Y., Amako, K., and Ohnishi, Y. (1967). *Virology* **31**, 166–168.
Tromans, W. J. and Horne, R. W. (1961). *Virology* **15**, 1–7.
Twort, F. W. (1915). *Lancet* **ii**, 1241–1243.
Vasquez, C., Granboulan, N., and Franklin, R. M. (1966). *J. Bacteriol.* **92**, 1779–1786.
Watanabe, I. (1964). *Nikon Rinsho* **22**, 243–251.
Weidel, W. and Kellenberger, E. (1955). *Biochim. Biophys. Acta* **17**, 1–9.
Williams, R. C. and Fraser, D. (1953). *J. Bacteriol.* **66**, 458–464.
Zinder, N. D., Valentine, R. C., Roger, M., and Stoeckenius, W. (1963). *Virology* **20**, 638–640.

CHAPTER 8 *Picornaviral Architecture*

*ROLAND R. RUECKERT**

* United States Public Health Service Career Development Awardee.

255

The picornaviruses are among the smallest animal viruses currently known. At present the group includes well over 200 serologically distinct agents. Several subgroups, notably poliovirus and foot-and-mouth disease virus (FMDV), have been extensively studied not only because of their considerable economic importance, but also because of their value as experimental tools for studying problems of macromolecular synthesis, assembly, and pathogenesis in animal cells.

Over half of the presently classified picornaviruses are derived from human hosts. This reflects man's obvious interest in recognizing and controlling human disease agents. The prevalence of these agents is indicated by the fact that there exist, in addition to the 150 odd serologically distinct entero- and rhinoviruses isolated from man, corresponding agents from monkeys, swine, cattle, horses, cats, birds, etc. (Andrewes, 1964). Thus it is probably safe to venture that thousands of picornaviral serotypes exist in nature in nonhuman hosts. In addition to FMDV some of the better known viruses of nonhuman origin include Teschen's disease of swine and the murine encephalomyocarditis (Jungeblut, 1958) and Theiler viruses (Theiler, 1937).

The purpose of this chapter is to summarize data relating to picornaviral structure, to point out that the presence of four nonidentical polypeptide chains with an aggregate weight of about 100 thousand daltons is characteristic of a number of different picornaviruses, and to suggest on the basis of current evidence that all mammalian picornaviruses have a common size and structure and may therefore derive from a common ancestral prototype.

I. Classification

In 1963 the International Enterovirus Study group, as part of a goal of classifying major groups of viruses on the basis of common biochemical and biophysical properties, defined a class of small RNA-containing viruses which they called the "picornaviruses" (*pico* meaning very small, and *RNA* referring to the type of nucleic acid in the virion). The picornavirus group was defined to include those viruses which (a) are small in size (150 to 300 Å in diameter), (b) are insensitive to inactivation by aqueous ether solution, i.e., bear no essential lipid envelope, and (c) contain (single-stranded) ribonucleic acid. On the basis of these simple criteria the picornavirus group clearly ought to have included the plant and bacterial ribonucleic acid-containing viruses. The Committee however subclassified only the animal viruses and left unclear the status and classification of the plant and bacterial viruses. As a consequence the term appears to have been adopted mainly by animal virologists and has come to be applied almost exclusively to a large group of polioviruslike agents. It is to these latter agents only that this chapter is devoted. A brief description of some of the clearly classified agents is summarized below. The reader will find more detailed biological and immunological criteria used in forming these classifications in the literature citations and in Horsfall and Tamm (1965).

A. ENTEROVIRUSES

Most of these agents were isolated from the human alimentary tract in the course of epidemiological research on poliovirus. They include (Committee on Enteroviruses, 1962) the polioviruses (three serotypes); the coxsackie viruses, type A (twenty-four serotypes) and type B (six serotypes); and the so-called echoviruses (Enteric Cytopathogenic Human Orphan, thirty serotypes). The latter viruses generally produce certain syndromes but no specific disease. Echovirus 10 is no longer classified as a picornavirus but as a reovirus because it contains double-stranded RNA; echovirus 28 is now classified as rhinovirus (1A) because it is acid labile (Ketler *et al.*, 1962).

B. RHINOVIRUSES

As might be expected from their name the rhinoviruses are commonly isolated from nasal secretions. They are rarely, if ever, isolated from the gastrointestinal tract. Also known as coryzavirus, ERC-group, murivirus, respirovirus, and Salisbury virus, these agents are most frequently isolated from adults with mild upper respiratory infections (Hamre, 1968; Tyrrell, 1968) and are thus responsible for one of the commonest and most costly viral infections of man, the common cold. About 90 serotypes have been officially recognized (Kapikian *et al.*, 1967; 1971). They are distinguished by their sensitivity to acid pH. In contrast to the enteroviruses the rhinoviruses quickly lose infectivity when incubated at pH 3 to 5 (Ketler *et al.*, 1962; Tyrrell and Chanock, 1963; Hamparian *et al.*, 1963). Interestingly all of the acid-sensitive rhinoviruses so far examined also have a buoyant density in CsCl solution of 1.38 to 1.41, compared to that of enterovirus which is typically 1.33 to 1.34. The significance of this correlation of acid sensitivity and increased buoyant density remains to be clarified. The rhinoviruses are divided into two categories, designated H and M. The H viruses grow in certain human diploid cell lines but not in monkey kidney cultures while the M viruses propagate in both types of cells.

C. Foot-and-Mouth Disease Virus (FMDV)

These agents cause a highly contagious infection of cloven footed animals and are of considerable economic importance in South America, Europe, and Africa where the disease is endemic. Seven serotypes are known: the so-called classical types O, A, and C; the South African Territories types SAT 1, SAT 2, and SAT 3, and a type which occurs in several Asian countries designated Asia 1. FMDV was the first animal disease agent to be identified as a viral (filterable) agent (Loeffler and Frosch, 1898). FMDV resembles the rhinoviruses in that it is rapidly inactivated at pH 3 and has a buoyant density of about 1.43. It has a sedimentation coefficient of 140 S, compared to 160 S for the rhinoviruses. For a more comprehensive treatment of FMDV the reader is referred to Bachrach (1968).

D. Unclassified Viruses

Most of the picornaviruses not mentioned above have been relegated to this category. We shall deal here with only one group, the *cardioviruses* which cannot be ignored because they are among the most thoroughly characterized of all the picornaviruses. These viruses, which have been reviewed by Jungeblut (1958), are also known as the Columbia-SK group or the encephalomyocarditis group. The name cardiovirus (Fenner, 1968) is adopted here because the name makes an unambiguous distinction between a group of viruses and a specific strain. Several members of this group including mengo, encephalomyocarditis (EMC), and Mouse-Elberfeld (ME) viruses have been intensively studied. According to Warren (1965) Columbia-SK, MM, EMC, and mengo viruses are indistinguishable by neutralization, complement fixation, or hemagglutination techniques (Dick, 1949; Warren *et al.*, 1949). ME virus was shown to belong to the cardiovirus group by Hausen and Schäfer (1962). The cardioviruses are closely similar to the enterovirus group but differ in pH stability under special conditions (see Section IV,B). Recent studies on the buoyant density and structural proteins of the cardioviruses indicate that they should probably be classified in a category with the enteroviruses.

E. SUMMARY

Broadly speaking the picornaviruses are divided into two distinct groups on the basis of their pH stability (Table I). Thus if a virus meets the qualification as a picornavirus (i.e., morphologically by its small size and lack of an evelope and chemically by presence of single-stranded RNA) its acid stability at pH 3 is examined by (more or less) "standard" procedures (e.g., Ketler et al., 1962; Tyrrell and Chanock, 1963). If the virus is labile it is a rhinovirus; if stable it is presumably either an enterovirus or it is unclassified.

TABLE I

PICORNAVIRUS SUBGROUPS

	Buoyant[a] density	Acid stability	Sedimentation[c] coefficient
Group 1			
A. Enterovirus	1.32–1.34	Stable pH 3–10	153–160
B. Cardiovirus	1.32–1.34	Stable pH 3–10[b]	151–162
Group 2			
A. Rhinovirus	1.38–1.43	Labile pH <3–7	158
B. FMDV	1.43	Labile pH <7	140[d]

[a] In cesium chloride solutions; see Table III.

[b] Many cardioviruses are unstable in the pH range 5–7 in the presence of 0.1 M chloride ion.

[c] Data from Table III.

[d] But FMDV cosediments with ME virus on a density gradient (G. F. Vande Woude, personal communication).

Both group 1 and group 2 can be further divided according to differences in pH stability, buoyant density, and sedimentation velocity. Thus the cardioviruses are similar to the enteroviruses in virtually all respects except that the former, unlike the latter, are unstable in the pH range 5 to 7 in the presence of physiological saline (Speir et al., 1962; Speir, 1962).

II. Morphology

A. Size

The picornaviral particle is roughly spherical with a diameter variously estimated between 210 and 300 Å (Table II, see later). This wide range of diameters reflects in large part the different techniques employed in making the measurements. Because it requires little material, electron microscopy is the most frequently used tool for studying morphology. Since in conventional electron microscopy the sample must be observed *in vacuo*, some distortion, flattening, and shrinkage may accompany removal of water from the hydrated structure. Though this artifact can be largely avoided by the freeze-drying method (Williams, 1953) or critical point procedure (Anderson, 1951) these methods are not always employed. Thus the wide range in diameters observed by electron microscopy vanishes when the diameters of different viruses are measured in the same laboratory by the same technique (Jamison and Mayor, 1966; Dales and Franklin, 1962).

Perhaps the most satisfactory measurements of diameter of the hydrated particle derive from low-angle X-ray studies of particles in free solution or from interparticle distances in wet crystals, where values of the order of 280 to 300 Å are obtained. Similar values have been obtained by electron microscopy using the negative staining technique and by hydrodynamic methods. The latter technique is, however, highly dependent upon accurate measurements of diffusion coefficients or of partial specific volumes. Until recently each of these parameters has been difficult to obtain because accurate measurements required relatively large amounts of material.

B. Fine Structure

Determining the number and arrangement of subunits in the picornaviral capsid is a challenging task both because of the small size of the subunits and also because the shell is evidently very compact. It seems to be impermeable to negative stains commonly used in electron microscopy and, unlike turnip yellow mosaic virus (Huxley and Zubay, 1960; Nixon and Gibbs, 1960) and bacteriophage ΦX174 (Maclean and Hall, 1962), shows little surface detail. Particles exhibiting real substructure are evidently rare. In 1959, Finch and Klug, on the basis of X-ray studies, suggested that the poliovirion is constructed of 60 identical subunits roughly 60 to 65 Å in size. Horne and Nagington (1959) observed a particle of poliovirus in which substructures, some 50 Å in diameter, were clustered in a group of five about a

TABLE II

Physical Properties of Selected Picornaviruses

Virion	s_{20}	D_{20} ($\times 10^{-7}$)	$\bar{V}$	Mol. wt. ($\times 10^6$)	Diameter (Å)			(%) RNA	E260/280
					EM	Hydrodynamic	X-ray		
Group 1									
A. Enterovirus									
Polio	160[a]	—	0.64[a]	6.8[a]	270[a]	—	304[b]	29[a]	1.69–1.74[a]
	154[c]	1.9[c]		5.5[c]			280[e]		
							305[e]		
Coxsackie	153	—	—	6.4–8.8[d]	280[d]	—	—	—	1.70–1.73[a]
Echo	157[f]	—	—	—	240–280	—	—	—	1.67[a]
B. Cardiovirus									
ME	155[g]	1.8[g]	0.64[a]	5.7[g]	240[g]	240[g]	—	31.5[h]	1.68[h]
EMC	156–160[i]	—	0.73[i]	10.0[i]	300[i]	—	—	30.5[i]	—
	162[j]	1.44[j]	0.678[j]	8.55[j]	300[j]	300[j]	—	31.7[j]	1.71[i]
Mengo	151[k]	1.47[k]	0.70[k]	8.3[k]	262[k]	293[k]	—	21[k]	1.62[k]
Group 2									
A. Rhinovirus									
type 14	158[l]	1.71[l]	0.682	7.1×10^6	230[l]	—	—	30[l]	1.74[l]
B. FMDV									
type A119	140[m]	n	n	(6 9 $\times 10^6$)[o]	230–250	230[m]	—	31.5[m]	—

[a] Schaffer and Schwerdt (1959); Schaffer and Frommhagen (1965).
[b] Finch and Klug (1959).
[c] Anderer and Restle (1964).
[d] Mattern (1962).
[e] Klug and Caspar (1960).
[f] Fabiyi et al. (1964).
[g] Hausen and Schäfer (1962).
[h] Rueckert and Schäfer (1965).
[i] Faulkner et al. (1961).
[j] Burness and Clothier (1970).
[k] Scraba et al. (1967).
[l] McGregor (1969).
[m] Bachrach et al. (1964); Schulze and Gralheer (1964).
[n] Not measured.
[o] Calculated on the assumption that 4.7×10^6 daltons protein (Bachrach and Vande Woude, 1968) represent 69% of the particle weight.

hollow center. Though they were unable to resolve the total number and arrangement of these subunits in the particle they interpreted their results in support of a 60 subunit shell.

In 1964, however, Mayor reported that poliovirus, several echoviruses, and a rhinovirus all exhibited capsomers distributed in a rhombic pattern inconsistent with a 60 subunit shell. She pointed out that such a pattern is more characteristic of a 32 or 42 capsomer shell and interpreted her photographs in favor of 32 clusters of polypeptide chains arranged at the vertices of a rhombic triacontahedron. Agrawal (1966) using the rotation technique of Markham *et al.* (1963) challenged Mayor's interpretation and concluded that poliovirus contained 42 rather than 32 capsomers. Hausen *et al.* (1963) also reported 42 capsomers in ME virus. In an instructive analysis of the fine structure of FMDV, Breese *et al.* (1965) emphasized the difficulty of applying the rotation technique in distinguishing between 32 and 42 subunit structures even when applied to models. They also pointed out that the difficulty is compounded when the technique is applied to virus structures which may be distorted by the dehydration and staining techniques used for electron microscopy. Nevertheless they reported that the structure of their particles best fit a 32 capsomer model. Other workers have confirmed 32 capsomer structures in echovirus (Jamison, 1969), rhinovirus (McGregor, 1969), and bovine enterovirus (Mattson *et al.*, 1969). A difficulty with interpretation of the electron micrographs is the absence of statistical information on the frequency with which such forms are observed. It is probably fair to say that in all reports claiming 60, 42, or 32 capsomers, the particles selected were "favorable" i.e., they occurred rarely.

In summary, picornaviral fine structure is exceedingly difficult to resolve by electron microscopy; the surface of the particle is evidently rather close packed. Nevertheless rare particles do exhibit fine structure and the weight of the evidence seems to favor a 32 capsomer arrangement.

III. Physicochemical Properties

A. THE VIRION

1. *Composition*

Picornaviruses are currently believed to consist only of RNA and protein. The presence of RNA has been demonstrated directly by analysis of purified preparations and also indirectly by metabolic inhibitor experiments. Attempts to detect DNA in purified virus preparations have been consistently negative. Among those examined for DNA are poliovirus (Schwerdt and Schaffer, 1955; Taylor and Graham, 1958), EMC and mengo virus (Faulkner *et al.*, 1961; Krug and Franklin, 1964), and FMDV (Bachrach *et al.*, 1964).

The picornaviruses have no envelope and there is no evidence that they contain any essential lipid. They are generally resistant to inactivation by aqueous ether solutions and many are relatively stable even in the presence of detergents such as deoxycholate and sodium dodecyl sulfate (Cartwright and Thorne, 1959; Mandel, 1962). Krug and Franklin (1964) have shown that purified ^{32}P-labeled mengo virus contains little or no phospholipid or phosphoprotein.

Although it is generally assumed that picornaviruses contain no carbohydrate there have been reports suggesting its presence (Rueckert, 1965; Lerner and Miranda, 1968; Stevens *et al.*, 1968; Halperen *et al.*, 1970). The hemagglutinin of coxsackie virus (Tillotson and Lerner, 1966) and EMC virus (Kunin, 1967) and the infectivity of ME virus (Mak, 1969) is inactivated by periodate. However there is no definitive evidence that any carbohydrate is an integral component essential for viral function and it remains to be demonstrated that carbohydrate is not an adventitious contaminant.

The RNA content of a number of picornaviruses is summarized in Table II. Most workers arrive at a value of 29 to 32% RNA (expressed as sodium ribonucleate). Recently Burness (1970) has made a very careful determination of the RNA content of EMC virus and found by several different types of measurement that the virion contains about 31.7% RNA. Thus it seems likely that the value of 21% (Scraba *et al.*, 1967) reported for the serologically related mengo virus is incorrect.

The extinction coefficients of several picornaviruses and of the RNA and protein isolated from them are summarized in Table III.

2. *Particle weight*

Given the sedimentation velocity, S, the partial specific volume, $\bar{v}$, and the diffusion coefficient, D, the molecular weight M, of a particle can be calculated from the Svedberg equation (1), where R is the gas constant, T, the absolute temperature, and p is the solvent density

$$M = \frac{RST}{D(1 - \bar{v}p)} \tag{1}$$

TABLE III

EXTINCTION COEFFICIENT OF SOME PICORNAVIRUSES AND OF THEIR RIBONUCLEIC ACIDS
AND CAPSID PROTEINS

Virus	1% $E(\lambda)^a$		
	Virion (260 mm)	RNA (260 mm)	Protein (280 mm)
Polio	82^b (1)	—	—
ME	78^e (2)	222^c (2)	14.9^d (3)
ME	85^e (4)	251^e (4)	16.4^d (4)
EMC	77.4^f (5)	221^f (6)	14.8^f (7)
Mengo	76.3^g (?)	—	16.9^g (?)
FMDV	76^h (8)	220^h (?)	11.1^i (?)

[a] The solvents used, noted in parentheses, were as follows: (1) 2% NaCl, 0.063 M potassium phosphate, pH 7.0; (2) 0.14 M NaCl, 0.02 M sodium phosphate, pH 7.2; (3) 0.002% acetic acid; (4) 0.01 M NaOH; (5) 0.1 M KCl, 0.02 M-sodium phosphate, pH 8.0; (6) 0.1 M tris-chloride, pH 7.2; (7) 0.1 M NaOH; (8) 0.05 M sodium phosphate, pH 7.2.

[b] Charney et al., 1961.

[c] Rueckert and Schäfer, 1965.

[d] Rueckert, 1965.

[e] Rueckert, 1968.

[f] Burness, 1970.

[g] Scraba et al., 1969; Kay et. al., 1970.

[h] Bachrach et al., 1964.

[i] Bachrach and Vande Woude, 1968.

There is a considerable spread in the apparent molecular weights measured for a number of picornaviruses (Table II). The problem which must be decided is, does this spread represent real differences between picornaviruses, or does it reflect experimental errors? Sedimentation coefficients are relatively easy to measure precisely; diffusion coefficients and partial specific volumes are not.

The reported sedimentation coefficients of most picornaviruses are very similar (153–162 S). Even small differences vanish in direct cosedimentation experiments in density gradients (Schaffer and Frommhagen, 1965; Medappa et al., 1971). Interestingly, FMDV, which is reported to sediment at 140 S in the analytical ultracentrifuge, cosediments with ME-virus (160 S) on sucrose gradients (G. F. Vande Woude, personal communication). The reason for this apparent discrepancy in sedimentation velocity of FMDV remains to be clarified. However, the weight of the evidence immediately suggests that different picornaviruses are similar, if not identical, in size.

Serious attempts to measure the partial specific volumes and diffusion coefficients of picornaviruses have been undertaken by only a few workers. Because of the limited amounts of available material all of these workers have used the same basic method, i.e., sedimentation in various concentrations of D_2O in H_2O and extrapolating the plot of sedimentation velocity versus D_2O concentration to zero velocity. Appropriate corrections must be made for binding of deuterium to the protein. The method requires a long extrapolation and yielded values (Table II) of 0.64 for poliovirus (Schaffer and Schwerdt, 1959), 0.70 for mengo virus (Scraba et al., 1967), 0.678 for EMC virus (Burness and Clothier, 1970), and 0.682 for rhinovirus 14 (McGregor, 1969). As pointed out by Schaffer and Schwerdt (1955) the value of 0.64, which was determined on only very limited amounts of poliovirus was probably too low; a value of the order of 0.68 is in better agreement with the chemical composition of the virus. Recalculation of the molecular weights of poliovirus and ME virus using the latter value for $\bar{V}$ and s_{20} and d_{20} values from Table II yields corrected particle weights of 6.2 and 6.4×10^6 daltons, respectively. Similarly the $\bar{V}$ of 0.73 used by Faulkner et al. (1961) in computing the weight of EMC virus as 10 million daltons is probably too high.

The values of diffusion coefficients of picornaviruses also show some variation (Table II). Thus Anderer and Restle (1964) and Hausen and Schäfer (1962) found values of 1.9 and 1.8×10^{-7} cm²/sec for poliovirus and ME virus, respectively. Scraba et al. (1967) and Burness and Clothier (1970) recently found somewhat lower values of 1.47 and 1.44×10^7 for mengo and EMC viruses. All of these values were determined by optical techniques. McGregor (1969) found an intermediate value of 1.71×10^{-7} for rhinovirus using a surface diffusion method.

Taken together it is clear that there is as much variation in the molecular weight values obtained with the immunologically related (presumably almost identical) cardioviruses as there is between the apparent molecular weights of unrelated viruses such as polio-, EMC, and rhinoviruses; with the possible exception of FMDV there is no reason to believe that the various picornaviruses differ in particle weight. On the grounds discussed above, and from the experimental measurements summarized in Table II it appears that the true particle weight of the picornaviruses falls somewhere between 6.2 and 8.6×10^6 daltons.

3. *Buoyant Density*

The enteroviruses and cardioviruses band in cesium chloride solutions at 1.32 to 1.34 while the rhinoviruses band at about 1.38 to 1.43 (Table IV). FMDV is the densest of the picornaviruses ($d = 1.43$). The empty capsids of FMDV however have about the same buoyant density (1.29 to 1.30) as those of the enteroviruses.

TABLE IV

BUOYANT DENSITY OF SOME PICORNAVIRUSES AND THEIR CAPSIDS IN CESIUM CHLORIDE SOLUTION

Group	Virion	Capsid	Reference[a]
A. *Enterovirus*			
Poliovirus type 1	1.34	1.29	1, 2
Poliovirus types 2, 3	1.34	—	1
Echovirus 4	1.35	—	2
Echovirus 6	1.33	1.30	3
Echovirus 7	1.34	—	1
Echovirus 12, 19	1.33	1.29	4, 5
Echovirus 16	1.32	—	2
Coxsackie virus A-9, 10, 21, 24	1.34	—	6, 7
Coxsackie virus B-2	1.33	—	2
B. *Cardiovirus*			
EMC	1.33	—	8, 9
EMC (density mutant)	1.36	—	9
ME, Columbia SK, MM	1.34	—	10, 12
Mengo	1.32[b]	—	11
C. *Rhinovirus*			
Rhino 1A, 1B, 2, 13, 14, 17	1.38–1.41	—	13, 14, 15
Rhino HGP	1.41	—	15
D. *Foot-and-mouth disease virus*			
Group A119	1.43	1.31	16, 17

[a] 1. Schaffer and Frommhagen (1965).   2. Jamison and Mayor (1966).   3. Forsgren (1968).   4. Halperen *et al.* (1964).   5. Fabiyi *et al.* (1964).   6. Mattern (1962).   7. Frommhagen and Martins (1961).   8. Faulkner *et al.* (1961).   9. Goodheart (1965).   10. Rueckert and Shäfer (1965).   11. Scraba *et al.* (1967).   12. Rueckert, unpublished observations.   13. Gerin *et al.* (1968).   14. McGregor and Mayor (1968).   15. Dans *et al.* (1966).   16. Trautman and Breese (1962).   17. Graves *et al.* (1968).

[b] Measured in cesium sulfate solution.

Buoyant density is determined to some extent by the nature of the solvent. For example Jamison and Mayor (1966) noted that a number of enteroviruses (poliovirus, type 1 and echoviruses 4, 6, 16, 19, and 23) band at a density of only 1.21 to 1.23 in potassium tartrate solution. Both virions and empty capsids had the same buoyant density in this solvent. Gerin *et al.* (1968) reported that rhinovirus 1A (echo 28) bands at a density of 1.26 in potassium citrate solution.

The reason for the higher buoyant density of the rhinoviruses and FMDV in cesium chloride solution is unclear. Some reasons which have been suggested (McGregor and Mayor, 1968) are (a) a greater percentage of RNA in the rhinoviruses than in the enteroviruses either because of a large genome or because they contain less structural protein in their capsids, (b) increased binding of heavy cesium ions, or (c) decreased hydration of the particles.

Early studies suggested that rhinoviruses may have a large genome (McGregor and Mayor, 1968); however subsequent work (McGregor, 1969) indicates that the composition of the rhinovirion is that same as that of the other enteroviruses (see also Table II). The buoyant density of poliovirus is very close to that expected for the hydrated particle (assuming 22% bound water, $d = 1.00$; 20% ribonucleic, $d = 0.55$; 58% protein, $d = 0.74$; Schaffer and Frommhagen, 1965). Burness (1970) came to a similar conclusion about EMC virus. This observation suggests that the particle binds relatively little cesium ion; if so then the ionizing groups of the RNA genome which represent about 7500 to 8000 anionic phosphate residues must be inaccessible for exchange with cesium ion. This suggests that the RNA may be rather tightly buried within the shell. A looser packing of the RNA between the protein subunits of the rhinoviruses might allow sufficient cation exchange of ceium ions to account for the greater buoyant density (McGregor *et al.*, 1966). However, a number of other possibilities could also explain this behavior and further investigation is required to settle the issue.

B. RIBONUCLEIC ACID

1. Composition

The picornaviruses contain a single strand of ribonucleic acid which can be extracted from the virion in infectious form by a variety of methods. These include treatment with phenol (Colter 1958; Colter *et al.*, 1957), acidic sodium dodecyl sulfate (Mandel, 1964), phenol-SDS mixtures (Bachrach, 1960; Scherer and Darnell, 1962), heat (Brown and Cartwright, 1961; Bachrach, 1961; Darnell *et al.*, 1961), urea (Sanders, 1964), and even treatment with mercurials (Philipson, 1964). Of these procedures extraction with phenol or phenol-SDS mixtures is generally most satisfactory for chemical studies.

The infectivity of extracted poliovirus RNA is highly sensitive to endonucleases such as snake venom phosphodiesterase or pancreatic ribonuclease, but is insensitive to alkaline phosphodiesterase (Holland *et al.*, 1960). The latter observation suggests that phosphate end groups are not required for infectivity of the RNA. In accord with this observation Leppla (1969) has shown that the 3'-terminal nucleotide of ME virus can be oxidized by periodate without a preliminary phosphatase digestion. This result implies that the 3'-terminus is unphosphorylated. Leppla also identified the 3'-terminal base as adenosine. The infectivity of ME virus RNA is unaffected by periodate treatment (Mak, 1969).

The single-stranded nature of picornaviral RNA is reflected by its base composition since the complementary base pairs A/U and G/C do not

TABLE V

BASE COMPOSITION OF PICORNAVIRAL RNA

Virion	Molar %				Method	Reference
	G	A	C	U		
Polio (3 types)	24	29	22	25	Acid	Schaffer *et al.* (1960)
Polio type 1	23.2	29.3	23.8	23.8	Alkali	Summers and Levintow (1965)[a]
Polio type 1	23.4	30.3	22.6	24.0	Alkali	Holland (1962)[b]
Coxsackie A9	28	27	20	25	Acid	Mattern (1962)
Coxsackie A10	28	28	21	23	Acid	Mattern (1962)
Coxsackie B1	24	29	23	24	Alkali	Holland (1962)
Mouse-Elberfeld	24	25	24	27	Acid	Rueckert and Schäfer (1965)
EMC	24	27	24	26	Acid	Faulkner *et al.* (1961)
EMC	24	26	26	24	Acid	Burness (1970)
Mengo	23	26	25	26	Alkali	Scraba *et al.* (1969)
Rhino, type 2	19.5	34.0	20.2	26.3	Alkali	Brown *et al.* (1970)
Rhino, type 14	19	35	19	27	Acid	McGregor (1969)
Foot-and-mouth; A-119	24	26	28	22	Acid	Bachrach *et al.* (1964)

[a] Based on thirteen determination with six virus preparations.
[b] Mean of six determinations.

generally follow the one-to-one stoichiometry required for double-stranded RNA. The base compositions of the three types of poliovirus are very similar (Table V); in fact, there is evidence that a one-quarter to one-third of the sequence may be common to each of them (Young *et al.*, 1968). The other enteroviruses and cardioviruses appear to differ somewhat from poliovirus but the significance of this difference is difficult to evaluate since somewhat different techniques are employed by various investigators. Even carefully performed analyses from two different laboratories on alkali hydrolyzed type 1 poliovirus RNA appear to differ in cytosine by 1.2 mole %. The variation is probably larger for many other entries for which fewer determinations were performed. There appears however little doubt that the base compositions of rhinovirus and FMDV differ significantly from each other and also from the enterovirus-cardiovirus group.

2. *Molecular Weight*

Given a particular size range of 6.2 to 8.6×10^6 daltons for the picornavirion and an RNA content of 29 to 32%, one computes a molecular weight of 1.8 to 2.8×10^6 daltons for the sodium salt of the RNA molecule. Until recently the accepted value has been about 2×10^6 daltons. This figure was supported by physical studies on RNA extracted from picornaviruses. Thus Anderer and Restle (1964) estimated the size of poliovirus RNA to be about 2×10^6 daltons on the basis of sedimentation and viscosity measurements. Scraba *et al.* (1967) obtained a value of 1.7×10^6 daltons for mengo virus RNA from sedimentation and diffusion coefficients.

Molecular weights of the RNA, estimated by the less rigorous procedure of comparative sedimentation with a standard RNA of known size under similar conditions, has also been obtained. For example the molecular weight of TMV RNA has been determined by careful sedimentation and viscosity measurements to be about 1.94×10^6 daltons (Boedtker, 1959; for a review see Fraenkel-Conrat, 1962). Using the empirical relation:

$$M = 1550(s_{20,w})^{2.1}$$

determined by Spirin (1963) for TMV RNA in 0.1 M NaCl, 0.01 M EDTA one calculates for a single-stranded poliovirus RNA with a sedimentation coefficient of 37 S in the same solvent (Holland *et al.*, 1960) a molecular weight of about 3×10^6 daltons. By variations of this comparative method molecular weights of the order of 2×10^6 were estimated for ME RNA (Hausen and Schäfer, 1962) and 3×10^6 for FMDV RNA (Strohmaier and Mussgay, 1959) and EMC RNA (Burness *et al.*, 1963; Montagnier and Sanders, 1963). Because, however, it is known (Strauss and Sinsheimer, 1963) that not all single-stranded RNA molecules possess similar configurations in solutions of moderate ionic strength such estimates have generally been considered to represent only approximations to the correct molecular weight.

Reexamination of viral RNA by newer techniques supports a molecular weight value of the order of 2.5 to 2.7 million daltons for picornaviral RNA. Thus Granboulon and Girard (1969) reported the contour length of both single- and double-stranded RNA of poliovirus to be equivalent to about 2.6×10^6 daltons; and a size of 2.7×10^6 daltons for EMC RNA (unpublished results of Granboulon and Montagnier, cited by the above authors). Fenwick (1968) estimated the size of EMC RNA by sedimentation studies in denaturing solvents to be about 2.4×10^6. Tannock et al. (1970) found a similar value for poliovirus RNA using the method of electrophoresis on polyacrylamide gel.

Poliovirus and rhinovirus when heated briefly at 45° to 50°C extrude long unbranched strands which are believed to represent ribonucleoprotein (RNP) because they are susceptible to pancreatic ribonuclease only after digestion with proteolytic enzymes (McGregor and Mayor, 1968). The strands extruded from rhinovirus are twice as long as those from poliovirus. This observation, at first interpreted as evidence that rhinovirus RNA is twice as large as that of poliovirus, is now believed to be an artifact since recent determination of the size and RNA content of rhinovirus (Table II) show that its RNA is probably about the same size as that of poliovirus (McGregor, 1969; McGregor and Mayor, 1971; Brown et al., 1970). The greater length of rhinoviral RNA strands is not understood but may be due to some type of specific end-to-end aggregation which does not occur with poliovirus.

C. Protein

1. Composition

The amino acid composition of whole virus is shown in Table VI. With the possible exception of histidine and methionine the amino acid compositions are rather similar. The differences in cysteine content of poliovirus probably reflect analytical difficulties.

A number of workers have also determined the amino acid composition of isolated virus protein (Table VII) freed of virus nucleic acid. There is a suggestion that rhinovirus and FMDV have higher threonine and histidine contents than the cardioviruses, ME, and mengo viruses, but the composi-

TABLE VI

AMINO ACID COMPOSITION OF PICORNAVIRIONS (WHOLE VIRUS)[a]

Amino acid	Moles per 100 moles of amino acid recovered				
	Polio (Type 1)		EMC	ME	FMDV A-119
	b	c	d	e	f
Aspartic	11.9	11.3	9.3	10.4	10 4
Threonine	9.1	9.2	8.6	9.9	10.9
Serine	7.0	8.1	7.9	8.3	7.0
Glutamic	7.7	8.0	8.7	8 0	8.8
Proline	7.2	7.1	7.5	8 0	5.4
Glycine	6.8	8.4	8.7	8.4	7.8
Alanine	7.8	7.6	7.1	7.9	8.3
Cysteine	(0.8)	(0.28)	1.5	1.1	0.53
Valine	7.2	6.0	6.8	6.0	6.9
Methionine	1.5	2.6	1.9	1.8	1.8
Isoleucine	4.8	4.4	3.9	4.0	3.6
Leucine	8.5	8.1	8.3	7.7	7.3
Tyrosine	3.9	3.8	3.8	4.2	4.8
Phenylalanine	4.4	4.0	5.1	5.6	3.7
Lysine	4.7	4.0	4.5	3.8	4.6
Histidine	2.4	2.1	1.4	1.9	3.3
Arginine	4.7	4.6	3.6	3.6	3.9
Tryptophan	—	—	—	—	0.9

[a] All analyses were performed by acid hydrolysis of whole virus, i.e., without first removing the RNA. Except for FMDV, the values for glycine are probably about 20% too high because of acid hydrolysis of nucleic acid purines. In the case of FMDV this correction has already been applied not only for glycine but also for threonine, serine, tyrosine, phenylalanine, and proline; the latter values were corrected by increasing the analytical values of a 41-hour hydrolyzate by 15, 7.5, 6, 4, and 4%, respectively. Cysteine is partially, and tryptophan completely, destroyed during acid hydrolysis but can be determined by special procedures (see cited literature).

[b] Levintow and Darnell (1960).
[c] Munyon and Salzman (1962).
[d] Faulkner et al. (1961).
[e] Rueckert and Shäfer (1965).
[f] Bachrach and Polatnick (1966).

TABLE VII

AMINO ACID COMPOSITION OF PICORNAVIRIONS (ISOLATED PROTEIN)

Amino acid	Moles per 100 moles of amino acid recovered			
	FMDV A-119[a]	ME[b]	Mengo[c]	Rhinovirus type 14[d]
Aspartic	10.3	10.0	10.3	10.6
Threonine	10.7	9.7	9.1	11.1
Serine	6.4	8.0	7.4	7.7
Glutamic	8.6	7.8	8.4	8.3
Proline	5.9	8.3	7.9	5.9
Glycine	8.1	7.0	7.3	7.8
Alanine	8.7	7.5	7.3	5.7
Cysteine	0.82	1.2	1.5	—
Valine	6.7	7.0	8.5	7.4
Methionine	1.5	1.8	1.1	2.0
Isoleucine	3.2	3.8	3.5	6.8
Leucine	7.4	7.6	7.4	9.4
Tyrosine	4.9	4.1	3.8	3.3
Phenylalanine	3.8	5.4	5.4	3.0
Lysine	4.6	3.8	4.1	4.2
Histidine	3.4	1.8	1.9	2.6
Arginine	3.0	3.5	3.8	3.1
Tryptophan	1.0	2.3	1.5	1.2

[a] Bachrach and Vande Woude (1968).
[b] Rueckert (1965).
[c] Scraba et al. (1969).
[d] McGregor (1969).

tions are not strikingly different. Comparison of amino acid analysis performed on the same virus by both procedures show satisfactory agreement (ME virus; FMDV). This observation is important because picornaviral proteins are known to be composed of multiple polypeptide chains (see below) some of which may be selectively lost during isolation by the phenol technique (Rueckert, unpublished observations). Thus it is probably preferable, and certainly simpler, to conduct amino acid analyses on whole virus rather than on isolated protein when only limited amounts are available for analysis.

2. *Polypeptide Elements*

The polypeptide chains of a few picornaviruses have been examined by ultracentrifugal and electrophoretic techniques. By sedimentation they appear to be homogeneous. The molecular weights of the polypeptides of poliovirus (Maizel, 1963; Anderer and Restle, 1964) and of ME virus (Rueckert, 1965) and EMC virus (Burness and Walters, 1967) fall between 26,000 and 30,000 daltons. A higher value of about 40,000 reported by Boeyé (1965) for poliovirus may have been due to difficulties in correcting for sodium dodecyl sulfate associated with the peptides. FMDV protein has an apparent molecular weight, by gel filtration of about 32,000 daltons after maleylation at pH 9 with maleic anhydride (Vande Woude and Bachrach, 1969).

Although it appears to be homogeneous in the ultracentrifuge enteroviral capsid protein can be resolved by electrophoresis into several types of nonidentical polypeptide chains (Maizel, 1963; Summers *et al.*, 1965; Rueckert, 1965; Rueckert and Duesberg, 1966; Jacobson *et al.*, 1970). Recent results indicate that the capsids contain four (Maizel and Summers, 1968) or five types of polypeptide chains (Rueckert *et al.*, 1969).

Multiple electrophoretic components have also been observed in purified preparations of FMDV. Vande Woude and Bachrach (1968) at first attributed this behavior to aggregation of a single polypeptide but subsequently suggested that this virus too has multiple polypeptide chains (Vande Woude and Bachrach, 1969; 1971) Wild *et al.* (1969) using a double-label technique, have shown that at least some of the multiple bands differ in amino acid composition. LaPorte (1969) reported that the virus contains three N-terminal amino acids, leucine, isoleucine, and threonine. Recent studies on rhinovirus 1A (Medappa *et al.*, 1971) and rhinoviruses 2 and 14 (Korant *et al.*, 1970) indicate that they too contain nonidentical polypeptide chains. Typical electrophoretic profiles of some labeled picornaviruses are illustrated in Fig. 1.

To a first approximation the electrophoretic mobility of polypeptides on polyacrylamide gels containing sodium dodecyl sulfate is directly proportional to the logarithm of its molecular weight (Shapiro *et al.*, 1967; Weber and Osborn, 1969; Dunker and Rueckert, 1969). The electrophoretic profiles of various picornaviruses differ slightly in relative mobility. The molecular weights of various picornaviral chains estimated in this way are summarized in Table VIII. The sum of the molecular weights of the nonidentical polypeptides ($\varepsilon + \alpha + \gamma$) or ($\alpha + \beta + \gamma + \delta$) of the enteroviruses, cardioviruses, and one rhinovirus is consistently about 96,000 daltons.

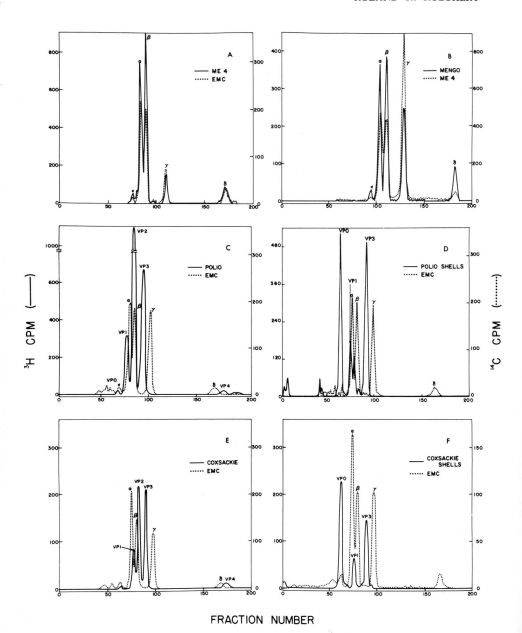

FRACTION NUMBER

FIG. 1. Radioactivity profiles of the polypeptide chains of several picornaviruses. Electrophoresis was in 10% polyacrylaminde gel according to Summers *et al.* (1965) except that 0.1 *M*-sodium beta-mercaptopropionate, which improves resolution of the alpha and beta chains of cardioviruses, was added to the electrode buffer (Hall, 1970; Radloff and Rueckert, unpublished results).

TABLE VIII

MOLECULAR WEIGHT OF PICORNAVIRAL CAPSID POLYPEPTIDES[a]

Chain nomenclature			Poliovirus					Cardiovirus			Coxsackie virus B1 (4)	Rhinovirus IA (7)
			Type 1				Type 2 (2)					
			Mahoney			CHAT (4)		ME (5)	ME, EMC, Mengo (4)	Mengo[b] (6)		
(1, 3)	(5)	(6)	(1)	(2)	(3)							
NCVP6(VPO)	ε	Early	41	—	40	41	—	41	40	42–54	40	37
VP1	α	I[c]	35	32	35	35	32	33	34	—	32	34
VP2	β	I[c]	28	28	28	30.5	25	30.5	30.5	31	30	29.5
VP3	γ	II	24	23	23	27	24	25	24	28	27	26
VP4	δ	III	6[d]	—	—	<10	—	10	7	20	<7	7

[a] Values in thousands of daltons; determined by electrophoresis in SDS-containing polyacrylamide gels.

[b] M mengo contains an additional component weighing 10,000 daltons (6).

[c] Peak I probably represents two unresolved components.

[d] May represent two unresolved components (3).

[e] References in parenthesis as follows: (1) Maizel and Summers, 1968; (2) Holland and Kiehn, 1968; (3) Jacobson et al., 1970; (4) Hall, 1970; Rueckert, unpublished experiments; (5) Dunker and Rueckert, 1969; (6) O'Callaghan et al., 1970; (7) Medappa et al., 1971.

There is good evidence that the electrophoretic components of poliovirus and ME virus represent nonidentical polypeptides. Thus the respective polypeptides of both viruses show distinctive tryptic mapping patterns (Maizel, 1963; Summers *et al.*, 1965; Rueckert and Duesberg, 1966; Jacobson *et al.*, 1970) and amino acid composition (Maizel and Summers, 1968; Burness and Walters, 1967; Stoltzfus and Rueckert, 1971). Preliminary data on the amino acid composition of several polypeptide chains from EMC and ME viruses are shown in Table IX. Direct comparison of the amino acid composition of the EMC chains with those in ME virus is not presently meaningful because the polypeptides were separated in two different electrophoretic systems and it is not clear which chains are analogous. However, these data support the notion that the chains are nonidentical.

Recent studies in this laboratory show that all five of the polypeptide chains from ME and EMC virus differ in primary sequence. The epsilon

TABLE IX

AMINO ACID COMPOSITION OF POLYPEPTIDE COMPONENTS OF ME AND EMC VIRUS

| Amino acid | Moles of amino acid per 100 moles of amino acid recovered | | | | | | |
| | EMC[a] | | | ME[b] | | | |
	1	2	3	alpha	beta	gamma	delta
Asp	8.9	10.1	8.6	9.2	10.1	7.4	19.6
Thr	7.0	10.0	9.6	7.7	9.7	9.9	4.2
Ser	13.0	10.6	8.7	9.9	8.1	9.9	16.6
Glu	10.7	9.5	9.0	9.7	10.1	6.9	8.5
Pro	5.3	8.8	7.1	7.0	5.1	8.1	4.3
Gly	15.1	10.5	9.8	10.7	8.9	9.1	9.0
Ala	6.7	7.3	8.4	6.0	7.7	8.7	7.6
Val	5.4	8.1	7.0	6.9	6.9	5.4	2.9
Ilu	3.1	3.6	4.2	2.9	3.5	5.1	3.7
Leu	6.3	8.1	7.0	6.9	7.8	6.6	8.7
Tyr	1.7	3.4	3.2	2.8	2.9	4.2	3.1
Phe	4.4	8.1	5.4	5.6	3.9	4.3	4.2
Lys	4.9	5.2	4.5	4.7	3.5	3.9	2.2
His	3.0	2.1	2.4	1.8	2.9	1.1	0.3
Arg	4.4	4.1	4.1	3.7	4.8	2.8	2.0
Trp	—	—	—	—	—	1.6	—
CySH	—	—	—	1.6	0.7	1.2	0
Met	—	—	—	0.9	1.5	2.1	1.0

[a] Burness, and Walters (1967).
[b] C. M. Stoltzfus, and R. R. Rueckert, 1971.

chain isolated from the virus produces cyanogen bromide cleavage fragments similar to those of the beta and delta chains (Stoltzfus and Rueckert, 1971). These results and those of Jacobson *et al.* (1970) leave little doubt that each of the chains in the electrophoretic patterns of ME virus and poliovirus corresponds to a different polypeptide. The fragmentation studies also imply that the epsilon chain is related in sequence to the two smaller chains, beta and delta, supporting the notion that the latter two chains are cleavage products of the epsilon chain.

IV. Antigens and Viral Derivatives

A. ANTIGENS

The picornaviruses exhibit at least four distinct antigens, three of which are capsid related. The other appears to be determined by viral RNA-replicating enzyme. As might be expected from their importance as disease agents the most thoroughly studied viruses are poliovirus and foot-and-mouth disease virus. We shall discuss first the entero- and cardiovirus antigens, then the antigens of FMDV.

1. *Enteroviruses*

Purified poliovirus preparations contain two distinct type-specific antigens which are characteristic of the ribonucleic acid containing virion (called N or D antigen) and of empty capsids (called H or C antigen) (Mayer *et al.*, 1957; Hummeler *et al.*, 1962). The virions can be completely converted from N to H antigenicity by a variety of denaturing conditions including heat, ultraviolet light, high pH, mercurials, phenol, and dessication (LeBouvier, 1955, 1959). Heating poliovirus at 56°C releases RNA from the particle and simultaneously converts it to H antigen (Hummeler and Hamparian, 1958; Roizman *et al.*, 1959; Hummeler *et al.*, 1962). Acridine orange stains the N but not the H antigen in agar precipitin tests (Ohlson, 1963) presumably because of loss of RNA from the H antigen.

The virion can also be converted to C antigenicity without loss of its encapsulated RNA by shorter heating at 56°C or by treatment with ultraviolet light (Roizman *et al.*, 1959; LeBouvier, 1959; Katagiri *et al.*, 1967). The surface rearrangement accompanying the N to H transition is accompanied by inability of the heated poliovirus to attach to host cells (Taylor and Graham, 1959; Katagiri *et al.*, 1968). Poliovirus preparations containing mixtures of the two antigens can be resolved chromatographically into fractions with a full complement of N reactive sites with a higher plating efficiency and a noninfectious fraction lacking N reactive sites (Tumilowicz and Hummeler, 1964; Schmidt, 1963, 1964). The N to H transition is accompanied by a change in the permeability of the shell to phosphotungstic acid (PTA). The N particles are impermeable to the stain while the H particles are readily penetrated and appear "empty" by electron microscopy (Hummeler et al., 1962). Dimmock (1967) concluded that the penetrability depends upon the reorientation of the subunits in the protein coat, not upon the nucleic acid content. He showed that it is possible under special conditions to produce apparently "full" particles from apparently "empty" ones.

The H or C antigen also occurs naturally in infected cell extracts. It is frequently encountered during the purification of enteroviruses including poliovirus (Mayer et al., 1957), various coxsackie viruses (Schmidt et al., 1963; Frommhagen, 1965), and echoviruses 6 (Forsgren, 1968), 12 (Halperen et al., 1964), and 19 (Fabiyi et al., 1964). The H antigen from infected cells corresponds to empty capsids which can be observed in electron micrographs (Horne and Nagington, 1959). It is devoid of RNA (Van Elsen and Boeyé, 1966), has a buoyant density in cesium chloride gradients of about 1.30, compared to 1.34 for the complete virion, and is assembled at about the same time as mature virus (Scharff and Levintow, 1963; Scharff et al., (1964) in proportions related to the type of host cell (Roizman et al., 1958a). Penman et al. (1964) detected these structures in cytoplasmic extracts from poliovirus-infected HeLa cells sedimenting in sucrose density gradients at about 73 to 75 S. The sedimentation coefficient has been variously estimated by others to be 80 to 90 S (Roizman et al., 1958b), 78 S (Boeyé and Van Elsen, 1967), and 90 S (Drees, 1965; Drees and Borna, 1965). It has been suggested that 74 S component may represent an intermediate in the morphogenesis of poliovirus (Jacobson and Baltimore, 1968b).

There is little or no immune cross-reaction between the C and D components (Roizman et al., 1957). Monotypic antibody against either of the two components precipitates only the homologous class of particles from a mixture (Hummeler et al., 1962). Thus the entire surface of these particles seems to possess either one or the other, but not both, antigenic specificities. Coxsackievirus is very similar to poliovirus in that its RNA-free empty capsids have H antigenicity. However the "coreless" particles from echo 12 virus, in contrast to those of poliovirus and coxsackievirus, appear to possess N antigenicity (Halperen et al., 1964).

In 1965, Watanabe et al., described a "10 S" particle, now known to sediment at about 14 S (Phillips, 1969) with a primary structure resembling that of the H and N particles but with an immunologic specificity unique from either, here called S antigen.

Still another antigen, designated here "G" antigen was described by Scharff *et al.* (1964). Poliovirus can be degraded by 6.35 *M* guanidine hydrochloride to a protein, probably polypeptide chains, sedimenting at about 2 S. This material produced an antibody which reacts with an antigen from infected cell extracts which sediments at about 5 S. There appears to be no 2 S G antigen in infected poliovirus extracts. The antigenic relationship between S and G antigen has not been clarified, i.e., it is not known whether they have the same or different antigenic specificity.

In a study of the polypeptide composition of these antigens, Phillips *et al.* (1968) showed that the 74, 14, and probably also the 5 S component, contained roughly equal amounts of three chains, NCVP6 (also known as VP0), VP1, and VP3 but lacked two chains, VP2 and VP4, which are present in the mature virion (Maizel *et al.*, 1967). Two capsid related antigens sedimenting at about 6 and 14 S, respectively, have been identified by serum-blocking power in extracts of EMC virus-infected Krebs ascites tumor cells. The 14 S material was estimated by an immunological diffusion technique to weigh about 420,000 daltons (Kerr *et al.*, 1965).

2. *Foot-and-Mouth Disease Virus*

Four virus-specific antigens, designated 140 S, 75 S, 12 S, and VIA have been associated with FMDV. The 140 S antigen corresponds to the virion which has a diameter of the order of 230 to 250 Å (Table II). The virus is readily inactivated at pH 6 (Randrup, 1954; Cartwright and Thorne, 1958) to generate infectious RNA (Mussgay, 1959) and smaller subunits about 70 to 80 Å in diameter (Bradish *et al.*, 1960; Bachrach and Breese, 1958; Breese and Bachrach, 1960) which sediment at 12 to 14 S (Cowan and Graves, 1966; Trautman *et al.*, 1959). The virus is similarly dissociated by heating to 56°C (Brooksby, 1953; Brown and Crick, 1958; Bachrach, 1960). The 12 S antigen is appreciably more stable to pH, temperature, and enzyme treatment than the virion (Polatnick, 1960). The 140 S and 12 S antigens elicit separate species of antibodies with different electrophoretic mobilities diameter (Bradish *et al.*, 1960; Bachrach and Breese, 1958; Breese and Bachrach, 1960) which sediment at 12 to 14 S (Cowan and Graves, 1966; Trautman *et al.*, 1959). The virus is similarly dissociated by heating to 56°C (Brooksby, 1953; Brown and Crick, 1958; Bachrach, 1960). The 12 S antigen is appreciably more stable to pH, temperature, and enzyme treatment than the virion (Polatnick, 1960). The 140 S and 12 S antigens elicit separate species of antibodies with different electrophoretic mobilities (Brown, 1958). Antiserum absorbed with homologous 12 S antigen contains a residual antibody which is specific for 140 S virions (Brown and Crick,

1958). Purified 140 S virions can combine with both of these antibodies (Cartwright, 1962). Only 140 S antigens are fully type specific; the 12 S antigen shows some cross-reaction with, but is not precipitated by, heterotypic antiserum.

Extracts from infected cells contain empty capsids (Breese, 1968) which sediment in sucrose density gradients at about 65 to 75 S (75 S antigen). These particles have a buoyant density in cesium chloride density gradients of about 1.31, compared with 1.43 for the virion, and unlike the virion are stable at pH 6 (Graves *et al.*, 1968; Planterose and Ryan, 1965). The amount of 75 S capsids produced varies with different virus strains; high yields are obtained with type A-1 virus, low yields with many other strains (Graves *et al.*, 1968). These particles are precipitated by homologous antibodies against both 140 S and 12 S antigens; thus it evidently has antigenic sites for both antibodies.

The virus-infection-associated (VIA) antigen appears to represent enzymically inactive FMDV-specific RNA replicase since its antibody inhibits polymerase activity (Polatnick and Arlinghaus, 1967). It sediments at less than 4.5 S on density gradients (Cowan and Graves, 1966; Cowan, 1966) and its complement fixing activity is destroyed by heating 30 minutes at 50°C or by treatment at pH 5.0. It is formed only in virus-infected cells prior to formation of virions. The VIA antigen is immunologically common to all seven serotypes of FMDV (Cowan and Graves, 1966).

B. Fragments Derived from the Virion

The stability of the picornavirion is markedly sensitive to ionic strength, pH, and temperature. Since its subunits are held together by noncovalent bonds many inactivation conditions lead to dissociation of the virion into smaller fragments which, in favorable cases, can be used to make inferences about the structure of the virion and the nature of the bonds holding it together. Perhaps the most useful variable is pH which generally seems to give the most specific products. The fact that the picornaviruses are classified by pH behavior already suggests that they vary considerably in their response.

1. *Enterovirus*

Poliovirus (Loring and Schwerdt, 1944) and coxsackie virus (Robinson, 1950) tolerate a very wide range of pH from 3 to about 9.5. At low temperature poliovirus survives treatment at pH 2.5 (Charney *et al.*, 1961). At alkaline pH it can be dissociated into RNA-free empty capsids in yields up to 90% under controlled conditions (Boeyé and Van Elsen, 1967). During this disruption the smallest polypeptide VP4 and the RNA can be quantitatively released in soluble form (Maizel *et al.*, 1967). Fibrillar structures of highly variable size and unknown composition are also produced during this process (Van Elsen *et al.*, 1968). Similar fibrils are produced by heat treating (50°C) poliovirus and rhinovirus (McGregor and Mayor, 1968). The fibers are digested by ribonuclease if the enzyme is present during the heat degradation step. However if heat degradation is completed in the absence of enzyme the fibrils are resistant to ribonuclease digestion unless they are first treated with pepsin. McGregor and Mayor have suggested that the fibrils may represent a ribonucleoprotein complex released from the virion.

Most enteroviruses are strongly stabilized against thermal inactivation by high ionic strength (Wallis and Melnick, 1961, 1962). Fujioka *et al.* (1969) noted that while hypertonic $MgCl_2$ solutions stabilize poliovirus against inactivation by heat and urea, they enhance its inactivation by guanidine hydrochloride. RNA can be released from poliovirus to yield empty shells by treatment with 8 M urea (Cooper, 1962). The virion can be degraded to 2 S subunits, probably polypeptide chains, by treatment with 6 M guanidine hydrochloride (Scharff *et al.*, 1964). In general the empty capsid of poliovirus seems to be very stable and it is the principle product of inactivation. The vigorous treatments required to dissociate it further usually lead to completely denatured proteins.

2. *Rhinovirus*

The rhinoviruses are by definition inactivated at pH 3–5. They are relatively stable at pH 7–8 (Ketler *et al.*, 1962). Except for FMDV relatively little seems to be known about the products of pH inactivation. Gerin *et al.* (1968) observed structures 80 Å in diameter at a buoyant density of 1.28 in CsCl gradients of purified rhinovirus. These structures appeared to be composed of five or six 20 Å subunits arranged around a hollow center 25 Å in diameter.

FMDV is stable at pH 7–9 but is unstable at pH 6.5 or at pH 10 (Bachrach *et al.*, 1957). At pH 5–6 it is dissociated into 12 to 14 S subunits and infectious RNA (see Section IV,A,2). However it has been reported that the infectivity of FMDV is quite stable at pH 3 in crude vesicular lymph (Pyl, 1933). This resistance was attributed to the use of a colloidal adsorbent but might conceivably represent a pH region of greater stability (see below). The isoelectric point of FMDV shifts from 7 to 5 with increasing ionic strength (Vande Woude, 1967). High ionic strength also stabilizes the virus at low pH. The stabilization of FMDV against heating at 50°C by 1 to 2 M MgCl$_2$ varies with virus strain. Some strains of type A, C, and O were stabilized by 1 or 2 M MgCl$_2$ or 2 M NaCl, while others were not or even showed enhanced inactivation (Wittman, 1967; Fellowes, 1967).

3. *Cardiovirus*

The infectivity of cardioviruses including Columbia-SK virus (Jungeblut and Sanders, 1940), EMC virus (Boesch and Drees, 1957), and MM virus

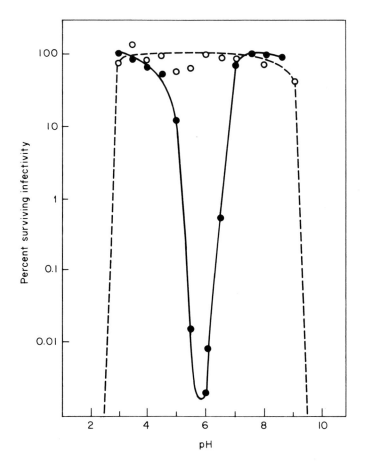

FIG. 2. pH stability profile of ME virus, strain S2, in the presence (●) and absence (○) of 0.1 M sodium chloride. Virus, diluted 20-fold into solutions buffered to the indicated pH with sodium citrate-phosphate buffer (Gomori, 1955), was incubated 60 minutes at 37°C. Treatment was terminated by 20-fold dilution into chilled buffer (pH 7.2) and samples were assayed immediately for infectivity (Rueckert, unpublished data).

(Schatz and Plager, 1948) is relatively unstable in the presence of unbuffered physiological saline solutions. In 1962, Speir *et al.* confirmed this behavior with mengo and MM viruses and noted further: (a) that two mouse encephalomyelitis viruses of the Theiler group, GVII and FA, were stable under the same conditions, and (b) that mengovirus was stable in distilled water and 1 *M* saline. They suggested that the peculiar salt instability might be a characteristic of the cardioviruses. In further studies Speir (1961a,b, 1962) showed that mengo virus is most labile at a NaCl concentration of about 0.15 *M* in the pH region 5–7 and that this instability is markedly and specifically enhanced by only two ions, chloride and bromide. Mak *et al.* (1970) repeated and completely confirmed these results.

The pH behavior of ME virus is illustrated in Fig. 2. The virus is stable in the absence of sodium chloride over the pH range from about 3 to 9; in the presence of 0.1 *M* sodium chloride there appears a sharp zone of instability in the pH region from about 4.5 to 7.0. With ME virus the inacti-

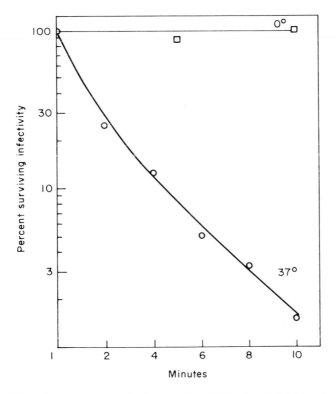

Fig. 3. Effect of temperature on the inactivation of ME virus in 0.1 *M* sodium chloride, 0.005 *M* sodium citrate, pH 5.7 (Rueckert, unpublished data).

vation rate is maximal at pH 5.7. The same behavior is obtained with equivalent chloride concentrations of the lithium, potassium, rubidium, cesium, magnesium, and calcium salts. There is little or no inactivation at this pH with 0.1 M sodium salts of nitrate, perchlorate, acetate, citrate, fluoride, iodide, or thiocyanate. The inactivation rate is temperature dependent (Fig. 3) so that the degree of inactivation can readily be controlled by suitably adjusting the time and temperature of incubation.

As seen in Fig. 4, the rate if inactivation is also strongly dependent upon the sodium chloride concentration; it is maximal at about 0.1 M. Stabiliza-

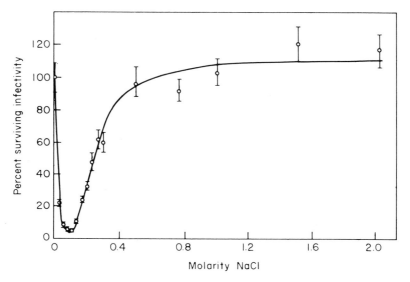

Fig. 4. Influence of sodium chloride concentration on the rate of inactivation of ME virus. Virus was incubated for 5 minutes at 37°C with the indicated concentration of sodium chloride in 0.05 M sodium citrate, pH 5.7, and assayed immediately thereafter for surviving infectivity. All pH's were adjusted after adding sodium chloride which depressed the pH of the buffer at high concentrations (Rueckert, unpublished data).

tion by sodium chloride concentrations exceeding 0.1 M reflects the characteristic thermostabilizing effect of high ionic strength typical of most picornaviruses. Thus the virus is completely stable at pH 5.7 in the presence of 0.1 M NaCl if 0.5 M sodium sulfate is also present.

The strong similarity in behavior between ME and mengo viruses make it likely that the chloride-dependent pH inactivation is indeed a characteristic of the cardiovirus group. In support of this conclusion the author has observed that poliovirus type 1 (CHAT) and coxsackie virus are completely stable for at least 30 minutes at 37°C in 0.10 M sodium chloride, 0.005 M

citrate, pH 5.7. ME virus has a half-life of roughly 0.9 minutes under these conditions.

ME virus (Hausen *et al.*, 1963; Rueckert *et al.*, 1969), EMC virus (Hall, 1970; Burness personal communication), and mengo virus (Mak *et al.*, 1970; 1971) are dissociated under these conditions into homogeneous water-soluble protein subunits which do not associate with the released RNA genome and which sediment on density gradients at 14 S. An insoluble precipitate (I protein) is also produced (Dunker and Rueckert, 1971). The 14 S subunit lacks delta and epsilon chains, contains equivalent molar proportions of alpha, beta, and gamma chains and weights about 400,000 (O'Callaghan *et al.*, 1970) to 420,000 (Rueckert *et al.*, 1969) daltons. It is a pentamer that can be further dissociated with 1 to 2 *M* urea into five smaller 5 S subunits. The latter subunit is a protomer composed of one each of three nonidentical chains (alpha, beta, and gamma) and is believed to represent one of the 60 subunits of the capsid.

The I protein contains the two minor components (delta and epsilon), which are always found in virus preparations; it also contains a definite proportion of the alpha, beta, and gamma chains. The stoichiometry of the nondelta portion of the I protein is consistent with that expected for a structure analogous to the 14 S pentamer containing one epsilon chain in place of one of the five beta chains. On the basis of these results Dunker and Rueckert (1970, 1971) proposed that the ME virion is composed of 60 protomers held together by two types of bonds, one type binding 12 pentamers to form a 60 subunit shell; a second type binding 5 protomers together in a pentamer. They suggested that mild acid dissociation of ME virus is due to specific rupture of the first type of interpentamer bond (Fig. 5).

V. Structure of the Virion

A. X-RAY STUDIES

In 1956, Crick and Watson proposed that isometric viruses were built, not of very large giant molecules, but rather from an assembly of smaller identical chains. Assembly of a superstructure from identical subunits implies repeated use of the same set of bonding contacts characteristic of, and presumably determined by, the protein subunit itself. Crick and Watson realized that a closed stable shell could be formed by such identical subunits only if the subunit were related in the completed structure by cubic symmetry.

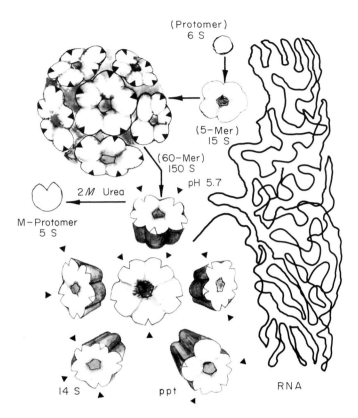

FIG. 5. A mechanism accounting for fragments derived by pH inactivation of ME virus. The virion is thought to contain 58 (α, β, γ, δ) protomers and 2 (α, ε, γ) protomers where epsilon represents an uncleaved precursor of the beta and delta chains. The 60 protomers are organized in twelve clusters of five which are held together by type 1 bonds. At pH 5.7 type 1 bonds are believed to be specifically dissociated to release 12 pentamers, viral RNA, and insoluble delta (▲) chains. Two pentamers (ppt) containing the uncleaved epsilon chains are thought to be insoluble while the (α, β, γ)$_5$ pentamers are soluble (14 S). The 5 S (M protomer) can be derived from the 14 S subunit by dissociation with 2 M urea (Dunker and Rueckert, 1970). The 6 S (i-protomer) is a hypothetical "immature" protomer (α, ε, γ) which is believed to be assembled through a 15 S pentamer (5-mer) during morphogenesis of the virion (Hall, McGregor, and Rueckert, 1971; Rueckert *et al.*, 1969; see also Section VI).

Of the possible classes of cubic symmetry, only three classes, tetrahedral, octahedral, and icosahedral (requiring 12, 24, or 60 asymmetric subunits, respectively) are relevant to structures built of asymmetric subunits (for an excellent summary see Mattern, 1969). Such symmetry in a particle can be detected, and its types distinguished, by the methods of X-ray crystallography. Experimental evidence confirming the existence of symmetry in various types of isometric viruses was quick in forthcoming (Caspar, 1956; see also Caspar and Klug, 1962).

1. *Poliovirus*

With the purification and crystallization of poliovirus (Schaffer and Schwerdt, 1955, 1959) came the opportunity of examining a picornavirus; poliovirus, like all other isometric viruses so far examined by this technique, had symmetry of the icosahedral (5:3:2) type implying that at least a part of the capsid is constructed of 60 or 60 *n* identical asymmetric structure units. At the time of the X-ray studies no information on the size of the polioviral polypeptide chains was available. Noting, however, the presence in the diffraction photographs of regular modulations of intensity corresponding to a repeat period of 60 to 65 Å, they proposed that the capsid was constructed of 60 identical structurally asymmetric subunits some 60 to 65 Å in diameter. Since poliovirus was believed to contain 25% RNA and to weigh about 6.7 million daltons (Schaffer and Schwerdt, 1959), one of the sixty identical subunits should weigh about 80,000 daltons; Finch and Klug noted that this is about the right size for a globular subunit about 60 Å in diameter. This model was corroborated by Horne and Nagington who found a poliovirus particle with pentagonally arranged knobs about 50 Å in diameter.

2. *Turnip Yellow Mosaic Virus*

That 5:3:2 symmetry and a 60 Å periodicity is insufficient to prove a 60 subunit model later became clear from studies on the structure of turnip yellow mosaic virus (TYMV). There is a close crystallographic similarity between poliovirus and TYMV. Diffraction patterns from TYMV also indicated icosahedral symmetry and a 60 Å repeat in the virion (Klug *et al.*, 1957). Just as for poliovirus, these results were at first thought to correspond to an icosahedral arrangement of 60 subunits 60 Å in diameter. However the 60 subunit model was later revised when chemical data (Harris and Hindley, 1961) indicated that the polypeptide chains of TYMV were only a third as large as expected and electron microscopy suggested that the capsid was organized in 32 rather than 60 capsomers (Huxley and Zubay, 1960; Nixon and Gibbs, 1960; Klug and Finch, 1960). Caspar and Klug (1962) showed that these observations could be reconciled without abandoning the simple physical principles proposed earlier by Crick and Watson if one relaxed the restriction that all subunits must be in exactly equivalent environments. By allowing slight distortions of bond lengths and angles they showed that it was possible to construct whole familes of shells containing 60 T identical subunits where the lowest values of T are 1, 3, 4, and 7. Subsequent work (Klug *et al.*, 1966; Finch and Klug, 1966) showed that the TYMV capsid is assembled of 180 identical polypeptide chains quasi-equivalently related in twelve clusters of 5 chains and twenty clusters of 6 chains. The modulations observed in the diffraction photographs were then reinterpreted in terms of approximately 60 Å spacings between the 32 capsomers (Klug and Finch, 1960). This remarkable and misleading geometric coincidence made it clear that further information would be required to settle the question of picornaviral structure.

B. DOES POLIOVIRUS HAVE 60 SUBUNITS?

The ensuing decade saw an increasing uncertainty about the 60 subunit model. This was due in part to a number of reports that poliovirus, echovirus, rhinovirus, and foot-and-mouth disease virus had 32 or 42 capsomers (see Section II on morphology). By classical structure theory based upon assembly from identical subunits (Caspar and Klug, 1962) the three types of morphology seemed to be incompatible because they represent three different surface lattice groups. Thus the proposed 60 subunit model, belonging to the simplest icosahedral surface lattice group, $T = 1$, would consist of 60 identical proteins; the 32 capsomer model, ($T = 3$), of 180 identical proteins; and the 42 capsomer model, ($T = 4$), of 240 identical chains.

Attempts to deduce the number of chains (from a knowledge of the size and composition of the virion and the size of the polypeptides) were complicated by the wide range of experimental variation in values for the composition of the virion and in values for the sizes of the virion and its polypeptide chains. For example, given a particle size range of 6.2 to 8.6 million daltons, a composition range of 20 to 32% RNA and a chain size range of 27 to 40 thousand daltons (see Section III.C.2) allowed a range of 120 to 270 chains per virion. Finally the presence of nonidentical polypeptide chains in picornaviruses compounded the difficulty since the theory of icosahedral structures has generally dealt only with surface lattices containing a single type of subunit.

Icosahedral symmetry in a shell comprised of identical polypeptide chains implies the presence of 60 or 60 T (where $T = 1$, 3, 4, 7, ..., Caspar and Klug, 1962) polypeptides; if however the shell is composed of not one, but

several different species of polypeptides then new and previously uncon-
sidered possibilities arise which still conform to the symmetry restrictions
required by the crystallographic work (see e.g. Fig. 6). Thus the presence
of nonidentical chains in an isometric virion was at once the most challenging to interpret and the most intriguing aspect of picornaviral architecture.
Indeed, it had been suggested that picornaviral architecture may differ
fundamentally from that of viruses containing identical polypeptide chains
(Burness and Walters, 1967). In 1964, Mayor speculated that the poliovirion
might be composed of twelve pentamers of one chain type and twenty
hexamers or trimers of another chain type(s). Similarly a 42 capsomer
model could be constructed out of twelve pentamers of one chain type and
thirty dimers, tetramers, or hexamers of another chain type. The number
of such possibilities is large in the absence of more definite information
about the molar proportions of polypeptide chains in the virion.

C. The Structure of ME Virus

In 1969, Rueckert *et al.* reported three important experimental observations relevant to the structure of the ME virion (a) the nonidentical alpha,
beta, and gamma chains are represented in the virion in equimolar proportions, and these three chains comprise about 93% of the mass of the
capsid; (b) the virion could be dissociated into homogeneous 14 S subunits
weighing about 420,000 daltons; and (c) the 14 S subunits could be dissociated with 1 to 2 *M* urea into homogeneous smaller 5 S subunits contained
equivalent molar proportions of alpha, beta, and gamma chains weighing
33,000, 30,000, and 25,000 daltons, respectively (Dunker and Rueckert,
1969). The sum of the molecular weights of these three chains (88,000) is
very close to the size (86,000) determined by sedimentation equilibrium
experiments for the 5 S subunit; the 5 S subunit thus appeared to be a
protomer composed of three nonidentical chains and the 14 S subunit
appeared to be an oligomer comprised of five of the protomers.

1. *Some Possible Morphological Structures*

The observation that the three nonidentical chains occurred in equimolar
proportions immediately ruled out a number of structural possibilities, which
need not be detailed here, and thereby defined the problem to a relatively
limited number of possible arrangements. Since these chains represent about
93% of the mass of the capsid it is reasonable as a first approximation to
assume that they represent the important structural proteins of the capsid.

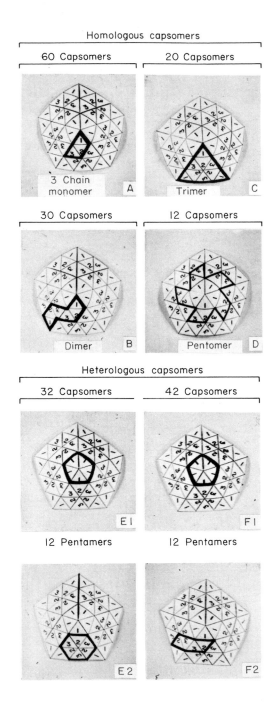

Homologous capsomers

60 Capsomers   20 Capsomers

A — 3 Chain monomer   C — Trimer

30 Capsomers   12 Capsomers

B — Dimer   D — Pentomer

Heterologous capsomers

32 Capsomers   42 Capsomers

E1   F1

12 Pentamers   12 Pentamers

E2   F2

Within the constraints of icosahedral symmetry there should be 60 or 60 n sets of the three chains (rejecting the unsatisfying possibility that the nonidentical chains are "pseudoequivalent," i.e., are interchangeable and behave as if they were all identical). Given a range of 120 to 270 chains per virion calculated from the experimental data (Section II) a shell composed of 180 chains appears to be the only likely possibility.

The capsid of isometric viruses is conveniently regarded as a monomolecular layer of subunits bonded together to form a closed surface lattice. A general way of arranging sixty sets of three nonidentical polypeptide chains in an icosahedral surface lattice is illustrated in Fig. 6. With the aid of the quasiequivalence concept, Caspar and Klug (1962) have enumerated all the possible ways of arranging *identical* polypeptide chains in such a capsid to generate icosahedral surface lattices. These can be classified according to triangulation numbers $T = 1, 3, 4, 7...$ The triangulation number of the lattices in Fig. 6 in its general form fits neatly into none of the categories, but represents a hybrid of $T = 1$ and 3.

Each of the eight panels represents the same basic icosahedral lattice; the polypeptide chains in such a lattice can be clustered in different ways depending upon which types of specific bonding interactions between polypeptide chains are strongest. The 32 capsomer pattern (Fig. 6b; E1, E2) is analogous to the clustering pattern of TYMV and represents the type of arrangement suggested by Mayor (1964) for poliovirus. Similarly a 42 capsomer structure (F1, F2) can be generated by clustering dimers across the

---

FIG. 6. Some chain clustering patterns (bold outline) which might be assumed by a shell composed of 60 sets of three nonidentical chains arranged in an icosahedral surface lattice (modified from Rueckert *et al.*, 1969). Each of the eight panels shows the fivefold vertex of an icosahedron folded from a "net" as described by Caspar and Klug (1962). The numbers and triangles identify bonding domains; they are mathematical symbols and imply nothing about the shapes, positions, or orientations of the three chains. The latter three parameters are not specified by symmetry relationships and would have to be determined experimentally for each virus. *a. Homologous capsomers.* An icosahedral shell has twelve vertices with a fivefold symmetry axis (D), twenty faces with a threefold symmetry axis (C), and thirty edges with a twofold symmetry axis (B). Each of the capsomers in panel *a* is composed of a single type of protomer (A). The protomer is a protein molecule composed of nonidentical polypeptide chains. *b. Heterologous capsomers.* The basic icosahedral surface lattice of the chains is the same as that shown in panel *a* but the chains are organized into two types of oligomers by means of a different clustering pattern. One oligomer is composed of five homologous chains of one type (E1 = F1). The remaining two types of chains can be arranged in two ways to form heterologous oligomers; one a trimer (E2); the other, a dimer (F2). Such homologous and heterologous oligomers might conceivably be synthesized separately and assembled to form structures satisfying the requirements of icosahedral symmetry.

twofold axes of symmetry. These 32 and 42 capsomer structures could conceivably be assembled from two different types of subunits; one type corresponding to the $(1)_5$ oligomer (twelve each), the other type to the $(2, 3)_3$ oligomer (twenty each for 32 capsomers), or the $(2, 3)_3$ oligomer (thirty each for 42 capsomers). Capsids built out of two different types of subunits are designated here as "heterologous" to distinguish them from capsids constructed from 60 identical protomers (Fig. 6a, A, B, C, D). The 12, 20, and 30 capsomer patterns are all variants of the same 60 protomer protomer shell and are called "homologous" capsids.

2. Structure of the ME Virion

The 14 and 5 S particles derived from ME virus (Section IV,B,3) are key fragments in deducing the organization of the polypeptides in the capsid; they are also very useful reference particles for evaluating the possible identity of biosynthetic intermediate particles in virus-infected cell extracts. Furthermore they provide experimental support for the hypothesis that the particle is held together by two different kinds of bonds. The latter hypothesis provides an integrating concept which suggests that assembly of the virion is probably also assembled in two steps corresponding to the step-wise utilization of two types of intersubunit bonds. These points will now be treated in greater detail.

a. Organization of Polypeptides in the Virion. As noted in Section IV the cardioviruses can be specifically dissociated at pH 5.7 into homogeneous

TABLE X

PROPERTIES OF SOME THEORETICAL FRAGMENTS FROM 32, 42 AND 60 SUBUNIT SHELLS

Capsomer model	Figure panel	Identity of possible subunits	Size range[a] (daltons $\times 10^3$)
32	E1	$\alpha 5$; $\beta 5$; $\gamma 5$	125–165
	E2	$(\beta, \gamma)_3$; $(\alpha, \gamma)_3$; $(\alpha, \beta)_3$	165–186
42	F1	$\alpha 5$; $\beta 5$; $\gamma 5$	125–165
	F2	$(\beta, \gamma)_2$; $(\alpha, \gamma)_2$; $(\alpha, \beta)_2$	110–126
60	A	(α, β, γ)	88
	B	$(\alpha, \beta, \gamma)_2$	176
	C	$(\alpha, \beta, \gamma)_3$	264
	D	$(\alpha, \beta, \gamma)_5$	440

[a] Assuming molecular weights of 33,000 for the alpha chain, 30,500 for the beta chain, and 25,000 for the gamma chain (see Table VIII).

14 S subunits weighing 420,000 daltons by sedimentation equilibrium studies; they contain equivalent molar proportions of each polypeptide chain, alpha, beta, and gamma. Summarized in Table X are the weights and polypeptide compositions of subunits which might be expected from the 32, 42, and 60 subunit models illustrated in Fig. 6. Inspection shows that the 5 S subunit uniquely fits the description of an (α, β, γ) protomer; the 14 S subunit uniquely matches the properties of the $(\alpha, \beta, \gamma)_5$ oligomer. It is difficult to see how these fragments could be produced by the dissociation of 32 or 42 subunit shells. Thus the sequential production of 14 and 5 S subunits is most simply explained by assuming that the capsid is constructed of 60 identical (α, β, γ) protomers. The fact that they are produced in two steps, the 14 S by dissociation at pH 5.7, the 5 S by treatment with 2 M urea, suggests that the protomers are held together in the shell by two different types of intersubunit bonds, one type holding twelve pentamers together to form the 60 protomer shell, the second type holding protomers together to form the pentamer (Fig. 7). In fact two different types of bonding contacts are both necessary and sufficient to specify a 60 subunit shell with icosahedral symmetry.

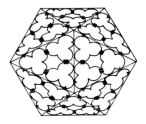

FIG. 7. A schematic model showing how an icosahedral surface lattice composed of 60 identical subunits can be determined by only two types of bonding contacts. In this example one type (▲) binds the protomers at the fivefold symmetry axis; a second type (■) binds the twelve pentamers together to complete the capsid. There are 60 bonds of each type since each protomer has its own set of two bonding surfaces. Note that bonding must occur on complementary, not identical surfaces of each partner, hence the bonds have "direction" or "polarity" indicated in Fig. 8 by arrows.

Two points remain to be considered. First, the (α, β, γ) protomer model fails to account for the delta and epsilon chains which are believed to be components of the virion. Second, analysis of the composition of fragments derived by dissociation of the virion cannot by itself be considered proof of the organization of chains in the virion since one cannot positively rule out the possibility, albeit unlikely, that the chains might reorganize after dissociation of the virion to form new fragments unrelated to their organization within the capsid. It is difficult however to see how this hypothesis could account for both the 14 and also the 5 S particles. Furthermore, we shall presently see a satisfying consistency in the simple concept of identical protomers in interpreting the identity and behavior of virus-specific particles believed to represent intermediates in the assembly of poliovirus.

b. Origin of the Delta and Epsilon Chains. Returning now to the issue of the delta and epsilon chains we note that the size of the epsilon chain (41,000) is close to the sum of the molecular weights of the beta (30,000) and delta (10,000) chains. Studies in the author's laboratory on the fragments produced from these three chains by cyanogen bromide cleavage show excellent correspondence of fragments from the epsilon chain with those derived from the beta and delta chains (Stoltzfus and Rueckert, 1971). Recent work on the biosynthesis of poliovirus suggests that maturation involves cleavage of a precursor polypeptide to form smaller chains (Maizel and Summers, 1968; Jacobson and Baltimore, 1968a; Holland and Kiehn, 1968; Jacobson *et al.*, 1970). The later authors have shown that the polioviral VP0 chain, which is structurally analogous to the epsilon chain of ME virus has sequence similarities to two smaller chains, VP2 and VP4; this observation provides strong support for the concept of maturation cleavage.

The notion that maturation of the virus involves cleavage of the epsilon chain combined with a 60 subunit model for the structure of the capsid leads quite naturally to the hypothesis that the (α, β, γ) protomer may be derived from an "immature" (α, ε, γ) protomer (i protomer) by cleavage of its epsilon chain to generate an (α, β, γ, δ) protomer (Rueckert *et al.*, 1969). In this event the capsid would be assembled from identical i protomers and the mature virion should consist of 60 (α, β, γ, δ) protomers. To account for the fact that ME virus contains two epsilon chains per average virion one assumes that two of the 60 i protomers fail to be cleaved. The significance of this observation is not presently clear. Recent studies in our laboratory suggest that the epsilon chain may be an integral component of the capsid. However, Maizel *et al.* (1967) have attributed the analogous NCVP6 chain which is also found in small amounts in poliovirions to contamination of the preparations with empty capsids since it could be significantly reduced in amount by rebanding the virus in CsCl gradients or by treatment with 1% SDS. Since we are unable to remove epsilon chain from ME virus by either rebanding in CsCl or by treatment with 1% SDS; and since the amount of this component is relatively constant in our virus preparations it is tempting to believe that incomplete cleavage of all 60 protomers may reflect some design the significance of which is not yet understood.

VI. Biosynthesis and Assembly

The studies just cited suggest that the ME virion, and by analogy probably also the poliovirion, is assembled from 60 identical "immature protomers" (α, ε, γ) weighing ($33+ 41 +25$) or about 100,000 daltons and that these protomers are cleaved at a specific site in the epsilon chain during maturation of the virus. A molecule of this size would be expected to sediment at about 5 to 6 S (Table XI). Such fragments have, in fact, been found in extracts from both EMC virus (Kerr *et al.*, 1965) and poliovirus- (Scharff *et al.*, 1964; Phillips *et al.*, 1968) infected cells.

TABLE XI

PROPERTIES OF FOUR SUBUNITS FROM ME AND EMC VIRUSES

Component	Molecular weight	Identity
4.9 S[c]	88,000[a]	(α, β, γ)
5.3 S[d]	100,000[b]	(α, ε, γ)?
14 S[c]	440,000[a]	(α, β, γ)$_5$
15 S[e]	490,000[b]	(α, ε, γ)$_5$

[a] Determined by sedimentation equilibrium studies and the sum of the weights of the polypeptide chains (Dunker and Rueckert, 1969; 1971).
[b] The sedimentation coefficient was calculated from the theoretical molecular weight according to the relation $S_1/S_2 = (M_1/M_2)^{2/3}$ which applies to globular or almost globular proteins with similar partial specific volumes (Martin and Ames, 1961) assuming (α, β, γ)$_5$ sediments at 14 S.
[c] Rueckert *et al.*, 1969; Dunker and Rueckert, 1971.
[d] Kerr *et al.*, 1965.
[e] Hall, McGregor, and Rueckert, 1971.

Kerr *et al.* (1965) reported the presence of a 14 S particle in extracts from EMC-infected Krebs-2 ascites cells, but it was uncertain at that time whether this fragment represented a precursor or a breakdown product. Recent studies in the authors laboratory show that this component contains roughly equimolar amounts of alpha, epsilon, and gamma chains and no trace of beta chain (Hall and Rueckert, 1971). Thus it cannot be a breakdown product from mature virus. Furthermore this component when cosedimented on a density gradient in a double-label experiment with 14 S subunits derived from mature virions (by pH dissociation) moved at a rate of about 15 S. This particle has just the sedimentation velocity and polypeptide composition expected for an $(\alpha, \varepsilon, \gamma)_5$ oligomer of an immature protomer (Table XI).

Two poliovirus-related fragments sedimenting at about 14 S [at first identified as 10 S (Watanabe *et al.*, 1962; Phillips *et al.*, 1968)] and 74 S have also been described in poliovirus-infected cell extracts. The latter two and probably also the 5 to 6 S particle in poliovirus in cell extracts contain the same three chains (Phillips *et al.*, 1968) which are electrophoretically analogous to the alpha, epsilon, and gamma chains of ME virus. Kinetic labeling experiments *in vivo* suggest that the 14 and 73 S particles are sequential precursors of the mature virion (Penman *et al.*, 1964; Jacobson and Baltimore, 1968b). Thus the polypeptides are labeled within a few minutes: the 14 S within 5 minutes, the 73 S within 20 minutes, and the virions after 30 minutes. The 14 S particles can be converted in cell-free extracts from infected cells into 74 S particles which cosediment with naturally occurring empty capsids (Phillips *et al.*, 1968; Phillips, 1969). Jacobson and Baltimore (1968b) found that under certain conditions *in vivo* label from empty capsids can be chased into virions and they suggested that the final step in poliovirus morphogenesis is combination of viral RNA with a protein shell and that concomitant with this union is a cleavage of one of the capsid chains to produce two smaller ones. This hypothesis accounts in a simple way for the known difference in polypeptide composition of mature virions and empty capsids of poliovirus (Maizel *et al.*,1967). These observations are consistent with the hypothesis (Rueckert *et al.*, 1969) that the enteroviruses and cardioviruses are assembled from an i-protomer. Incorporating the recent identification of 14 S particles in infected cell extracts (Phillips, 1969; Hall, McGregor, and Rueckert, 1971) this hypothesis may be summarized as follows:

$$\text{precursor} \xrightarrow{\quad} \underset{\text{i-protomer}}{\overset{5-6\,\text{S}}{(\alpha, \varepsilon, \gamma)?}} \xrightarrow{\ A\ } \underset{\text{i-pentamer}}{\overset{14-15\,\text{S}}{(\alpha, \varepsilon, \gamma)_5}} \xrightarrow{\ B\ } \underset{\text{procapsid}}{\overset{65-75\,\text{S}}{(\alpha, \varepsilon, \gamma)_{60}}} \xrightarrow[\text{invest RNA}]{\text{cleave } \varepsilon} \underset{\text{virion}}{\overset{160\,\text{S}}{(\alpha, \beta, \gamma, \delta)_{60}}}$$

The cleavage sequence from precursor to the pentamer remains to be clarified. Recent work on the biosynthesis of poliovirus suggests that the entire enterovirus ribonucleic acid genome is translated into a single giant polypeptide of the order of 200 to 260 thousand daltons (Jacobson and Baltimore, 1968a; Kiehn and Holland, 1970) which is then cleaved to form smaller chains (Maizel and Summers, 1968; Jacobson and Baltimore, 1968b; Holland an Kiehn, 1968). An uncleaved capsid precursor chain weighing roughly 100,000 daltons has been tentatively identified in extracts from Hela cells infected with poliovirus (Jacobson *et al.*, 1970; Cooper *et al.*, 1970) or with EMC virus (Butterworth *et al.*, 1971).

As noted in Section V,C,2 the ME virion behaves as if it were held together by two different kinds of intersubunit bonds. This suggests the rationale for condensation steps A and B in the above assembly scheme. According to this notion each i-protomer has a characteristic set of two specific bonding sites. Step A represents utilization of the first bonding site which leads to formation of pentamers. Completion of the pentamer is then followed by utilization of the second set of bonding sites through which twelve pentamers are symmetrically fused to form the 60 protomer capsid. This simple two-bond hypothesis accounts for the assembly mode of the virion and suggests

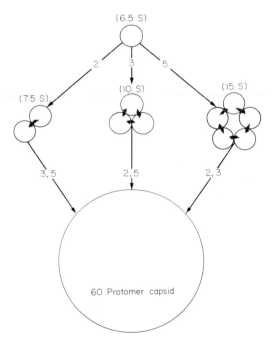

FIG. 8. Alternative two-step pathways in the assembly of an icosahedrally symmetric capsid from 60 identical protomers. Assembly by two sets of bonds can occur in three ways, through the dimer, the trimer, or the pentamer. This is so because symmetry in the capsid must be generated by the intersubunit bonds. The bonds have been selected by evolution to establish such symmetry axes, or else they could not direct assembly of an icosahedral shell. There are only three symmetry axes, hence only three possible two-step assembly modes. EMC virus and poliovirus appear to be assembled via the pentamer pathway. The numbers in the arrows designate the symmetry axes generated by that particular bonding step.

that all of the stable intermediates in capsid assembly may have been identified. The fact that only pentamers and not (say) dimers or trimers, are observed also suggests (see Figs. 6 and 8) that the two bonds are not used at random, but rather are used in a specific order; i.e., bond set A is used first, then and only then bond set B. Such an ordered process might be brought about, for example, if bond set B were latent in the i-protomer and became activated only after completion of the pentamer. Activation might be brought about by changes in configuration to generate previously unexposed complementary surfaces on each protomer. That such surface changes actually occur in poliovirus synthesis is indicated by differences in the antigenic determinant of the 5 and/or 10 S (G and S antigens), the 74 S empty shell (H antigen), and the intact virion (N antigen) (see Section IV,A,1).

VII. Conclusion

The picornaviruses can be classified in two major subgroups: (a) the enterovirus-cardiovirus group and (b) the rhinovirus-FMDV group. The physical basis for the greater buoyant density and acid lability of the latter remains unclear. However the bulk of the evidence indicates that the virions of both groups are chemically, physically, and morphologically similar, and there are currently no strong grounds for believing that virions of these two groups differ fundamentally in structure. The presence of several nonidentical polypeptide chains appears to be a characteristic feature of the picornaviral capsid.

Mounting evidence favors the following model of picornaviral architecture. The virion is assembled of a single strand of RNA and 60 identical protein subunits weighing about 96,000 daltons each. Each subunit is thought to be synthesized as a single polypeptide chain which is subsequently cleaved at three virus-specific sites at some time during viral morphogenesis; two of these cleavages appear to precede virus maturation while the third accompanies the maturation step. Thus the mature virion contains equimolar proportions of four nonidentical polypeptide chains while capsid-related precursor particles typically contain only three chains, one of which is a precursor of two smaller chains in mature virions. In each case the net weight of the three or four nonidentical polypeptide chains is about 96,000 daltons.

From this model the size of the protein capsid is calculated to be 60 ×96,000, or about 5.7 million daltons. Given an RNA content of 30%

TABLE XII

PHYSICAL PROPERTIES OF REPRESENTATIVE ISOMETRIC RNA VIRUSES

Virus	s_{20}	Diameter (Å)	RNA		Protein		Particle weight ($\times 10^6$)
			RNA (%)	RNA (mol. wt)	Subunit no.	$\times 10^3$ (mol. wt.)	
Ribophage R17	80	230–266	31.7	1.1–1.2	180	14.7	3.5–4.0[a]
Bromegrass mosaic	86	260	21.4	1.0	180	20.3	4.6–4.7[a]
Turnip yellow mosaic	117	280	33–35	1.9–2.0	180	20.1	5.4–6.0[a]
Tomato bushy stunt	132	300	15–20	—	—	40[e]	7.8–9.8[a]
Tobacco necrosis (satellite)	50	170	20	0.41	60	27[g]	2.0[a]
ME, EMC, mengo; polio	160	270–300	30–32	2.5–2.7	60	96	8.3–8.6[b]
Rhinovirus, FMDV,	140–160	270–300	30–32	2.2–2.8	60	96[h]	7[b,d];8.4[h]

[a] Kaper (1966).
[b] Dunker and Rueckert (1971).
[c] McGregor (1969).
[d] Bachrach (1968).
[e] Michelin-Lausarot et al. (1970).
[f] may be distributed between two shells (Finch et al., 1970).
[g] Roy et al. (1969).
[h] Medappa et al. (1971).

to 32%, and assuming the rest (69%) to be protein, the size of the virion is computed to be 5.7/0.69, or about 8.4 million daltons; the size of the RNA genome is then 8.4×0.31, or about 2.6 million daltons (computed as the sodium salt).

Since the average nucleotide residue (sodium salt) weighs 340 to 345 daltons (calculated from the base compositions in Table V), the RNA genome of such a virion contains about 7500 nucleotides. Assuming a coding ratio of 3 nucleotides per amino acid, this is sufficient to encode the sequence of polypeptides containing 2500 amino acid residues. Given a weight of 110 daltons per average amino acid residue (calculated values are 111 for lysozyme, myoglobin, and tobacco mosaic virus coat protein, 106 for the coat protein of bacteriophage R17, 109 for chymotrypsinogen, 110 for ribonuclease, and 119 for cytochrome c), the genome has a coding capacity of about 270,000 daltons of protein. Thus the capsid protomer would account for about 96/270 or 35% of the viral genome.

It is desirable that the 4-chain protomer model of picornaviral architecture be tested by further investigations with other picornaviruses, particularly those in the rhinovirus-FMDV group. It remains to be seen, for example, if all picornaviruses contain 96,000 dalton subunits, and, if so, whether all are cleaved three times. It is conceivable that some types of viruses may bear uncleaved subunits or subunits cleaved only once or twice. It is also conceivable new evolutionary relationships with other viruses will emerge from such studies.

Table XII summarizes the basic properties of picornaviruses and compares them with some well characterized types of plant and bacterial RNA viruses. This table reveals no obvious size relationships between either the RNA or the protein subunit of any of the major virus groups.

REFERENCES

Agrawal, H. O. (1966). *Arch. Gesamte Virusforsch.* **19**, 365.
Anderer, F. A. and Restle, H. (1964). *Z. Naturforsch.* **19B**, 1026.
Anderson, T. F. (1951). *Trans. N. Y. Acad. Sci.* **13**, 130.
Andrewes, C. H. (1964). "Viruses of Vertebrates." Williams and Wilkins, Baltimore, Maryland.
Bachrach, H. L. (1960). *Virology* **12**, 258.
Bachrach, H. L. (1961). *Proc. Soc. Exp. Biol. Med.* **102**, 610.
Bachrach, H. L. (1968). *Ann. Rev. Microbiol.* **22**, 201.
Bachrach, H. L. and Breese, S. S., Jr. (1958). *Proc. Soc. Exp. Biol. Med.* **97**, 659.
Bachrach, H. L. and Polatnick, J. (1966). *Proc. Soc. Exp. Biol. Med.* **124**, 465.
Bachrach, H. L. and Vande Woude, G. F. (1968). *Virology* **34**, 282.

Bachrach, H. L., Breese, S. S., Jr., Callis, J. J., Hess, W. R., and Patty, R. E. (1957). *Proc. Soc. Exp. Biol. Med.* **95**, 147.

Bachrach, H. L., Trautman, R., and Breese, S. S., Jr. (1964). *Amer. J. Vet. Res.* **25**, 333.

Boedtker, H. (1959). *Biochim. Biophys. Acta* **32**, 519.

Boesch, P. and Drees, O. (1957). *Naturwissenschaften* **44**, 544.

Boeyé, A. (1965). *Virology* **25**, 550.

Boeyé, A. and Van Elsen, A. (1967). *Virology* **33**, 335.

Bradish, C. J., Henderson, W. M., and Kirkham, J. B. (1960). *J. Gen. Microbiol.* **22**, 379.

Breese, S. S., Jr. (1968). *J. Gen. Virol.* **2**, 465.

Breese, S. S., Jr. and Bachrach, H. L. (1960). *4th Intern. Conf. Electron Microscopy, Berlin* **2**, 619.

Breese S. S., Jr., Trautman, R., and Bachrach, H. L. (1965). *Science* **150**, 1303.

Brooksby, J. B. (1953). *Symp. Nature Virus Multiplication: Soc. Gen. Microbiol.*, p. 246.

Brown, F. (1958). *Nature (London)* **181**, 1130.

Brown, F. (1963). *Nature (London)* **197**, 590.

Brown, F. and Cartwright, B. (1961). *Nature (London)* **192**, 1163.

Brown, F. and Crick, J. (1958). *Virology* **5**, 133.

Brown, F., Newman, J. F. E., and Stott, E. J. (1970). *J. Gen. Virol.* **8**, 145.

Burness, A. T. H. (1970). *J. Gen. Virol.* **6**, 373.

Burness, A. T. H. and Clothier, F. W. (1970). *J. Gen. Virol.* **6**, 381.

Burness, A. T. H. and Walters, D. S. (1967). *Nature (London)* **215**, 1350.

Burness, A. T. H., Vizoso, A. D., and Clothier, F. W. (1963). *Nature (London)*, 197 1177.

Butterworth, B. E., Stoltzfus, C. M., and Rueckert, R. R. (1971). *Fed. Proc. (Biochem.).*

Cartwright, B. (1962). *J. Immunol.* **88**, 128.

Cartwright, S. F. and Thorne, H. V. (1958). *Nature (London)* **182**, 717.

Cartwright, S. F. and Thorne, H. V. (1959). *J. Gen. Microbiol.* **20**, 61.

Caspar, D. L. D. (1956). *Nature (London)* **177**, 476.

Caspar, D. L. D. and Klug, A. (1962). *Cold Spring Harbor Symp. Quant. Biol.* **27**, 1.

Charney, J., Machlowitz, R., Tytell, A. A., Sagin, J. F., and Spicer, D. S. (1961). *Virology* **15**, 269.

Colter, J. S. (1958). *Progr. Med. Virol.* **1**, 1.

Colter, J. S., Bird, H. H., Moyer, A. W., and Brown, R. A. (1957). *Virology* **4**, 522.

Committee on Enteroviruses (1962). *Virology* **16**, 501.

Cooper, P. D. (1962). *Virology* **16**, 485.

Cooper, P. D., Summers, D. F., and Maizel, J. V. (1970). *Virology* **41**, 408.

Cowan, K. M. (1966). *J. Immunol.* **97**, 647.

Cowan, K. M. and Graves, J. H. (1966). *Virology* **30**, 528.

Dales, S. and Franklin, R. M. (1962). *J. Cell. Biol.* **14**, 281.

Dans, P. E., Forsyth, B. R., and Chanock, R. M. (1966). *J. Bacteriol.* **91**, 1605.

Darnell, J. E., Jr., Levintow, L., Thoren, M. M., and Hooper, J. L. (1961). *Virology* **13**, 271.

Dick, G. W. A. (1949). *J. Immunol.* **62**, 375.

Dimmock, N. J. (1967). *Virology* **31**, 338.

Dimmock, N. J. and Harris, W. J. (1967). *Virology* **31**, 715.

Drees, O. (1965). *Arch. Gesamte Virusforsch.* **15**, 344.

Drees, O. and Borna, C. (1965). *Z. Naturforsch.* **20B**, 870.

Dunker, A. K. and Rueckert, R. R. (1969). *J. Biol. Chem.* **244**, 5074.

Dunker, A. K. and Rueckert, R. R. (1970). *Biophys. Soc. Abstr.*, p. 158a.

Dunker, A. K. and Rueckert, R. R. (1971). *J. Mol. Biol.*, in press.

Fabiyi, A., Engler, R., and Martin, D. C. (1964). *Arch. Gesamte Virusforsch* **14**, 621.

Faulkner, P., Martin, E. M., Sved, S., Valentine, R. C., and Work, T. S. (1961). *Biochem. J.* **80**, 597.

Fellowes, O. N. (1967). *Appl. Microbiol.* **14**, 206.

Fenner, F. (1968). *In* "The Biology of Animal Viruses." Vol. 1, pp. 20–21. Academic Press, New York.

Fenwick, M. L. (1968). *Biochem. J.* **107**, 851.

Finch, J. T. and Klug, A. (1959). *Nature (London)* **183**, 1709.

Finch, J. T. and Klug, A. (1966). *J. Mol. Biol.* **15**, 344.

Finch, J. T., Klug, A., and Leberman, R. (1970). *J. Mol. Biol.* **50**, 215.

Forsgren, M. (1968). *Arch. Gesamte Virusforsch.* **28**, 19.

Fraenkel-Conrat, H. (1962). *Surv. Biol. Progr.* **4**, 59.

Frommhagen, L. H. (1965). *J. Immunol.* **95**, 818.

Frommhagen, L. H. and Martins, M. J. (1961). *Virology* **15**, 30.

Fujioka, R., Kurtz, H., and Ackermann, W. W. (1969). *Proc. Soc. Exp. Biol. Med.* **132**, 825.

Gerin, J. L., Richter, W. R., Fenters, J. D., and Holper, J. C. (1968). *J. Virol.* **2**, 937.

Gomori, G. (1955). *In* "Methods in Enzymology" (S. P. Colowick, and N. O. Kaplan, eds.), Vol. I, p. 138. Academic Press, New York.

Goodheart, C. R. (1965). *Virology* **26**, 466.

Granboulan, N. and Girard, M. (1969). *J. Virol.* **4**, 475.

Graves, J. H., Cowan, K. M., and Trautman, R. (1968). *Virology* **34**, 269.

Hall, L. (1970). Ph. D. Thesis, Univ. of Wisconsin, Madison, Wisconsin.

Hall, L. and Rueckert, R. R. (1971). *Virology*, **43**, 152.

Hall, L., McGregor, S., and Rueckert, R. R. (1971). In preparation.

Halperen, S., Eggers, H. J., and Tamm, I. (1964). *Virology* **23**, 81.

Halperen, S., Stone, H. O., and Korant, B. D. (1970). *Bacteriol. Proc.*, p. 156.

Hamparian, V. V., Hilleman, M. R., and Ketler, A. (1963). *Proc. Soc. Exp. Biol. Med.* **112**, 1040.

Hamre, D. (1968). "Monographs in Virology, 1: Rhinoviruses." Karger, Basel.

Harris, J. J. and Hindley, J. (1961). *J. Mol. Biol.* **3**, 117.

Hausen, P. and Schäfer, W. (1962). *Z. Naturforsch.* **17B**, 15.

Hausen, P., Hausen, H., Rott, R., Scholtissek, C., and Schäfer, W. (1963). *In* "Viruses, Nucleic Acids and Cancer," p. 282. Williams and Wilkins, Baltimore, Maryland.

Holland, J. J. (1962). *Proc. Nat. Acad. Sci. U.S.* **48**, 2044.

Holland, J. J. and Kiehn, E. D. (1968). *Proc. Nat. Acad. Sci. U.S.* **60**, 1015–1022.

Holland, J. J., McLaren, L. C., Hoyer, B. H., and Syverton, J. T. (1960). *J. Exp. Med.* **112**, 841.

Horne, R. W. and Nagington, J. (1959). *J. Mol. Biol.* **1**, 333.

Horsfall, F. L. and Tamm, I. (1965). "Viral and Rickettsial Infections of Man," 4th ed. Lippincott, Philadelphia, Pennsylvania.

Hummeler, K., Anderson, T. F., and Brown, R. A. (1962). *Virology* **16**, 84.

Hummeler, K. and Hamparian, V. V. (1958). *J. Immunol.* **81**, 499.

Huxley, H. E. and Zubay, G. (1960). *J. Mol. Biol.* **2**, p. 189.

International Enterovirus Study Group (1963). *Virology* **19**, 114.

Jacobson, M. F. and Baltimore, D. (1968a). *Proc. Nat. Acad. Sci. U.S.* **61**, 77.

Jacobson, M. F. and Baltimore, D. (1968b). *J. Mol. Biol.* **33**, 369.

Jacobson, M. F., Asso, J., and Baltimore, D. (1970). *J. Mol. Biol.* **49**, 657.

Jamison, R. M. (1969). *J. Virol.* **4**, 904.

Jamison, R. M. and Mayor, H. D. (1966). *J. Bacteriol.* **91**, 1971.

Jungeblut, C. W. (1958). *In* "Handbuch der Virusforschung," Vol. 4, p. 459. Vienna, Springer-Verlag.

Jungeblut, C. W. and Sanders, M. (1940). *J. Expt. Med.* **72**, 407.

Kaper, J. M. (1966). *In* "Molecular Basis of Virology" (H. Fraenkel-Conrat, ed.), pp. 1–128. Reinhold, New York.

Kapikian, A. Z. *et al.* (1967). *Nature (London)* **213**, 761.

Kapikian *et al.* (1971). *Virology* **43**, 524.

Katagiri, S., Hinuma, Y., and Ishida, N. (1967). *Virology* **32**, 337.

Katagiri, S., Hinuma, Y., and Ishida, N. (1968). *Virology* **34**, 797.

Kay, C. M., Colter, J. S., and Oikawa, K. (1970). *Canad. J. Biochem.* **48**, 940.

Kerr, M. F., Martin, E. M., Hamilton, M. G., and Work, T. S. (1965). *Biochem. J.* **94**, 3137.

Ketler, A., Hamparian, V. V., and Hilleman, M. R. (1962). *Proc. Soc. Exp. Biol. Med.* **110**, 821.

Kiehn, E. D. and Holland, J. J. (1970). *J. Virol.* **5**, 358.

Klug, A. and Caspar, D. L. D. (1960). *Advan. Virus Res.* **7**, 280.

Klug, A. and Finch, J. T. (1960). *J. Mol. Biol.* **2**, 201.

Klug, A., Finch, J. T., and Franklin, R. M. (1957). *Biochim. Biophys. Acta* **25**, 242.

Klug, A., Longley, W., and Leberman, R. (1966). *J. Mol. Biol.* **15**, 315.

Korant, B. D., Lonberg-Holm, K. K., and Halperen, S. (1970). *Biochem. Biophys. Res. Commun.* **41**, 477.

Krug, R. M. and Franklin, R. M. (1964). *Virology* **22**, 48.

Kunin, C. M. (1967). *J. Virol.* **1**, 274.

LaPorte, J. (1969). *J. Gen. Virol.* **4**, 631.

LeBouvier, G. L. (1955). *Lancet* **ii**, 1013.

LeBouvier, G. L. (1959). *Brit. J. Exp. Pathol.* **40**, 605.

Leppla, S. (1969). Ph.D. Thesis, Univ. of Wisconsin, Madison, Wisconsin.

Lerner, A. M. and Miranda, Q. R. (1968). *Virology* **36**, 277.

Levintow, L. and Darnell, J. E., Jr. (1960). *J. Biol. Chem.* **235**, 70.

Loeffler, F. and Frosch, P. (1898). *Zentrabl. Bakteriol. Parasitenk. Aft. Orig.* **23**, 371.

Loring, H. S. and Schwerdt, C. E. (1944). *Proc. Soc. Exp. Biol. Med.* **57**, 173.

McGregor, S. (1969). Ph.D. Thesis, Baylor University, Houston, Texas.

McGregor, S. and Mayor, H. D. (1968). *J. Virol.* **2**, 149.

McGregor, S. and Mayor, H. D. (1971). *J. Virol.* **7**, 41.

McGregor, S., Phillips, B. A., and Mayor, H. D. (1966). *Proc. Soc. Exp. Biol. Med.* **122**, 118.

Maclean, E. C. and Hall, C. E. (1962). *J. Mol. Biol.* **4**, 173.

Maizel, J. V., Jr. (1963). *Biochem. Biophys. Res. Commun.* **13**, 483.

Maizel, J. V., Jr. and Summers, D. F. (1968). *Virology* **36**, 48.

Maizel, J. V., Jr., Phillips, B. A., and Summers, D. F. (1967). *Virology* **32**, 692.

Mak, T. W. (1969). M. S. Thesis, Univ. of Wisconsin, Madison, Wisconsin.

Mak, T. W., O'Callaghan, D. J. and Colter, J. S. (1970). *Virology* **40**, 565.

Mak, T. W., O'Callaghan, D. J., Kay, C. M., and Colter, J. S. (1971). *Virology* **3**, 579.

Mandel, B. (1962). *Virology* **17**, 288.

Mandel, B. (1964). *Virology* **22**, 360.

Markham, R., Frey, S., and Hills, G. J. (1963). *Virology* **20**, 88.

Martin, R. G. and Ames, B. N. (1961). *J. Biol. Chem.* **236**, 1372.

Mattern, C. F.T. (1962). *Virology* **17**, 520.

Mattern, C. F. T. (1969). *In* "The Biochemistry of Viruses" (H. B. Levy, ed.), p. 55. Dekker, New York.

Mattson, D. E., Moll, T., and Barya, M. A. (1969). *Amer. J. Vet. Res.* **30**, 1577.

Mayer, M. M., Rapp, H. J., Roizman, B., Klein, S. W., Cowan, K. M., Lukens, D., Schwerdt, C. E., Schaffer, F. L., and Charney, J. (1957). *J. Immunol.* **78**, 435.

Mayor, H. D. (1964). *Virology* **22**, 156.

Medappa, K. C., McLean, C., and Rueckert, R. R. (1971). *Virology*, in press.

Michelin-Lausarot, P., Ambrosino, C., Steere, R. L., and Reichman, M. E. (1970). *Virology* **41**, 160.

Montagnier, L. and Sanders, F. K. (1963). *Nature* (*London*) **197**, 1178.

Munyon, W. and Salzman, N. P. (1962). *Virology* **18**, 95.

Mussgay, M. (1959). *Monatsh. Tierheilk.* **11**, 185.

Nixon, H. L. and Gibbs, A. J. (1960). *J. Mol. Biol.* **2**, 197.

O'Callaghan, D. J., Mak, T. W., and Colter, J. S. (1970). *Virology* **40**, 572.

Ohlson, M. (1963). *Acta Path. Microbiol. Scand.* **57**, 494.

Penman, S., Becker, Y., and Darnell, J. E. (1964). *J. Mol. Biol.* **8**, 541.

Philipson, L. (1964). *Biochem. Biophys. Res. Commun.* **17**, 352.

Phillips, B. A. (1969). *Virology* **39**, 811.

Phillips, B. A., Summers, D. F., and Maizel, J. V., Jr. (1968). *Virology* **35**, 216.

Planterose, D. N. and Ryan, J. R. C. (1965). *Virology* **26**, 372.

Polatnick, J. (1960). *Proc. Soc. Exp. Biol. Med.* **103**, 27.

Polatnick, J. and Arlinghaus, R. B. (1967). *Virology* **31**, 601.

Pyl, G. (1933). *Physiol. Chem.* **218**, 249.

Randrup, A. (1954). *Acta Pathol. Microbiol. Scand.* **34**, 366.

Robinson, L. K. (1950). *Proc. Soc. Exp. Biol. Med.* **75**, 580.

Roizman, B. (1957). *J. Immunol.* **82**, 19.

Roizman, B., Hopken, W., and Mayer, M. M. (1957). *J. Exp. Med.* **78**, 386.

Roizman, B., Hopken, W., and Mayer, M. M. (1958a). *J. Immunol.* **80**, 386.

Roizman, B., Mayer, M. M., and Rapp, H. J. (1958b). *J. Immunol.* **81**, 419.

Roizman, B., Mayer, M. M., and Roone, P. R., Jr. (1959). *J. Immunol.* **82**, 19.

Roy, D., Fraenkel-Conrat, H., Lesnaw, J., and Reichman, M. E. (1969). *Virology* **38**, 368.

Rueckert, R. R. (1965). *Virology* **26**, 345.

Rueckert, R. R. (1968). *In* "Methods in Cancer Research," (H. Busch, ed.), p. 455. Academic Press, New York.

Rueckert, R. R. and Duesberg, P. H. (1966). *J. Mol. Biol.* **17**, 490.

Rueckert, R. R. and Hall, L. (1971). In preparation.

Rueckert, R. R. and Schäfer, W. (1965). *Virology* **26**, 333.

Rueckert, R. R., Dunker, A. K. and Stoltzfus, C. M. (1969). *Proc. Nat. Acad. Sci. U.S.* **62**, 912.

Sanders, F. K. (1964). *In* "Techniques in Experimental Virology" (R. J. C. Harris, ed.), p. 277. Academic Press, New York.

Schaffer, F. L. and Frommhagen, L. H. (1965). *Virology* **25**, 662.

Schaffer, F. L. and Schwerdt, C. E. (1955). *Proc. Nat. Acad. Sci. U.S.* **41**, 1020.

Schaffer, F. L. and Schwerdt, C. E. (1959). *Advan. Virus Res.* **6**, 159.
Schaffer, F. L., Moore, H. F., and Schwerdt, C. E. (1960). *Virology* **10**, 530.
Schaffer, F. L. and Schwerdt, C. E. (1965). *In* "Viral and Rickettsial Diseases of Man."
 (F. L. Horsfall and J. Tamm, eds.), 4th ed., p. 94. Lippincott, Philadelphia, Penn-
 sylvania.
Scharff, M. D. and Levintow, L. (1963). *Virology* **19**, 491.
Scharff, M. D., Maizel, J. V., Jr., and Levintow, L. (1964). *Proc. Nat. Acad. Sci. U.S.*
 51, 329.
Schatz, A. and Plager, H. (1948). *Proc. Soc. Exp. Biol. Med.* **67**, 452.
Scherer, K. and Darnell, J. E. (1962). *Biochem. Biophys. Res. Commun.* **7**, 486.
Schmidt, N. J., Dennis, J., Frommhagen, L. H., and Lenette, E. H. (1963). *J. Immunol.*
 90, 654.
Schmidt, W. A. K. (1963). *Arch. Gesamte Virusforsch.* **13**, 368.
Schmidt, W. A. K. (1964). *Arch. Gesamte Virusforsch.* **14**, 508.
Schulze, P. and Gralheer, H. (1964). *Acta Virol.* **8**, 473.
Schwerdt, C. E. and Schaffer, F. L. (1955). *Ann. N. Y. Acad. Sci.* **61**, 740.
Scraba, D. G., Kay, C. M., and Colter, J. S. (1967). *J. Mol. Biol.* **26**, 67.
Scraba, D. G., Hostvedt, P., and Colter, J. S. (1969). *Can. J. Biochem.* **47**, 165.
Shapiro, A. L., Viñuela, E., and Maizel, J. V., Jr. (1967). *Biochem. Biophys. Res. Commun.*
 28, 815–820.
Speir, R. W. (1961a). *Proc. Soc. Exp. Biol. Med.* **106**, 402.
Speir, R. W. (1961b). *Virology* **14**, 382.
Speir, R. W. (1962). *Virology* **17**, 588.
Speir, R. W., Aliminosa, K. V., and Southam, C. M. (1962). *Proc. Soc. Exp. Biol. Med.*
 109, 80–82.
Spirin, A. S. (1963). *Progr. Nucl. Acid Res.* **1**, 301.
Stevens, C. L., Kerner, A. M., Bryant, C. P., and Poiley, E. J. (1968). *Bacteriol. Proc.*,
 p. 168.
Stoltzfus, C. M. and Rueckert, R. R. (1971). In preparation.
Strauss, J. H. and Sinsheimer, R. L. (1963). *J. Mol. Biol.* **7**, 43.
Strohmaier, K. and Mussgay, M. (1959). *Science* **130**, 217.
Summers, D. F. and Levintow, L. (1965). *Virology* **27**, 44.
Summers, D. F., Maizel, J. V., Jr., and Darnell, J. E., Jr. (1965). *Proc. Nat. Acad. Sci.*
 U.S. **54**, 505.
Tannock, G. A., Gibbs, A. J., and Cooper, P. (1970). *Biochem. Biophys. Res. Commun.*
 38, 298.
Taylor, J. and Graham, A. F. (1958). *Virology* **6**, 488.
Taylor, J. and Graham, A. F. (1959). *Trans. N. Y. Acad. Sci.* **21**, 242.
Theiler, M. (1937). *J. Exp. Med.* **65**, 705.
Tillotson, J. R. and Lerner, A. M. (1966). *Proc. Nat. Acad. Sci. U.S.* **56**, 1143.
Trautman, R. and Breese, S. S., Jr. (1962). *J. Gen. Microbiol.* **27**, 231.
Trautman, R., Savan, M., and Breese, S. S., Jr. (1959). *J. Amer. Chem. Soc.* **81**, 4040.
Tumilowicz, J. J. and Hummeler, K. (1964). *J. Bacteriol.* **87**, 1105.
Tyrrell, D. A. J. (1968). "Virology Monographs 2: Rhinoviruses." Springer-Verlag,
 New York.
Tyrrell, D. A. J. and Chanock, R. M. (1963). *Science* **141**, 152.
Vande Woude, G. F. (1967). *Virology* **31**, 436.
Vande Woude, G. F. and Bachrach, H. L. (1968). *Arch. Gesamte Virusforsch.* **23**, 353.

Vande Woude, G. F. and Bachrach, H. L. (1969). *Bacteriol. Proc.* p. 167.

Vande Woude, G. F. and Bachrach, H. L. (1971). *J. Virol.* **7**, 250.

Van Elsen, A. and Boeyé, A. (1966). *Virology* **28**, 481.

Van Elsen, A., Boeyé, A., and Teuchy, H. (1968). *Virology* **36**, 511.

Wallis, C. and Melnick, J. L. (1961). *Texas Rept. Biol. Med.* **19**, 683.

Wallis, C. and Melnick, J. L. (1962). *Virology* **16**, 504.

Warren, J. (1965). *In* "Virus and Rickettsial Infections of Man" (F. L. Horsfall, and I. Tamm, eds.), p. 562. Lippincott, Philadelphia, Pennsylvania.

Warren, J., Smadel, J. E., and Russ, S. B. (1949). *J. Immunol.* **62**, 387.

Watanabe, Y., Watanabe, K., and Hinuma, Y. (1962). *Biochim. Biophys. Acta* **61**, 976.

Watanabe, Y., Watanabe, K., Katagiri, S., and Hinuma, Y. (1965). *J. Biochem. (Tokyo)* **57**, 733.

Weber, K. and Osborn, M. (1969). *J. Biol. Chem.* **244**, 4406.

Wild, T. F., Burroughs, J. N., and Brown, F. (1969). *J. Gen. Virol.* **4**, 313.

Williams, R. C. (1953). *Exp. Cell Res.* **4**, 188.

Wittman, G. (1967). *Zentrabl. Bakteriol. Parasitenk. Infektionskr. Hyg. Abt. I. Orig.* **202**, 133.

Young, N. C., Hoyer, B. H., and Martin, M. A. (1968). *Proc. Nat. Acad. Sci. U.S.* **61**, 548.

CHAPTER 9 **Arboviruses: Incorporation in a General System of Virus Classification**

JORDI CASALS

I. Introduction

The arboviruses constitute a set defined by the epidemiologic fact, that they are transmitted between vertebrate hosts through the agency of biting, blood-sucking arthropods (Casals, 1957; WHO Study Group, 1967). Attempts have been made to incorporate them into universal systems of classification of viruses (Lwoff *et al.*, 1962); the attempts have resulted at times in confusion arising, in part, from the use of the term, arbovirus, to designate both an epidemiologic set and a set in a system based on properties of the virion.

Efforts toward clearing this confusion were initiated at least as far back as 1966 (Casals, 1966). It has been all along evident that the arboviruses constitute a heterogeneous set when the properties of the virion and the activity of the virion at the cellular level are considered. Since in the various systems of viral classification advanced (Lwoff *et al.*, 1962; Committee, 1965; Melnick and McCombs, 1966; Gibbs *et al.*, 1966; Bellet, 1967) the properties of the virion are the only, or main, basis for classifying, it follows that the arboviruses cannot as an undivided body be incorporated in those systems. The solution was to dismantle the arbovirus set and redistribute its parts into the pertinent groups of the universal system.

The International Committee on Nomenclature of Viruses (Wildy *et al.*, 1967) designated a Subcommittee on Viruses of Vertebrates which has made certain proposals (Andrewes, 1970) to end the confusion that had heretofore prevailed. These proposals are: to use the term arbovirus only as an epidemiologic designation; to distribute the arboviruses into the classes of the universal system to which they belong; and to create a name to designate the set of viruses that have RNA, cubic symmetry, and an envelope, regardless of whether they are or not arboviruses.

II. Arboviruses

A. DEFINITIONS

The word arbovirus is an abbreviation for "arthropod-borne virus of vertebrates"; it defines a concept not related to the chemical, physical, or morphological properties of the virion. That these properties may in the last resort determine all activities of a virus goes without saying; this, however, is irrelevant to the present definition.

The definition to which we suscribe (WHO Study Group, 1967) is: "arboviruses are viruses which are maintained in nature principally, or to an important extent, through biological transmission between susceptible vertebrate hosts by hematophagous arthropods." In the natural cycle are involved, in addition to virus, vertebrate host, and vector, vertebrate reservoir and amplifier; possibly the vector acts in some instances also as reservoir.

Essential to the definition is the expression "biological transmission," by which it is meant that a period of time, from 5 or 6 to 10 or 12 days, elapses between the moments when the vector becomes infected by biting a viremic host and when it can transmit the virus to a new vertebrate host. During this period, designated extrinsic incubation, the arthropod though infected cannot transmit the virus by bite; the virus multiplies in the tissues of the arthropod causing—with hardly an exception—no damage, ill effects or recognizable lesions; and finds its way to the salivary glands of the arthropod which can then transmit it by biting a new vertebrate host. Viremia ensues in the vertebrate setting up conditions for infection of a new arthropod and continuation of the cycle.

The definition excludes mechanical transmission—usually immediate—by an arthropod whose mouth parts have become contaminated on biting an infected host. It must not be concluded, however, that an arbovirus can be transmitted in nature only through biological transmission; infection of man by direct contact with patients, probably through droplet infection, or by ingestion of contaminated food are known to occur (Downs, 1970).

The recognized range of natural vectors has been so far confined to mosquitoes, ticks, *Phlebotomus* and *Culicoides*; mites, though on occasion suspected, have not been conclusively proved.

B. ANTIGENIC GROUPS

There are at present between 250 and 300 virus serotypes which with various degrees of legitimacy are assembled in the arbovirus set. The currently published, most comprehensive listing and description of these agents (Taylor, 1967) has 204 entries to which an additional 44 are being added (Berge, 1970, personal communication); another publication (WHO Study Group, 1967) lists, either by name or strain designation, 252 viruses some of which are not included in the former.

About 4/5 of the arboviruses are assembled in antigenic groups (Table I), on the basis of serological overlaps (Casals, 1957; WHO Study Group, 1967); according to the established concept, all cross-reacting viruses constitute a group. Several groups have been bound in a supergroup owing to the fact that occasional viruses from a group reproducibly show low titered antigenic overlaps with viruses from another group (Whitman and Shope, 1962; Casals, 1963). Under a strict application of the definition of antigenic group all such groups and viruses thus related should be assembled in one single group. Practical considerations have, however, made it advisable thus far to maintain the separate groups as given in Table I.

TABLE I

An Antigenic Classification of Arboviruses

Group	Number of viruses	Group	Number of viruses
A	18	Changuinola	2
African Horsesickness	9	Congo	2
Anopheles A	3	EHDD	2
Anopheles B	2	Ganjam	2
B	39	Hughes	3
Bakau	2	Kaisodi	3
Bluetongue of sheep	12	Kemerovo	5
Bunyamwera Supergroup		Mapputta	2
Bunyamwera	13	Mossuril	2
Bwamba	2	Nyando	2
C	11	Phlebotomus Fever	11
California	9	Qalyub	2
Capim	7	Quaranfil	2
Guama	6	Timbo	2
Koongol	2	Turlock	3
Patois	4	Uukuniemi	4
Simbu	15	Vesicular stomatitis	5
Tete	2	Ungrouped	43
Unassigned	1	Tacaribe[a]	6

[a] Evidence for an arthropod cycle in nature is scant for all the viruses in the group.

Number of viruses in the groups is in some instances, approximate. There are at this time well over 40 additional serotypes as yet incompletely identified and unreported, not included in the table.

Some groups include viruses for which there is little or no evidence that they are arboviruses. One of the groups, Tacaribe, is listed separately; while it has been assumed in the past that the viruses in the group may be arthropod-borne in nature or that they had a cycle, perhaps not the main one, involving an arthropod, current evidence seems not to support this assumption (see Section II,C,4).

C. Criteria Used in Practice for Defining an Arbovirus

In the strict sense the definition of arbovirus requires that the cycle of biological propagation from arthropod to vertebrate and back to arthropod be observed under controlled conditions as it occurs in nature.

Properties of a virus which are not directly related to the transmission cycle, though helpful at times for orientation, should not be entertained as defining criteria. Easy isolation and propagation by intracerebral inoculation into newborn or adult mice, or inactivation of the virus by lipid solvents are properties shared with other viruses not remotely suspected of being arthropod-borne, such as herpes and rabies. It is perhaps time even to ask critically whether antigenic relationship with well-established arboviruses is an acceptable norm.

Only properties relating to the transmission cycle should therefore be considered in attempting to determine whether a virus is an arbovirus. The amount of information available concerning the natural transmission of the arboviruses varies greatly from one to the other with the result that viruses are considered to be arthropod-borne with varying degrees of conformity with, or fulfillment of, the defining criteria (WHO Study Group, 1967; Downs, 1970). Even disregarding that the mere fact of investigating transmission as it occurs in nature may disrupt events, consequently transmission is no longer natural: it is a matter of record that complete observation and reproduction of the cycle, relatively undisturbed, in nature has been secured only in few instances, considering the large number of arboviruses accepted.

The degrees of information required to fulfill the criteria for definition of an arbovirus fall in three categories:

1. Criteria fulfilled:

   a. Observation of the complete natural cycle achieved.

   b. Observation of the natural cycle achieved in a less complete or uninterrupted manner; overwhelming epidemiological and laboratory evidence is available.

2. Criteria less adequately fulfilled:

   a. Artificial reproduction of the complete cycle in laboratory animals.

   b. Artificial reproduction of parts of the cycle, usually coupled with detection of virus in animal tissues by inoculation of experimental hosts.

   c. Serial propagation by experimental inoculation of blood-sucking arthropods.

3. Circumstances of isolation of the virus.
4. Other criteria independent from association with arthropods.

Whatever experimental procedures or ecological considerations are employed, it is generally agreed that multiplication of the virus in an arthropod and its transmission by bite to a vertebrate are the minimum criteria necessary before it can be accepted as biologically arthropod-borne.

1.a. Three viruses or virus diseases fall in this class; urban yellow fever, sandfly fever, and dengue. While in the early studies the actual viral sero-type is unknown, the observations with these three diseases remain as classical epidemiological examples that established a complete natural cycle. Undoubtedly the fact that they are diseases of man, one of them very serious, may account for early efforts made to solve the cycle for, by inter-rupting it, epidemics might be stopped. It must not be assumed from these examples, however, that human involvement is an essential feature of the arboviruses' ecology. Rather the opposite; infection of man is generally accidental, not essential to the perpetuation of the cycle and, moreover, usually leads to a deadend (Downs, 1970).

The natural cycle for urban yellow fever was definitely established by the work of the U.S. Yellow Fever Commission, in 1900-1901. Well-con-trolled observation showed that the disease was transmitted to healthy persons by mosquitoes, *Aedes aegypti*, that had fed on patients in the acute stage of their illness; these patients had contracted the disease through natural exposure. An interval of some 10–12 days—somewhat variable de-pending on circumstances—must elapse from the time a mosquito takes an infected blood meal to the moment when it can transmit it by bite to a well person, extrinsic incubation. Inoculation of blood taken from a patient during the first days of illness to a healthy person resulted in disease. Finally, it was later shown that control of *Aedes aegypti* prevented urban epidemics; this observation completed the case for the arthropod-vertebrate cycle with this illness.

A similar clear-cut proof of the natural cycle was observed (Doerr *et al.*, 1909) with sandfly fever. Transmission by the bite of a midge, *Phlebotomus papatasi*, was authenticated through the use of human volunteers. Midges were fed during the first days of illness on patients who had acquired the disease through natural exposure in an endemic area; 7–10 days later the arthropod transmitted the disease, by bite, to human volunteers in an area far removed from the endemic zone.

The association between dengue fever and a vector, *Aedes aegypti*, was established (Siler *et al.*, 1926) indicating the capacity of the mosquito to become infected by feeding on a patient in the acute stage of illness and its ability to transmit it from 10 to 11 days later, but not before, to a healthy volunteer.

1.b. There are a number of viruses for which the evidence that they are arthropod-borne in the natural infections that they cause is nearly as com-plete as that shown above. With these agents, the complete cycle in nature has not been observed in one continuous operation, mainly due to the fact that no such attempts (which are always difficult to carry out) were made, or found necessary, in view of other, overwhelming evidence that established them as arthropod-borne. With the agents in question the association in nature between a vector, the virus, and the disease in man or lower animals has been so well and repeatedly established; and the experimental transmission in the laboratory between vertebrate hosts by the vector, with a required extrinsic incubation, so well documented, that one must consider the natural cycle as adequately proved.

Within this category are included: Western equine encephalitis (WEE) and St. Louis encephalitis (SLE) and their vector, *Culex tarsalis*, in the western United States; Japanese encephalitis (JE) and *Culex tritaeniorhynchus*, in Japan; Russian tick-borne encephalitis of group B(RSSE) and the tick, *Ixodes persulcatus*, in the far eastern regions of the Soviet Union; Colorado tick fever (CTF) and *Dermacentor andersoni* in the Mountain States.

In this class may also be included, with in some instances less complete evidence, the following viruses and some of their associated vectors: Kyasanur Forest disease (KFD) and *Haemaphysalis spinigera* in Mysore, India; Nairobi sheep disease and *Ripicephalus appendiculatus*, in Kenya, Uganda; Murray Valley encephalitis (MVE) and *Culex annulirostris* in Australia; Rift Valley fever (RVF) and *Eretmapodites chrysogaster*, also *Aedes caballus*, in East Africa; West Nile (WN) and *Culex univittatus* in Egypt; Eastern equine encephalitis (EEE) and *Culiseta melanura* (in birds), in eastern United States; o'nyong-nyong (ONN) and *Anopheles funestus*, in eastern Africa; chikungunya and *Aedes* mosquitoes in Africa and Thailand.

A few additional viruses may also be included in the present category on the basis of good epidemiological association between virus, disease, and particular arthropods, even though studies on laboratory transmission may be lacking or unsatisfactory; the reason for the latter being the difficulty in colonizing the arthropod. The diseases are: African horsesickness and bluetongue of sheep which have as vectors *Culicoides* species; bovine ephemeral fever is also considered to have *Culicoides* as vectors. Recently (Plowright *et al.*, 1969) evidence has been adduced to suggest that African swine fever may be a vector-transmitted disease, *Ornithodoros moubata*, in Tanzania.

2. A large proportion of the arboviruses are so considered with evidence less complete than that shown with the viruses in Section 1.; the criteria for definition have been less adequately fulfilled. Within this partial fulfillment there are various degrees.

2.a.b. With some viruses the evidence consists in the reproduction in the laboratory of an artificial transmission cycle between vertebrate hosts by means of an arthropod. Mosquitoes are infected by inoculation of a viral suspension, or by feeding on a viremic animal experimentally inoculated, or are wild-caught infected; after a sufficient number of days to allow for an extrinsic incubation, the arthropod is made to feed on a new animal which is then observed for signs of illness, viremia, or antibody development. The cycle is completed, at least once, by allowing new mosquitoes to feed on the vertebrate at the time it circulates virus and determining whether the mosquito is infected, either by further transmission by bite or by inoculation of its tissues into susceptible animals.

With some viruses the experimental evidence does not reproduce the complete cycle; thus, after feeding on an infected vertebrate host and allowing for time to elapse, the arthropod is tested for evidence of viral multiplication by inoculation into a host system. Or, conversely, an artificially infected arthropod is set upon a susceptible vertebrate host and the tissues of the latter tested for presence of virus.

The artificiality of the system consists in that the host employed is seldom the natural one, but often newborn mice; the mosquito may have no or little epidemiological significance in nature; and in interrupting the cycle and testing for virus by inoculation of tissue suspensions.

There are in this class at least 29 viruses and among them, Bunyamwera, California encephalitis, Ilheus, Mayaro, Middelburg, Nodamura, Oriboca, Semliki, Sindbis, vesicular stomatitis Indiana (VSV-I) and New Jersey (VSV-NJ), Wesselsbron, and Zika. All have also been isolated from wild-caught arthropods and eight or ten have, in addition, been propagated serially by mosquito inoculation.

2.c. With other viruses the evidence for their arboviral nature, while still pertaining to the transmission cycle, is more artificial than thus far considered; it consists on the observation that a virus can be maintained serially by parenteral inoculation of arthropods, usually mosquitoes. Inoculation is done intrathoracically and passage to the next arthropod is done a few days later either by grinding the entire mosquito or, better yet, its salivary glands only. Maintenance of a virus in this fashion for several consecutive passages is considered good evidence that it is an arbovirus. Strictly considered, this evidence is valid only if positive; a true arbovirus may fail to multiply in the wrong kind of arthropod.

Also included in this group are a few viruses which on inoculation to mosquitoes or fed to mosquitoes have been found to multiply and persist for some time in the body of the arthropod, determined by mouse inoculation.

At least 50 viruses have been proved to propagate by serial passage in mosquitoes and one in ticks; or have shown to multiply and persist. With few exceptions—8 or 10—all have also been isolated from wild-caught arthropods. Representative viruses are: Anopheles A and B, Apeu, Bussuquara, Bwamba, Cocal, Corriparta, epizootic hemorrhagic disease of deer (EHD-NJ), Eubenangee, Guaroa, Hart Park, Marituba, Melao, Oropouche, Quaranfil, Tacaiuma, Uganda S, Venkatapuran, Witwatersrand, and Wyeomyia.

3. With a large number of viruses, over 100, the only criterion at the moment available for consideration as arboviruses is circumstances of isolation. Viruses are classed as arboviruses that have been isolated, some once, others repeatedly, from wild-caught mosquitoes, ticks, midges, or other blood-sucking arthropods which when ground up and inoculated into susceptible hosts had digested their last blood meal. Viruses isolated from sentinel animals exposed in such a manner that their only likely source of infection are hematophagous arthropods are provisionally considered or suspected of being arboviruses.

4. In another category, viruses are included as arboviruses solely on the grounds of an antigenic relationship with established arboviruses. From the total number, 240, considered in this analysis, 26 distinct serotypes belong in this category; in addition, there are 6 to 8 viruses of the antigenic group Tacaribe that deserve special consideration.

Attempts to passage or detect multiplication of the virus in arthropods have been reported as negative with some: bat salivary gland and Modoc viruses of group B and Candiru and Icoaraci of phlebotomus fever group. No reported attempts to test for arthropod susceptibility are available with 22 among which are included seven group B agents, four from Simbu group, and three from phlebotomus fever group.

An antigenic group, Tacaribe, presents special problems. It includes the etiological agents of several important diseases of man: Junin (Argentinian hemorrhagic fever), Machupo (Bolivian hemorrhagic fever), and Lassa (Lassa fever) viruses. In addition (Rowe et al., 1970b), a virus not known to have an arthropod cycle in nature, lymphocytic choriomeningitis, is antigenically part of the group; as are other agents not known to be associated with human disease, Amapari, Pichinde, Tacaribe, Tamiami, and two other viruses isolated in South America, as yet incompletely characterized (Johnson, personal communication, 1970). Junin virus has been isolated from mites on a few occasions (Mettler, 1969); Tacaribe from mosquitoes once (Downs et al., 1963); and Amapari on a few occasions from mites taken off rodents that were aviremic at the time (Woodall, personal communication,

1968). With these exceptions, there is nothing in the epidemiology of the diseases that indicates arthropod transmission to man or between lower animals; determined efforts to isolate Machupo virus from arthropods in the course of a large epidemic failed (Kuns, 1965), while its presence in the urine of chronically infected rodents points to a different mode of transmission than by the bite of an arthropod (Johnson, 1965). In view of the importance of the human diseases involved, inclusion of this antigenic group with the arboviruses is of more than passing importance and should no longer be done unless there is definite evidence in its favor. As it now stands the evidence fails to support an arthropod cycle in the maintenance of the viruses, nor do they appear to fit the definition of an arbovirus by any of the currently accepted criteria.

Two viruses, Kern Canyon and Lagos bat are, so far, antigenically unrelated to any established arboviruses; attempts to passage them in mosquitoes have failed. These agents fulfill none of the requirements for consideration as arboviruses and should, therefore, be excluded from the set, until otherwise established.

The existence of viruses antigenically related to proved arboviruses but which are themselves not able to propagate in arthropods, raises a doubt with respect to the adequacy of antigenic relatedness as a criterion for inclusion; the solution is to accept the fact that nonarboviruses can be related to arboviruses. In this connection Baker's comments (1943) on the evolution of the arthropod-borne viruses may be relevant; successive steps in the evolution may lead from an originally latent virus of an arthropod to a virus that has a primary arthropod-vertebrate cycle, then to another with a side, secondary arthropod-vertebrate cycle and, finally, to a virus with a vertebrate-vertebrate propagation. In the latter case the virus would eventually lose its potential to multiply in arthropod vectors. There are instances in which arthropod-borne viruses can maintain themselves, in limited conditions, in nature in the absence of an arthropod vector, for example, EEE in pheasants (Holden, 1955); VEE in man (Briceño-Rossi, 1964), and group B tick-borne encephalitis (CETB) virus between cow and man by the ingestion of contaminated milk (Blaskovic, 1967).

III. Properties of the Virions

In systems of universal viral classification currently under consideration the properties of the virion are the criteria used for establishing different taxons; those of the arboviruses are analyzed in this section.

Since arboviruses are defined on epidemiological-ecological grounds, there is no a priori reason why the agents included in the set should all be alike in other respects, for example, in the properties of their virions; nor why nonarboviruses could not share basic properties with arboviruses.

In the following analysis, African horsesickness and bluetongue viruses are each considered as one virus, disregarding their numerous antigenic types; tick-borne virus of group B (Russian spring-summer encephalitis and central European tick-borne encephalitis) is also considered as one agent. Viruses of the antigenic group Tacaribe are included even though there are reasons (see Section II,C,4) for excluding them from the arbovirus set. The total number of viruses considered is 240.

A. NUCLEIC ACID TYPE

Nucleic acid examination has been reported in at least 40 arboviruses; in a number, infectious nucleic acid has been extracted and the type determined by the effect of the corresponding nucleases; in others the type is inferred from the effect of bromodeoxyuridine (BUDR) on viral multiplication in cell cultures.

All these viruses contain RNA with the exception of African swine fever, from which infectious DNA has been extracted (Plowright et al., 1966). Infectious RNA has been extracted from at least the following 16 viruses: bluetongue, CETB, chikungunya, dengue 1, dengue 2, EEE, JBE, MVE, Semliki, Sindbis, SLE, VEE, VSV-I, WEE, West Nile, and yellow fever.

In another 23 viruses the presence of RNA has been deduced from the lack of inhibition by BUDR of virus multiplication in cell cultures. The viruses are: African horsesickness, Cocal, CTF, Congo (Crimean hemorrhagic fever), Corriparta, bovine ephemeral fever, Gamboa, Guaroa, Juan Diaz, Junin, Kemerovo, Lipovnik, Lone Star, Louping ill, Machupo, Matucare, Mayaro, Mermet, Omsk hemorrhagic fever, Tamiami, Tribec, Uukuniemi, and VSV-NJ.

B. MORPHOLOGY

Considerable advances have been accomplished in this sector during the last 3 or 4 years (see Casals, 1966; WHO Study Group, 1967). Electron microscopy of infected cells in thin sections and negative contrast staining of sedimented viral suspensions has been extended to arboviruses of several antigenic groups as well as to ungrouped ones. As a consequence several patterns are now discerned in these viruses with respect to symmetry of capsid, shape, size, envelope, and morphogenesis. The emergence of these patterns amply documents the claim (Casals, 1966; WHO Study Group, 1967) that the arboviruses are a mixed set that does not fit as a whole in any of the taxons of the proposed systems of classification based on properties of the virion.

Even with these recent contributions, data are not available on the large majority of arboviruses. Results of electron microscopy have been reported in 63 of the 240 viruses considered in this survey; in 20, a statement is made on the type of symmetry as well as giving information on size, shape, envelope, and site of development and maturation of the viron. The reports on the remaining 43 agents fail to state type of symmetry or the statement is to the effect that none was discernible; the description of these viruses is limited to the other morphological characteristics.

1. *Symmetry*

The capsid's symmetry has been clearly stated in 15 viruses; stated less definitely in 4; and hinted at in one instance.

Clearly, definitely stated symmetries are: cubic, in African horsesickness (Breese *et al.*, 1969), African swine fever, bluetongue, Eubenangee, Kemerovo, and Sindbis viruses; helical, in Batai, Inkoo, and Uukuniemi; helical-complex, in Cocal, bovine ephemeral fever, Hart Park-Flanders, Kern Canyon, VSV-I, and VSV-NJ viruses.

The type of symmetry has been stated less categorically but, it appears, still convincingly in four viruses: Chenuda, CTF (Murphy *et al.*, 1968a), and Middelburg and Semliki (Simpson and Hauser, 1968); these four viruses show cubic symmetry.

Observations with Powassan virus (Abdelwahar *et al.*, 1964) showed under the envelope a surface geometric arrangement compatible with cubic symmetry.

With an additional virus, VEE, it was reported (Klimenko *et al.*, 1965) that helical structures presumably representing tightly packed nucleoprotein components of the virus were seen; no claim was made on symmetry.

2. *Other Morphological Characteristics*

Morphological details other than symmetry have been detected in 43 additional arboviruses: shape, size, envelope, and density or lightness of different areas—core or halo. These details were also noted with the viruses in the preceeding subsection (Section III,B,1).

a. Shape. The virion's shape is given for nearly all the viruses examined by electron microscopy.

The six viruses with helical-complex symmetry are bullet-shaped, stubby rods with a round and a blunt end. The 11 viruses with cubic symmetry are reported as being round, polygonal or, more precisely, icosahedral in shape.

The shape of 41 viruses, including those with helical symmetry, is described by one, or several, of the following words: round, oval, spherical, or polygonal. Two viruses, Machupo and Tacaribe (Murphy *et al.*, 1969), are reported to be round, oval, and pleomorphic.

b. Size. Reported size determinations by electron microscopy show that the arboviruses are heterogeneous (Table II). The viruses with small di-

TABLE II

SIZE OF ARBOVIRUSES DETERMINED BY ELECTRON MICROSCOPY

Virus	Diameter, or width × length (nm)
African swine fever	200
Machupo, Tacaribe[a]	60–260; mean 110
Bunyamwera, California, Guaroa, Gumbo Limbo, Keystone, La Crosse, Lone Star, Marituba, San Angelo, Shark River, Uukuniemi	90–100 (Shark River, 104)
Chenuda, Colorado tick fever	80
African horsesickness, Eubenangee, Kemerovo, Middelburg, Rift Valley fever, Semliki, Sindbis, Tribec, VEE	60–80 (AHS, also 94; Eubenangee, enveloped, 101)
Nodamura	55–75
Aura, Batai, bluetongue, Cache Valley, chikungunya, EEE, Getah, Inkoo (Tahyna-like), Kairi, Manzanilla, Melao, Oriboca, Pacui, Restan, WEE, Wesselsbron, Whataroa, Wyeomyia	45–58
Mayaro	40
Dengue 1, dengue 2, CETB-RSSE, Cowbone ridge, JBE, Langat, MVE, Omsk HF, Powassan, SLE, West Nile, yellow fever	30–40 (a few extremes 17–50)
Bullet-shaped: bovine ephemeral fever, Cocal, Hart Park, Kern Canyon, VSV-I, VSV-NJ	60–73 × 120–220

[a] See footnote Table I.

ameter, between 30 and 40 nm, all belong thus far in group B. The distribution of other grouped viruses in size categories does not correspond precisely with their antigenic groups. Inconsistencies in reported sizes may, in part, be due to differences in technique; also, some reports give by preference the internal diameter of enveloped viruses (Bastardo *et al.*, 1968; Bergold *et al.*, 1969).

In addition to the more precise determinations by electron microscopy, estimates have been reported on the size of other arboviruses based on their filterability through commercially available graded membranes. The diameter of these viruses, on the assumption that they are spherical, is considered to be between the average pore diameter (APD) of a membrane that allows passage with no loss of infectivity and that of the next, tighter membrane that substantially reduces infectivity; a correction factor gives the estimated size range (Casals, 1968). The estimates between relatively wide margins for a number of viruses are: Mt. Elgon bat, between membranes of APD 450 and 220 nm; Bahig, Bhanja, Bwamba, Caraparu, Congo, Gamboa, Ganjam, Grand Arbaud, Junin, Kaisodi, Mapputta, Murutucu, Quaranfil, Silverwater, Tensaw, Thogoto, and Wad Medani, between 220 and 100 nm; Corriparta, Koongol, Matucare, and Ntaya, between 100 and 50 nm.

Other size estimates, also by filtration, have been given for additional viruses: Zika, between 18 and 25 nm; Chagres, between 55 and 110; and Turlock, between 120 and 128 nm.

c. Envelope. The presence or absence of an envelope surrounding the capsid has been settled with those arboviruses whose symmetry has been determined (see Section III,B,1) and with nearly all in which electron microscopy showed the shape, if not the structure, of the viral particle. As with respect to other properties of the virion, the arboviruses fall in distinct categories.

Statements that an envelope was present have been made for 46 viruses of the 59 examined, including viruses from several antigenic groups and ungrouped ones. That no envelope was visible has been reported for: African horsesickness, bluetongue, and Rift Valley fever viruses. It has also been stated that, on the whole, Chenuda, CTF, Eubenangee, and Kemerovo viruses had no envelopes, although an occasional extracellular enveloped virion could be detected. Finally, no statements are made for 6 more viruses, dengue 1, dengue 2, Lone Star, MVE, Tribec, and Wesselsbron; presumably, the ones belonging in group B are enveloped.

d. Morphogenesis; Maturation. Electron microscopy of thin sections of infected cell cultures or mouse brain tissue reveals a diversity in the details of development and maturation of arboviruses.

Based on available descriptions on about 35 viruses there appears to be a common property: all replicate in the cytoplasm; beyond this several distinct patterns have been observed.

Group A nucleoids appear to form at cytoplasmic membranes, sometimes are scattered throughout the cytoplasm; crystalline arrays have been observed. The nucleoids migrate to the plasma membrane where complete virions are formed by budding and extrusion into intercellular spaces (Morgan *et al.*, 1961; Acheson and Tamm, 1967; Lascano *et al.*, 1969).

Viruses of group B appear to replicate and bud almost exclusively on thickened membranes of cytoplasmic organelles which envelope the viral particle; these accumulate in the lumina of the distended endoplasmic reticulum membranes and are released into intercellular spaces either by migration or cell disruption (Ota, 1965; Murphy *et al.*, 1968b; Filshie and Rehacek, 1968).

Several members of the Bunyamwera and California groups have been examined (Southam *et al.*, 1964; Murphy *et al.*, 1968c,d) and observed to have similar morphology and mode of maturation. The virion matures by budding from cytoplasmic membranes into cisternae or vacuoles, predominantly in the golgi area; no budding is prominent at the cell surface membrane. Virus release is by cell disruption or through transport by migration of the vacuole to the cell margin, fusion, and expulsion.

Certain viruses, CTF and Chenuda, share some developmental characteristics which are similar to those of reoviruses (Murphy *et al.*, 1968a). The virus particles are associated with intracytoplasmic granular matrices, and with arrays of intracytoplasmic filaments and kinky threads; these formations are distributed throughout the cytoplasm. The virus is, for quite a while, almost exclusively intracellular; later, it is released by cell disruption. No budding was observed, although a few enveloped particles could be seen in vacuoles; there were no crystalline arrays.

The arboviruses with bullet shapes characteristically mature at marginal cytoplasmic membranes and bud through the cell membrane to accumulate extracellularly (Howatson and Whitmore, 1962).

C. LIPID SOLVENT SUSCEPTIBILITY

Andrewes and Horstmann (1949) proposed that the susceptibility of viruses to the action of lipid solvents be used as a criterion for classification. Ethyl ether and sodium deoxycholate (SDC) and, to a lesser extent, chloroform have been used (Theiler, 1957; Sunaga *et al.*, 1960; Feldman and Wang, 1961). Inactivation of viral infectivity by these chemicals indicates the presence of essential lipids in the virion and is generally accepted as equivalent to possession of an envelope by the virus particle.

Determination of the action of these chemicals on the virus is technically simple; a large proportion of the arboviruses have been studied. Of the 240 viruses surveyed, 191 have been tested; 44 of the remaining belong in antigenic groups most of whose members have been tried and it is assumed that they would behave like their group mates. Information is lacking in 5 ungrouped viruses.

The majority of the viruses have been tested with SDC at 0.1% final dilution, others at a dilution of 0.5%. It has been shown (Sunaga *et al.*, 1960) that the effect of SDC is dependent on concentration; also that chloroform has a stronger inactivating effect than either ether of SDC. The effects of the latter are considered similar.

The dosage-dependent effect of SDC may lead to contradictory conclusions concerning the resistance of a virus; resistance ought to be stated in connection with the concentration of chemical used. A clear demonstration of differing degrees of susceptibility is given in Table III, based on tests carried out in our laboratory. The procedure used in all cases was to incubate at 37°C mixtures of equal volumes of a virus suspension, 10^{-2}, and of an adequate dilution of SDC; the residual virus was titrated by the intracerebral route of inoculation in newborn mice. The amount of inactivated virus by comparison to that present in the control suspension is expressed in dex (Haldane, 1960).

As shown in Table III, mouse polioencephalitis and encephalomyocarditis viruses, picornaviruses, and mouse hepatoencephalitis virus, reovirus type 3, are resistant to the action of 0.5% SDC. Some arboviruses, Chenuda, CTF, EHD-NJ, and Kemerovo are inactivated by 0.5% SDC but resist a 0.1% solution; the remaining listed viruses are inactivated by SDC diluted 0.1%. It appears, therefore, that on the basis of the effect of SDC viruses should, properly, be classified as resistant, partially resistant, and susceptible.

All the 191 arboviruses tested are reported to be susceptible to the action of one or more of the chemicals, usually SDC, with the following exceptions: Nodamura virus, tested in the form of a suspension of infected mouse brain tissue, is resistant (Scherer, 1968); African horsesickness and bluetongue of sheep are resistant or partially resistant (Howell, 1962; Studdert, 1965); Chenuda, CTF, EHD-NJ, and Kemerovo (Table III; also Borden *et al.*, 1971), Corriparta (Carley and Standfast, 1969), Eubenangee (Schnagl *et al.*, 1969), and Tribec are partially resistant.

TABLE III

INACTIVATION OF SELECTED VIRUSES BY SODIUM DEOXYCHOLATE

Virus, strain	Sodium deoxycholate $(\%)^a$	
	0.5	0.1
Polioencephalitis, mouse (GD7)	−0.4	
Encephalomyocarditis, mouse (CDC)	−0.3	
Reovirus 3 (hepatoencephalitis, mouse)	0.8	
Chenuda (Eg Ar 1152)	+4.6	−0.4
Colorado tick fever	+4.5	0.9
EHD of deer, New Jersey	4.1	0.1
Kemerovo (R10)	+4.5	−0.5
Sindbis (Eg Ar 339)	+5.2	2.2
Bunyamwera (Smithburn)	+4.7	4.7
Junin (XJ)	+4.0	+4.0
Nyamanini (Eg Ar 1304)	+4.4	+4.3
Oriboca (Be An 17)	5.1	5.4
Uukuniemi (S-21)	+2.6	+3.6
West Nile (Eg 101)		+7.0

^{a} Dex of virus inactivated; −, indicates that treatment by SDC increased the virus titer; +, indicates that no end-point of inactivation was reached.

IV. Arboviruses in a General System of Classification

A. CRITERIA FOR SYSTEMS OF CLASSIFICATION

Systems for differentiation of major groups of viruses and classification are, essentially, particle oriented. An International Committee on Nomenclature of Viruses (1963) recommended that a system be established with a hierarchy of criteria in which were considered: nucleic acid type, symmetry of capsid, presence of envelope, and size. Subdivisions within the major groups were to be made on cytopathological, immunological, biochemical, and other criteria. In general, most of the systems proposed (see Chapter 1, by Dr. Lwoff; also Section I), have accepted these criteria at least in part.

The principle on which these systems are based has been adversely criticized on the grounds that since the visible characteristics of the virion are a phenotypic expression they are less important than knowledge of the genotype. A system based on properties of the virion may not represent phylogeny, it assumes an unproved sequence of values or hierarchy, creates artificial groups with no valid relationships discernible among them (Cooper, 1967; Gibbs *et al.*, 1966).

While recognizing the possible flaws of currently proposed systems they are, nevertheless, the only ones available; an attempt will be made in the next section (see Section IV,B) to fit the arboviruses in a virion-based general system.

B. Incorporation of the Arboviruses in a Universal System

Analysis of the virion's properties (Section III) makes it abundantly clear that the arboviruses are heterogeneous. Any attempt to fit them into generally accepted taxonomic groups of a universal system requires that the arbovirus set be disassembled and the resulting subsets or individual viruses be then placed in the corresponding taxonomic groups. One such attempt is represented schematically in Fig. 1 and given in some detail in Table IV.*

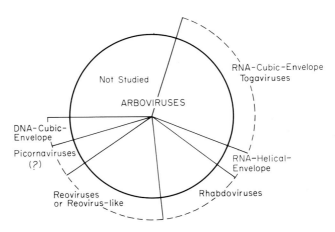

Fig. 1. Distribution of the arboviruses among groups of a universal system of viral classification.

* Figure 1 and Table IV, as well as the discussion that follows, represent in a general manner, if not in detail, the substance of a report by an *ad hoc* Study Group for the Arboviruses to Dr. C. H. Andrewes, Chairman, Vertebrate Virus Subcommittee, International Committee on Nomenclature of Viruses (Andrewes, personal communication, 1969). The Study Group consisted of: Dr. J. S. Porterfield, Chairman, Dr. J. Casals, Prof. M. P. Chumakov, Dr. Claude Hannoun, and Prof. M. Mussgay.

TABLE IV

INCORPORATION OF THE ARBOVIRUSES IN A UNIVERSAL SYSTEM OF CLASSIFICATION

Arboviruses		Togaviruses (RNA-cubic-enveloped)	
Not togaviruses	Togaviruses	Arboviruses	Not arboviruses
African horsesickness	Group A	Antigenically, group A	Antigenically, group B[a]
Bluetongue	Sindbis	Sindbis	Modoc[c]
Kemerovo	Middelburg[b]	Middelburg	Bat salivary gland[c]
Tribec	Semliki[b]	Semliki	
Colorado tick fever	Other (?)	Other	Antigenically, Tacaribe group:
Chenuda (? NA)			Arenoviruses
Corriparta (? NA)			LCM[c]
Other	Group B	Antigenically, group B	Machupo[c]
RNA-cubic-naked	Tick-borne enc.[b]	Tick-borne enc.	Tacaribe[c]
Reovirus, reoviruslike; diplornavirus	Omsk HF[b]	Omsk HF	Lassa[c]
	St. Louis[b]	St. Louis	Other[c]
Nodamura (cubic ?)	Powassan[b]	Powassan	
RNA-cubic-naked	Other (?)	Other (?)	Antigenically, other arbovirus groups
Picornavirus			?
	Other groups	Other antigenic groups	
Batai (? NA)	?	Some arboviruses	
Inkoo (? NA)		Some not arboviruses	Rubella

TABLE IV (*Continued*)

Arboviruses		Togaviruses (RNA-cubic-enveloped)	
Not togaviruses	Togaviruses	Arboviruses	Not arboviruses
Uukuniemi (RNA)	Ungrouped	Ungrouped	Other RNA-cubic-enveloped (?) or
RNA-helical-enveloped	?	Some arboviruses	some of the RNA-undetermined
Myxoviruslike		Some not arboviruses	symmetry-enveloped (?)
Cocal			
VSV-NJ			
VSV-I			
Bovine ephemeral fever			
Hart Park (? NA)			
Other			
RNA-helical: complex-enveloped			
Rhabdovirus			
African swine fever			
DNA-cubic-enveloped			
Iridovirus			
Cotia			
Poxviruslike			

[a] Serial propagation by inoculation in mosquitoes unsuccessfully attempted with Modoc and bat salivary gland viruses. There are at least 6 additional viruses in antigenic group B which may be togaviruses for which there is no proof of transmission by arthropods in nature nor recorded attempts to serial passage in the laboratory.

[b] Likely to be a togavirus; however, no categorical statement on its symmetry available.

[c] No evidence that it is arthropod-borne; possibly a togavirus but there is no knowledge on its symmetry.

Figure 1, not drawn to scale, is given in order to show that arboviruses belong in different taxons in a classification based on properties of the virion; these taxons in turn comprise viruses that are not arboviruses. The designation "togavirus" was suggested by the Study Group as a temporary expedient to prevent the confusion arising from the use of a word, arbovirus, to designate both an epidemiological set (arthropod-borne viruses of vertebrates) and a taxon in the general system (viruses with RNA-cubic or presumed cubic symmetry-envelope). As the figure also shows, there are arboviruses that cannot be classified for lack of available information.

In Table IV an effort is made to present a synthesis even though much information is still not at hand. Considering first the known or accepted arboviruses (left half of the table) they fall into the following divisions.

1. Arboviruses that are not togaviruses (RNA-cubic-enveloped) but fit in other established or proposed taxons:

a. *Reovirus, reoviruslike* (Borden *et al.*, 1971), *diplornavirus* (Verwoerd, 1969). Absence of an envelope, resistance or partial resistance to SDC and characteristic virion structure place the listed viruses, probably additional ones not listed for lack of information, in this group. Chenuda and CTF viruses have, in addition, some features in their development and maturation that are similar to those of reoviruses (Murphy *et al.*, 1968a).

b. *Picornavirus.* Nodamura virus has been maintained serially in mosquitoes by parenteral inoculation and transmitted by bite; is resistant to ether and SDC, contains RNA, and is relatively small (Scherer, 1968). Since, however, there are no published reports on the type of its symmetry, inclusion in the picornaviruses is tentative.

c. *Myxoviruslike.* The viruses in this set, Batai, Inkoo, and Uukuniemi have helical symmetry (von Bonsdorff *et al.*, 1969); the aspect of the helix differs from that of the myxoviruses. They have, presumably, RNA although no report on the type of nucleic acid is available for all.

d. *Rhabdovirus.* Vesicular stomatitis viruses (New Jersey and Indiana serotypes), Cocal, Hart Park, and, probably, a few others are bullet-shaped, enveloped viruses containing RNA and showing a complex structure with an internal helix characteristic of the rhabdoviruses. The criteria for considering bovine ephemeral fever an arbovirus are not totally fulfilled; however, the epidemiological circumstances justify inclusion in the set.

e. *Iridovirus.* This name had been suggested (Committee, 1965) to designate large DNA viruses, with an envelope and cubic symmetry of the compound type; to this description answer tipula iridescent and African swine fever viruses (Almeida *et al.*, 1967). It may be premature to consider

African swine fever virus an arbovirus; there is, however, good epidemiological evidence to this effect, including mention of transmission by the bite of a tick, *Ornithodoros erraticus* (Plowright *et al.*, 1969).

f. *Poxviruslike*. A virus, Cotia, isolated repeatedly from wild-caught mosquitoes has been observed to have a virion morphology similar to the poxviruses (Borden, Shope, and Murphy, personal communication, 1970).

2. Arboviruses that are togaviruses. This is a set that has been usually referred to as the "typical or true" arboviruses. As indicated in Table IV, lack of definite information on capsid's symmetry prevents at this time inclusion in this set of many arboviruses that, presumably, will in time be included. Sindbis virus is definitely a member of this taxon (Horzinek and Mussgay, 1969); so are, most likely, Middelburg and Semliki viruses (Simpson and Houser, 1968) and, probably, the listed viruses of antigenic group B. At the moment it is not justified to assign other arboviruses, grouped or ungrouped, to this taxon for lack of knowledge of capsid symmetry.

Having distributed a number of arboviruses in accepted divisions of a general system and having placed others in a taxon, togavirus, heretofore recognized but unnamed, it remains to see whether viruses exist which answer the general description of togaviruses but are not arboviruses. (This analysis appears on the right half of Table IV.)

3. Togaviruses that are arboviruses. This set should be, of course, an exact duplicate of the converse, i.e., arboviruses that are togaviruses. If electron microscopy reveals in the future that the symmetry of the listed viruses of antigenic groups A and B, as well as of other unlisted grouped or ungrouped viruses is of cubic type, the viruses so characterized being arboviruses will also be included in this set.

4. Togaviruses that are not arboviruses. This is a particularly difficult set to consider due to lack of information on virion properties, particularly capsid symmetry, and on the natural epidemiological cycle of the agents involved.

The two viruses of antigenic group B listed, Modoc and bat salivary gland, have not been isolated from wild-caught arthropods; attempts to passage them by parenteral inoculation of mosquitoes have failed (Whitman, personal communication, 1969); nor are there any epidemiological associations that favor the view that they have an arthropod cycle in nature. There is no evidence by any of the criteria used that these two viruses are arboviruses. There are 6 additional viruses, antigenically belonging in group B, for which there is a similar lack of evidence that they are arthropod-borne;

no reported information is available as to whether attempts have been made to passage them by parenteral inoculation in arthropods. The viruses are: Cowbone ridge, Dakar bat, Entebbe bat, Israel turkey meningoencephalitis, MML, and Negishi.

It cannot be stated whether Modoc, bat salivary gland, and the other viruses are togaviruses. Their nucleic acid type has not been reported, nor is there knowledge about their morphology; should these viruses by analogy to other group B agents be considered or proved to be togaviruses, they might well be examples of togaviruses that are not arboviruses.

The viruses of antigenic group Tacaribe are included in a newly defined taxon for which the name arenoviruses has been proposed (Rowe et al., 1970a). The arenoviruses have antigens in common, contain RNA, are susceptible to lipid solvents, and have an envelope; no capsid symmetry has been discerned. If future investigations were to reveal cubic symmetry a decision would be needed concerning the position of the arenoviruses with respect to the larger, inclusive taxon of the togaviruses. The lack of convincing evidence that the viruses of the Tacaribe group are arthropod-borne has been reviewed (see Section II,C,4).

Rubella virus appears to be a togavirus that is not arthropod-borne. It contains RNA; by electron microscopy it shows an electron-dense core surrounded by a clear halo, interpreted as an envelope; it develops by budding through the cytoplasmic membrane or sometimes into cytoplasmic vacuoles; its diameter is 60 nm and it is susceptible to ether and SDC (Holmes and Warburton, 1967; Holmes et al., 1969). Recent studies (Mussgay, personal communication, 1970) indicate that rubella virus has an isometrical core strongly suggesting icosahedral symmetry. There is no evidence that rubella virus is arthropod-borne; nor has it been found antigenically related to any of the arboviruses (Mettler et al., 1968).

The position in the present scheme of classification of nonarthropod-borne viruses that have RNA, an envelope, and undetermined type of symmetry, perhaps even no symmetry, is much too vague in all respects to warrant a discussion.

V. Conclusions

There are various reasons why it may be premature to try to incorporate such a large and heterogeneous group, the arboviruses, into a general system of viral classification. As the preceeding sections have shown there is on the whole scant information on the properties of their virion; in hardly 10% of the arboviruses is there complete knowledge of the type of nucleic acid, capsid symmetry, and envelope, properties whose knowledge is indispensable for current viral taxonomy.

Furthermore, a system of classification has not yet been fully agreed upon; while, in general, a system based only on properties of the virion is accepted, there is no complete agreement concerning the divisions and subdivisions in the scheme, nor on the proper value to assign to each of the virion's properties in the hierarchy.

In the group consisting of viruses with RNA, cubic symmetry, and no envelope there are two main divisions, picornaviruses and reoviruses. On the basis of recent work with African horsesickness and bluetongue viruses (Verwoerd, 1969; Breese et al., 1969) and with Chenuda, Colorado tick fever, Kemerovo, and other viruses (Borden et al., 1971) it has been proposed that new taxons be recognized, diplornavirus and reoviruslike. If these proposals are found acceptable, should there be or not an all encompassing label for all the viruses that have these three basic properties, RNA, cubic symmetry, and no envelope: picornaviruses (including enteroviruses and rhinoviruses), reoviruses, reoviruslike viruses, and diplornaviruses?

Still more complex and, seemingly, more unresolved is the situation with RNA, helical (or helical component) symmetry, enveloped viruses (Waterson and Almeida, 1966). Agents so defined are a mixed collection for which there is no overall designation; the term myxovirus, obviously, does not apply to the collection, at least as the term was first used (Andrewes et al., 1955). Possibly an overall designation is meaningless except as a shortcut for "RNA-helical-enveloped."

The morphology and other properties of certain arboviruses, Batai, Inkoo, and Uukuniemi seem to place them in the RNA-helical-enveloped collection; these viruses are distinct from myxoviruses (von Bonsdorff et al., 1969) and probably from other similar agents. The taxonomic problem arising here is whether these arboviruses should be assigned to a new taxon in the RNA-helical-enveloped collection, on a par with myxoviruses, paramyxoviruses, coronaviruses, and rhabdoviruses.

Similar complications may appear in the set of viruses defined by RNA, cubic symmetry, and envelope, to which the designation togavirus has been provisionally given. In this taxon are now included, or proposed, arboviruses from groups A and B and a nonarbovirus, rubella virus. Arboviruses of the Bunyamwera supergroup differ sufficiently in morphology from those in groups A and B so that even if some were shown to possess cubic symmetry, they might not be placed in the togavirus group, unless the latter is interpreted as equivalent to RNA-cubic-enveloped with no regard to other characteristics. A similar comment could be made with respect to the arenoviruses (Rowe *et al.*, 1970a); if they were resolved into cubic symmetry a decision would be needed as to whether they are a sub-division of the togaviruses or a taxon of equal rank with them.

The problem to solve is whether an all-inclusive group, RNA-cubic-enveloped, serves any meaningful purpose in a general system or should taxons be defined by the above plus other properties—size, number of capsomeres, presence of inclusionlike bodies in the core; togavirus would then be used in the more restricted sense.

Difficulties of detail inherent in an attempt to incorporate the arboviruses in a general system of classification, therefore, exist and derive mainly from lack of knowledge of the virion's properties and lack of a well-systematized, generally accepted scheme of taxonomy. There is little doubt, however, that the following applies to the classification of the arboviruses:

1. The term arbovirus designates an epidemiological concept which is irrelevant to the criteria on which current general systems of viral classification are based; the term arbovirus should not appear in these systems of classification

2. The set of the arboviruses cannot be incorporated as a whole into a division of the universal system; the set must be disassembled and individual viruses, or antigenically related ones, should be distributed among the pertinent taxons of the system.

3. A taxon defined by virions with RNA, cubic symmetry, and envelope is recognized in the general systems. Some arboviruses, at the moment mainly from antigenic groups A and B, belong in the taxon; it appears that viruses that are not arthropod-borne also belong in it.

ACKNOWLEDGMENT

The author gratefully acknowledges the helpful advice and constructive criticism given by Dr. Wilbur G. Downs and Dr. Robert E. Shope; the latter, in addition, supplied me with much information as yet unpublished.

REFERENCES

Abdelwahar, K. S. E., Almeida, J. D., Doane, F. W., and McLean, D. M. (1964). *Can. Med. Ass. J.* **90**, 1068.
Acheson, N. H. and Tamm, I. (1967). *Virology* **32**, 128.

Almeida, J. D., Waterson, A. P., and Plowright, W. (1967). *Arch. Gesamte Virusforsch.* **20**, 392.

Andrewes, C. H. (1970). *Virology* **40**, 1070.

Andrewes, C. H. and Horstmann, D. M. (1949). *J. Gen. Microbiol.* **3**, 290.

Andrewes, C. H., Bang, F. B., and Burnet, F. M. (1955). *Virology* **1**, 176.

Baker, A. C. (1943). *Amer. J. Trop. Med.* **23**, 559.

Bastardo, J. W., Bergold, G. H., and Munz, K. (1968). *Amer. J. Trop. Med. Hyg.* **17**, 115.

Bellett, A. J. D. (1967). *J. Virol.* **1**, 245–259.

Bergold, G. H., Graf, T., and Munz, K. (1969). "Arboviruses of the California Complex and the Bunyamwera Group, Proceedings of a Symposium" (V. Bardos, ed.), page 41. Publishing House of the Slovak Academy of Sciences. Bratislava.

Blaskovic, D. (1967). *Bull. W.H.O.* **36** [Suppl. 1], 5.

Borden, E. C., Murphy, F., Shope, R., and Harrison, A. (1971). To be published.

Breese, S. S., Jr., Ozawa, Y., and Dardiri, A. H. (1969). *J. Amer. Vet. Med. Ass.* **155** [2], 391.

Briceño-Rossi, A. L. (1964). *Rev. Venezolana Sanidad Asistencia Social* **29**, 351.

Carley, J. G. and Standfast, H. A. (1969). *Amer. J. Epidemiol.* **89**, 583.

Casals, J. (1957). *Trans. N. Y. Acad. Sci.* **19** [Ser. 2], 219.

Casals, J. (1963). *Anais Microbiol.* **11A**, 13.

Casals, J. (1966). *9th Int. Congr. Microbiol. Moscow*, p. 441, Ivanovsky Institute of Virology, Moscow.

Casals, J. (1968). *Nature (London)* **217**, 648.

Committee. (1965). *Ann. Inst. Pasteur*, Paris, **109**, 625.

Cooper, P. D. (1967). *Brit. Med. Bull.* **23**, 155.

Doerr, R., Franz, K., and Taussig, S. (1909). "Das Pappatacifieber." Deuticke, Leipzig.

Downs, W. G. (1970). "Arboviruses: Epidemiological Considerations" (S. Mudd, ed.), page 538. Saunders, Philadelphia, Pennsylvania.

Downs, W. G., Anderson, C. R., Spence, L., Aitken, T. H. G., and Greenhall, A. H. (1963). *Amer. J. Trop. Med. Hyg.* **12**, 640.

Feldman, H. A. and Wang, S. S. (1961). *Proc. Soc. Exp. Biol. Med.* **106**, 736.

Filshie, B. K. and Rehacek, J. (1968). *Virology* **34**, 435.

Gibbs, A. J., Harrison, B. D., Watson, D. H., and Wildy, P. (1966). *Nature (London)* **209**, 450.

Haldane, J. B. S. (1960). *Nature (London)* **187**, 879.

Holden, P. (1955). *Proc. Soc. Exp. Biol. Med.* **88**, 607.

Holmes, I. H. and Warburton, M. F. (1967). *Lancet* **ii**, 1233.

Holmes, I. H., Wark, M. C., and Warburton, M. F. (1969). *Virology* **37**, 15.

Horzinek, M. and Mussgay, M. (1969). *J. Virol.* **4**, 514.

Howatson, A. F. and Whitmore, G. F. (1962). *Virology* **16**, 466.

Howell, P. G. (1962). *Onderstepoort J. Vet. Res.* **29**, 139.

Johnson, K. M. (1965). *Amer. J. Trop. Med. Hyg.* **14**, 816.

Klimenko, S. M., Yershov, F. I., Gofman, Y. P., Nabatnikov, A. P., and Zhdanov, V. M. (1965). *Virology* **27**, 125.

Kuns, M. L. (1965). *Amer. J. Trop. Med. Hyg.* **14**, 813.

Lascano, E. F., Berria, M. I., and Oro, J. G. B. (1969). *J. Virol.* **4**, 271.

Lwoff, A., Horne, R. W., and Tournier, P. (1962). *Cold Spring Harbor Symp. Quant Biol.* **27**, 51.

Melnick, J. L. and McCombs, R. M. (1966). *Progr. Med. Virol.* **8**, 400.

Mettler, N. E. (1969). *Pan Amer. Health Organ., Sci. Publ. No.* **183**.

Mettler, N. E., Petrelli, R. L., and Casals, J. (1968). *Virology* **36**, 503.

Morgan, C., Howe, C., and Rose, H. M. (1961). *J. Exp. Med.* **113**, 219–234.

Murphy, F. A., Coleman, P. H., Harrison, A. K., and Gary, G. W., Jr. (1968a). *Virology* **35**, 28.

Murphy, F. A., Harrison, A. K., Gary, W. G., Jr., Whitfield, S. G., and Forrester, F. T. (1968b). *Lab. Invest.* **19**, 652–662.

Murphy, F. A., Harrison, A. K., and Tzianabos, T. (1968c). *J. Virol.* **2**, 1315.

Murphy, F. A., Whitfield, S. G., Coleman, P. H., Calisher, C. H., Rabin, E. R., Jenson, A. B., Melnick, J. L., Edwards, M. R., and Whitney, E. (1968d). *Exp. Mol. Pathol.* **9**, 44.

Murphy, F. A., Webb, P. A., Johnson, K. M., and Whitfield, S. G. (1969). *J. Virol.* **4**, 535.

Ota, Z. (1965). *Virology* **25**, 372.

Plowright, W., Brown, F., and Parker, J. (1966). *Arch. Gesamte Virusforsch.* **19**, 289.

Plowright, W., Parker, J., and Peirce, M. A. (1969). *Nature (London)* **221**, 1071.

Rowe, W. P., Murphy, F. A., Bergold, G. H., Casals, J., Hotchin, J., Johnson, K. M., Lehmann-Grube, F., Mims, C. A., Traub, E., and Webb, P. A. (1970a). *J. Virol.*, **5**, 651.

Rowe, W. P., Pugh, W. E., Webb, P. A., and Peters, C. J. (1970b). *J. Virol.* **5**, 289.

Scherer, W. F. (1968). *Proc. Soc. Exp. Biol. Med.* **129**, 194.

Schnagl, R. D., Holmes, I. H., and Doherty, R. L. (1969). *Virology* **38**, 347.

Siler, J. F., Hall, M. W., and Hitchens, A. P. (1926). *Manila Bur. Sci. Monogr. No.* **20**, 62 and 170.

Simpson, R. W. and Hauser, R. E. (1968). *Virology* **34**, 358.

Southam, C. M., Shipkey, F. H., Babcock, V. I., Bailey, R., and Erlandson, R. A. (1964). *J. Bacteriol.* **88**, 187.

Studdert, M. J. (1965). *Proc. Soc. Exp. Biol. Med.* **118**, 1006.

Subcommittee. (1963). *Virology* **21**, 516.

Sunaga, H., Taylor, R. M., and Henderson, J. R. (1960). *Amer. J. Trop. Med. Hyg.* **9**, 419.

Taylor, R. M. (1967). *U.S. Dept. Health, Education and Welfare Public Health Service Publ. No.* **1760**, Washington, D.C.

Theiler, M. (1957). *Proc. Soc. Exp. Biol. Med.* **96**, 380.

Verwoerd, D. W. (1969). *Virology* **38**, 203.

von Bonsdorff, C.-H., Saikku, P., and Oker-Blom, N. (1969). *Virology* **39**, 342.

Waterson, A. P. and Almeida, J. D. (1966). *Nature (London)* **210**, 1138.

Whitman, L., and Shope, R. E. (1962). *Amer. J. Trop. Med. Hyg.* **11**, 691.

WHO Scientific Group. (1967). *W.H.O. Tech. Rept. Ser. No.* **369**.

Wildy, P., Ginsberg, H. S., Brandes, J., and Maurin, J. (1967). *Progr. Med. Virol.* **9**, 476.

CHAPTER 10   *Comparative Properties of Rod-Shaped Viruses*

L. HIRTH

I. Introduction

Recent findings have resulted in a new interest in the classification of viruses as a unified group, as discussed in the first chapter in this book by Lwoff and Tournier. In 1962, Lwoff and his co-workers proposed a system of classification in which elongated plant viruses were classified as the group of the rod-shaped viruses (Lwoff *et al.*, 1962). These viruses are non-enveloped, contain RNA, and have a helical symmetry. Some are rigid particles (Rigidoviridales); others are flexible (Flexiviridales; Hirth, 1965). Some viruses are not easily classified in the system of Lwoff *et al.* (LHT system 1962). An example is the case of alfalfa mosaic virus (AMV), a multicomponent virus. Some of its components appear to be elongated

335

and their organization is unique, but certainly nearer to cubic, than to helical symmetry.

The objective of this chapter is to discuss the available information concerning the general structure and particular characteristics of the rod-shaped viruses and AMV.

II. Rigidoviridales

A. TOBACCO MOSAIC VIRUS (TMV)

The structure of tobacco mosaic virus has been well established by the work of Franklin *et al.* (1957). TMV is a rod-shaped virus 3000 Å in length and 180 Å in width. The nucleic acid is RNA with a molecular weight of 2×10^6 daltons. TMV is constituted by 2200 morphological subunits, each with a molecular weight of approximately 17,000.

The protein subunits are packed in a helical rod. Each turn contains 16 1/3 subunits and the same disposition of the subunits is reproduced every three turns. It is established that RNA is included between the subunits and not contained in the central hole of the virus. The configuration of the RNA inside the capsid is a helix, but this configuration is dependent on the mode of aggregation of TMV protein. In solution, TMV RNA is a random coil, of which 70% is in double-stranded form, and 30% in single-stranded one.

1. *Disassembly of TMV*

An excellent review on the disassembly and assembly of elongated and spherical viruses, was published by Leberman (1968). Some details of disassembly and assembly of TMV not described in that review will be mentioned here.

Disaggregation of TMV into its constituents may be easily obtained by various methods; however it is difficult to simultaneously extract both viral nucleic acid and protein in undenatured form. Preparation of TMV protein is obtained by means of acetic acid (Fraenkel-Conrat, 1957) and RNA by the bentonite–phenol method (Gierer and Schramm, 1956; Fraenkel-Conrat *et al.*, 1961). Several other procedures reveal RNA-protein interactions, which were discussed from a theoretical point of view by Caspar (1963). It is evident that salt linkages between the free OH-phosphate group and free amino group of dibasic amino acids play an important role in the stabilization of the nucleocapsid, but it also appears that there are other factors involved in the interaction between RNA and protein. It has been established that each protein subunit interacts with three nucleotides and that these three nucleotides are different for each subunit. Experiments using urea (Buzzel, 1962), and more recently sodium dodecyl sulfate (SDS) (Symington, 1969a,b), have shown that stripping seems to begin at the 5'-linked end. As treatment by SDS proceeds, stripping of the other end of the particles begins in several cases and some shortened particles with two RNA tails are observed. In addition, stripping by SDS occurs in different steps, indicating that some stable regions alternate with less stable ones. Since it appears that the strength of interaction due to phosphate groups has the same importance at the level of each protein subunit, it is likely that the linear variation in the stability observed along the TMV rod is due to the interactions between bases and amino acids. This point of view is substantiated by a recent observation by Kaper (1969) that hydrogen bonding may exist between amino groups of the bases and dicarboxylic acid of the subunits of the coat in the case of the turnip yellow mosaic virus (TYMV). From polar stripping it may be deduced that stability of the 5'-linked end of TMV particle is weaker than that of the 3'-linked end. These observations indicate that viral RNA not only plays an informational role but also contributes to the determination of stable and less stable regions in the TMV particle.

2. Assembly of TMV Components

TMV protein is capable of assembly in different forms in the absence of RNA. This was described in great detail by Caspar (1963) and Lauffer (1966). At 0.5°C, pH between 6.5 and 10.5, and at low protein and salt concentrations, TMV protein aggregates in trimers. The stability of this aggregation product, however, is relatively low. When the pH decreases, for example, aggregation in two-turn discs occurs; these discs seem to be packed head-to-head, but this was not entirely demonstrated. Another stable aggregate is the rodlike structure, which may be obtained by stacking each of 17 protein-subunits discs or by a helical arrangement of subunits. In the latter case the organization of the rod is very similar to that of TMV particles; however, some differences occur, for example, the length of the protein aggregate is indeterminate, and the interactions between protein subunits are weaker than in TMV and in stacked-disc rods; helical polymers are less stable and give rise to stacked-disc rods. The conditions for obtaining such a helical arrangement with TMV protein subunits are not well known. It should be noted that stacked-disc rods are obtained from viruses extracted from tobacco leaves, in which TMV multiplies at non-optimal temperatures (Lebeurier and Hirth 1966). Assembly of TMV protein occurs with a protein entirely free of contaminating viral RNA; this contrasts greatly with observations made with spherical viruses and AMV. In this case, RNA-free protein seems to be unable to aggregate (see below).

The *in vitro* assembly of TMV protein and nucleic acid was obtained for the first time by Fraenkel-Conrat and Williams (1955). A number of experiments have been done by numerous authors since this initial observation, and a critical review of the results obtained was recently published by Leberman (1968).

At this point the problem of specificity of the association between TMV protein and RNA of different origins will be discussed. It has been well established by Holoubek (1962) that cross-reconstitution between protein of one strain of TMV can take place with RNA of other TMV strains,

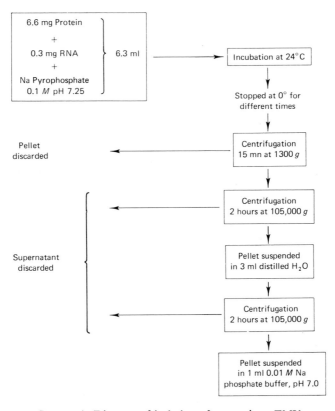

SCHEME 1. Diagram of isolation of reconstitute TMV.

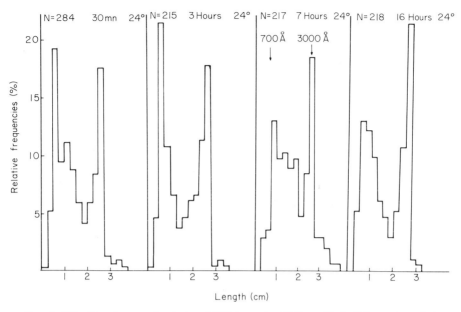

FIG. 1. Kinetics of reconstitution at 24°C (protein: RNA = 22:1). Histograms of the length of reconstituted TMV particles. The relative frequencies of each class are given by the ratio $K_i/N \cdot K_i$ = number of particles in the ith class; N = total number of particles measured. The length of the interval of class is 0.25 cm (~280 Å) at 90,000 magnification (1 cm = 1100 Å ± 100). The length of the particles is given with a precision of ± 0.1 cm (~110 Å).

but conflicting results were obtained when different groups of researchers attempted to obtain aggregation between TMV protein and different types of RNA. For example, Fraenkel-Conrat and Singer (1959) were unable to get rodlike particles by mixing TMV protein with wheat germ RNA or ascites cell RNA. On the other hand, Hart and Smith (1956) were able to "reconstitute" with yeast RNA. Matthews (1966) was able to recombine high molecular weight TYMV RNA with TMV protein, the best yield being obtained at pH 4.8. With polyribonucleotides the reconstitution process occurs, according to Fraenkel-Conrat and Singer (1963), only with poly A or with copolymers which have a high content in A. It is worth noting that with polyribonucleotides aggregation occurs only when the pH is significantly lower than 7, and the best yield seems to be obtained under conditions where the protein begins to aggregate by itself. Under these conditions, one cannot be sure that the aggregation process is specific; some RNA molecules may be "trapped" by aggregating proteins.

It would be interesting to know whether under standard conditions TMV proteins recognize TMV RNA, because TMV reconstitution might prove a useful system in studying the general problem of specific RNA-protein interactions; indeed, this type of interaction already plays an important role in the viral-RNA replication process (replicase-viral RNA interaction).

In my laboratory, we have reinvestigated some aspects of the reconstitution of TMV, specifically the problem of a polar reconstitution. Several experiments suggest this polar reconstitution. TMV was reconstituted as shown in Scheme 1. The reconstituted material was analyzed using electron microscopy, sucrose gradient centrifugation, and inoculation of hypersensitive plants. Kinetic studies of TMV reconstitution show that each suspension of reconstituted TMV seems to be a mixture of two distinct populations whose maxima are 700 and 3000 Å, respectively. To ascertain whether the 700 Å corresponds to a preferential length of the particles is due to the particular ratio of protein to RNA used, we investigated the reconstitution with different protein-RNA ratios. Figures 1 and 2 show that the histograms are of the same types. If we try to separate the two populations of particles by density gradient centrifugation in a linear sucrose

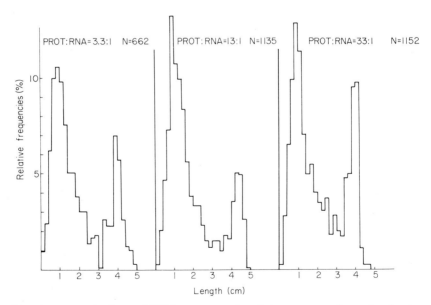

FIG. 2. Reconstitution at 24°C in presence of various amounts of protein after incubation of 1 hour. Histograms of the length of reconstituted TMV particles. The relative frequencies of each class are given by the ratio $K_i/N \cdot K_i =$ number of particles in the ith class; $N =$ total number of particles measured. The length of the interval of class is 0.2 cm ($\sim$140 Å) at 140,000 magnification (1 cm = 700 Å $\pm$ 70). The length of the particles is given with a precision of $\pm$0.1 cm ($\sim$70 Å).

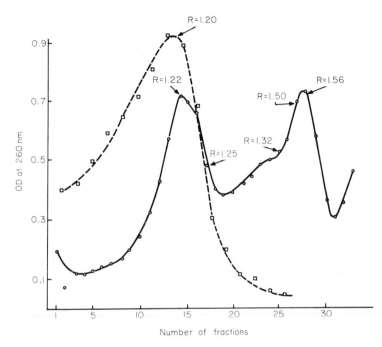

FIG. 3. Separation on a linear sucrose gradient (8-20%, aqueous solution of sucrose) of reconstituted TMV. Time of reconstitution, 1 hour at 24°C; centrifugation for 105 minutes at 25,000 rpm. □-----□ Standard TMV; O———O, reconstituted TMV. $R = OD_{260nm}/OD_{280nm}$.

gradient, we observe two distinct peaks (Fig. 3). The heavy peak is virus and is highly infectious. The light peak is composed of short particles: the ratio of absorption at 2600/2800 Å is near 1.5, indicating a high RNA content. Following the action of RNase, the short particles have the same length as before, but their ratio is 1.20. Electron micrographs, after platinum shadowing of the untreated fractions, showed RNA molecules which are coated for a mean length of 700 Å (Fig. 4). Thus there was an incomplete coating of RNA molecules. It was interesting to see whether the incompletely coated molecules were able to achieve encapsidation after the addition of more native protein. The light material fraction corresponding to incompletely coated virus RNA and coming from TMV reconstituted for 30 minutes at 24°C, were pooled together and divided into identical lots. The first was used as a control and the second was combined with fresh native protein at pH 7.25. Three hours after the addition of protein, the samples were layered on a sucrose gradient. The control shows a homogeneous peak corresponding to the light material and the ratio A_{2800}/A_{2600} Å is near 1.50. The sample to which native protein had been added showed a more heterogeneous tracing (Fig. 5). Less light material was present and heavy material with a ratio near 1.20 appeared. Infectivity of the control was very low in the different fractions. On the other hand, infectivity was very much increased at the level of the heavy fractions in the treated sample. This demonstrates the formation of infectious particles from incompletely coated RNA molecules. Thus, an important part of the 700 Å particles had a free RNA fraction which was able to yield infectious particles after reconstitution.

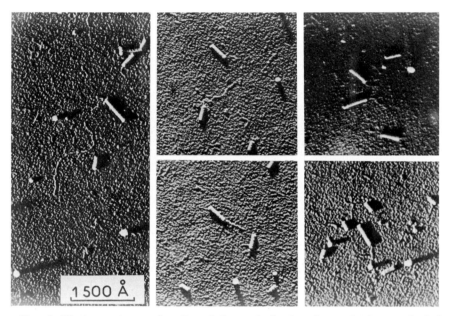

FIG. 4. Electron micrographs after platinum shadowing. Incompletely coated viral RNA molecules corresponding to the light material of Fig. 3 × 120,000. RNA tails are easily observed.

Different aspects of TMV reconstitution are revealed by these experiments (Stussi *et al.*, 1969). Electron microscopic studies did not reveal short particles with two tails of RNA, nor reconstituted particles of complementary length (2300 Å), suggesting that reconstitution is a polar process: it seems unlikely that RNA chains contain two regions located at equal distances from both ends, where the reconstitution process stops. This type of experiment substantiates stripping experiments by revealing that affinity of the TMV protein for RNA varies considerably along the polynucleotide chains and is so weak at one particular point that it becomes rate limiting.

Polar reconstitution, if any, might be demonstrated indirectly by the influence of the action of exonucleases on *in vitro* morphogenesis of TMV. Previously, Singer *et al.* (1965) demonstrated that the release by means of spleen phosphodiesterase nucleotides of the 5'-OH end (left side) of the TMV RNA chain is followed by an important decrease of the infectivity of the viral RNA and by a low yield of particles when the digested RNA was reconstituted. The snake venom phosphodiesterase seems to have a less drastic influence on infectivity of RNA and on the obtention of nucleoprotein particles, respectively (Singer and Fraenkel-Conrat, 1963).

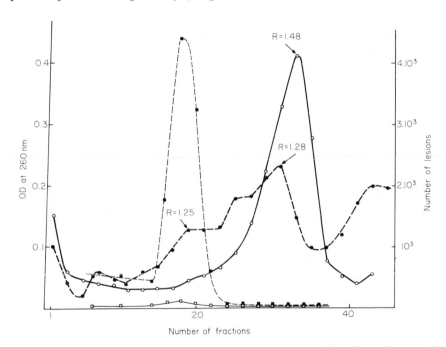

FIG. 5. Study of the infectivity of incompletely coated particles after native protein was added. Protein 2.2 mg, was added to the incompletely coated RNA (here about 100 μg). After incubation for 5 hours at 24°C, the mixture was separated on a sucrose gradient (8-20%). Lot without added protein: ○———○, OD; □———□, infectivity. Lot with added protein: ●-----●, OD; ■-----■, infectivity. The infectivity of each fraction is given by the number of lesions induced on each leaf of *N xanthi* n.c. × dilution.

Some recent experiments (Guilley *et al.*, 1971) in our laboratory show that viral RNA digested (10 minutes) by spleen phosphodiesterase (SPDE) (Scheme 2) gave rise to heavy molecules (30 S) and large pieces of RNA (17–25 S) (Fig. 6). By adding TMV protein a few very short (300–350 Å) rod-shaped particles are obtained (Fig. 7). The reconstituted material is

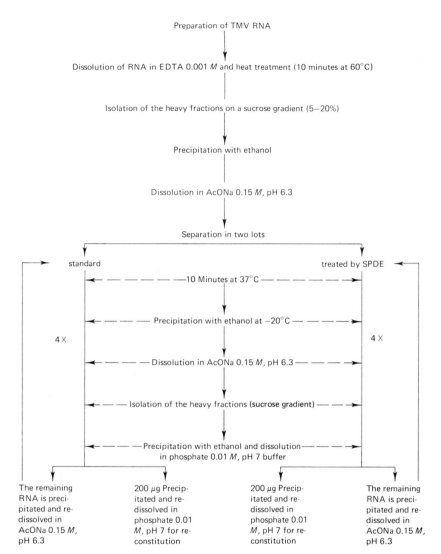

SCHEME 2. Experimental procedure for the study of the action of spleen phosphodiesterase (SPDE) on TMV RNA.

not infectious (Fig. 8). It seems, in this case, that spleen phosphodiesterase has, in addition to its exonuclease activity, an endonuclease one, but that incomplete heavy molecules (removing of 5–7 nucleotides from 5'-OH end) and lighter, but relatively large molecules of RNA, respectively, are not recognized by TMV protein. The origin of the short particles is not clear. Treatment of TMV RNA by purified snake venom phosphodiesterase has much less influence on the reconstitution process, and the yield of re-constituted particles decreases relatively slowly under treatment of the viral RNA by the enzyme. From this observation it can be deduced that TMV protein is able to recognize some specific conformation apparently located at the 5'-OH end of the viral RNA. Removing of a number of nucleotides of the 5'-OH end changes its conformation and no reconstitution occurs. TMV reconstitution may be a good system to use in studying specific RNA-protein interaction.

If a given RNA in solution has the particular configuration which is recognized by TMV protein, reconstitution may occur. Recently Breck and Gordon (1970) described the assembly of potato virus X RNA with TMV protein. TMV RNA and potato virus X RNA, both extracted from helical viruses, may contain regions of similar conformation, which allow the recognition of nucleic acid by TMV protein.

Other viruses have some structural properties very similar to that of tobacco mosaic virus and differ only by the thickness of the particles and by the number of subunits per turn of the helix. For example, barley stripe mosaic virus is a rod-shaped particle of 1250 Å length and a diameter of 200 Å. The protein subunit arrangement around the long axis of the particle is helical. The pitch of the helix is 26 Å and the number of subunits in five turns is close to an integer (Finch, 1965). Reconstitution *in vitro* of this virus was not described, but aggregation of protein subunits *in vitro* was

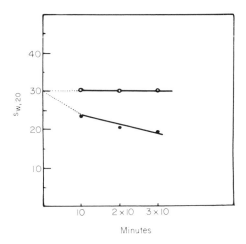

FIG. 6. Evolution of s_{W20}, of TMV RNA with time of treatment with SPDE. After 10 minutes approximatively 20% of the molecules have a s_{W20}, ranging from to 26 and 30 S; the lightest molecules sediment at 17 S; the average value is 23 S. ○———○, standard RNA; ●———●, SPDE-treated RNA.

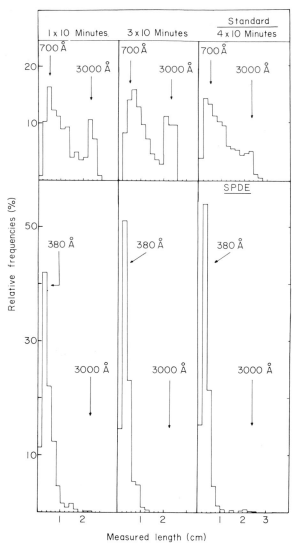

Fig. 7. Histograms of the lengths of reconstituted TMV particles. TMV RNA was treated or not (standard) by SPDE. The histogram were established under the conditions described in Fig. 2.

described by Atabekov *et al.* (1968a). It is worth mentioning also that plants infected with "winter wheat mosaic virus" contain paracrystals which contain protein but no nucleic acid. The protein may be disaggregated *in vitro* under conditions very near to those utilized for TMV disassembly, and reassociation of the protein subunits in rod-shaped particles is easily obtained under conditions similar to those of TMV protein reassociation (Atabekov *et al.*, 1968b). The arrangement of the protein subunits is helical, but the different parameters of the helix have not yet been described.

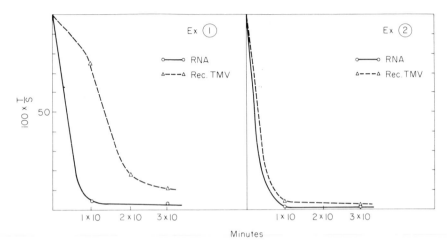

FIG. 8. Evolution of the infectivity of TMV RNA treated for different times by SPDE, and of the particles reconstituted with treated RNA. Reconstitution experiments have been performed as indicated in Scheme 1. S and T, specific infectivities of the standard and treated RNA and the corresponding reconstituted TMV particles.

B. Tobacco Rattle Virus

Tobacco rattle virus is another interesting example of Rigidoviridales. The complete infectious cycle is obtained by inoculating the two types of rod-shaped particles found in the preparations of tobacco rattle virus (TRV) (Lister, 1966; Frost *et al.*, 1967).

In this case the complete genetic information is divided into two parts. Long particles (190 nm) found in the preparations contain information for TRV RNA replication, and the inoculation of the long particles alone (or of their extracted RNA) give rise, on tobacco leaves, to necrotic lesions which do not contain virus particles but apparently only free viral RNA. For all strains of TRV the length of the long particles is identical; on the other hand, the length of the short particles is dependent on the strain and varies from 50 to 115 nm. Inoculation of homogeneous short particle preparations does not produce any lesions. If however, short rods are added to long ones and if the mixture is inoculated onto tobacco leaves, the major part of the lesions formed contain virus particles of both lengths. It seems that short particles are not infectious by themselves but contain information for coat protein synthesis; but complementation of the function of long particles by short ones was considered during many years as specific, that is to say, the function of long particles of a given strain can be complemented only by the short particles of the same strain. Recently Semancik and Kajiyama (1968) have shown that short particles of different strains are capable of complementing functions of long particles of others strains. This shows that there is one common function for all short particles (protein coat synthesis) and also suggests that some specific properties of different strains come from short particles.

The structure of TRV is relatively well known (Finch, 1965; Offord, 1966). Although some differences exist between TMV and TRV, they are similar in structure.

The organization of the long and short particles of TRV seems to be the same. The diameter of TRV is 256 Å and the pitch of the helix is 26 Å. The arrangement of the subunits is helical and a repeated period of three turns containing $(75 + 1)$ subunits is observed. As in the TMV particle a central hole of 27 Å radius exists and the RNA is embedded in protein subunits. The molecular weight of the protein subunit, calculated by physical methods, is approximately 23,000–24,000. Calculations made from the different data obtained for TRV show that the long particles contain approximately 1830 protein subunits; this corresponds to a molecular weight of 44×10^6 daltons. The geometry of the structure suggests that four nucleotides interact with each of the protein subunits; the RNA of the long particles contains 7100 nucleotides (mol. wt., 2.1×10^6 daltons). The molecular weight of the whole particle is 46×10^6 daltons. The molecular weight of the short particles varies greatly with different strains but, for the European common strain, is approximately 17×10^6 daltons (the molecular weight of the RNA is close to 1×10^6 daltons). It is worth observing that the entire genetic information for TRV corresponds theoretically to a greater number of proteins than for that of TMV. In fact, the long particles of TRV contain somewhat more nucleotides than TMV particles which contain all the information necessary for its multiplication. It is worth noticing also that TMV itself contains two times more information than brome grass mosaic virus, which, nevertheless, is able to multiply without a "helper." The problem of this important excess of information is a general one in the case of "multicomponent viruses" and its significance is not very clear. *In vitro* morphogenesis of TRV was studied recently by Semancik and Reynolds (1969). The self-assembly process occurs when viral RNA extracted from whole virus preparations is mixed with protein subunits isolated by means of acetic acid under suitable conditions. In fact, it seems that the two types of particles do not reconstitute at the same rate, and the short RNA particles seem to be easily recognizable from the protein subunits which are identical for the two types of particles. This is true especially in the case where the ratio protein/RNA is not very distant from the stochiometry. In this case no formation of long particles occurs, although the corresponding RNA is present. Reconstitution of long particles occurs only when an excess of viral RNA is present in the mixture; but it is not clear from the publication quoted whether the RNA added to the protein is a mixture of the two RNA species or contains only heavy molecules. No explanation was proposed for the differences observed in the case of reconstitution between short and long particles. It is not known whether, in the case of TRV, the association between RNA and protein is a specific process. If the process is specific, it is reasonable to assume that long and short RNA molecules have some specific configuration in common which permits recognition by protein subunits.

III. Flexiviridales

The Flexiviridales group consists of a large number of plant viruses, many of which cause important plant diseases. Relatively little information is available concerning their structure and especially the arrangement of the protein subunits. Helical symmetry is observed in all viruses belonging to this group. The origin of the sinuous structure of these viruses was discussed by Caspar and Klug (1962) and was attributed to the fact that the protein subunits are not strictly "equivalent" but "quasi-equivalent." In the case of quasi-equivalent arrangement the different bonds between one subunit and the other surrounding subunits do not have exactly the same strength and are more or less specific. This contrasts greatly with TMV, where the bonding between the subunits is exactly the same. The consequence of this "quasi-equivalence" is the nonexistence of a straight axis of symmetry for the Flexiviridales. Varma *et al.* (1968) have determined by means of electron microscopy that filamentous viruses can be divided in three groups, according to the appearance of the substructure observed after staining with uranyl formate. Groups I and III contain very flexuous particles showing clear and indistinct substructures, respectively. Group II is constituted by viruses which are relatively rigid and show a definite structure. The length of Flexiviridales varies from 500 to 1.250 nm.

It is difficult to define the groups on the basis of particle length, width, and substructure. Different proposals were made to classify Flexiviridales, but it seems that definitions of family and genus in this case are still a matter of discussion as much of the data are relatively uncertain in the case of these viruses. However, one characteristic seems to be common to all filamentous viruses: the pitch of the helix. This characteristic was calculated for ten viruses using optical diffraction diagrams and it is bracketed between 33 and 37 Å. Recently the pitch of the helix of the sugar cane mosaic virus was calculated in my laboratory as approximately 34 Å. These values are very different from those obtained for Rigidoviridales which approximate 23–26 Å. This difference justifies the distinction between Rigidoviridales and Flexiviridales in the LHT system. It is worth noticing that such viruses as potato virus X, Y, and S, white clover mosaic virus, and sugar beet yellows, belong to the Flexiviridales.

IV. Alfalfa Mosaic Virus

Although the main purpose of this paper is to compare some characteristics of the rod-shaped viruses, we thought it would also be interesting to discuss some properties of a multicomponent virus, alfalfa mosaic virus (AMV), the heavy particles of which have an elongated form and have been considered in the past by some virologists as more or less similar to rod-shaped viruses. In fact such viruses do not yet have a definite place in the LHT system. It is worth noticeing that the various AMV components, completely purified, or with very low contamination by other components, are not infectious by themselves. It was established by the work of Van Vloten-Doting and Jaspars (1967) and Van Vloten-Doting et al. (1968) that two RNA types, corresponding to a molecular weight of 1.3, and 0.33×10^6 daltons, respectively, must be inoculated together in order to induce local lesions or generalized infection. The complementary functions of the components of AMV have been described recently in detail in a general review by Hull (1969). It seems well established from recent work that the organization of the protein subunits of this virus is very similar to that of icosahedral viruses, but some characteristics of AMV appear worth discussing from the viewpoint of the several types of particles observed, and from that of the specificity of RNA-protein association.

Purified preparations of AMV studied by electron microscopy and analytical ultracentrifugation show four major and several minor components; these latter components are not always present in the preparations. The physical characteristics of these different components and their chemical composition have been discussed widely. The recent review of Hull (1969) indicates that all components have a nucleic acid content of 17% and that the protein subunits corresponding to each component have the same chemical composition and the same serological specificity. Table I gives the S

TABLE I

VALUES OF AMV COMPONENT

Components	S value	Molecular weight ($\times 10^6$)	Length of the particles (nm)
Top a	68	3.13	32
Top b	76	3.72	37
Middle	88	4.89	47
Bottom	99	6.06	56.5

values, the molecular weight, and the size of the different major components. Additional details concerning the physicochemical properties of these components may be obtained from the excellent article by Hull (1969). It is worth noticing that the width of the different particles is the same: 180 Å. From the molecular weight of each component, the molecular weight of their total RNA may be easily calculated but it seems that, except for the bottom component, the RNA of all the other components is split into several pieces. The data concerning the different types of RNA extracted from whole and fractionated preparations of AMV are contradictory. Gillaspie and Bancroft (1965) extracted only three types of RNA 25, 17, and 12 S; Hull et al. (1969a) claim that four types of RNA exist, 24, 20, 17, and 12 S, whereas, Jaspars found only two. Recently, Pinck (1969) described four types of RNA (Mw 1.35; 0.98; 0.72; 0.32×10^6). Whatever each component contains, it is clear that the morphology is different for light and for heavy components; the latter are elongated and have round ends. All the particles are built of protein subunits having a molecular weight of 32,000 or 25,000 (Kruseman et al., 1971) and partially organized according to the icosahedral system; this is clear for a minor component "top z" which seems to be organized as phage $\varphi \times 174$ (12 pentamers = 60 subunits). The detailed organization of the different minor and major components was described by Gibbs et al. (1963), and Hull et al. (1969b), all of whom propose, in fact, the same model. Hull's model consists of an

icosahedron cut across its threefold axis and with hexagons added to form
the tubular portion. Thus the elongation of the starting icosahedron (top z)
proceeds by successive addition of three hexagons (18 morphological subunits) (Fig. 9). The calculation of the number of the subunits made from the
proposed structure for each compound seems to be in agreement with the
data obtained by electron microscopy and by physicochemical analysis
although the discrepancy in the estimation of the molecular weight of the
protein subunit (32,000–24,000) is rather confusing. In fact the structure
proposed by Gibbs *et al.* (1963) and by Hull *et al.* (1969b) describes more

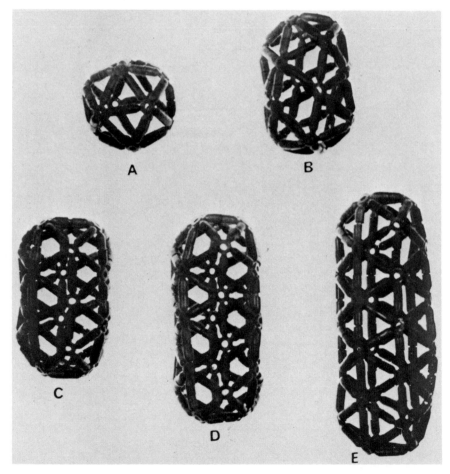

FIG. 9. Models showing the structure proposed by Hull *et al.* (1969b) for AMV
components. A, top z component; B, top a component; C, top b component; D, middle
component; E, bottom component.

the lattice of the particles than the disposition of the subunits. More work is necessary to obtain a definitive idea of the true organization of protein subunits in the case of AMV.

From the point of view of morphogenesis, and especially from that of the specificity of RNA-protein interactions, AMV may present some interesting characteristics. We have previously seen that, for rod-shaped viruses with helical symmetry, the organization of protein subunits permits their association in a helix, but that, under proper conditions only, the corresponding RNA can be embedded in the rod. In fact, in this case, the RNA plays a role only in the stabilization and in the determination of the length of the particles. The case of icosahedral viruses seems to be different, even if we consider only that of unstable spherical viruses.

Bancroft and Hiebert (1967), and Hiebert et al. (1968) have recently demonstrated that native protein of different viruses (obtained by disaggregation of the virions by high salt concentrations) may be assembled in the presence of different types of RNA or of polyelectrolytes, such as, polynucleotides, polyvinylsulfates. For example, CCMV (cowpea chlorotic mottle virus) protein subunits may associate with TMV RNA to give spherical particles which are infectious. It is worth mentioning that association of protein subunits alone gives rise, under proper conditions, to empty shells which seem not to exist *in vivo*; but it is not clear from the work of Bancroft and his co-workers whether the protein is contaminated by a large or by a small amount of residual RNA. AMV, which is very easily degraded by salts and which presents several possibilities of organization of its subunits into spherical and more or less elongated particles, has recently been investigated to determine its reassociation *in vitro*.

It is worth noticing that a true reconstitution of AMV was not obtained until now. Recently Lebeurier et al. (1969b) and Hull (1970) reported experiments in which they obtained the different types of virus particles by adding protein subunits to undegraded RNA of AMV. However, the yield of heavy particles was low and the infectivity absent (Lebeurier et al., 1969b) or low (Hull, 1970) and some other components which did not exist in the initial preparation were found in relatively large amounts in the reconstituted ones. On the other hand, Lebeurier et al. (1969a) and Bol and Kruseman (1969) have succeeded in performing "dissociation-reassociation" experiments and were able to obtain infectious preparations having the main properties of the initial material. These experiments have recently been confirmed by Hull (1970). "Dissociation-reconstitution" experiments differ from "true" reconstitution in that probably the dissociation is not a complete one and that some protein subunits still interact strongly with their corresponding whole RNA. This complex may play the role of template for the reassociation of the free subunits present in the medium, provided that suitable conditions of ionic strength, pH, etc., are realized. This type of experiment suggests that some regions of the different types of RNA have more affinity for the subunits than others, and that this may be considered as an indirect proof of a relative specificity of RNA-protein interactions. New information on the problem of specificity of interactions was recently arrived at in my laboratory by comparing the different types of particles obtained from the association of protein subunits alone under different conditions. Contrary to the affirmations of Hull, we were able to obtain an association of protein subunits without adding RNA, Lebeurier *et al.* (1969b), but in all cases, the protein contained a low, yet significant amount of residual RNA. Under the best conditions possible, the amount of residual RNA was approximatively 0.5%, as estimated with ^{32}P. As shown in Fig. 10

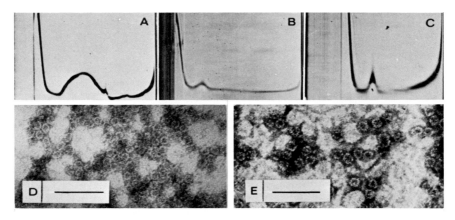

FIG. 10. Schlieren patterns and electron micrographs of AMV protein retaining little RNA (0.5%) originally in 1.5 *M* NaCl, 0.02 *M* tris pH 7.4, 0.001 *M* Cleland's reagent back-dialyzed, for 4 hours at 22°C against: (A) 0.1 *M* NaCl, 0.01 *M* phosphate, pH 6.6, 0.001 *M* Cleland's reagent, 0.005 *M* MgCl$_2$, 0.01 *M* KCl; (B) same buffer as in A, pH 7.4; (C) same buffer as A, pH 8.0. All photographs were taken 10 minutes after the rotor reached a speed of 44,770 rpm; sedimentation is from left to right. (D) and (E) Small spheres formed at pH 6.6 and 8.0, 180 Å diameter. Stained in 1 % uranyl acetate at pH 4.2. Scale: 1000 Å.

association of subunits of AMV may occur, but gives rise to apparently empty spherical particles, the organization of which seems to be similar to that described by Hull (1969) in natural preparations as top z. It is worth noticing that associated particles have the same specific radioactivity as the initial dissociated protein; thus they contain residual RNA. By dialyzing virions against buffers containing various concentrations of NaCl, it is possible to obtain AMV protein contaminated by various amounts of residual RNA. The reassociation of such different types of proteins gives rise to various types of particles. Figure 11 shows Schlieren patterns of reassociated protein contaminated by 1.5% of residual RNA; the main peaks correspond to

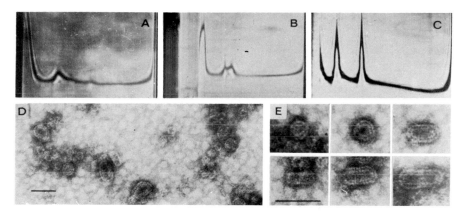

FIG. 11. Schlieren patterns of AMV protein containing 1.5% of residual RNA originally in 1 M NaCl, 0.02 M tris, pH 7.4, 0.001 M Cleland's reagent, back-dialyzed for 4 hours against: 0.1 M NaCl, 0.01 M phosphate, 0.005 M MgCl$_2$, 0.01 M KCl, 0.001 M Cleland's reagent. (A) pH 8.0, 22°C; (B) pH 6.6, 22°C; (C) pH 7.4, 36°C. All photographs were taken 10 minutes after the rotor reached a speed of 44,770 rpm; sedimentation is from left to right. (D) and (E) Electron micrographs of selected particles corresponding to conditions A: small spheres of 180 Å, double-shelled spheres of 310 and 370 Å. Bacilli-form particles of diameter 180 Å and length: 250–370, and 440 Å. Double-shelled bacilli-form particles of 250 and 310 Å diameter and lengths: 310, 370, 440, and 500 Å. Stained in 1% uranyl acetate. Scale, 500 Å.

spheres of 35 and 77 S (180 and 310 Å in diameter). Among these empty spherical particles are found some unusual double shells and double-shelled ellipsoid particles (Fig. 11). With increasing RNA contamination, protein subunits aggregate very easily and give rise to numerous double-shelled spheres and some double-shelled ellipsoids (Fig. 12). The ellipsoids are of different lengths, the longest double ellipsoids obtained have a length close to the length of the natural bottom component. It should be noted that the residual RNA is degraded, as observed from analytical ultracentrifugation analysis after extraction by phenol. On the other hand, addition of unde-graded AMV RNA to protein containing small amounts of residual RNA gives rise only to simple spheres and ellipsoids, as well as to some rare long bacilliform particles. It seems also that AMV protein is unable to encapsi-date RNA having a molecular weight higher than 1.3×10^6 daltons (for

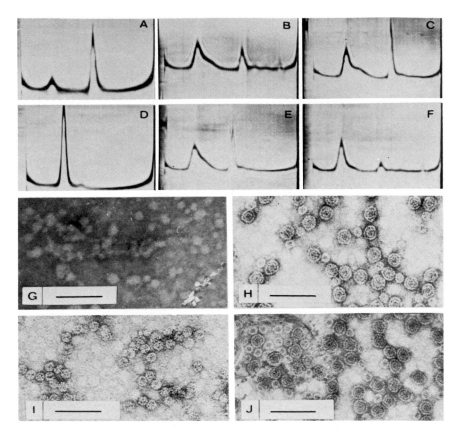

FIG. 12. Schlieren patterns of protein retaining about 3% of residual RNA originally in 0.8 M NaCl, 0.02 M tris, pH 7.4, 0.001 M Cleland's reagent, back-dialyzed for 4 hours against, 0.1 M NaCl, 0.01 M phosphate, 0.005 M MgCl$_2$, 0.01 M KCl, 0.001 M Cleland's reagent at: (A) pH 6.6, 36°C; (B) pH 7.4, 22°C; (C) pH 7.4, 2°C; (D) pH 8.0, 36°C; (E) pH 8.0, 22°C; (F) pH 8.0, 2°C. All photographs were taken 10 minutes after the rotor reached a speed of 44,770 rpm; sedimentation is from left to right. (G) Electron micrograph of material dialyzed against buffer described in the legend of Fig. 11. (H) Small and double-shelled particles corresponding to Fig. (A). (I) Electron micrograph corresponding to Fig. (D). (J) Electron micrograph of material dialyzed against the same buffer at pH 6.6, 22°C. Stained in 1% uranyl acetate. Scale, 1000 Å.

example, TMV and TYMV RNA's) (Lebeurier *et al.*, 1969b; Hull, 1970). In contrast to these results, very long filamentous particles were obtained by replacing TMV RNA with *Haemophilus influenzae* DNA having a molecular weight of 3.10^6 daltons. The filaments seem to be composed of stacked discs. If degraded AMV RNA (3%) is added to DNA, filaments are also found, but in smaller number. As can be seen in electron micrographs

(Fig. 13), in addition to the filaments, the same spheres and ellipsoids were discovered as when degraded RNA by itself was used. Therefore in the presence of an excess of *Haemophilus* DNA (50%), the AMV protein seems able to recognize its specific RNA and also to encapsidate other nucleic acids.

Several matters, particularly in the case of AMV protein association, are especially worthy of attention:

1. It seems that association needs the presence of a small amount of RNA; the higher the content of contaminating RNA in the protein, the better the yield in reassociated products. It seems, in this case, that RNA plays an activating role by catalyzing the process of association, which is probably due to the formation of salt linkages between the ionized phosphate groups and the basic amino acids (37 arginine and lysine residues; Hull *et al.*, 1969a). The association of AMV protein with *Haemophilus* DNA can also be explained by the same type of bonding.

2. On the other hand, the addition of undegraded AMV RNA to protein

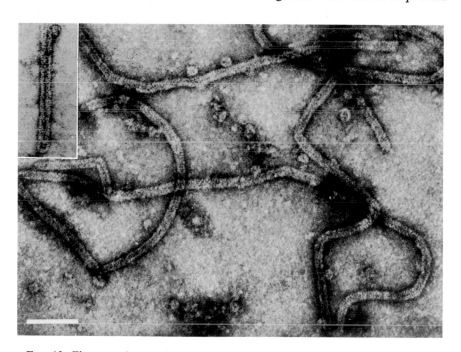

FIG. 13. Electron micrograph of a mixture of AMV protein + DNA of *Haemophilus influenzae* (50%) + degraded AMV RNA (3%), dialyzed against: 0.1 M NaCl, 0.01 M phosphate, 0.0025 M MgCl$_2$, 0.01 M KCl, 0.001 M Cleland's reagent at pH 8.0, 4 hours at 22°C. Stained in 1% uranyl acetate. Scale, 1000 Å.

gives rise to particles resembling those of the native preparation of virus; some specific forms (double-shelled particles) observed in the preparation obtained by association of more or less contaminated subunits alone were not found.

We may therefore consider that the residual RNA contains nucleotide sequences interacting more strongly with AMV protein than with the other parts of the RNA chain; the occurrence of specific interactions between some bases and some amino acids, suspected for TMV (Caspar, 1963), and recently well established in the case of TYMV (Kaper, 1969), may explain that the presence of residual RNA would confer a special configuration to the AMV protein subunits, allowing their association to give rise to particles of particular morphologies. It is worth noticing that small pieces of RNA can do things which a large one can not.

The case of AMV seems to be a special one if we compare it with the rod-shaped viruses and with true spherical viruses. Protein subunits seem to have very weak interpeptide bonds; RNA, even in low amounts is necessary for association. This relatively low affinity of subunits for each other is indirectly confirmed by the fact that gaps between the subunits exist especially in the case of middle and bottom components. The existence of these gaps may explain the penetration into the particles of basic proteins such as ribonuclease (Bol and Veldstra, 1969) and cytochrome (Hull *et al.*, 1969b). Some nucleotide sequences of AMV RNA probably alter both the configuration and the affinity of protein subunits for each other, and allow the formation of particular compounds. Other RNA or polyelectrolytes, even of high molecular weight, seem unable to induce the formation of elongated particles. Thus in the case of AMV a relative specificity of interaction between RNA and viral protein exists in any case for the formation of heavy particles. It would be interesting to know what nucleotide sequences interact strongly with AMV protein.

V. Conclusion

Some conclusions may be drawn from the comparative properties of the viruses studied in this paper.

The helical arrangement of the protein-subunits around RNA is a characteristic of the rod-shaped viruses but it seems that in certain cases the subunits of spherical viruses (Bancroft *et al.*, 1969) or of AMV (Lebeurier *et al.*, 1970) are able to aggregate around double-stranded nucleic acids, either in a stacked disc or in a helical arrangement (Pfeiffer and Witz, unpublished results). It is worth noting that, in the latter case, the morphology and shape of the particles are determined by the nucleic acid itself, which is double-stranded and rigid. With some unstable spherical viruses, protein subunits show a relative plasticity which permits aggregating in different ways. This however does not apply to the subunits of the rod-shaped viruses; until now no other types such as rod-shaped aggregates were described for these viruses.

Rod-shaped viruses seem more efficient in protecting the viral nucleic acid than unstable spherical viruses; even in the case of very stable icosahedral viruses such as TYMV, protection of RNA is poor: the latter is often cut inside the virion.

On the other hand, rod-shaped viruses encapsidate *in vitro* their corresponding RNA andit is worth considering whether the relative non specificity of *in vitro* encapsidation for viruses belonging to the type of BBMV exists also *in vivo*; if such was true a part of the virions might contain plant RNA and a part of viral RNA might escape encapsidation.

In the group of rod-shaped viruses, the Rigidoviridales seem to be the more efficient system for conserving the biological properties of the RNA. The Flexiviridales, which are very often long particles, may be easily degraded by breaking. Thus apparently Rigidoviridales seem to be the best system for assuring perpetuation of the viral properties. It is worth noticing that Caspar and Klug (1962) have observed that spherical viruses represent a "cheaper" system than rod-shaped viruses, because much less protein is necessary to protect a nucleic acid having a very high molecular weight. They claim that spherical viruses are more efficient for conserving and propagating the viral genetic information than rod-shaped viruses. But this point of view seems difficult to generalize, because at least some spherical RNA-viruses are relatively unstable, relatively sensitive to nuclease, and devoid of a true specificity for *in vitro* encapsidation.

REFERENCES

Atabekov, J. G., Novikov, V. K., Kiselev, N. A., Kaftonova, A. S., and Egorov, A. M. (1968a). *Virology* **36**, 620.
Atabekov, J. G., Popova, G. A., Kiselev, N. A., Kaftonova, A. S., and Petrovsky, G. V. (1968b). *Virology* **35**, 458.
Bancroft, J. B. and Hiebert, E. (1967). *Virology* **32**, 354.
Bancroft, J. B., Hiebert, E., and Bracker, C. E. (1969). *Virology* **39**, 924.
Bol, J. F., and Kruseman, J. (1969). *Virology* **37**, 385.
Bol, J. F. and Veldstra, H. (1969). *Virology* **37**, 74.
Breck, L. O., and Gordon, M. P. (1970). *Virology* **40**, 397.

Buzzel, A. (1962). *J. Amer. Chem. Soc.* **82**, 1636.

Caspar, D. L. D. (1963). *Advan. Protein Chem.* **18**, 37.

Caspar, D. L. D. and Klug, A. (1962). *Cold Spring Harbour Symp. Quant. Biol.* **27**, 1.

Finch, J. T. (1965). *J. Mol. Biol.* **12**, 612.

Fraenkel-Conrat, H. (1957). *Virology* **4**, 1.

Fraenkel-Conrat, H. and Singer, B. (1959). *Biochim. Biophys. Acta* **33**, 359.

Fraenkel-Conrat, H. and Singer, B. (1963). *Virology* **23**, 354.

Fraenkel-Conrat, H. and Williams, R. C. (1955). *Proc. Nat. Acad. Sci. U.S.* **41**, 690.

Fraenkel-Conrat, H., Singer, B., and Tsugita, A. (1961). *Virology* **14**, 54.

Franklin, R. E., Klug, A., and Holmes, K. C. (1957). *Ciba Found. Symp. Nature Viruses.*

Frost, R. R., Harrison, B. D., and Woods, R. D. (1967). *J. Gen. Virol.* **1**, 57.

Gibbs, A. J., Nixon, H. L., and Woods, R. D. (1963). *Virology* **19**, 441.

Gierer, A. and Schramm, G. (1956). *Nature (London)* **177**, 102.

Gillaspie, A. G. and Bancroft, J. B. (1965). *Virology* **27**, 391.

Guilley, H., Stussi, C., and Hirth, L. (1971). *C. R. Acad. Sci. Paris*, **272**, 1181.

Hart, R. G. and Smith, J. D. (1956). *Nature (London)* **178**, 739.

Hiebert, E., Bancroft, J. B., and Bracker, C. E. (1968). *Virology* **34**, 492.

Hirth, L. (1965). *C.R. Acad. Sci. Paris* **261**, 4556.

Holoubek, V. (1962). *Virology* **18**, 401.

Hull, R. (1969). *Advan. Virus Res.* **15**, 365.

Hull, R. (1970). *Virology* **40**, 34.

Hull, R., Rees, M. W., and Short, M. N. (1969a). *Virology* **37**, 404.

Hull, R., Rees, M. W., Markham, R., and Hills, G. J. (1969b). *Virology* **37**, 416.

Kaper, J. M. (1969). *Science* **166**, 248.

Kruseman, J., Kraal, B., Jaspars, E. M. J., Bol, J. F. Brederode, F. T., and Veldstra, H. (1971). *Biochemistry*, **10**, 447.

Lauffer, M. A. (1966). *Biochemistry* **5**, 2440.

Leberman, R. (1968). *Symp. Soc. Gen. Microbiol.* **18**, 183.

Lebeurier, G. and Hirth, L. (1966). *Virology* **38**, 16.

Lebeurier, G., Wurtz, M., and Hirth, L. (1969a). *C.R. Acad. Sci. Paris* **268**, 1897.

Lebeurier, G., Wurtz, M., and Hirth, L. (1969b). *C.R. Acad. Sci. Paris* **268**, 2002.

Lebeurier, G., Fraenkel-Conrat, H., Wurtz, M., and Hirth, L. (1971). *Virology*, **43**, 51.

Lister, R. M. (1966). *Virology* **28**, 350.

Lwoff, A., Horne, R., and Tournier, P. (1962). *Cold Spring Harbour Symp. Quant. Biol.* **27**, 51.

Matthews, R. E. F. (1966). *Virology* **30**, 82.

Offord, R. E. (1966). *J. Mol. Biol.* **17**, 370.

Pinck, L. (1969). Ph. D. Thesis, University of Strasbourg.

Semancik, J. S. and Kajiyama, M. R. (1968). *Virology* **34**, 170.

Semancik, J. S., and Reynolds, D. A. (1969). *Science* **164**, 559.

Singer, B. and Fraenkel-Conrat, H. (1963). *Biochim. Biophys. Acta* **72**, 534.

Singer, B., Sherwood, M., and Fraenkel-Conrat, H. (1965). *Biochim. Biophys. Acta* **108**, 306.

Stussi, C., Lebeurier, G., and Hirth, L. (1969). *Virology* **38**, 16.

Symington, J. (1969a). *Virology* **38**, 209.

Symington, J. (1969b). *Virology* **38**, 317.

Varma, A., Gibbs, A. J., Woods, R. D., and Finch, J. T. (1968). *J. Gen. Virol.* **2**, 107.

Van Vloten-Doting, L. and Jaspars, E. M. J. (1967). *Virology* **33**, 684.

Van Vloten-Doting, L., Kruseman, J., and Jaspars, E. M. J. (1968). *Virology* **34**, 737.

CHAPTER 11 **Bullet-Shaped Viruses**

KLAUS HUMMELER

I. Introduction

The group of bullet-shaped viruses are of particular interest because they parasitize vertebrates, invertebrates, and plants. Their morphology is characteristic and distinct from other viruses and, therefore, it has been suggested that they should be grouped into a new genus. New members are reported with an increasing frequency. Beyond structural details, however, little is known about biological, physical, and biochemical properties of many of them, although some of them have been investigated in detail. It remains to be seen whether all of these agents belong properly to a single genus, but enough similarities as well as differences in their properties are known at this time to make a comparative analysis useful.

The taxonomy of these viruses is still unsatisfactory and presents some difficulties. The vesicular stomatitis virus (VSV) is the most extensively investigated member of this group and is generally considered the prototype. Originally, therefore, it was suggested to name the genus *stomatoviridae*. This is obviously inappropriate in view of the biological activities of other members of this group. The criterion for relegating viruses into this genus is, at the moment, based merely on structure and hence a term expressing those characteristics would be more valid.

Most of the vertebrate viruses of this group resemble a bullet. They are elongated with one rounded and one planar end. Therefore, the term bullet-shaped virus has been used extensively. This is hardly satisfactory because it appears that a number of members of this genus, especially those parasitizing plants, have two rounded ends and are therefore bacilliform. Melnick and McCombs (1966) suggested *rhabdoviridae* for the vertebrate members of this group. This term is now widely used and is the recommended name by a subcommittee of the International Committee on Nomenclature of Viruses. The final decision on this recommendation was made at the International Microbiological Congress held in Mexico City in 1970. This is unfortunate because rhabdovirus (from rhabdos=rod) expresses only one, and not the most important structural feature, of these viruses. It has furthermore been preempted, and properly so, by plant virologists as the genus name for viruses which are true rods in their mature form such as tobacco mosaic and others (see the chapter by Hirth in this publication). Important differences among these viruses have become apparent and the lack of data available for many of them make a common terminology problematic.

The purpose of this article is solely to render a comparative analysis of these viruses including such data as are available on structure, morphogenesis, biochemical, biophysical, and antigenic properties. Other aspects such as historical data, epidemiology, transmission, pathogenesis, have been ignored. Some of the viruses listed in Table I will be considered. The references in the table indicate merely the first description permitting clearly the inclusion of these agents provisionally in this genus. No claim can be made that the list is complete since the rapid development in this field precludes any such attempt.

TABLE I

Virus	Abbreviation	Host[a]	References[b]
Vertebrate viruses			
Vesicular Stomatitis	VSV	Arthropods, mammals	Howatson and Whitmore, 1962
Rabies		Mammals	Hummeler et al., 1967
Lagos		Bat	Shope et al., 1970
IbAn 27377		Shrew	Shope et al., 1970
Kern Canyon	KCV	Myotis bat	Murphy and Fields, 1967
Flanders-Hart Park	FHPV	Arthropods, birds	Murphy et al., 1966
Mt. Elgon	MEBV	Bat	Murphy et al., 1970
Piry		Opossum	Murphy and Shope, 1969
Chandipura		Man	Murphy and Shope, 1969
M-1056		Microtus, rodent	Murphy and Shope, 1969
Viral hemorrhagic septicemia	VHS	Rainbow trout	Zwillenberg et al., 1965
Infect. hematopoietic necrosis	IHN	Salmon	Amend and Chambers, 1970
Oregon sockeye disease	OSD	Salmon	Amend and Chambers, 1970
Sacramento River chinook disease	SRCD	Salmon	Amend and Chambers, 1970
Bovine ephemeral fever	BEF	Cattle	Ito et al., 1969
Marburg		Monkey	Siegert et al., 1967

TABLE I (*Continued*)

Virus	Abbreviation	Host[a]	References[b]
Invertebrate and plant viruses			
Sigma		*Drosophila*	Berkaloff *et al.*, 1965
Potato yellow dwarf	PYDV	Leaf hopper	McLeod *et al.*, 1966
Sowthistle yellow vein	SYVV	Aphid	Richardson and Sylvester, 1968
Gomphrena	GV		Kitajima and Costa, 1966
Lettuce necrotic yellows	LNYV	Aphid	Harrison and Crowley, 1965
Wheat striate mosaic	WSMV	Leaf hopper	Lee, 1964
Melilotis latent	MLV		Kitajima *et al.*, 1969
Plantain			Hitchborn *et al.*, 1966
Corn mosaic	CMV	Leaf hopper	Herold *et al.*, 1960
Rice transitory yellowing	RTYV		Shikata and Chen, 1969
Northern cereal mosaic	NCMV		Shikata and Lu, 1967
Broccoli necrotic yellows	BNYV		Hills and Campbell, 1968
Eggplant mottled dwarf	EMDV		Martelli, 1969
Russian winter wheat mosaic	RWWMV	Leaf hopper	Razvjazkina *et al.*, 1968
Clover enation	CEV	Various	Rubio-Huertos and Bos, 1969
Cereal striate mosaic	CSMV	Plant hopper	Conti, 1969

[a] Not necessarily indicative of host range. Viruses originally isolated from these hosts.
[b] First morphological description to allow inclusion in genus.

II. Structure

The appearance of these viruses in either thin sections or negative contrast is characteristic although there are differences among them. Besides the classic bullet shape of VSV and rabies, they may be bacilliform, as some of the plant viruses, or conically shaped as some strains of BEF (Lecatsas *et al.*, 1969). The Marburg virus bears the least resemblance to these agents. It seems to exist as long flexible cylindrical filaments and has a more complex internal structure (Peters *et al.*, 1970).

Recently it has been suggested (Peters and Kitajima, 1970) that all these viruses may be bacilliform and that the bullet shape is an artifact produced by the negative stain employed. Prefixation with glutaraldehyde maintained the bacilliform shape. Without fixation one rounded end broke off leaving a bullet which, at least in the case of SYVV, may be the internal component and not the virus particle itself. This seems to be the case for a number of plant viruses but evidently not for the vertebrate agents. Fixation *in situ* of infected cells revealed in thin sections bacilliform viruses in infected plants, but not with animal or fish viruses. Here the appearance in section is that of a bullet, identical with those in negative contrast, with or without prefixation. The structural discrepancies may be due to differences in the morphogenesis of these viruses. While it seems agreed that all viruses which have been investigated in more detail mature on preexisting cell membranes, vertebrate viruses utilize cytoplasmic membranes, whereas many of the plant viruses mature on nuclear membranes. As a consequence, these viruses have a complex structure and their mode of replication must be taken into account in attempts at a comparative analysis of their structure. As a matter of fact with some of these viruses drastic differences in morphogenesis may also become apparent depending on the host cell infected or on the strain employed.

A. Dimensions

The width of these viruses show a reasonable uniformity ranging from about 65 nm (FHPV) to about 90 nm (RTYV). The measurements are greatly influenced by the mode of preparation. Thin sections using the dehydration process will decrease the diameter whereas negative stanning with its surface tension will increase it. The length of these viruses, on the other hand, varies greatly, disregarding aberrant forms. Apart from the Marburg virus which shows a median of 665 nm but can be found up to

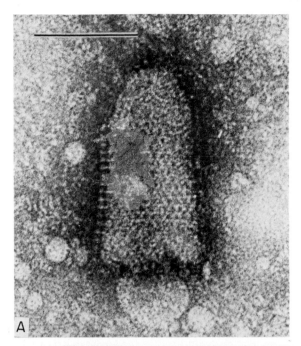

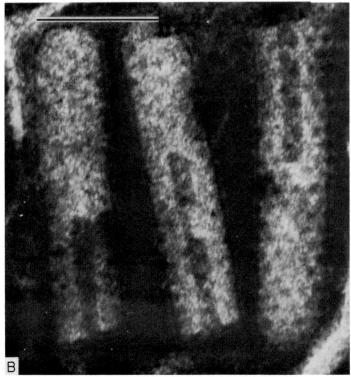

8 μm in length, vertebrate viruses range from 130 (Kern Canyon) to 230 nm (Mt. Elgon), and the plant viruses from 120 (RTYV) to 500 nm (NCMV). The vertebrate viruses, in general, show a greater uniformity in their dimensions than the plant agents.

B. Surface Structure

Differences in the surface structures become apparent after negative staining. Most viruses permit penetration of the contrast material, revealing the internal component, the nucleocapsid. This makes it difficult to gain insight into the arrangement of the surface structure of VSV, for instance. Rabies, on the other hand, has a more stable surface and penetration of contrast material is the exception rather than the rule. Under certain circumstances the rabies virion envelope exhibits a distinct surface arrangement which appears as hexagons (Hummeler *et al.*, 1967). This type of surface arrangement into hexamers has also been described for some plant viruses such as BNYV (Hills and Campbell, 1968), and possibly SYVV (Peters and Kitajima, 1970) (Fig. 1). The hexamer surface arrangement, however, is not unique to this group of viruses having also been described for influenza virus (Almeida and Waterson, 1967). Whether these surface arrangements are artifacts produced by the negative contrast or represent the basic structure of the envelope remains to be seen. In terms of structural rigidity the latter explanation seems more reasonable. Why some virus envelopes are easily penetrated whereas others are not cannot be explained at this point. More information on the chemical composition of the coat may provide the answer.

C. Internal Structure

The envelope of the viruses encloses an internal helical structure which is the nucleoprotein. Penetration of the envelope by negative contrast material permits the helical nucleoprotein to appear clearly as transverse striations (Fig. 2A). With rabies virus, penetration is rare but when observed the cross-striations do not differ from those of other agents of this genus. The periodicity of these transverse striations is remarkably uniform among all members of this group where it has been possible to measure it. The

Fig. 1. Surface arrangement. (A) Rabies; (B) BNYV (courtesy G. J. Hills). The bar represents 0.1 μm.

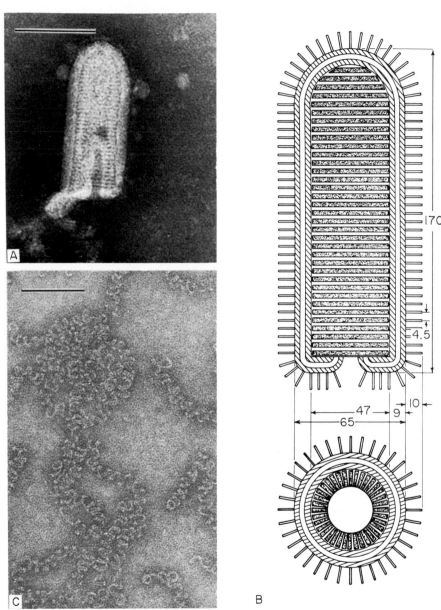

FIG. 2. Nucleocapsid. (A) Helical arrangement of the nucleocapsid in the virion (VSV) apparent as transverse striations. Periodicity 4.5 nm. (Courtesy A. F. Howatson.) (B) Model of VSV. (Dimensions in nm.) In longitudinal section the nucleoprotein helix appears as transverse striations. In cross section arrangement of individual protein subunits can be seen. (Courtesy A. F. Howatson.) (C) Nucleocapsid from purified rabies virions. Isolated in sucrose gradient. Note recoiling of strands. The bar represents 0.1 μm.

range is from 4 to 5 nm. The cross-striations indicate the helical configuration of the nucleoprotein enclosed by the coat (Fig. 2B). A detailed structural analysis of the nucleoprotein has been accomplished for VSV and rabies (Nakai and Howatson, 1968; Sokol et al., 1969) and the differences are minimal. In the mature virion of both viruses there are about 30 turns of the strand in the enveloped space with a few more in the rounded end. In contrast to the paramyxoviruses the bonds between the helical turns are exceedingly labile. The configuration is readily destroyed by minimal forces and destruction of the envelope results in release of the nucleoprotein as a single wavy strand which frequently recoils loosely at a smaller diameter than that of the *in situ* helix (Fig. 2C). Isolation of the nucleocapsid from purified virions in sucrose gradients also causes coiling (Sokol et al., 1969). This led to the assumption in the case of rabies (Sokol et al., 1969) that the helix in the virion may be a tertiary structure, but in thin sections the nucleoprotein appears as single strands in the cytoplasm (Zajac and Hummeler, 1970). Cesium chloride density gradients produce single strands which contain all the nucleic acid of the virus (Sokol et al., 1969; Zajac and Hummeler, 1970). These isolated strands are identical with those released from virions (Zajac and Hummeler, 1970). The strands exhibit subunits whose dimensions for VSV are approximately $9 \times 3 \times 3$ nm and for rabies $6.5 \times 2 \times 2$ nm. The subunits are threaded on a single strand of nucleic acid. The rabies virion, for instance, contains a strand approximately 4.2 μm long with approximately 1500 such subunits. The strand in VSV contains about 1000 subunits. Among other viruses of this group, Mt. Elgon bat virus shows evidence of such a single helical arrangement of its nucleoprotein (Murphy et al., 1970) and evidence presented for plant viruses suggest that the same arrangement exists in SYVV (Peters and Kitajima, 1970), LNYV (Wolanski et al., 1967), and PYDV (MacLeod, 1968). Since transverse striations are apparent in all viruses of this genus it is reasonable to suggest that they all contain a single helical arrangement of their nucleoprotein, the length of the strand possibly being the only variation. It is of interest to point out here that Peters and Kitajima (1970) in their study of SYVV, which appears bacilliform, found the single helix not to fill the whole of the envelope but to terminate short of one rounded end which is absence of prefixation breaks off forming the bullet which contained all of the genome. Thus the bacilliform shape may be only a minor variation in morphology. There seems to be no difference in the arrangement of the nucleocapsid between vertebrate and invertebrate viruses of this group.

D. Aberrant Structures

Aberrant structures are frequently found in cells infected with viruses which utilize cell membranes for their maturation. In rabies (Hummeler *et al.*, 1967), and less frequently in VSV (Galasso, 1967), long rods become evident which are identical in diameter with the virions and probably contain nucleocapsid cores in sequence with indentations at the proper virion length. They occur later in the growth cycle. At the time when the synthetic mechanisms of the cell are exhausted, bizarre forms are found in rabies-infected cells (Hummeler *et al.*, 1967). This phenomenon points up difficulties encountered in interpretation of morphogenetic processes. Brain tissues of experimental animals infected with rabies were investigated after onset of disease when an orderly morphogenesis had ceased and interpretation of these aberrant forms were difficult. It was thought for instance that the long rods belong in the proper sequence of morphogenesis, and that the virion segmented from them (Matsumoto, 1963). Only precisely timed growth studies, preferably *in vitro*, in combination with tests for biological properties, permit conclusive interpretation of these structures.

While no biological functions have been found for aberrant forms in rabies and other members of this group, in VSV such forms occur naturally and can be produced experimentally by serial undiluted passage of the virus. Under these conditions, low yields of infectious virus are obtained. This interference is due to the production of noninfectious particles which are shorter than mature VSV virions and are probably derived from them. They have been named T (truncated) in comparison to the longer mature forms which are called B (bullet shaped) (Hackett, 1964; Huang *et al.*, 1966). Interference with B particles by T particles under these circumstances is VSV specific. T particles are smaller than B and also contain accordingly less nuclei acid but are antigenically and chemically indistinguishable from B (Huang *et al.*, 1966). The mechanism of the interference is still poorly understood.

Truncated forms have also been found in rabies (Hummeler *et al.*, 1967) but no efforts have been made as yet to determine their biological properties.

III. Biochemical and Biophysical Properties

A. Nucleic Acids

Ribonucleic acid (RNA) has been found in viruses of this genus either by direct isolation and identification or indirectly by the lack of effects of inhibitors of DNA replication on virus multiplication. In all cases the RNA was found to be single-stranded. The RNA protein, or nucleocapsid, and the RNA of VSV and rabies have been investigated in detail and the results showed only minor differences as can be seen in Table II (Huang and Wagner, 1966; Sokol *et al.*, 1969).

TABLE II

COMPARISON OF NUCLEOCAPSID AND RNA OF VSV AND RABIES

	VSV	Rabies
Nucleoprotein		
Length	3.5 μm	4.2 μm
Number subunits	1000	1300
Sedimentation constant	140	200
RNA Protein ratio		1:25
Density in CsCl gm/cm³	1.32	1.32
RNA		
Sedimentation constant	40–45	45
Molecular weight	4.0 × 10⁶ daltons	4.6 × 10⁶ daltons
Infectivity	None	None
Single-strand	Yes	Yes

The VSV data are based on values obtained from purified B particles. The smaller amount of RNA in T particles would give heterogeneous results. It may be of interest to point out that despite the presence of a small amount of T-like particles in rabies progeny, the rabies RNA proved to be remarkably homogeneous in contrast to VSV preparations containing T particles. Base composition of the RNA has been determined for VSV (Brown *et al.*, 1967) and the percentage distribution of ^{32}P for B particle RNA obtained were A = 29.3, C = 21.1, G = 20.9, and U = 28.7. Base composition of T particle RNA was slightly different but it was not thought to be significant. The base composition of rabies RNA has not been determined, as yet. Both VSV and rabies RNA are not infectious (Huang and Wagner, 1966; Sokol *et al.*, 1969).

B. PROTEINS

1. *VSV*

Electrophoretic patterns of purified VSV showed four separate proteins (Kang and Prevec, 1969; Wagner *et al.*, 1969). Their respective molecular weights were found to be (1) 275,000, (2) 81,500, (3) 60,000, and (4) 34,500. The first protein, which is a minor component, may possibly be an aggregate. Proteins nos. 2 and 4 are contained in the viral coat and have been named G (glycoprotein) and S (surface), respectively. Protein no. 3, or N, is the protein of the nucleocapsid. It is assumed that they are single peptides (Wagner *et al.*, 1969).

Wagner *et al.* (1970) studied the kinetics and cellular sites of synthesis of these proteins and related that N was synthesized in large pools in the cytoplasm as early as 2 hours after infection with subsequent aggregation, presumably with progeny RNA, to form the nucleocapsid. Proteins G and S were synthesized near and inserted into cytoplasmic membranes from which the virions matured.

2. *Rabies*

Sokol *et al.* (1971) investigated the proteins of rabies by electrophoresis and found four major components. In order of their mobility they were numbered 1–4. Numbers 1, 3, and 4 are coat proteins, number 1 presumably the glycoprotein similar to the one found in VSV. Number 2 is the nucleocapsid protein. The molecular weights of these virus constituents were (1) the glycoprotein, 80,000; (2) the nucleocapsid protein, 62,000; and (3) and (4) small coat components, 40,000 and 25,000, respectively. According to these results it seems that rabies contains one more major protein component than VSV.

IV. Antigenic Properties

A. SEROLOGICAL REACTIVITIES

Antigenic differences among strains of these viruses have been described for VSV. Two serological types of VSV are known, the Indiana and New Jersey strains. Several other VSV strains isolated showed a relationship to the Indiana serotype which has three subtypes, Indiana 1, the prototype, 2 (including cocal and Argentina viruses), and 3 (Brazil virus) (Federer *et al.*, 1967). Antigenic relationships among some vertebrate viruses is illustrated in Tables III and IV (kindly supplied by Drs. F. Murphy and R. E. Shope). Most members are antigenic entities except possibly Piry and Chandipura which in neutralization tests showed considerable cross-reaction and some reactions with VSV strains.

TABLE III

NEUTRALIZATION TEST RESULTS[a]

Virus	Titer log LD$_{50}$	Mouse hyperimmune sera										Guinea pig serum Marburg
		VS Indiana	Cocal	VS-New Jersey	Piry	Chandipura	Hart Park	Flanders	Rabies	Kern Canyon	Mt. Elgon	
VS-Indiana	7.1	>6.2[b]	4.2	1.9	0	2.6	0	0	0	0	0	0
Cocal	7.7	4.2	>7.0	2.7	1.7	1.8	0	0	0	0	0	0
VS-New Jersey	5.5	2.2	0	>5.0	0	0	0	0	0	0	0	0
Piry	7.9	2.3	2.6	2.5	5.7	3.4	0	0	0	0	0	0
Chandipura	7.3	2.5	2.7	0	4.6	>5.7	0	0	0	0	0	0
Hart Park	4.6	0	0	0	0	0	>4.1	>4.1	0	0	0	0
Flanders	4.4	0	0	0	0	0	>3.9	>3.9	0	0	0	0
Rabies	5.9	0	0	0	0	0	0	0	>5.4	0	0	0
Kern Canyon	5.4	0	0	0	0	2.2	0	0	0	>4.9	0	0
Mt. Elgon	5.8	0	0	0	0	0	0	0	0	0	>5.3	0

[a] Table supplied by F. Murphy and R. E. Shope.
[b] Log neutralization index; 0 = <1.5.

TABLE IV

RESULTS OF COMPLEMENT FIXATION TESTS[a]

Antigens	Hyperimmune mouse sera										Guinea pig serum Marburg
	VS-New Jersey	VS-Indiana	Cocal	Piry	Chandipura	Flanders	Hart Park	Rabies	Mt. Elgon	Kern Canyon	
VS-New Jersey	256/512[b]	0	0	0	0	0	0	0	0	0	0
VS-Indiana	0	256/512	32/128	0	0	0	0	0	0	0	0
Cocal	0	32/512	256/512	0	0	0	0	0	0	0	0
Piry	0	0	0	128/32	8/4	0	0	0	0	0	0
Chandipura	0	0	0	0	128/64	0	0	0	0	0	0
Flanders	0	0	0	0	0	64/16	32/4	0	0	0	0
Hart Park	0	0	0	0	0	32/16	256/64	0	0	0	0
Rabies	0	0	0	0	0	0	0	128/512	0	0	0
Mt. Elgon	0	0	0	0	0	0	0	0	64/64	0	0
Kern Canyon	0	0	0	0	0	0	0	0	0	256/512	0
Marburg	0	0	0	0	0	0	0	0	0	0	64/32
Normal	0	0	0	0	0	0	0	0	0	0	0

[a] Table supplied by F. Murphy and R. E. Shope.

[b] Titer of serum/antigen; 0 = no reaction at 1:4 serum dilution.

Recently Shope *et al.* (1970) found that the Lagos and IbAn viruses are related to rabies by complement fixation and neutralization tests. These viruses, including rabies, may eventually be designated as a subgroup of this genus.

Antigenic relationships among the salmonid viruses have been studied by McCain (1970). He found OSD, IHN, and SRC to be related; OSD and IHN being indistinguishable. Thus these two agents may be identical. The South African and Australian isolates of BEF are serologically related, but they have not been compared to the Japanese strains (Doherty *et al.*, 1969).

Antigenic relationships among the described proteins of VSV have been studied by Kang and Prevec (1970). Isolated and purified coat and nucleo-capsid proteins were used for immunization of rabbits. The immune sera were tested by various immunological procedures such as neutralization, complement fixation, and antibody blocking tests for their antibody content. The results can be summarized as follows: Sera prepared against coat proteins showed high titers of neutralizing antibodies whereas those against nucleoprotein were devoid of them. Complement fixation tests revealed that coat proteins and nucleoproteins were distinct antigens within one sero type of VSV. A comparison between the proteins of the Indiana and the New Jersey strains, which do not cross-react in neutralization tests, showed that while coat antigens were specific for each sero type, antinucleoprotein sera cross-reacted.

These results are similar to those described long since for myxoviruses, particularly influenza, with the reactivity of their nucleoprotein (S) and coat antigen (V).

Cartwright *et al.* (1969) found in VSV the antigen responsible for neutralizing antibodies in the projection, or spikes, of the virus whereas the lipo-proteins of the coat induced host-specific antibodies, an indication of the presence of host material in the viral envelope as a consequence of the budding process.

A similar careful analysis of the proteins of rabies has not been attempted, as yet. The recent isolation and identification of the proteins of rabies, however, will facilitate a proper analysis of their antigenic reactivities in the near future.

A serological comparison of the plant viruses has not been done as extensively as with vertebrate viruses. Different strains within one type have been compared and the Australian isolates of LNYV were found to be related (Harrison and Crowley, 1965). No cross-reactions were found between SYVV and VSV (Sylvester *et al.*, 1968). The two strains of PYDV the New York (AC) strain and the New Jersey (SC) cross-reacted in neutralization tests but were separate from SYVV (Liu, 1969).

B. HEMAGGLUTININS

Hemagglutinating activity has been described for rabies, VSV, and Kern Canyon viruses (Halonen *et al.*, 1968). The agents agglutinate goose erythrocytes at low temperatures. In rabies the HA is associated with a coat of intact virions and there is no evidence of a soluble HA (Kuwert *et al.*, 1968). Disintegration of the viral envelope by deoxycholate leaves subunits which have no HA activity but inhibit the reaction. Red cells can be used repeatedly for absorption and elution with this virus indicating lack of neuraminidase in the coat. The relationship between infectivity and HA is linear, the ratio of plaque-forming units and HA activity being about 10^6 similar to myxoviruses. The exact location of the HA on the virus coat is not known (Sokol *et al.*, 1969).

V. Morphogenesis

The morphogenesis of these viruses plays a key role in their eventual structure and chemical composition since host membranes are so obviously involved. While viruses of this genus presumably mature from preexisting cell membranes, although it has not been observed as yet for all of them, the type of membrane involved is not uniform. All vertebrate viruses involve cytoplasmic membranes or the plasmalemma in the process of maturation. A number of plant viruses, on the other hand, utilize nuclear membranes. Thus there is a drastic difference in the site of synthesis of their nucleoprotein and presumably in the mechanism of replication of their nucleic acid. The latter process cannot be considered here because little information is available. The development of only a few of these viruses has been followed *in vitro* in controlled time studies. Most morphogenetic insights have been gained indirectly at the peak or past the peak of development. It is difficult therefore to compare the various viruses in their developmental stages. Among the vertebrate viruses, VSV (Howatson and Whitmore, 1962; Zajac and Hummeler, 1970) and rabies (Hummeler *et al.*, 1967, 1968) have been investigated in detail and will be discussed here. They will be compared with three plant viruses where the nucleoprotein is synthesized in the nucleus and maturation of particles occurs on nuclear membranes. They are SYVV, PYDV, and Gomphrena virus. SYVV has recently been shown to multiply in primary cell cultures of its aphid vector with evidence of synthesis in the nucleus (Peters and Black, 1970) and PYDV has been grown in continuous lines of its leaf hopper vector showing identical morphogenetic features as in plant cells (Chiu *et al.*, 1970). These experimental conditions will permit controlled growth studies. These viruses to be discussed are representatives of different modes of replication, but other members of this genus, as yet not investigated, may show still others.

A. VERTEBRATE VIRUSES

The nucleoprotein of vertebrate viruses is synthesized in the cytoplasm. The nucleoprotein does not exist in helical form at the site of synthesis but rather as single strands. It is difficult to visualize, therefore, unless it accumulates and forms inclusions. An exception seems to be the trout virus (HSV) where the nucleoprotein can be found in the cytoplasm as helixes with a diameter of 18 nm (Zwillenberg *et al.*, 1965) (Fig. 3A). The rate of accumulation of the nucleoprotein in the cytoplasm seems to depend on the host cell or the virus strain used for infection. These influences on morphogenesis are particularly evident in rabies but can be shown also with VSV if experimental conditions are suitably altered.

Rabies is the most venerable member of this group. The dramatic clinical illness in man led to experimental work since Pasteur's days. Its particular selectivity for certain neurons in the central nervous system pinpointed investigations and characteristic cytoplasmic inclusions, or bodies, were found with regularity in these cells. They were first described by Negri in 1903 and considered pathognomonic. The fine structure of the Negri bodies in nerve tissue of infected animals has been elucidated by alternate thick and thin sections of neurons containing Negri bodies (Miyamoto and Matsumoto, 1965). They consist of a ground substance with virus particles frequently embedded in it. Following infection with fixed virus strains, or by serial passage of street virus in animal brain tissue, the size and number of these virus-specific cytoplasmic foci are greatly reduced but not eliminated. To determine the exact nature of these inclusions it became necessary to investigate the replication of the virus in tissue cultures because the events in the growth of the virus in infected central nervous systems of experimental animals were studied at a late stage of intracellular development at a point when illness became apparent in the experimental animals, thus the very early stages of replication were missed (Hummeler *et al.*, 1967, 1968).

The first morphological evidence of replication of rabies virus in tissue culture is the appearance of a finely granular material which replaces normal constituents of the cytoplasm and is not limited by any membrane structure. Eventually this granular material becomes filamentous and its appearance

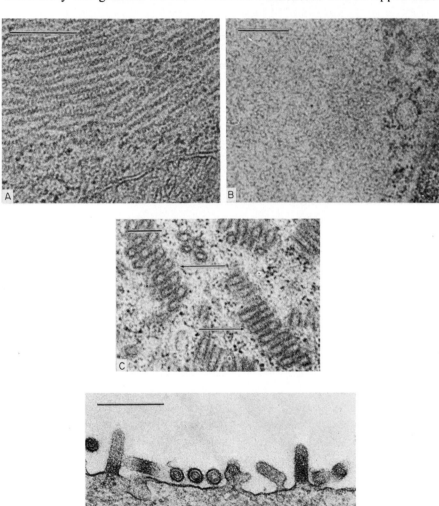

FIG. 3. Morphogenesis of vertebrate viruses. (A) Cytoplasmic inclusion formed by helical nucleoprotein (VHSV - trout). (B) Cytoplasmic inclusion formed by single-stranded nucleoprotein (rabies). (C) Viral envelope synthesized *de novo* in cytoplasm (rabies). Note continuity of membranes (arrows). (D) Formation of virus envelope by budding off preexisting cell plasma membrane (VHSV - trout). The bar represents 0.25 μm.

coincides with that of specific fluorescence seen with conjugated antisera (Fig. 3B). These early changes were observed uniformly with all virus strains used. However, the maturation of the virions showed decided differences depending on the strain of virus used for these experiments. Particles may emerge on the periphery of these filamentous inclusions or inside them without obvious involvement of preexisting cell membranes (Fig. 3C) and only later virions are formed by budding from membranes of cytoplasmic vacuoles and the plasmalemma. With other strains, however, virus particles form solely on preexisting cell membranes by budding with no evidence of *de novo* synthesis of the envelope in the cytoplasm (Fig. 3D). This drastic difference in morphogenesis, depending on the virus strain, has also been observed in experimental animals. Street virus strains showed the characteristic *de novo* synthesis of viral membranes whereas vaccine strains matured solely by budding from cytoplasmic membranes (P. Atanasiu, personal communication).

The filamentous inclusions were shown to consist of viral nucleoprotein strands which became the nucleocapsid of the virion upon maturation. When little budding of virus particles occurred large masses of ribonucleoprotein accumulated in the cell leading to the formation of the characteristic inclusions. When morphogenesis of the virion was an efficient process, however, the synthesis of nucleoprotein occurred at a rate equivalent to its incorporation into virions and little accumulation occurred.

These experimental data demonstrate the influence of various strains of one virus on the morphogenetic process. In this rabies is unique among the members of this group investigated thus far.

Despite the considerable body of literature on the intracellular replication of VSV, little is known about the morphogenesis of the ribonucleoprotein of this virus. Irregular arrays of virus—specific material in the cytoplasm of VSV—infected cells have been described (David-West and Labzoffsky, 1968), but the nature of the material was unknown. The finding of "rabies-like" inclusions in VSV-infected calf kidney cells (Schultze and Liebermann, 1966) led us to investigate the dependence of the morphogenesis of the ribonucleoprotein on the type of host cell used (Zajac and Hummeler, 1970).

Several tissue culture cell lines of human and animal origin were infected with VSV at a high multiplicity of input and accumulation of the nucleoprotein of VSV in the cytoplasm of three of them could readily be demonstrated. The results of growth experiments are shown in Table V. Of the three cell lines showing inclusion formation, the human lymphoblastoid cell lines followed predictable patterns. Little maturation of the infectious

TABLE V

INCLUSION FORMATION AND VIRUS REPLICATION IN VARIOUS CELL LINES 18 HOURS AFTER
INFECTION WITH VSV

Cell	PFU/ml ($\times 10^6$)	Fluoresc. cells (%)	Inclusion
Human foreskin	50.0	40	—
Human amnion	30.0	70	—
Human lymphoblastoid	5.0	40	+
Human lymphoblastoid	8.5	70	+
Hamster kidney	1260.0	95	+

virus occurred on preexisting cell membranes, indicated by the low titers
of infectivity, and nucleoprotein accumulated in the cytoplasm. The baby
hamster kidney cell line BHK-21, on the other hand, exhibited still another
pattern. Large amounts of infectious virus were produced and yet inclusions
were found readily in the cytoplasm of these cells indicating that the accumulation of nucleoprotein was not dependent on the maturation of virions on
cell membranes. It is not known as yet whether the formation of inclusions
in VSV-infected cells is also dependent on virus strains employed as in rabies
because these results were obtained with a single strain of VSV. The filaments in these cytoplasmic inclusions were identified as the ribonucleoprotein of VSV by various means. Their appearance and progression paral-
leled development of virus-specific immunofluorescence and production of
virus progeny. They were found in proximity to budding virions indicating
that they become the helical nucleocapsid upon maturation of virions on
cell membranes. They do not exist as helical structures in the cytoplasm
since only unwound strands were found. This was also emphasized by the
fact that the diameter of the strands in thin sections was approximately
9 nm whereas the *in situ* helical nucleocapsid of VSV has the diameter of
approximately 50 nm. When the inclusion material was extracted from infected cells and purified in cesium chloride density gradients the strands
resembled the ribonucleoprotein isolated from purified virions. They incorporated uridine-^{3}H indicating the presence of RNA and they exhibited
virus-specific complement fixing activity, had a density of 1.32 gm/cm^3 in
CsCl, and appeared as single wavy strands with a variation in width from
2.5 to 8.5 nm (Fig. 4). This is identical to the ribonucleoprotein strand found
in virions (Nakai and Howatson, 1968).

The results obtained with VSV showed the striking similarity in the

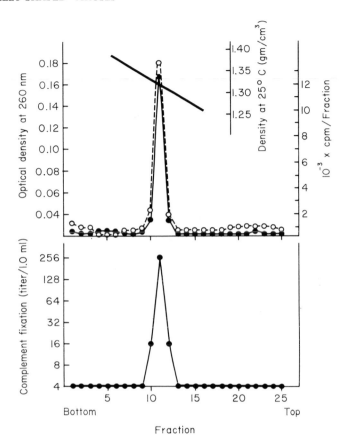

FIG. 4. Equilibrium centrifugation in CsCl of VSV nucleoprotein isolated from infected cells. Density (———), total radioactive counts of uridine-³H (●———●), and optical densities at 260 nm (○———○) are shown in upper panel. Virus-specific complement fixing activity (●———●) in lower panel.

morphogenetic process with that of rabies except for the *de novo* synthesis of envelope material with certain strains of rabies which are unique and resemble that of poxviruses (Dales and Siminovitch, 1961). The dependency of morphogenesis on the host cell and possibly the strain of virus warrants further investigation.

B. PLANT VIRUSES

Two plant viruses, PYDV and SYVV, have been propagated successfully in cell cultures of their respective vectors, the PYDV in continuous cells of its leaf hopper vector (Chiu *et al.*, 1970) and SYVV in primary cultures of its aphid vector (Peters and Black, 1970). These results not only demonstrated multiplication of these plant viruses in their insect vectors but also that the morphogenesis, especially PYDV in leaf hopper cells, is indistinguishable from that in its plant host (Chiu *et al.*, 1970). Infected cells showed specific fluorescence in their nuclei 18–24 hours after infection. The discrete stain became more intense and diffuse at 48 hours, extending into the cytoplasm. Electron microscopy verified the morphogenetic process previously described in infected plants. All basic developmental features observed in the cytoplasm of cells infected with vertebrate viruses were observed in the nuclear replication of these plant viruses. It is evident from the illustrations in various publications that the nucleoprotein exists in nuclei infected with PYDV or SYVV as discrete single filaments (Fig. 5A). Budding off the inner nuclear membrane occurred in close proximity to these inclusions. Of particular interest here are observations with Gomphrena virus (Kitajima and Costa, 1966). The described nuclear "tubular particles" appear to be the ribonucleoprotein in tightly coiled helixes similar to the trout virus (VHSV) in the cytoplasm (Zwillenberg *et al.*, 1965) (Fig. 5B). The virus coat was acquired during maturation on the inner membrane of the nucleus and virus particles accumulated in the intermembrane space. Other plant viruses showing nuclear development, among those listed in Table I, are MLV, EMDV, and CEV, whereas the remainder replicate in the cytoplasm, except WSMV which appears to utilize both the nucleus and the cytoplasm (Lee, 1970).

The nuclear development of some of these plant viruses is of interest in relation to their nucleic acid content. The type of nucleic acid has, unfortunately, not been determined unequivocally*. In some of them it may indeed be single-stranded RNA. Definite knowledge of this would be of particular interest because of differences in the site of replication of the nucleic acids of these viruses as compared to vertebrate viruses. No comparison can be made at this point since it would be pure speculation and the elucidation of these pathways will have to await more biochemical data. These should become available as these viruses become more amenable to experimental manipulation in continuous cell lines of their respective vectors.

The preceding studies on the morphogenesis of the various viruses showed the distinct influence of the mode of replication on the structure. Matura-

---

* Dr. R. MacLeod recently informed the author that he has isolated RNA from PYDV with a MW of 4.3×10^6 daltons. He also obtained evidence of four major proteins and hemagglutinin.

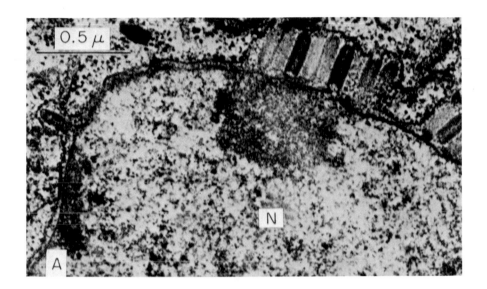

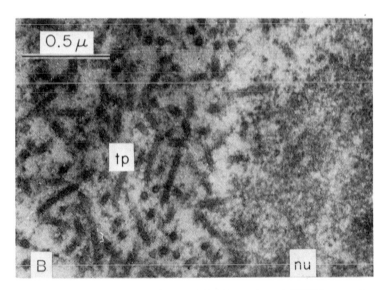

FIG. 5. Morphogenesis of plant viruses. (A) Maturation of PYDV on nuclear membranes. Nucleoprotein appears as strands (Chiu *et al.*, 1970). N = nucleus. (B) Helical nucleoprotein of Gomphrena virus in nucleus. (Kitajima and Costa, 1966). nu, nucleolus; tp = tubular particles. The bar represents 0.5 μm.

tion on nuclear membranes seems to lead preferentially to bacilliform virions whereas cytoplasmic membrane and plasmalemma maturation results in bullet-form shapes. Fixed thin sections used in the study of morphogenetic processes show this clearly. An occasional bacilliform VSV particle is surely an aberration, as may be the bullet shape of some plant viruses that are actually bacilliform. Aberrant structures are particularly frequent at late stages of development and *in vitro* investigations of early events of the morphogenesis of these viruses are paramount to arrive at valid conclusions as to the mechanism of their replication and their final structure.

VI. Summary

A new and important family of viruses has been described. These agents are of particular interest because they are associated with disease of vertebrates, invertebrates, as well as plants. They are separable from other viral families solely on the basis of morphology. They may also share the same type of nucleic acid and despite some divergence in their structure the nucleocapsid exists as a single-stranded helix, the periodicity of which shows good agreement for all members thus far investigated. The nucleocapsid varies only in its length, as do the particles themselves. Minor differences in structure such as bacilliform versus bullet shape may be the result of differences in morphogenesis.

Some of the vertebrate viruses are known to multiply in insects, as do several of the plant viruses. VSV when injected into *Drosophila* seems to elicit the same symptomatology as is caused by the sigma virus (Perier *et al.*, 1966). Insects thus are a bridge between vertebrates and plants but no viruses are known to be able to multiply in all three types of hosts.

Whether all viruses listed in Table I will eventually be classified in this genus remains to be seen. It is possible that some of them are different strains of the same virus, and more immunological studies are indicated. The drastic differences in morphogenesis may also be a cause for reclassification. The nuclear involvement in plant virus replication is not found in vertebrate viruses which may speak for differences in the biochemical replication of their nucleic acid. Whether these differences warrant a separation remains to be investigated. More information about the properties of many of these viruses must be obtained so that their relationship or their differences can be more clearly defined.

REFERENCES

Almeida, J. D. and Waterson, A. P. (1967). *J. Gen. Microbiol.* **46**, 107.

Amend, D. F. and Chambers, V. C. (1970). *J. Fish. Res. Bo., Canada*, **27**, 1285.

Berkaloff, A., Bregliano, J. C., and Chanessian, A. (1965). *C. R. Acad. Sci. Paris* **260**, 5956.

Brown, F., Cartwright, B., Crick, J., and Smale, C. J. (1967). *J. Gen. Virol.* **1**, 479.

Cartwright, B., Smale, C. J., and Brown, F. (1969). *J. Gen. Virol.* **5**, 1.

Chiu, R. M., Liu, H. Y., MacLeod, R., and Black, L. M. (1970). *Virology* **43**, 87.

Conti, M. (1969). *Phytopathol. Z.* **66**, 275.

Dales, S. and Siminovitch, L. (1961). *J. Biophys. Biochem. Cytol.* **10**, 475.

David-West, T. S. and Labzoffsky, N. A. (1968). *Arch. Gesamte Virusforsch.* **23**, 105.

Doherty, R. L., Standfast, H. A., and Clark, I. A. (1969). *Australian J. Sci.* **31**, 365.

Federer, K. E., Burrows, R., and Brooksby, J. F. (1967). *Res. Vet. Sci.* **8**, 103.

Galasso, G. J. (1967). *Proc. Soc. Exp. Biol. Med.* **124**, 43.

Hackett, A. J. (1964). *Virology* **24**, 51.

Halonen, P. E., Murphy, F. A., Fields, B. N., and Reese, B. R. (1968). *Proc. Soc. Exp. Biol. Med.* **127**, 1037.

Harrison, B. D. and Crowley, N. C. (1965). *Virology* **26**, 297.

Herold, F., Bergold, G. H., and Weibel, J. (1960). *Virology* **12**, 335.

Hills, V. M. and Campbell, R. N. (1968). *J. Ultrastr. Res.* **24**, 134.

Hitchborn, J. H., Hills, G. J., and Hull, R. (1966). *Virology* **28**, 768.

Howatson, A. F. and Whitmore, G. F. (1962). *Virology* **16**, 466.

Huang, A. S. and Wagner, R. R. (1966). *Virology* **30**, 161.

Huang, A. S., Greenawalt, J. W., and Wagner, R. R. (1966). *Virology* **30**, 161.

Hummeler, K., Koprowski, H., and Wiktor, T. J. (1967). *J. Virol.* **1**, 152.

Hummeler, K., Tomassini, N., Sokol, F., Kuwert, E., and Koprowski, H. (1968). *J. Virol.* **2**, 1191.

Ito, Y., Tanaka, Y., Inaba, Y., and Omori, T. (1969). *Nat. Inst. Animal Health, Quart.* **3**, 35.

Kang, C. Y. and Prevec, L. (1969). *J. Virol.* **3**, 404.

Kang, C. Y. and Prevec, L. (1970). *J. Virol.*, **6**, 20.

Kitajima, E. W. and Costa, A. S. (1966). *Virology* **29**, 523.

Kitajima, E. W., Lauritis, J. A., and Swift, H. (1969). *J. Ultrastr. Res.* **29**, 141.

Kuwert, E., Wiktor, T. J., Sokol, F., and Koprowski, H. (1968). *J. Virol.* **2**, 1381.

Lecatsas, C., Theodorides, A., and Erasmus, P. J. (1969). *Arch. Gesamte Virusforsch.* **28**, 390.

Lee, P. E. (1964). *Virology* **23**, 145.

Lee, P. E. (1970). *J. Ultrastr. Res.* **31**, 282.

Liu, H. (1969). Ph.D. Thesis, Univ. of Illinois, Urbana, Illinois.

McCain, B. B. (1970). Ph.D. Thesis, Oregon State Univ., Corvallis, Oregon.

MacLeod, R. (1968). *Virology* **34**, 771.

MacLeod, R., Black, L. M., and Moyer, F. H. (1966). *Virology* **29**, 540.

Martelli, C. P. (1969). *J. Gen. Virol.* **5**, 319.

Matsumoto, S. (1963). *J. Cell Biol.* **19**, 565.

Melnick, J. J. and McCombs, R. M. (1966). *Progr. Med. Virol.* **8**, 400.

Miyamoto, K. and Matsumoto, S. (1965). *J. Cell Biol.* **27**, 677.

Murphy, F. A. and Fields, B. N. (1967). *Virology* **33**, 625.

Murphy, F. A. and Shope, R. E. (1969). Personal communication.

Murphy, F. A., Coleman, P. H., and Whitfield, S. G. (1966). *Virology* **30**, 314–317.

Murphy, F. A., Shope, R. E., Metselaar, D., and Simpson, D. I. H. (1970). *Virology* **40**, 288.

Nakai, T. and Howatson, A. F. (1968). *Virology* **35**, 268.

Perier, J., Printz, P., Canivet, M., and Chwat, J. C. (1966). *C. R. Acad. Sci. Paris* **262**, 2106.

Peters, D. and Black, L. M. (1970). *Virology* **40**, 847.

Peters, D. and Kitajima, E. W. (1970). *Virology* **41**, 135.

Peters, D., Muller, G., and Slenczka, W. (1970). "Marburg Virus Symposium," **7**, 241. Springer, Berlin.

Razvjazkina, G. M., Poliakova, G. P., Stein-Margolina, V. A., and Cherny, N. E. (1968). *1st Int. Congr. Plant Pathol., London*, p. 162.

Richardson, J. and Sylvester, E. S. (1968). *Virology* **35**, 347.

Rubio-Huertos, M. and Bos, L. (1969). *Neth. J. Plant Pathol.* **75**, 329.

Schultze, P. and Liebermann, H. (1966). *Arch. Exp. Veterinaermed.* **27**, 13.

Shikata, E. and Chen, M. (1969). *J. Virol.* **3**, 261.

Shikata, E. and Lu, Y. T. (1967). *Proc. Jap. Acad.* **43**, 918.

Shope, R. E., Murphy, F. A., Harrison, A. K., Causey, O. R., Kemp, G. E., Simpson, D. I. H., and Moore, D. L. (1970). *J. Virol.* **6**, 690.

Siegert, R., Shu, H. L., Slenczka, W., Peters, D., and Muller, G. (1967). *Deut. Med. Wochschr.* **92**, 2341.

Sokol, F., Stancek, D., and Koprowski, H. (1971). *J. Virol.*, **7**, 241.

Sokol, F., Schlumberger, H. D., Wiktor, T. J., Koprowski, H., and Hummeler, K. (1969). *J. Virol.* **38**, 651.

Sylvester, E. S., Richardson, J., and Wood, P. (1968). *Virology* **36**, 693.

Wagner, R. R., Schnaitman, T. A., and Snyder, R. M. (1969). *J. Virol.* **3**, 395–403.

Wagner, R. R., Snyder, R. M., and Yamazaki, S. (1970). *J. Virol.* **5**, 548.

Wolanski, B. S., Francki, I. B. and Chambers, T. C. (1967). *Virology* **33**, 287.

Zajac, B. A. and Hummeler, K. (1970). *J. Virol.* **6**, 243.

Zwillenberg, L. O., Jensen, M. H., and Zwillenberg, H. H. L. (1965). *Arch. Gesamte Virusforsch.* **17**, 1.

CHAPTER 12 *Structure and Transcription of the Genomes of Double-Stranded RNA Viruses**

STEWART MILLWARD AND A. F. GRAHAM

In 1963 two viruses, reovirus and wound tumor virus, were shown by Gomatos and Tamm (1963b) to have genomes comprised of double-stranded RNA (dsRNA), and for some time these were the only RNA viruses known to have double-stranded genomes. More recently several other such viruses have been found: cytoplasmic polyhedrosis virus (Miura *et al.*, 1968), rice dwarf virus (Sato *et al.*, 1966), and the bluetongue virus of sheep (Verwoerd, 1969). In anticipation of other such viruses being found, Vorwoerd has suggested that the group be christened the "diplorna" viruses.

There is general agreement that the structural proteins of reovirus are arranged in the form of a double capsid. The inner capsid or core is approximately 450 Å in diameter and contains the viral genome. An outer layer of capsomeres, probably 92 in number, is arranged in a regular manner on

* This work was supported by U.S. Public Health Service Research Grant AI-02454 from the National Institute of Allergy and Infectious Disease, and by the Medical Research Council of Canada.

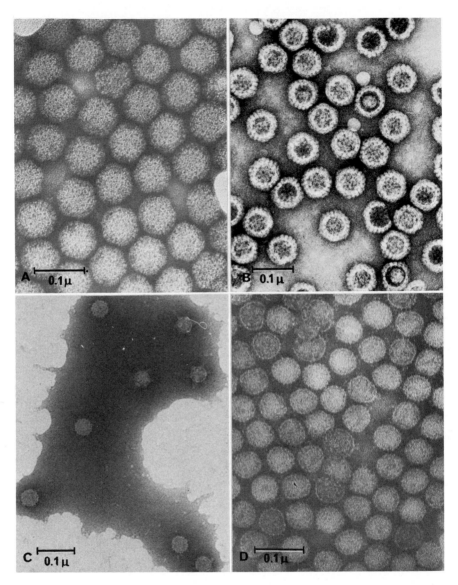

FIG. 1. Electron micrographs of: (a) reovirus, (b) wound tumor virus, (c) cytoplasmic polyhedrosis virus, and (d) reovirus cores. All preparations were negatively stained with neutral 1% phosphotungstate. Electron micrographs of reovirus (a) and reovirus cores (d) were kindly prepared for us by Dr. H. A. Blough; those of wound tumor virus (b) and cytoplasmic polyhedrosis virus (c) were obtained through the courtesy of Drs. G. Streissle and L. J. Lewandowski, respectively.

the surface of the core, giving an overall diameter of approximately 550 to 600 Å (Vasquez and Tournier, 1962, 1964; Loh et al., 1965; Dales et al., 1965; Mayor et al., 1965; Gomatos et al., 1962). The cores can be released by treatment of the virions with various reagents (Loh and Shatkin, 1968; Shatkin and Sipe, 1968; Smith et al., 1969). Wound tumor virus has a type of structure very similar to the reoviruses (Streissle and Granados, 1968; Wood and Streissle, 1970), as shown in Fig. 1a,b. On the other hand, cytoplasmic polyhedrosis virus (Fig. 1c) (L. Lewandowski, personal communication) and bluetongue virus (Els and Verwoerd, 1969) do not have the double capsid type of structure. They both appear to have the size and, to some extent, surface structure of reovirus cores (Fig. 1d).

One consistent and striking feature characterizes all the known dsRNA viruses: when the dsRNA is removed from the purified virions, it invariably "fragments" in a reproducible manner. Until recently the reason for this ready fragmentation of the genome has been a mystery; we shall discuss it from the point of view of reovirus where the phenomenon has been most thoroughly studied.

I. Structure of the Reovirus Genome

On the strength of chemical analyses, the molecular weight of the reovirus genome was originally put at a minimum value of 10^7 daltons by Gomatos and Tamm (1963a). However, two experimental parameters were soon found that were inconsistent with this molecular weight. First, the sedimentation rate of viral RNA was lower than the value expected for a dsRNA molecule with a weight of 10^7 daltons. Second, when RNA was extracted from purified virions and examined in the electron microscope, the majority of lengths was much too short. Careful estimates (Gomatos and Stoeckenius, 1964; Kleinschmidt et al., 1964; Dunnebacke and Kleinschmidt, 1967) showed that the dsRNA strands fell into a trimodal distribution of sizes. The average lengths of the three classes (Vasquez and Kleinschmidt, 1968) are shown in Fig. 2, along with the molecular weights calculated from these lengths (Millward and Graham, 1970). It is clear that none of the pieces has a molecular weight close to 10^7 daltons.

The three size classes can be shown in another way, by velocity centrifugation through a sucrose gradient (Fig. 3). If molecular weights are calculated from the sedimentation rates, they fall close to the values obtained for the three classes by electron microscopy. Iglewski and Franklin (1967)

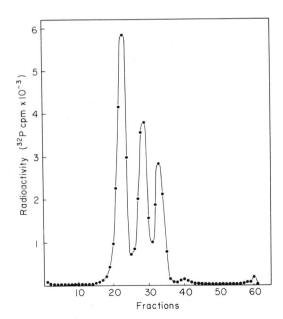

0.38μ mol. wt. = 0.89 x 10^6
ds RNA – 1
0.66μ mol. wt. = 1.55 x 10^6
ds RNA – 2
1.13μ mol. wt. = 2.66 x 10^6
ds RNA – 3

FIG. 2. Average lengths of the three size classes of reovirus dsRNA. The contour lengths are those reported by Vasquez and Kleinschmidt (1968). The molecular weights were calculated as reported by Millward and Graham (1970). The dsRNA-1 class consists of four segments while dsRNA-2 and dsRNA-3 consist of three segments each (Loh and Shatkin, 1968; Watanabe *et al.*, 1968a).

and Bellamy *et al.* (1967) did extensive studies of various methods for extracting dsRNA from purified virions in order to get out the intact genome. Invariably they obtained the three size classes of dsRNA in the same relative proportions, but never the complete genome. Nevertheless, Dunnebacke and Kleinschmidt (1967) and Granboulan and Niveleau (1967) had seen occasional strands up to 7 μ in length during electron microscopy of viral RNA. Such a length would be equivalent to a molecular weight of

Radioactivity (32P cpm x 10^-3)
1
2
3
4
5
6
10
20
30
40
50
60
Fractions

FIG. 3. Velocity sedimentation of reovirus dsRNA. The dsRNA, labeled *in vivo* with ^{32}P, was layered on a 5 to 20% linear sucrose gradient and centrifuged for 8 hours at 20°C in the SW 41 rotor of the Beckman, Model L-2 ultracentrifuge. The direction of sedimentation is from right to left and the three size classes are clearly resolved (S. Millward, unpublished data).

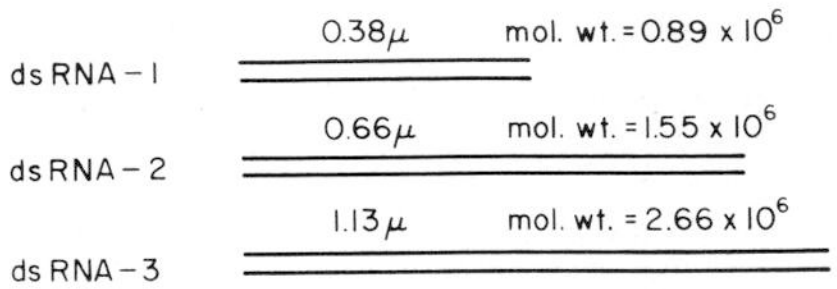

approximately 18×10^6 daltons. One might infer from these results that the genome existed as a linear thread within the virion, but was extremely fragile and broke up reproducibly into segments upon removal of the protein coat.

A major question now was whether the breaks were introduced at random or not. The frequency distribution and lengths of the three size classes seen with the electron microscope strongly suggested that the breaks were not random. More direct proof was produced by two groups of workers (Watanabe and Graham, 1967; Bellamy and Joklik, 1967). Essentially, the proof consisted of separating the three size classes of dsRNA from each other and determining whether there was a base sequence homology among them in hybridization tests. No such homology was found and, therefore, the genome must have broken at specific weak points during extraction from the virion and not at random.

From the total molecular weight of the genome and the weights of the three size classes (Fig. 2), it is clear that the genome must contain more than one fragment of each size. The number of fragments was estimated by analyzing ^{3}H-labeled viral RNA on columns of polyacrylamide gel, as shown in Fig. 4 (Watanabe *et al.*, 1968). The smallest size class (dsRNA-1) separated into three fractions, the medium size (dsRNA-2) into two, and the largest class (dsRNA-3) was unresolved. On the basis of the average molecular weights of each class and the amount of ^{3}H in each, it was calcu-

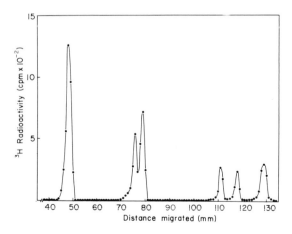

FIG. 4. Electrophoretic analysis of reovirus dsRNA in 2.5% polyacrylamide-agar gel. dsRNA from purified virus labeled with uridine-^{3}H was extracted by treatment with phenol, precipitated with ethanol and analyzed. (Reprinted from Watanabe *et al.*, 1968a by courtesy of *J. Mol. Biol.*)

lated that dsRNA-1a contains two segments, —1b and —1c are single, —2a contains two segments, —2b is single, and —3 contains three segments. This calculation can be checked by directly visualizing the distribution of segments. Figure 5 shows a polyacrylamide gel analysis in which the various fractions were stained with methylene blue (Millward, unpublished). Similar results have been obtained by Shatkin *et al.* (1968). Thus, the viral genome contains ten segments of dsRNA with a combined molecular weight of approximately 16×10^6 daltons.

The molecular weights of the individual segments can be estimated from analysis such as that shown in Fig. 5. If the average molecular weights of the three classes of dsRNA are plotted against the average distance migrated on a gel, they fall on a straight line, as shown in Fig. 6. When the distance

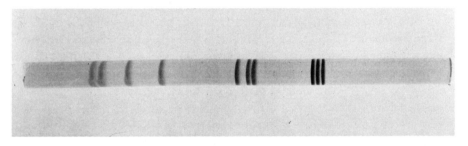

Fig. 5. Reovirus dsRNA (about 20 μg) was subjected to electrophoresis in 5% polyacrylamide gel for 60 hours at 8 mA. The ten segments were visualized by staining the gel with methylene blue. The direction of migration is from right to left (S. Millward, unpublished data).

migrated by each individual segment is then plotted on this line, its molecular weight can be obtained. Recently, it has been shown that the dsRNA genomes of wound tumor virus (Wood and Streissle, 1970; Kalmakoff *et al.*, 1969), cytoplasmic polyhedrosis virus (Kalmakoff *et al.*, 1969), and bluetongue virus (Verwoerd *et al.*, 1970) also yield a number of segments when analyzed by electrophoresis on polyacrylamide gels. Figure 7 is a chart summarizing our present knowledge on the number and molecular weights of the dsRNA segments in these various viruses.

The extraordinary tendency for the genomes of dsRNA viruses to fragment poses an important question—what is the structure of the genome at the break-points? We are starting to get some answers to this question in the following way.

Four general models of the reovirus genome can be proposed, as shown

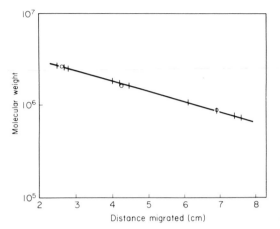

Fig. 6. The linear relationship between molecular weight and distance migrated by the reovirus dsRNA segments in polyacrylamide gel. The average molecular weight of each size class was determined as described for Fig. 2 and plotted against the average distance each size class migrated (open circles). The molecular weight of each segment can now be obtained from its position in the gel (S. Millward, unpublished data).

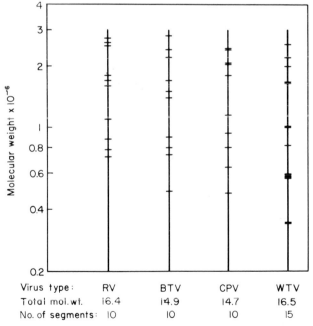

Fig. 7. Summary chart of dsRNA viruses showing the number and estimated molecular weight of the dsRNA segments. RV, reovirus; BTV, bluetongue virus (Verwoerd et al., 1970); cytoplasmic polyhedrosis virus, CPV (Kalmakoff et al., 1969); wound tumor virus, WTV (Wood and Streissle, 1970). Data for rice dwarf virus not available.

in Fig. 8. In trying to decide among these models, a technique has been employed to specifically label the 3′-terminals. This is done by oxidizing the RNA with periodate, which forms a dialdehyde at the free 2′-, 3′-OH positions. The dialdehyde is then reduced with tritiated borohydride of very high specific activity which introduces, theoretically, two tritium atoms per 3′-terminal. Assume the 3′-terminals of the dsRNA segments can be labeled with equal efficiency whether the oxidation step is carried out on the intact virion or on extracted dsRNA. After reduction, what would be the distribution of the tritium label in light of the proposed models (Fig. 8)?

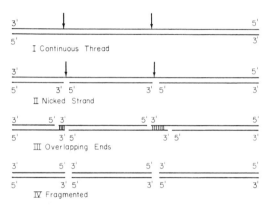

Fig. 8. Schematic models for the dsRNA genome of reovirus. The arrows represent the weak points in the viral genome. (Reprinted from Millward and Graham, 1970 by courtesy of *Proc. Nat. Acad. Sci. U.S.*)

When the dsRNA is extracted from purified virus, ten segments are found and each will have two 3′-terminals to be labeled by oxidation and reduction. Thus, each segment should be equally labeled with tritium. If the structure of the genome *within* the virion is either model III or IV, the same number of ends, twenty, are available to be labeled and all segments should be equally labeled if the oxidation–reduction reaction is carried out on *intact virions*. At the other extreme we have model I. Here only the two extreme ends are available for oxidation and reduction and when the oxidized dsRNA is extracted from the virions, reduced, and analyzed, only two of the ten segments should contain tritium. Suppose model II represents the structure of viral RNA: one strand is continuous, the other is nicked in nine places. Oxidation of this structure *within the virion* will modify only the left-hand segment at both 3′-terminals, while the other nine segments will be modified at one 3′-terminal only. This means that during the reduction step nine of the ten subunits will incorporate half as much tritium as when the oxidation is done on extracted dsRNA. In the actual experiment (Millward and Graham, 1970) reovirus was labeled with ^{32}P, to provide an RNA marker, and purified, and the virus preparation was divided into two. RNA was extracted from one part, oxidized with periodate, and reduced with borohydride-^{3}H. The other part of the virus was oxidized and the RNA was then extracted and reduced. Both specimens of RNA were analyzed by gel electrophoresis and the results are shown in Fig. 9. The ten segments are separated into five fractions. One cannot get as high a resolution of the genomic segments by this technique as with

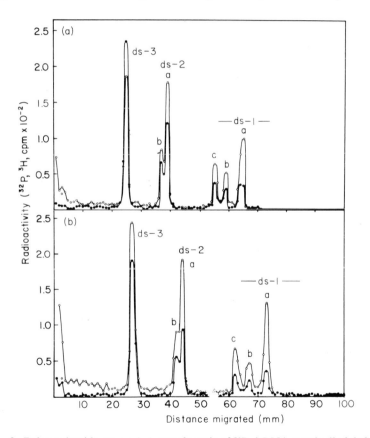

FIG. 9. Polyacrylamide-agar gel electrophoresis of ^{32}P dsRNA terminally labeled with ^{3}H. (a) Purified reovirus was oxidized with periodate, deproteinized and the resulting dsRNA reduced with ^{3}H-NaBH$_4$. (b) Oxidation with periodate and reduction with ^{3}H-NaBH$_4$ were carried out on isolated reovirus dsRNA. Open circles, ^{3}H; closed circles, ^{32}P. Fraction dsRNA-1a contains two segments, -1b and -1c are single segments; dsRNA-2a contains two segments, -2b is a single segment; dsRNA-3 contains three segments. (Reprinted from Millward and Graham, 1970 by courtesy of *Proc. Nat. Acad. Sci. U.S.*)

stained gels. Nevertheless, it is clear that all fractions are labeled with tritium; therefore, we can eliminate model I.

According to model II, oxidation of intact virions would modify about one-half as many 3'-terminals as when the oxidation is carried out on deproteinized dsRNA. Therefore, the dsRNA from oxidized virions can accept approximately one-half as much tritium during the reduction step. But, quantitative analysis of the tritium distribution in Fig. 9 showed that both samples of dsRNA incorporated tritium to about the same extent in all classes and we thereby eliminate model II (Millward and Graham, 1970).

If model III were correct, one expects to have single-stranded tails at the ends of each segment. When dsRNA labeled with ^{32}P to high specific activity is treated with snake venom phosphodiesterase, no ^{32}P is released suggesting that there are no tails at the 3'-terminals. Also, when the 3'-terminals of dsRNA were oxidized with periodate and then reduced with tritiated borohydride, no more than 50% of the label introduced could be removed by ribonuclease and this is a further indication that there are no single-stranded tails at the 3'-ends of the dsRNA. The loss of up to 50% of the label can be explained by the altered conformation at the 3'-end as a result of oxidation and reduction of the 3'-terminal residue. On the other hand, when dsRNA uniformly labeled with ^{32}P to high specific activity is treated with spleen phosphodiesterase in the presence of phosphomonoesterase, a significant amount of label was released. This observation suggested there may be a short region of nonhydrogen bonded nucleotides on the 5'-terminals of the dsRNA segments. To test this possibility the 5'-terminals of dsRNA were labeled with ^{32}P using the polynucleotide kinase catalyzed transfer of phosphate from (γ-^{32}P)-ATP (Richardson, 1965), and the lability of the labeled ends was tested by reaction with T_1-and pancreatic RNase. The results of these tests indicated that the 5'-tails were composed solely of purines (S. Millward, unpublished results). This result means there can be no complementary hydrogen bonding, in the Watson-Crick sense, between the tails of the various segments and therefore we can exclude the overlapping tail model (model III, Fig. 8). We conclude that model IV represents the structure of the reovirus genome with one or more purine nucleotides extending beyond the bihelix at the 5'-end. Whether or not the segments are linked, or how they are bound inside the virus core is still unknown.

While none of the other dsRNA viruses has yet been analyzed in this way, it seems very likely that they will be found to have a similar type of segmented genomic structure. Preliminary results on the analysis of the 3'-terminals of intact cytoplasmic polyhedrosis virions support this suggestion (Millward and Lewandowski, unpublished data).

II. Transcription of the Reovirus Genome

Several years ago it was found that two types of virus-specific RNA were being synthesized in reovirus-infected cells (Kudo and Graham, 1965). One species was dsRNA, which was shown to be viral progeny RNA. The second species was single-stranded RNA (ssRNA). It was made in an amount approximately equal to that of dsRNA and was found to be virus-induced messenger RNA (mRNA) (Prevec and Graham, 1966; Shatkin and Rada, 1967). Upon sedimentation of this mRNA through a sucrose gradient, three size classes were observed, as shown in the pattern of Fig. 10. In fact, the sedimentation rates of these messenger RNA's were found to be the same as the sedimentation rates of viral dsRNA after denaturation (Watanabe and Graham, 1967). This result meant that for each size class of dsRNA in the virion, a mRNA of similar size was synthesized in infected cells and raised the possibility that each mRNA was transcribed exclusively from a segment of genome of corresponding length. This possibility could be tested by determining whether each class of mRNA would hybridize uniquely with the corresponding size class of dsRNA. Watanabe *et al.* (1967b) provided the following experimental evidence in answer to this question.

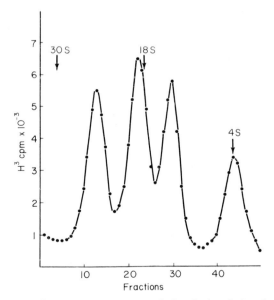

FIG. 10. Sucrose gradient sedimentation analysis of virus-induced mRNA from L cells infected with reovirus. (Reprinted from Watanabe *et al.*, 1968a by courtesy of *J. Mol. Biol.*)

Figure 11 shows an acrylamide gel analysis of viral dsRNA (open circles, ^{3}H label) and the three size classes of dsRNA were cleanly separated. Prior to electrophoresis, some ^{14}C-labeled viral RNA had been denatured and annealed and then mixed with the native ^{3}H-labeled RNA. This material (closed circles) migrated exactly with the native RNA. This result indicated that after denaturation and annealing the dsRNA regained its native configuration. A second control experiment is shown in Fig. 12. Here, ssRNA labeled with ^{3}H was mixed with dsRNA-^{14}C and subjected to electrophoresis

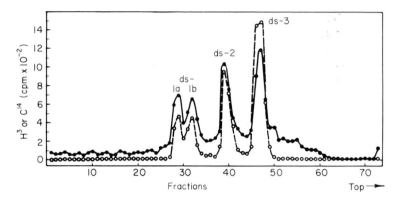

FIG. 11. Electrophoretic analysis of dsRNA and of dsRNA which had been denatured then annealed. Native reovirus dsRNA (^{3}H-labeled, open circles) was mixed with annealed ^{14}C-labeled dsRNA (closed circles). The mixture was analyzed by electrophoresis in polyacrylamide agar gel. (Reprinted from Watanabe *et al.*, 1967b by courtesy of *Proc. Nat. Acad. Sci. U.S.*)

on a polyacrylamide gel column (upper panel). The three classes of ssRNA (open circles) were separated from each other and from the three classes of dsRNA. The lower panel shows the pattern obtained when the mixture was denatured and annealed before analysis. All the ssRNA had now moved out of its normal position and migrated under the double-stranded RNA peaks. This result showed that hybridization was highly efficient and that the hybrids had a structure similar to that of the native dsRNA segments.

With these two preliminary results, the critical experiment could now be done. The three classes of ssRNA labeled with ^{3}H were separated from each other by sucrose gradient sedimentation. Each class of ssRNA was then mixed with viral dsRNA labeled with [14]C. Each of the three mixtures was denatured, annealed, and subjected to gel analysis (Fig. 13). The upper panel shows the hybridization of the smallest class of ssRNA with viral RNA. There was no ssRNA in the normal position; it had all hybridized and the hybrid was formed exclusively with the smallest class of dsRNA. In the middle panel is shown the hybrid formed by the medium class of ssRNA: it is exclusively with the middle class of dsRNA. The largest ssRNA hybridizes uniquely with the largest class of viral RNA, as shown in the bottom panel. The conclusion from this experiment is unequivocal; each class of ssRNA is transcribed exclusively from that class of dsRNA of corresponding size. Using a somewhat different analytical approach, Bellamy and Joklik (1967) came to exactly the same conclusion.

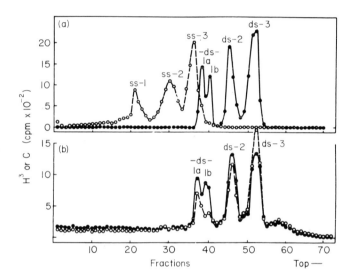

FIG. 12. Electrophoretic analysis of a mixture of ssRNA and dsRNA before and after hybridization. ^{3}H-labeled ssRNA, approximately 1 μg, 50,000 cpm (○——○), was mixed with ^{14}C-labeled dsRNA, 40 μg, 67,000 cpm (●——●), in 0.1 ml of 0.01 M STE buffer. (a) Part of the mixture was analyzed immediately; (b) an equal part was denatured and annealed, then analyzed. ss-1 is ssRNA-1; ss-2 is ssRNA-2; and ss-3 is ssRNA-3. (Reprinted from Watanabe et al., 1967b by courtesy of Proc. Nat. Acad. Sci. U.S.)

Thus, the various segments of the viral genome in the infected cell are transcribed from one end to the other into mRNA's. Apparently, the discontinuities between the segments have the effect of full stops during transcription. Recent analyses of the reovirus capsid proteins (Loh and Shatkin, 1968; Smith et al., 1969) suggest that there are seven species of polypeptides which fall into three size classes; the molecular weights of these classes have been estimated. For each of the three size classes of polypeptides, there is a size class of dsRNA with approximately the equivalent amount of genetic information encoded in it (Smith et al., 1969). The in-

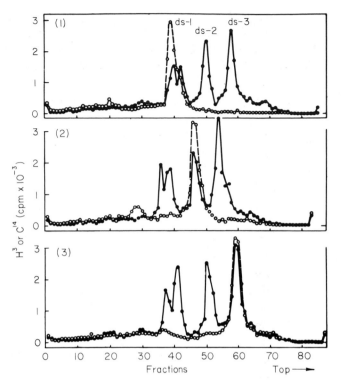

FIG. 13. Hybridization of each of the three size classes of ssRNA with reovirus dsRNA. Each of ³H-labeled ssRNA-1, ssRNA-2, and ssRNA-3, approximately 1 μg, 60,000 cpm, and 0.1 ml of 0.01 *M* STE buffer (O——O) was mixed separately with a sample of ¹⁴C-labeled dsRNA, 40 μg, 67,000 cpm, in 0.2 ml of 0.01 *M* STE buffer (●——●). These mixtures were denatured, annealed, and analyzed by electrophoresis. (a) ³H-ssRNA-1 with ¹⁴C-dsRNA; (b) ³H-ssRNA-2 with ¹⁴C-dsRNA; (c) ³H-ssRNA-3 with ¹⁴C-dsRNA. (Reprinted from Watanabe *et al.*, 1967b by the courtesy of *Proc. Nat. Acad. Sci. U.S.*)

ference from these results is twofold: seven of the ten segments of the reovirus genome are coded for viral capsid polypeptides, and the seven corresponding mRNA's are also monocistronic. If the remaining three mRNA's are also monocistronic, only three functions remain to be associated with the viral genome. One of these three functions is surely the synthesis of the viral RNA replicase (Watanabe *et al.*, 1968b).

In fact, there is still some question as to how many of the ten genomic segments are transcribed during infection. At least six are. For example, when ssRNA is extracted from L cells late in the infectious cycle and hybridized with viral RNA and the products are analyzed by gel electrophoresis, the pattern shown in Fig. 14 is obtained. The ssRNA was labeled with ³H (broken lines) and it is clear that each of the six classes of dsRNA (solid lines) has a ssRNA hybridized with it, that is, a corresponding mRNA. Although all ten components were not resolved in the analysis, this hybridization technique has enabled us to approach the question whether there is any control over transcription of the viral genome during the infectious cycle of the virus. Putting the question in its most simple form, we would ask whether the mRNA's made early in infection are the same as those made later. In principle, the experiment is a simple one (Watanabe *et al.*, 1968a). An infected culture is divided in two. One portion is labeled with uridine-³H between 4 to 6 hours after infection, the earliest time that a reasonable amount of mRNA can be seen. The second portion is labeled between 10 to 12 hours after infection, when viral progeny is being made at its maximum rate. The ssRNA is extracted from both cultures, hybridized with viral RNA and the products are analyzed by gel electrophoresis. Figure 15 shows the results: the early mRNA is shown in the top panel, late mRNA in the bottom panel. Clearly, the two patterns are different. At early times there is much less mRNA in the 1b, 1c, and 2b positions than at late times, suggesting that the corresponding segments of the genome are

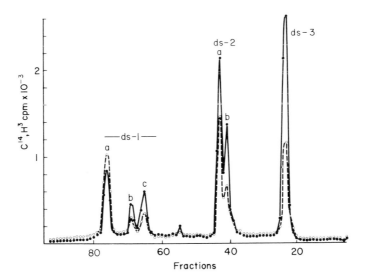

FIG. 14. Separation of the hybridization products of ssRNA and dsRNA by electrophoresis on polyacrylamide-agar gel. ssRNA was extracted from a culture of L cells labeled with uridine-³H 10–12 hours after infection with reovirus. The ssRNA was mixed with ¹⁴C-dsRNA, denatured with dimethyl sulfoxide then annealed. These results show that each of the six classes of dsRNA is transcribed for mRNA in infected cells. (Reprinted from Watanabe *et al.*, 1968a by courtesy of *J. Mol. Biol.*)

responsible for late functions, i.e., they are transcribed late in infection. This method can be made quantitative, and by pulse labeling at different times during infection and analyzing the mRNA's, shifts in the relative amounts of the various mRNA components can be seen, indicating that there is control over expression of the various viral functions at the transcriptional level (Watanabe *et al.*, 1968a). Perhaps the clearest demonstration of regulated transcription in this system is given by the following experiment.

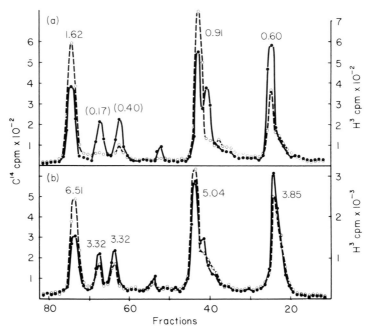

Fig. 15. Electrophoretic analysis of ssRNA extracted from cells labeled with uridine-³H at different times after infection with reovirus. (a) Infected cells labeled 4 to 6 hours after infection and (b) 8–10 hours after infection. The ssRNA was hybridized with ¹⁴C-labeled dsRNA from purified reovirus prior to analysis on polyacrylamide-agar gels. The numbers marked on each peak of the pattern represent the ratio ³H/¹⁴C for that component; ○——○, ³H; ●——●, ¹⁴C. (Reprinted from Watanabe *et al.*, 1968a by courtesy of *J. Mol. Biol.*)

The principle of the experiment is based on the observation that when inhibitors of protein synthesis are added soon after infection, there is no later synthesis of viral dsRNA (Shatkin and Rada, 1967; Kudo and Graham, 1966; Watanabe *et al.*, 1967a). Some virus-induced protein synthesis is required before dsRNA can be made and, therefore, corresponding mRNA's should be transcribed from the parental genome. From experience with various phage systems, these same mRNA's would be expected to form when protein synthesis is blocked.

As a test of this idea, L cells were infected and at time zero cycloheximide was added to block protein synthesis. ³H-labeled uridine was then added to the culture for several hours and the total RNA of the culture was extracted. This RNA was hybridized with dsRNA and the products were analyzed by electrophoresis on gel columns. A typical result is seen in Fig. 16. No hybrid is formed with segments dsRNA-1b, -1c, and -2b of the genome, but hybrids are formed with other segments. Thus certain segments of the parental genome can be transcribed in the absence of protein synthesis. At least three segments cannot be transcribed until some virus-induced protein synthesis occurs and late functions would be ascribed to these segments, 1b, 1c, and 2b. Resolution of these hybrids by electrophoresis in polyacrylamide gel has now improved to the extent that all ten hybrids can be resolved in a single analysis. This makes it possible to determine the kinetics of synthesis of viral mRNA throughout the infectious cycle and to assign early or late functions to all ten segments of the reovirus genome (Millward and Nonoyama, 1970).

Transcription of the information for early functions must be carried out by a preexisting enzyme since it goes on in the absence of protein synthesis.

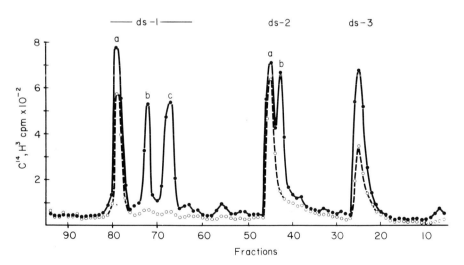

Fig. 16. Analysis of ssRNA formed in cells infected with reovirus and in the absence of protein synthesis. Cells were infected and cycloheximide was added at time zero. ssRNA was labeled with uridine-³H 4 to 8 hours after infection, extracted, hybridized with ¹⁴C-dsRNA, and analyzed by polyacrylamide-agar gel electrophoresis. ○——○, ³H; ●——●, ¹⁴C. (Reprinted from Watanabe et al., 1968a by courtesy of J. Mol. Biol.)

Most likely this transcription is carried out by an RNA polymerase that is an integral part of the virion. Brief heat shock or chymotrypsin digestion of purified virions "activates" an enzyme that, when provided with nucleoside triphosphates in the conventional system, synthesizes ssRNA (Shatkin and Sipe, 1968; Borsa and Graham, 1968; Skehel and Joklik, 1969). This ssRNA does not hybridize with itself, but hybridizes with high efficiency to denatured dsRNA. Clearly, the enzyme is transcribing mRNA from one strand of the viral genome *in vitro*. The first observation of a transcriptase in reovirus-infected cells was made by Gomatos (1968); he showed that, *in vitro*, the enzyme would synthesize correctly the three size classes of ssRNA. More recently, Skehel and Joklik (1969) have obtained similar results with the virion RNA polymerase which is associated with the cores of reovirus; they have, moreover, shown that each of the ten genomic segments is transcribed *in vitro*. Very likely, the polymerase observed by Gomatos (1968) is also contained in the cores of progeny reovirus, since he isolated the enzyme late in the infectious cycle by methods that would release the cores from infectious particles.

In summary, the reovirus genome consists of ten discrete segments within the virion and each of the segments is transcribed independently from end to end and from one strand only, both *in vivo* and *in vitro*. A limited number of segments code for early viral functions and these can be transcribed in the absence of protein synthesis, probably by the virion-associated polymerase.

Detailed studies on the structure and replication of the other dsRNA viruses are just beginning to emerge. The genomes of cytoplasmic polyhedrosis virus (Lewandowski *et al.*, 1969) and bluetongue virus (Verwoerd *et al.*, 1970), like that of reovirus, each consist of ten subunits of dsRNA. The genome of wound tumor virus has been reported to consist of twelve (Kalmakoff *et al.*, 1969) or fifteen (Wood and Streissle, 1970) subunits. As described above for reovirus cores, both cytoplasmic polyhedrosis virus (Lewandowski *et al.*, 1969) and wound tumor virus (unpublished data) contain a RNA polymerase associated with purified virions. It remains to be seen if bluetongue virus and rice dwarf virus also contain this polymerase activity.

III. Biological Significance of a Segmented Genome

Why some viruses have discontinuities in their genetic material, while others do not, is a question of considerable biological interest not only with respect to the dsRNA viruses but also certain of the ssRNA viruses. For example, Rous sarcoma virus (Duesberg, 1968), and influenza virus (Pons and Hirst, 1968), have ssRNA genomes which may also be comprised of discrete segments. Possibly mixing of the segments could occur during viral replication and thus provide a mechanism for genetic variation and a selective advantage for survival of the virus. In fact, such a mixing of segments has been suggested to account for the high recombination frequencies of certain mutants of reovirus (Fields and Joklik, 1969) and influenza virus (Hirst, 1962). We suggested recently that the functional significance of the reovirus fragmented genome might be to provide the cell with mono-cistronic mRNA's (Millward and Graham, 1970). This hypothesis was based on the suggestion that mammalian cells may be able to initiate protein synthesis only at the beginning of a message and not at intermediate points along the chain (Jacobson and Baltimore, 1968). Thus, in infected cells the polycistronic mRNA's of poliovirus, mengovirus, and Coxsackie virus are apparently translated into continuous protein chains which are subsequently cleaved into functional polypeptides (Jacobson and Baltimore, 1968; Summers and Maizel, 1968; Holland and Kiehn, 1968). On the other hand, dsRNA viruses could have adapted to the situation at the transcription level, discontinuities in the genome providing initiation and termination signals for a series of monocistronic messages. There is evidence to suggest that this might be the case. Messenger RNA's in reovirus-infected cells correspond exactly in length to the dsRNA segments in the viral genome (Bellamy and Joklik, 1967; Watanabe et al., 1968a). Certain of these messengers are approximately the right size to have been translated directly into the several structural protein units found in reovirus virions by Loh and Shatkin (1968) and Smith et al. (1969). This would mean that the lengths of virus-specific polypeptide chains are determined by the lengths of the genomic segments of the virus, a hypothesis that ultimately can be checked experimentally.

REFERENCES

Bellamy, A. R. and Joklik, W. K. (1967). *J. Mol. Biol.* **29**, 19–26.
Bellamy, A. R., Shapiro, L., August, J. T., and Joklik, W. K. (1967). *J. Mol. Biol.* **29**, 1–17.
Borsa, J. and Graham, A. F. (1968). *Biochem. Biophys. Res. Commun.* **33**, 895–901.
Dales, S., Gomatos, P. J., and Hsu, K. C. (1965). *Virology* **25**, 193–211.
Duesberg, P. H. (1968). *Proc. Nat. Acad. Sci. U.S.* **60**, 1511–1518.
Dunnebacke, T. H. and Kleinschmidt, A. K. (1967). *Z. Naturforsch.* **22B**, 159–164.
Els, H. J. and Verwoerd, D. W. (1969). *Virology* **38**, 213–219.
Fields, B. N. and Joklik, W. K. (1969). *Virology* **37**, 335–342.
Gomatos, P. J. (1968). *J. Mol. Biol.* **37**, 423–439.

Gomatos, P. J. and Stoeckenius, W. (1964). *Proc. Nat. Acad. Sci. U.S.* **52**, 1449–1455.
Gomatos, P. J. and Tamm, I. (1963a). *Proc. Nat. Acad. Sci. U.S.* **49**, 707–714.
Gomatos, P. J. and Tamm, I. (1963b). *Proc. Nat. Acad. Sci. U.S.* **50**, 878–885.
Gomatos, P. J., Tamm, I., Dales, S., and Franklin, R. M. (1962). *Virology* **17**, 441–454.
Granboulan, N. and Niveleau, A. (1967). *J. Microsc.* **6**, 23–30.
Hirst, G. K. (1962). *Cold Spring Harbor Symp. Quant. Biol.* **27**, 303–309.
Holland, J. J. and Kiehn, E. D. (1968). *Proc. Nat. Acad. Sci. U.S.* **60**, 1015–1022.
Iglewski, W. J. and Franklin, R. M. (1967). *J. Virol.* **1**, 302–307.
Jacobson, M. F. and Baltimore, D. (1968). *Proc. Nat. Acad. Sci. U.S.* **61**, 77–84.
Kalmakoff, J., Lewandowski, L. J., and Black, D. R. (1969). *J. Virol.* **4**, 851–856.
Kleinschmidt, A. K., Dunnebacke, T. H., Spendlove, R. S., Schaffer, F. L., and Whitcomb, R. F. (1964). *J. Mol. Biol.* **10**, 282–288.
Kudo, H. and Graham, A. F. (1965). *J. Bacteriol.* **90**, 936–945.
Kudo, H. and Graham, A. F. (1966). *Biochem. Biophys. Res. Commun.* **24**, 150–155.
Lewandowski, L. J., Kalmakoff, J., and Tanada, Y. (1969). *J. Virol.* **4**, 857–865.
Loh, P. C. and Shatkin, A. J. (1968). *J. Virol.* **2**, 1353–1359.
Loh, P. C., Hohl, H. R., and Soergel, M. (1965). *J. Bacteriol.* **89**, 1140–1144.
Mayor, H. D. and Jordan, L. E. (1968). *J. Gen. Virol.* **3**, 233–240.
Mayor, H. D., Jamison, R. M., Jordan, L. E., and Mitchell, M. V. (1965). *J. Bacteriol.* **89**, 1548–1556.
Millward, S. and Graham, A. F. (1970). *Proc. Nat. Acad. Sci. U.S.* **65**, 422–429.
Millward, S. and Nonoyama, M. (1970). Cold Spring Harbor Symp. Quant. Biol. (in press).
Miura, K., Fujii, I., Sakaki, T., Fuke, M., and Kawase, S. (1968). *J. Virol.* **2**, 1211–1222.
Pons, M. W. and Hirst, G. K. (1968). *Virology* **34**, 385–388.
Prevec, L. and Graham, A. F. (1966). *Science* **154**, 522–523.
Richardson, C. C. (1965). *Proc. Nat. Acad. Sci. U.S.* **54**, 158–165.
Sato, T., Kyogoku, Y., Higuchi, S., Mitsui, Y., Iitaka, Y., Tsuboi, M., and Miura, K. (1966). *J. Mol. Biol.* **16**, 180–190.
Shatkin, A. J. and Rada, B. (1967). *J. Virol.* **1**, 24–35.
Shatkin, A. J. and Sipe, J. D. (1968). *Proc. Nat. Acad. Sci. U.S.* **59**, 246–253.
Shatkin, A. J., Sipe, J. D., and Loh, P. (1968). *J. Virol.* **2**, 986–991.
Skehel, J. J. and Joklik, W. K. (1969). *Virology* **39**, 822–831.
Smith, R. E., Zweerink, H. J., and Joklik, W. K. (1969). *Virology* **39**, 791–810.
Summers, D. F. and Maizel, J. V., Jr. (1968). *Proc. Nat. Acad. Sci. U.S.* **59**, 966–971.
Streissle, G. and Granados, R. R. (1968). *Arch. Gesamte Virusforsch.* **25**, 369–372.
Vasquez, C. and Kleinschmidt, A. K. (1968). *J. Mol. Biol.* **34**, 137–147.
Vasquez, C. and Tournier, P. (1962). *Virology* **17**, 503–510.
Vasquez, C., and Tournier, P. (1964). *Virology* **24**, 128–130.
Verwoerd, D. W. (1969). *Virology* **38**, 203–212.
Verwoerd, D. W., Louw, H., and Oellermann, R. A. (1970). *J. Virol.* **5**, 1–7.
Watanabe, Y. and Graham, A. F. (1967). *J. Virol.* **1**, 665–677.
Watanabe, Y., Kudo, H., and Graham, A. F. (1967a). *J. Virol.* **1**, 36–44.
Watanabe, Y., Prevec, L., and Graham, A. F. (1967b). *Proc. Nat. Acad. Sci. U.S.* **58**, 1040–1046.
Watanabe, Y., Millward, S., and Graham, A. F. (1968a). *J. Mol. Biol.* **36**, 107–123.
Watanabe, Y., Gauntt, C. J., and Graham, A. F. (1968b). *J. Virol.* **2**, 869–877.
Wood, H. A. and Streissle, G. (1970). *Virology* **40**, 329–334.

CHAPTER 13 **_The Structure and Assembly of Influenza and Parainfluenza Viruses_**

RICHARD W. COMPANS AND PURNELL W. CHOPPIN

I. Introduction

The name myxovirus was originally proposed for a group of viruses which then included influenza, mumps, and Newcastle disease viruses (Andrewes *et al.*, 1955). These viruses had in common an affinity for certain mucins, as well as physical and biological properties such as the ability to adsorb to red blood cells of chickens and various other species, causing hemagglutination. The viruses also possess an enzyme, neuraminidase, which cleaves neuraminic acid from mucoproteins. With increasing knowledge of the structure and replication of these viruses, it has become apparent that they fall into two groups with distinct properties. Furthermore, as indicated in Table I, additional viruses have properties which indicate that they are members of the group of large myxoviruses. These include the

TABLE I

Occurrence in Nature of Influenza and Parainfluenza Viruses

Group	Virus	Source
Influenza	Type A	Man, horses, swine, birds
	Type B	Man
	Type C	Man
Parainfluenza	Types 1, 2, 3, 4	Man, mice (type 1), cattle (type 3)
	SV5	Monkey, dog, man(?), hamster(?)
	Newcastle disease	Birds
	Mumps	Man
	Measles	Man
	Rinderpest	Cattle, sheep, goats, water buffalo
	Canine distemper	Dogs

parainfluenza viruses, simian virus 5, measles virus, and tentatively, rinderpest and canine distemper viruses. There has been no official designation of names for these two groups of myxoviruses by the Committee for the Nomenclature of Viruses, and we refer to them here as the *influenza virus group* and the *parainfluenza virus group*. The two groups have also been referred to as myxoviruses and paramyxoviruses, respectively. The two groups have many common structural features, with an internal helical ribonucleoprotein surrounded by a lipid-containing envelope.

As indicated in Table I, members of both the influenza and parainfluenza virus groups are found in a variety of vertebrate species. Many of these viruses infect the respiratory tract, although avian influenza viruses and several of the parainfluenza viruses cause severe generalized diseases. There is no example as yet of a virus from plants, bacteria, or invertebrates which can be included definitely in the influenza or parainfluenza groups. One plant virus, tomato spotted wilt virus, has some properties which suggest a possible similarity to myxoviruses (Best, 1968). It is an enveloped, RNA-containing virus, but information about its internal structure, which is necessary to classify this virus, has not been obtained. It should be pointed out that bullet-shaped enveloped RNA-containing viruses with helical nucleocapsids are found in plants and invertebrates, so there would be no obvious reason why the structural properties of influenza or parainfluenza virions are incompatible with such hosts.

In this chapter we summarize the current information about the morphology, composition, and assembly of influenza and parainfluenza virus particles. Macromolecular synthesis in cells infected by these viruses has been reviewed elsewhere (Barry and Mahy, 1970; Blair and Duesberg, 1970). Much of this information has been obtained very recently, and detailed studies on only a few members of each group have been made. Nevertheless, from the information that has been obtained thus far, it appears that each of the two groups is relatively homogeneous with respect to the major structural features of the virus particle.

II. Structure of Influenza Viruses

A. The Influenza Virus Group

Three distinct virus-specific components with antigenic activity have been recognized in influenza virions, and used in their classification. One is associated with the internal nucleoprotein component, and two with the viral envelope. One of the envelope antigens (hemagglutinin) is the component involved in the specific agglutination of red blood cells by virus particles, and the other is the enzyme involved in their subsequent elution (neuraminidase).

Human influenza viruses fall into three antigenic types, designated A, B, and C, whereas all influenza viruses isolated thus far from other species are type A. The three types are distinguished by antigenically distinct nucleoprotein components as recognized by complement fixation. Type A viruses can be further divided into antigenic subtypes. Three of these are designated A_0, A_1, and A_2, and another (Hong Kong 1968) has not been classified as a distinct subtype. All four of these groupings have antigenically distinct hemagglutinin components, and they can be distinguished by neutralization tests, since antibody to hemagglutinin neutralizes virus infectivity. The antigenic composition of the neuraminidase also varies independently of the hemagglutinin. Subtypes A_0 and A_1 have closely related neuraminidases, as do the A_2 and Hong Kong viruses. Minor differences in surface antigens have been detected within the influenza type B group, but not the major variation found with type A.

B. Morphology of the Virion

Some of the morphological properties of influenza virions are listed in Table II. Influenza virions are pleomorphic, and two morphological types are generally found: roughly spherical virus particles with a diameter of

TABLE II

Morphology of Influenza and Parainfluenza Viruses

	Influenza viruses (Å)	Parainfluenza viruses (Å)
Diameter of virion	800–1000	1200–4500
Diameter of nucleocapsid	~90	~180
Length of nucleocapsid	500–1300	10,000
Length of surface projections	80–100	~80
Occurrence of filamentous virions	+	+

~1000 Å (Fig. 1), and long filamentous virions with a similar diameter and a length up to ~4 μ. The production of filamentous virions is a genetic property of virus strains which can be exchanged in recombination experiments (Kilbourne and Murphy, 1960; Choppin, 1963; Kilbourne, 1963). In recently isolated strains of influenza virus, filaments are frequently observed (Chu et al., 1949; Choppin et al., 1960), and it is possible that the filament is the predominant form of virus in man. The specific infectivity of the filament has been reported to be higher than that of spherical virions, which may be related to the fact that filaments contain more nucleic acid per particle (Ada et al., 1958).

When virus particles are examined by the negative staining method, an outer layer of closely spaced projections 80–100 Å in length is seen clearly (Fig. 1). These spikes are found on spherical (Horne et al., 1960) and filamentous particles (Choppin *et al.*, 1961). Evidence has been obtained that two different types of spikes may be present on the influenza virion, one associated with hemagglutinin and the other with neuraminidase (Laver and Valentine, 1969; Webster and Darlington, 1969). In these studies the subunits with hemagglutinin activity were found to be rod-shaped, measuring 40×140 Å. The neuraminidase-containing subunits had a more complex morphology, with an 80×50 Å cylinder located at the tip of a fiber which had a small knob of about 40 Å at its other end. The subunits aggregated in the absence of detergents, with only one end of each subunit in mutual contact. This finding led Laver and Valentine to suggest that subunits have a hydrophobic end and a hydrophilic end, and that the hydrophobic end functions in binding of the subunit to the viral membrane.

In the occasional virus particles which are penetrated by negative stain, internal strands about 90 Å in diameter are detected. This internal component has been isolated by disrupting virus particles with ether (Hoyle, 1952; Lief and Henle, 1956; Rott and Schäfer, 1964). Such treatment also yields fragments of the envelope which have hemagglutinin and neura-

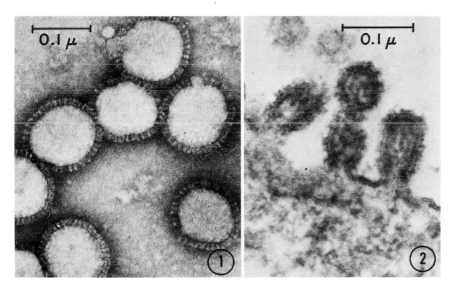

FIG. 1. Influenza virions negatively stained with sodium phosphotungstate. The particles have a relatively uniform diameter of about 1000 Å, and show clearly the layer of spikes on their surfaces. × 160,000.

FIG. 2. Influenza virions in the process of budding at the cell surface. The particle on the right shows dense internal strands about 60 Å in width. ×200,000. (From Compans and Dimmock, 1969.)

minidase activity, and have the appearance of rosettes (Hoyle *et al.*, 1961; Choppin and Stoeckenius, 1964a). More recently, the internal component has been isolated and characterized from detergent-treated virus particles (Duesberg, 1969; Kingsbury and Webster, 1969; Pons *et al.*, 1969). The nucleoprotein isolated in this way had a diameter of about 150 Å and length ranging from 500 to 1300 Å; it appeared to be composed of chains of protein subunits wound into a helical coil with terminal loops (Pons *et al.*, 1969). Several distinct size classes were resolved by velocity sedimentation on sucrose gradients, and these classes contained distinct RNA molecules (Duesberg, 1969; Kingsbury and Webster, 1969). As discussed below (Section II,C), the RNA extracted from influenza virus is found to consist of several distinct pieces, and appears that the different size classes of nucleoprotein reflect the size of the RNA's they contain.

When virus particles are examined in thin sections, the envelope appears as a complex membrane with projections on the outer surface. The appearance of the viral membrane depends both on the staining method used and the host cell of origin (Compans and Dimmock, 1969; Compans *et al.*, 1970a). The relationship between viral and cellular membrane is discussed below (Section IV). In the interior of virus particles are dense strands about 60 Å in diameter (Fig. 2), which presumably correspond to the internal strands seen occasionally by negative staining.

The arrangement of the internal strands within the virion is uncertain, primarily because internal structure can be seen in only a small percentage of virions by negative staining. An occasional virion shows a 60 Å strand thought to be arranged in a continuous, tight coil about 600 Å in diameter (Apostolov and Flewett, 1965; Almeida and Waterson, 1970). However, in thin sections of virus particles, where internal strands are seen in many virions, no such arrangement was detected (Compans and Dimmock, 1969). It appeared that most virions were slightly elongated, with internal strands oriented parallel to the long axis; whether the internal component is continuous or in pieces was uncertain.

There are no obvious morphological differences between most influenza virus strains, apart from the frequency of filamentous virions. However, it has been suggested that influenza type C virions may differ morphologically from types A and B (Apostolov and Flewett, 1969). The type C viral envelope had different staining properties than envelopes of A and B virions, and spikes were often absent. A hexagonal pattern was seen on the surface of influenza C filaments, and the internal strands seen in sectioned type C virions had a width of 90 Å, compared with 60 Å for types A and B.

On serial undiluted passage, the yield of infectious influenza virus declines but a high level of noninfectious, hemagglutinating particles is produced (von Magnus, 1954). Such noninfectious virus has been termed "incomplete virus." The morphology of incomplete virus differs from that of standard, highly infectious virus in that more large pleomorphic forms are observed, and the dense internal structure is less evident in incomplete virus (Birch-Andersen and Paucker, 1959; Morgan et al., 1962; Moore et al., 1962). Differences in the surface structure of standard and incomplete influenza virus have also been noted (Almeida and Waterson, 1970). On the envelopes of incomplete virus, a square array of subunits was often found, while on standard virus a hexagonal pattern was evident.

Recently it has been found that the amount of incomplete virus formation is in part host-cell dependent (Choppin, 1969). A high yield of infective virus and little incomplete virus is produced by a line of bovine kidney cells (MDBK) infected at high multiplicity with the WSN strain of influenza A virus. Three serial high multiplicity passages were required before a significant amount of incomplete virus was detected. This system thus permits single cycle experiments with influenza virus in a continuous cell line with the production of a high yield of complete, infective virus. The absence of such a system in the past has impeded biochemical studies of virus replication.

C. CHEMICAL COMPOSITION

The influenza virion contains about 1% RNA (Table III), which is associated with the internal ribonucleoprotein within the virion. Recently RNA from virions has been extracted and characterized in several laboratories. On centrifugation in sucrose density gradients, RNA molecules with relatively low estimated molecular weights of around 500,000 were observed (Davies and Barry, 1966; Pons, 1967; Duesberg and Robinson, 1967). Using polyacrylamide gel electrophoresis, the extracted RNA could be resolved into five components (Duesberg, 1968; Pons and Hirst, 1968). These observations have suggested that, unlike other viruses with single-stranded RNA genomes, the genome of influenza virus may exist as several discrete pieces. Such a structure for the influenza genome had been postulated earlier on the basis of genetic experiments (Hirst, 1962). The finding of several pieces of RNA in influenza virions may explain the failure in most cases to recover infectious RNA from virions (Ada et al., 1959; Sokol and Szurman, 1959), although it is also possible that a virion protein is essential for infectivity.

The RNA's of standard and incomplete influenza virus have been com-

TABLE III

CHEMICAL COMPOSITION OF INFLUENZA AND PARAINFLUENZA VIRUS PARTICLES

	RNA[a]	Protein[a]	Lipid[a]	Carbo-hydrate[a]	
Influenza virion	0.9	60–70	~25	5–8	Ada and Perry, 1954; Frommhagen et al., 1959
Influenza nucleocapsid	10	90	—	—	Pons et al., 1969
Parainfluenza virion	0.9	73	20	6	Klenk and Choppin, 1969a
Parainfluenza nucleocapsid	4	96	—	—	Compans and Choppin, 1967a; Kingsbury and Darlington, 1968; Hosaka, 1968

[a] Values in percent.

pared by polyacrylamide gel electrophoresis (Duesberg, 1968; Pons and Hirst, 1968; Choppin and Pons, 1970). Whereas five components were present in standard virus, there was a progressive decrease in the amount of the largest piece of RNA with serial undiluted passage. Depending on the host cell, by the second to the fourth passage, the largest RNA piece was essentially absent in the incomplete virus. Thus there appears to be a common defect in the virus particles produced by serial undiluted passage. With virus grown in the chick embryo as inoculum, the abortive infection of HeLa cells by influenza virus is also characterized by the absence of the largest piece of RNA, suggesting that the mechanism responsible for this abortive infection is similar to that in incomplete virus formation in other cells (Choppin and Pons, 1970).

Studies of the acrylamide gel patterns of influenza virus RNA's have been made with type A viruses only. However, the base composition has been determined for a variety of A and B strains (data summarized by Bellett, 1967). In all instances a high content of uracil is observed. Influenza B viruses appear to have a consistently higher uracil content than type A viruses. Scholtissek (1970) has begun comparative studies of the genetic relatedness of different influenza viruses by hybridization experiments between viral "plus-strand" RNA and complementary "minus-strand" RNA from different strains. High levels of cross-hybridization were observed between different type A viruses, including human, swine, and avian strains. However, there was little cross-hybridization between A and B strains.

About 60 to 70% of the mass of the influenza virion is protein, and there is little or no protein of the host cell detectable in purified virus particles (Laver and Webster, 1966; Holland and Keihn, 1970). Seven polypeptides have been resolved by polyacrylamide gel electrophoresis of disrupted virus particles (Compans et al., 1970b; Schulze, 1970). Four of these polypeptides appear to be covalently linked with carbohydrate, and the projections on the surface of the virion are composed of these glycoproteins (Compans et al., 1970b; Schulze, 1970). One of the polypeptides which contains no carbohydrate is associated with the RNA to form the nucleocapsid of the virus (Pons et al., 1969; Joss et al., 1969). Another non-glycoprotein, which is the major protein in the virion, appears to be a non-spike protein in the viral membrane (Compans et al., 1970b; Schulze, 1970). The antigenic properties of this protein, which was previously unrecognized, have not been investigated.

The dimensions of the hemagglutinin subunits isolated from detergent-disrupted virus particles suggest a molecular weight in excess of 150,000, but the molecular weight of the polypeptide in such subunits is only ~47,000 indicating that several chains are present in the intact spike (Webster, 1970). There are differences in the peptide maps of hemagglutinin subunits of antigenically different virus strains (Laver, 1964), and antigenic mutants selected in the laboratory also have differences in the amino acid sequence of the hemagglutinin protein (Laver and Webster, 1968). These observations suggested that the marked antigenic variation which is characteristic of influenza virus is related to changes in the amino acid sequence of the viral proteins.

The neuraminidase subunits isolated from detergent-disrupted virus particles have been analyzed by acrylamide gel electrophoresis, and two closely spaced bands were observed (Webster, 1970). The molecular weights of the polypeptides were estimated at about 58,000. Antigenic variation has been found with influenza neuraminidases, and is independent of the variation in antigenicity of the hemagglutinin (Schulman and Kilbourne, 1969; Meier-Ewert et al., 1970).

The influenza viral envelope contains a large amount of lipid (Table III). The source of the lipids in the viral envelope has been a subject of interest for some time. Wecker (1957) obtained evidence that some preformed cellular lipids were incorporated into the virion, and Kates and co-workers (1962) analyzed the lipids of influenza virion grown in chick embryo and calf kidney cells and found that the virion lipids resembled those of the host cell, particularly those of the nuclear fraction. However these workers did not obtain a plasma membrane fraction, and it is likely that with the method of cell fractionation used, the plasma membranes were in the nuclear fraction. Recently, it has been found in studies with purified plasma membranes obtained from four different types of cultured host cells, and myxoviruses grown in these cells, that with a few exceptions, the lipids of the plasma membranes of the host cell are incorporated quantitatively into the virions (Klenk and Choppin, 1969b; 1970a,b; 1971). Furthermore, changes in the culture medium or in the lipid composition of the infected cells as a result of viral cytopathic effects were also reflected in the virions (Klenk and Choppin, 1971). Additional evidence that the viral lipid composition depends on the host cell was obtained by Simpson and Hauser (1966), who observed that sensitivity to phospholipase depends on the cell in which influenza virus is grown. On the other hand, based on differences in the fatty acid composition of three strains of influenza virus grown in the chick embryo, it has been postulated that the viral envelope proteins play a major role in determining the lipid composition of the viral membrane (Blough, 1968; Tiffany and Blough, 1969). However, these strains were grown under multiple cycle conditions and factors such as different batches of eggs, different growth cycles, and different cytopathic effects and alterations in host cell metabolism may have affected the lipid composition. Although in certain cells some limited rearrangement of plasma membranes lipids specified by the viral envelope proteins may be possible, the available evidence suggests that the lipid composition of the plasma membrane of the host cell is the major determinant of the lipid composition of the virion.

In addition to the ribose in the viral RNA, the influenza virion contains 5–8% carbohydrate, all of which is associated with the envelope (Frisch-Niggemeyer and Hoyle, 1956). A host cell carbohydrate antigen has been found to be associated with the virion (Harboe, 1963; Laver and Webster, 1966; Lee et al., 1969). Although most of the carbohydrate in the virion is covalently bound to protein, some is probably present in glycolipid (Klenk and Choppin, 1970b).

III. Structure of Parainfluenza Viruses

A. The Parainfluenza Virus Group

The antigenic relationships between the structural components of parainfluenza viruses have not been extensively investigated, but there may be some limited antigenic cross-reaction between Newcastle disease (NDV), mumps, SV5, and parainfluenza viruses types 1-4. As described below, one of the characteristic features of this group is the structure of the helical nucleocapsid, and on this basis measles virus appears to belong to the group. The nucleocapsids of rinderpest and canine distemper viruses have not been characterized, but these viruses are antigenically related to measles virus.

B. Morphology of the Parainfluenza Virion

The virions of the parainfluenza group are markedly pleomorphic when examined by negative staining, with diameters ranging from 1200 to over 3000 Å (Table II). The virion (Figs. 3 and 4) consists of a membrane covered with spikes about 80 Å in length, and a well-defined helical nucleocapsid about 180 Å in diameter (Horne and Waterson, 1960; Choppin and Stoeckenius, 1964b). Regular filamentous virions of the type seen with influenza virus are not found in unfixed negatively stained preparations of parainfluenza virions, and it was generally thought that such filaments did not occur in the parainfluenza virus group. However, numerous filaments have been found in sections of fixed parainfluenza virus-infected cells (Prose et al., 1965; Compans et al., 1966; Howe et al., 1967; Yunis and Donnelly, 1969; Feller et al., 1969). Therefore, filamentous virions are a common feature of the influenza and parainfluenza viruses. It is possible that when unfixed parainfluenza filaments are negatively stained they are disrupted or distorted into large pleomorphic forms.

The nucleocapsid of the parainfluenza virion (Figs. 4 and 5) is a single-stranded helix with a periodicity of about 50 Å, composed of subunits measuring $\sim$25 × 70 Å (Choppin and Stoeckenius, 1964b). Although all of the viruses in this group have nucleocapsids with the same basic structure, there is some variability in their rigidity. Sendai (parainfluenza type 1) virus has a very rigid nucleocapsid, and that of NDV is very flexible; SV5 nucleocapsid appears to be intermediate in rigidity. The nucleocapsids of SV5, NDV, mumps, Sendai, and measles viruses all appear to have a uniform length of about 1 μ (Hosaka et al., 1966; Compans and Choppin, 1967a,b; Hosaka and Shimizu, 1968; Kingsbury and Darlington, 1968; Nakai et al., 1969; Finch and Gibbs, 1970). Longer nucleocapsids, which are multiples of the 1 μ unit length, are found occasionally and are thought to be end-to-end aggregates of 1 μ nucleocapsids. From optical diffraction studies on electron micrographs of negatively stained Sendai nucleocapsids, Finch and Gibbs (1970) have calculated that the structural subunits are

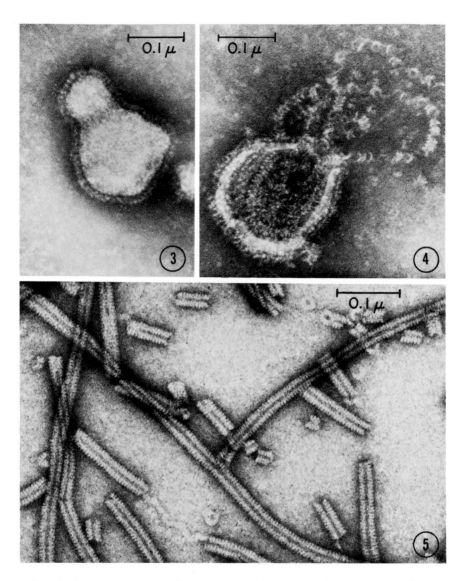

Figs. 3–5. Parainfluenza (SV5) virions negatively stained with sodium phosphotungstate. The uniform layer of spikes can be seen on the irregularly-shaped virion in Fig. 3, which is not penetrated by the stain. Fig. 4 shows a virion whose membrane has ruptured, with the helical nucleocapsid partially extruded. Fig. 3, × 180,000; Fig. 4, × 160,000. Figure 5 shows purified SV5 nucleocapsid stained with uranyl acetate, showing the ∼50 Å periodicity and central core characteristic of the helical structure. Some segments of the helix which have fragmented are viewed end-on, showing the central ∼40 Å hole. × 175,000.

hour glass-shaped, and are arranged in the helix with their long axes inclined at an angle of about 60° to the long axis of the nucleocapsid. They estimated that there are 11 or 13 subunits in each turn of the helix, giving a value between 2400 and 2800 for the number of structural units in the 1 μ nucleocapsid. The nucleocapsids of mumps and measles virus appeared to have a similar substructure.

The size of the virions containing a single 1 μ nucleocapsid has been estimated at about 1200 Å (Klenk and Choppin, 1969a), and in negatively stained preparations, Sendai virions with diameters of 1600 to 2500 Å were present in a fraction with 1 μ nucleocapsids (Hosaka et al., 1966). Larger virions contain a greater length of nucleocapsid (Hosaka et al., 1966; Compans et al., 1966). Hosaka and co-workers (1966) separated classes of Sendai virions with multiple nucleocapsids, and inactivation experiments indicated that they contained multiple genomes. A class of rapidly sedimenting Newcastle disease virions with multiple genomes has also been described (Dahlberg and Simon, 1969).

C. CHEMICAL COMPOSITION

The chemical composition of highly purified parainfluenza virions has been determined (Klenk and Choppin, 1969a), and the results are listed in Table III. The virus particles contain 0.9% RNA. Earlier analyses of NDV (Cunha et al., 1947) which gave a much higher RNA content were performed on preparations whose purity was unknown. A rapidly sedimenting species of RNA has been isolated from several members of the parainfluenza virus group (Duesberg and Robinson, 1965; Kingsbury, 1966; Barry and Bukrinskaya, 1968; Compans and Choppin, 1968), and on the basis of its sedimentation coefficient a molecular weight of about 6 to 7 $\times$ 10^6 daltons has been estimated. A similar estimate was made for the amount of RNA contained in 1 μ length of the helical nucleocapsid, which is thought to contain one viral genome (Compans and Choppin, 1967a). Some low molecular weight RNA is also associated with the virion, but its significance is uncertain. The base composition of the RNA from several members of the parainfluenza virus group is listed in Table IV; a common feature is the high content of uracil. On the basis of their base composition, ribonuclease sensitivity, and sedimentation properties, parainfluenza viral RNA's appear to be single-stranded. Attempts to demonstrate infectivity of isolated RNA molecules have been unsuccessful. Since the high sedimentation coefficient and homogeneity of the viral RNA suggest that intact molecules have been

TABLE IV

BASE COMPOSITION OF RNA FROM PARAINFLUENZA VIRIONS

Virus	C^a	A^a	G^a	U^a	Reference
Newcastle disease					
L-Kansas strain	23.0	23.8	23.8	29.4	Duesberg and Robinson, 1965
Beaudette strain	23.2	20.1	25.4	31.2	Kingsbury, 1966
SV5	20.1	27.8	21.1	30.8	Compans and Choppin, 1968
Sendai	24.3	24.2	21.5	30.0	Iwai et al., 1966
	23.4	24.1	22.9	29.6	Blair and Robinson, 1968

[a] Values in mole %.

recovered, there is no obvious reason why the RNA is noninfectious. One possibility is that a protein of the virion is essential for infectivity.

The viral RNA is contained within the helical nucleocapsid, which has an RNA content of about 4%, with the remainder being protein (Compans and Choppin, 1967a; Kingsbury and Darlington, 1968; Hosaka, 1968). The nucleocapsid protein isolated from virions is a single species with a molecular weight of ～60,000, and is the major protein of the virion (Evans and Kingsbury, 1969; Bikel and Duesberg, 1969; Haslam et al., 1969; Mountcastle et al., 1970). When SV5 nucleocapsid was isolated from infected cells, the protein subunit had a molecular weight of 43,000 (Caliguiri et al., 1969). The subunits of NDV and Sendai nucleocapsids isolated from trypsinized infected cells were also very similar in size, 46,000–47,000 mol. wt. (Mountcastle et al., 1970). The smaller subunits of the nucleocapsids isolated from cells are probably the result of a specific proteolytic cleavage. A typical parainfluenza nucleocapsid helix can be formed from either the large or small subunits; however the helix composed of the smaller subunits appears to be more tightly coiled and less flexible (Mountcastle et al., 1970). The amino acid composition of SV5 nucleocapsid protein isolated from cells has been determined, and the subunit is rich in leucine, arginine, aspartic acid, and glutamic acid (Compans et al., 1971).

The remaining viral proteins are presumably associated with the viral envelope, which contains neuraminidase and the components responsible for hemagglutination, cell fusion, and hemolysis. The proteins of NDV, SV5, and Sendai virions have been studied by SDS-polyacrylamide gel electrophoresis (Evans and Kingsbury, 1969; Haslam et al., 1969; Bikel and Duesberg, 1969; Caliguiri *et al.*, 1969; Mountcastle *et al.*, 1971). Although there are differences in the patterns observed, each of the viruses probably contain at least five polypeptides. There is a major protein with a molecular weight of about 40,000, and two minor proteins between this and the nucleocapsid protein (60,000). There are also one or two proteins with a molecular weight of 67,000–76,000. With NDV it has been reported (Haslam *et al.*, 1969) that the hemagglutinin polypeptide has a molecular weight of 80,000, and the neuraminidase polypeptide, 38,000. However, as yet neither component has been isolated in pure form prior to analysis by gel electrophoresis. The protein composition of the parainfluenza virion does not vary when virions are grown in four different cell types, suggesting that the proteins are coded by the viral genome (Caliguiri *et al.*, 1969). Further evidence that the virion proteins are virus-coded was obtained in experiments in which cells were prelabeled with radioactive amino acids, then infected, and virion proteins analyzed. A low level of radioactivity, due to turnover of the label, was found uniformly distributed among the various proteins. No individual protein(s) showed a relatively higher amount of label, as would be expected if a cellular protein was incorporated into the virus (Klenk, Caliguiri, and Choppin, unpublished).

Although all the proteins in the virion appear to be newly synthesized under the direction of the viral genome, the lipid composition depends largely on the host cell. The lipids of SV5 virions grown in four different host cell types have been analyzed, and in nearly all cases the composition was very similar to that of the isolated purified plasma membrane of the host cells (Klenk and Choppin, 1969b; Klenk and Choppin, 1970a,b). Within narrow limits, some selection of lipids may be possible by the virus (Klenk and Choppin, 1970a), but this is the exception. In general, the lipids of the plasma membrane appear to be incorporated quantitatively into the virion. A comparison of these lipid analyses and of studied of biological behavior of the various cell membranes and their response to the virus (Holmes and Choppin, 1966; Compans *et al.*, 1966; Choppin *et al.*, 1968; Holmes *et al.*, 1969) suggests that there is a correlation between membrane lipid composition and virus yield and sensitivity to virus-induced cell fusion and immune cytolysis.

One significant exception to the incorporation of cell membrane lipids into the parainfluenza virion is the absence of neuraminic acid-containing glycolipids (gangliosides) in the virions (Klenk and Choppin, 1970b). No protein-bound neuraminic acid was found in the virions either, presumably due to the action of the viral neuraminidase which is incorporated into those specific regions of the cell membrane which will become the viral envelope.

The virion contains about 6% of nonnucleic acid carbohydrate, much of which is probably covalently linked to viral proteins. Two of the proteins of SV5, with estimated molecular weights of 56,000 and 67,000, have been shown to be glycoproteins on the basis of the incorporation of labeled carbohydrate precursors (Klenk et al., 1970). In addition to glycoproteins, neutral glycolipids are present in the virions (Klenk and Choppin, 1970b). As indicated above, these neutral glycolipids resemble those of the plasma membrane of the host cell.

Whether the amounts and sequence of sugars in the carbohydrate portion of the virion is influenced by the viral genome or completely determined by the host remains to be elucidated. However, the apparent absence of host cell proteins in the virion and the presence in the virion of host cell carbohydrate bound to viral protein and of glycolipids, which may be antigenic, suggest that any host cell antigens present in myxovirus particles will be carbohydrate rather than protein.

IV. The Assembly Process

The assembly of influenza and parainfluenza virions is a multistep process which is common to most enveloped viruses. The nucleocapsid or internal component is assembled in the interior of the cell, and then becomes associated with a segment of a cell membrane which contains viral envelope proteins. By a process of budding, the nucleocapsid is enclosed within the modified membrane and a complete virion is formed.

The formation of virus particles by a budding process was first recognized in early electron microscopic studies of influenza virus-infected cells (Murphy and Bang, 1952). The internal component of influenza virus is difficult to recognize in the interior of the infected cell, although strands about 60 Å in width are seen clearly in sectioned virions; occasionally strands of similar appearance can be detected in clusters in the cytoplasm of infected cells (Compans et al., 1970a). When cells are stained by procedures which show a three-layered unit membrane, a similar membrane is found in the envelope of the budding virion, and during budding there is continuity between the leaflets of the cellular and viral membranes; however, the viral envelope also contains an additional innermost layer which is not found in the cell membrane (Compans and Dimmock, 1969; Bächi et al., 1969).

The distinct layer of viral projections is seen on the outer surface of the viral envelope, but not on the adjacent cell membrane (Compans and Dimmock, 1969; Bächi et al., 1969). Thus both the outer and inner surfaces of the viral envelope appear to differ morphologically from cell membrane by the presence of virus-specific structures.

Although spikes can be seen clearly on budding influenza virions and not on adjacent cell membranes, chicken erythrocytes appear to adsorb specifically to morphologically normal membrane (Compans and Dimmock, 1969). It was suggested that such areas of membrane may contain viral envelope proteins in the form of a precursor to the spikes seen on the virion.

Most influenza virions are observed budding at the surface membrane of the infected cell, but occasionally some particles appear to be budding into vacuoles. In a study of the A_0/WSN strain of influenza virus in chick embryo fibroblasts, labeling with colloidal thorium dioxide (Thorotrast) was used to investigate the origin of such apparent vacuoles (Compans and Dimmock, 1969). It was found that all "vacuoles" were labeled with thorotrast, indicating that they were either continuous with or derived from the cell surface. Thus all influenza virus maturation appeared to take place at a single type of membrane, the cell surface membrane. The cell surface membrane differs from intracellular membranes in structure and composition (Benedetti and Emmelot, 1968; Sjöstrand, 1968). The structure of the envelope proteins or the site at which they are synthesized may determine which of the cellular membrane systems will serve as a site for viral maturation.

The helical nucleocapsids of parainfluenza viruses can be seen clearly in infected cells (Prose et al., 1965; Compans et al., 1966; Howe et al., 1967; Duc-Nguyen and Rosenblum, 1967). The nucleocapsid appears to be assembled in the cytoplasmic matrix, and then to associate with segments of the cell membrane that show viral surface projections (Fig. 6). Often the nucleocapsid is seen in a very regular arrangement at the cell surface. Subsequently, the maturation of virions occurs by a process of budding, and either spherical or filamentous particles are produced (Figs. 6 and 8).

The nucleocapsids of parainfluenza viruses often accumulate in large amounts in infected cells, and such accumulations represent the eosinophilic cytoplasmic inclusions that are typical of these viruses (Holmes and Choppin, 1966; Compans et al., 1966). This accumulation was most pronounced in cells producing a low yield of infectious virus. It was suggested that in such cells, which disintegrate after extensive cell fusion, there was a block in viral maturation at the cell membrane. In a different cell type which yielded high levels of infectious virus, there appeared to be a bal-

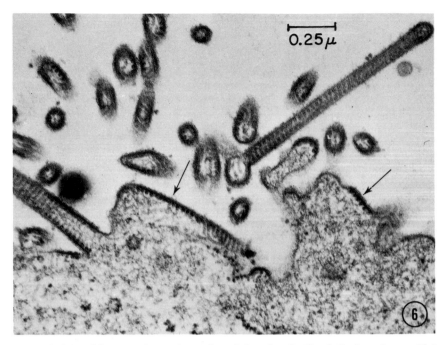

FIG. 6. SV5 virions at the surface of an infected cell. The helical nucleocapsid is regularly coiled within the long filamentous virions, and at some parts of the cell surface (arrows), the nucleocapsid is aligned under the cell membrane. × 52,000. (From Compans *et al.*, 1966.)

ance between synthesis of nucleocapsid and its incorporation into mature virions.

Erythrocytes are adsorbed specifically to surface membranes of parainfluenza-infected cells (Baker *et al.*, 1965; Duc-Nguyen, 1968) and often there is no morphological alteration detectable at the plasma membrane in the area of hemadsorption (Fig. 7). Furthermore, as shown in Figs. 9 and 10, ferritin-conjugated antiviral antibody is often found attached to patches of membrane where no nucleocapsid is present. These areas of membrane therefore appear to contain viral components but differ from the regions of membrane which show spikes and attached nucleocapsid. It is likely that these areas contain viral envelope proteins, and they may be precursors to the modified membrane which becomes the envelope of the virion.

As described above, the genome of parainfluenza virions appears to be a single molecule, and if this is the case assembly of these viruses can proceed by the envelopment of a single nucleocapsid. However, as described above (Section II,C), the available evidence indicates that the influenza viral genome exists in pieces. Furthermore, since the nucleocapsid also is recovered in pieces corresponding to those of the RNA, the assembly of influenza virions does not appear to occur by envelopment of a single structure, and a mechanism must exist for the various RNA pieces to be enclosed in a single envelope. Since the nucleocapsid appears to contain only a single protein species, it does not seem likely that nucleocapsid pieces could associate in a specific aggregate by interaction between proteins. It has been pointed out elsewhere (Compans et al., 1970a) that a high proportion of infectious particles could be formed by random incorporation of RNA pieces, provided that the total number of pieces per virion is sufficiently high. For example, if five different pieces of RNA are required for infectivity, about 22% of virus particles could be infectious if virions contained a total of seven pieces of RNA. Alternatively, some specific interaction between the RNA's of the nucleocapsids may be possible, since the RNA is sensitive to nuclease and may therefore be exposed at the surface of the structure (Duesberg, 1969; Kingsbury and Webster, 1969; Pons et al., 1969).

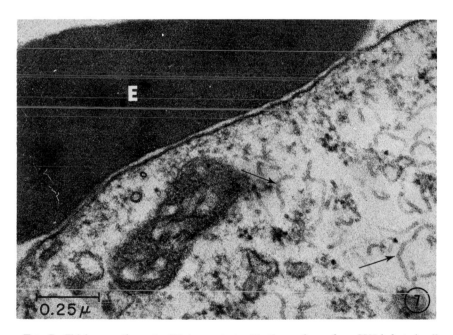

FIG. 7. Chicken erythrocyte (E) in contact with the surface of an SV5-infected cell. Nucleocapsid (arrows) is present in the cytoplasm, but is not associated with the cell membrane where the erythrocyte appears to be adsorbed. × 70,000.

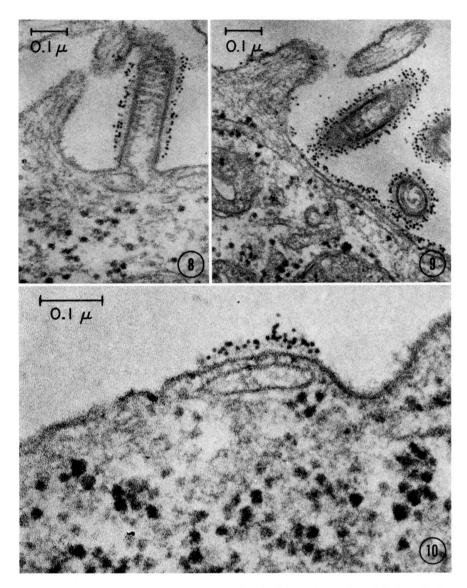

FIGS. 8–10. SV5-infected cells treated with ferritin-conjugated antiviral antibody. Figure 8. An SV5 filament in the process of budding is tagged with antibody molecules, but the cell surface just adjacent to the filament is devoid of antibody. × 90,000. Figure 9. Two virions as well as the adjacent cell surface are tagged with antibody. × 85,000. Figure 10. Antibody molecules localized over a discrete patch of the cell surface. There is no nucleocapsid detectable in association with the membrane, but the tagging indicates the presence of viral antigen. A small vesicle is present just below the area where antibody is attached. × 160,000.

The release of progeny influenza and parainfluenza virions from the cell surface is thought to involve the viral neuraminidase. By using recombinant influenza viruses containing hemagglutinin and neuraminidase from different parents, antibody specific for neuraminidase has been prepared (Webster and Laver, 1967; Kilbourne et al., 1968). Antineuraminidase antibody does not neutralize infectivity, but markedly reduces the level of released progeny virus when added to infected cells (Seto and Rott, 1966; Webster and Laver, 1967; Kilbourne et al., 1968). The antibody does not appear to have a direct effect on viral maturation, and the release of virus is enhanced if an antigenically distinct neuraminidase from Vibrio cholera is added to infected antibody-treated cells (Compans et al., 1969). These observations suggest that the viral neuraminidase functions in permitting the elution of virions from receptors at the cell surface after maturation is completed, and that antineuraminidase antibody inhibits release by interfering with this process.

V. Other Enveloped Viruses with Helical Nucleocapsids

The influenza and parainfluenza viruses represent two of the best-studied groups of enveloped viruses with helical symmetry, but other types of viruses which do not fit into either group nevertheless are similar in structure and are probably assembled by a similar process. Two viruses, pneumonia virus of mice (PVM) (Compans et al., 1967) and respiratory syncytial virus (Zakstelskaya et al., 1967; Joncas et al., 1969), have helical nucleocapsids similar in appearance to those of the parainfluenza viruses, but with a diameter of only about 135 Å. They have spike-covered envelopes and their assembly process is apparently very similar to that of influenza and parainfluenza viruses; long filamentous virions are frequently observed. On the basis of inhibitor studies (Harter and Choppin, 1967) PVM appears to contain RNA. No neuraminidase has been detected, but PVM virions will cause hemagglutination (Mills and Dochez, 1944). These two viruses do not appear to fit into either group of myxoviruses, but may form a third group with many similar morphological properties. No investigation of an antigenic relationship between PVM and respiratory syncytial virus has been reported.

As discussed elsewhere in this volume, a large group of plant and animal viruses have an elongated, bulletlike shape, and a helical internal structure. The nucleocapsid of one of these viruses, rabies virus, has been studied in detail (Sokol et al., 1969), and it is composed of 4% RNA and 96% protein like the nucleocapsid of parainfluenza virus; however, there are differences in the morphology of the nucleocapsids. When cells are doubly infected with the parainfluenza virus SV5 and vesicular stomatitis virus (VSV), a bullet-shaped virus, phenotypically mixed virus particles are produced (Choppin and Compans, 1970). Some virions contain VSV genomes within SV5 envelopes, and more frequently, both VSV and SV5 antigens are present in the same envelope of a virus containing the VSV genome. These results indicate that the VSV nucleocapsid can interact with segments of cell membrane which contain SV5 envelope proteins in the manner which leads to viral maturation. The base composition of SV5 RNA (Compans and Choppin, 1968) and VSV RNA (Brown et al., 1967) are very similar, and their nucleocapsids both have protein subunits with molecular weights of about 60,000 (Mountcastle et al., 1970; Wagner et al., 1969). These observations raise the possibility that parainfluenza viruses and the bullet-shaped viruses may be evolutionarily related.

VI. Summary and Conclusions

There are many similarities in the morphology of influenza and parainfluenza virions, and in the process of their assembly at the cell surface. Virions of both groups consist of a spike-covered envelope and a helical internal component. The envelope contains the viral hemagglutinin and neuraminidase, and in the case of the parainfluenza viruses, the component(s) responsible for cell fusion and hemolysis. The virus particles of both groups vary in size and shape, within smaller spherical particles presumably containing a single genome, and long filamentous virions probably containing multiple genomes. Viruses of both groups occur in many vertebrates, but no plant or bacterial virus has been described which fits clearly into either group.

Most of the detailed studies on the structure of influenza virions have been done with type A viruses, and similar information has not yet been obtained for viruses of the B or C antigenic type. The influenza nucleocapsid has been isolated in the form of short strands, and several distinct sizes of nucleocapsid correspond to RNA molecules of different sizes. Seven polypeptides have been detected in influenza virions; one of these is found in the nucleocapsid, and the remainder are presumably associated with the envelope. The hemagglutinin and neuraminidase have been isolated and they appear to be two morphologically distinct types of spikes on the viral envelope. There is marked antigenic variation in the hemagglutinin of influenza A viruses, and it is probably due to variation in the amino acid sequence. Some degree of antigenic variation is also found with the viral neuraminidase.

The helical nucleocapsids of several parainfluenza viruses have been characterized, and there is good agreement between the various observations, indicating that nucleocapsid structure is an easily identifiable characteristic of this group of viruses. The nucleocapsid is a single-stranded helix with a length of about 1 μ. The viral RNA is a single molecule with an estimated molecular weight of about 6×10^6 daltons. The nucleocapsid protein is a single species with a molecular weight of about 60,000, and four or five other proteins are present in the intact virion. Two of the viral envelope proteins are glycoproteins.

The available evidence suggests that the proteins of both influenza and parainfluenza virions are virus-coded. In contrast, with a few significant exceptions, the lipids of the host cell plasma membrane are incorporated quantitatively into the virion during maturation at the cell membrane. Neuraminic acid-containing glycolipids (gangliosides) are not found in the virion, nor is protein bound neuraminic acid, presumably due to the action of the viral neuraminidase. The virions of both groups also contain carbohydrate covalently linked to protein and in neutral glycolipids. Host carbohydrate moieties may represent host antigens incorporated into virions.

Assembly of influenza and parainfluenza viruses occurs in the following sequence. Prior to viral morphogenesis, virion envelope proteins are incorporated into patches of the plasma membrane of the infected cell. Such areas cannot be distinguished from normal membrane by their morphology, but show specific adsorption of erythrocytes and tagging with ferritin-conjugated antiviral antibody. The nucleocapsid, which is assembled in the interior of the cell, appears to associate specifically with such areas of membranes, and the characteristic viral spike layer can then be detected on the outer surface of the membrane. By a process of budding, the nucleocapsid is then enveloped by the spike-covered membrane, and a complete virion is produced.

ACKNOWLEDGEMENTS

Research by the authors was supported by Contract AT (30-1)-3983 from the United States Atomic Energy Commission and Research Grant AI-05600 from the National Institutes of Allergy and Infectious Diseases, United States Public Health Service. The electron microscope was obtained through equipment grant GB-7681 from the National Science Foundation.

REFERENCES

Ada, G. L. and Perry, B. T. (1954). *Aust. J. Exp. Biol. Med. Sci.* **32**, 453–468.

Ada, G. L., Perry, B. T., and Abbot, A. (1958). *J. Gen. Microbiol.* **19**, 23–39.

Ada, G. L., Lind, P. E., Larkin, L., and Burnet, F. M. (1959). *Nature (London)* **184**, 360–361.

Almeida, J. D. and Waterson, A. P. (1970). *In* "The Biology of Large RNA Viruses" (R. D. Barry and B. W. J. Mahy, eds.), pp. 27–51. Academic Press, New York.

Andrewes, C. H., Bang, F. B., and Burnet, F. M. (1955). *Virology* **1**, 176–184.

Apostolov, K. and Flewett, T. H. (1965). *Virology* **26**, 506–508.

Apostolov, K. and Flewett, T. H. (1969). *J. Gen. Virol.* **4**, 365–370.

Bächi, T., Gerhard, W., Lindenmann, J., and Mühlethaler, K. (1969). *J. Virol.* **4**, 769–776.

Baker, R. F., Gordon, I., and Stevenson, D. (1965). *Virology* **27**, 441–445.

Barry, R. D. and Bukrinskaya, A. G. (1968). *J. Gen. Virol.* **2**, 71–79.

Barry, R. D. and Mahy, B. W. J. (eds.) (1970). "The Biology of Large RNA Virus." Academic Press, New York.

Bellett, A. J. D. (1967). *J. Virol.* **1**, 245–259.

Benedetti, E. L. and Emmelot, P. (1968). *In* "The Membranes" (A. J. Dalton and F. Haguenau, eds.), pp. 33–120. Academic Press, New York.

Best, R. (1968). *Advan. Virus Res.* **13**, 64–146.

Bikel, I. and Duesberg, P. H. (1969). *J. Virol.* **4**, 399–393.

Birch-Andersen, A. and Paucker, K. (1959). *Virology* **8**, 21–40.

Blair, C. D. and Duesberg, P. H. (1970). *Advan. Virus Res.*, in press.

Blair, C. D. and Robinson, W. S. (1968). *Virology* **35**, 537–549.

Blough, H. A. (1968). *Wistar Inst. Symp. Monogr.* **8**, 55–59.

Brown, F., Martin, S. J., Cartwright, B., and Crick, J. (1967). *J. Gen. Virol.* **1**, 479–486.

Caliguiri, L. A., Klenk, H.-D., and Choppin, P. W. (1969). *Virology* **39**, 460–466.

Choppin, P. W. (1963). *Virology* **21**, 278–281.

Choppin, P. W. (1969). *Virology* **39**, 130–134.

Choppin, P. W. and Compans, R. W. (1970). *J. Virol.* **5**, 609–616.

Choppin, P. W. and Stoeckenius, W. (1964a). *Virology* **22**, 482–492.

Choppin, P. W. and Stoeckenius, W. (1964b). *Virology* **23**, 195–202.

Choppin, P. W., Murphy, J. S., and Tamm, I. (1960). *J. Exp. Med.* **112**, 945–952.

Choppin, P. W., Murphy, J. S., and Stoeckenius, W. (1961). *Virology* **13**, 549–550.

Choppin, P. W., Compans, R. W., and Holmes, K. V. (1968). *In* "Proceedings of the Second International Symposium on Medical and Applied Virology" (M. Sanders and E. H. Lennette, eds.), pp. 16–30. W. Green, St. Louis, Missouri.

Choppin, P. W. and Pons, M. W. (1970). *Virology* **42**, 603–610.

Chu, C. M., Dawson, I. M., and Elford, W. J. (1949). *Lancet* i, 602–603.

Compans, R. W. and Choppin, P. W. (1967a). *Proc. Nat. Acad. Sci. U. S.* **57**, 949–956.

Compans, R. W. and Choppin, P. W. (1967b). *Virology* **33**, 344–346.

Compans, R. W. and Choppin, P. W. (1968). *Virology* **35**, 289–296.

Compans, R. W. and Dimmock, N. J. (1969). *Virology* **39**, 499–515.

Compans, R. W., Holmes, K. V., Dales, S., and Choppin, P. W. (1966). *Virology* **30**, 411–426.

Compans, R. W., Harter, D. H., and Choppin, P. W. (1967). *J. Exp. Med.* **126**, 267–276.

Compans, R. W., Dimmock, N. J., and Meier-Ewert, H. (1969). *J. Virol.* **4**, 528–534.

Compans, R. W., Dimmock, N. J., and Meier-Ewert, H. (1970a). *In* "The Biology of Large RNA Viruses" (R. D. Barry and B. W. J. Mahy, eds.), pp. 87–108. Academic Press, New York.

Compans, R. W., Klenk, H. D., Caliguiri, L. A., and Choppin, P. W. (1970b). *Virology* **42**, 880–889.

Compans, R. W., Lackland, H., and Choppin, P. W. (1971). In preparation.

Cunha, R., Weil, M. L., Beard, D., Taylor, A. R., Sharp, D. G., and Beard, J. W. (1947). *J. Immunol.* **55**, 69–89.

Dahlberg, J. E. and Simon, E. H. (1969). *Virology* **38**, 666–678.

Davies, P. and Barry, R. D. (1966). *Nature (London)* **211**, 384–387.

Duc-Nguyen, H. (1968). *J. Virol.* **2**, 494–506.

Duc-Nguyen, H. and Rosenblum, E. N. (1967). *J. Virol.* **1**, 415–429.

Duesberg, P. H. (1968). *Proc. Nat. Acad. Sci. U. S.* **59**, 930–937.

Duesberg, P. H. (1969). *J. Mol. Biol.* **42**, 485–499.

Duesberg, P. H. and Robinson, W. S. (1965). *Proc. Nat. Acad. Sci. U. S.* **54**, 794–800.

Duesberg, P. H. and Robinson, W. S. (1967). *J. Mol. Biol.* **25**, 383–405.

Evans, M. J., and Kingsbury, D. W. (1969). *Virology* **37**, 597–604.

Feller, U., Dougherty, R. M., and DiStefano, H. S. (1969). *J. Virol.* **4**, 753–762.

Finch, J. T. and Gibbs, A. J. (1970). *J. Gen. Virol.* **6**, 141–150.

Frisch-Niggemeyer, W., and Hoyle, L. (1956). *J. Hyg.* **54**, 201–212.

Frommhagen, L. H., Knight, C. A., and Freeman, L. K. (1959). *Virology* **8**, 176–197.

Harboe, A. (1963). *Acta Pathol. Microbiol. Scand.* **57**, 488–492.

Harter, D. H. and Choppin, P. W. (1967). *J. Exp. Med.* **126**, 251–266.

Haslam, E. A., Cheyne, I. M., and White, D. O. (1969). *Virology* **39**, 118–129.

Hirst, G. K. (1962). *Cold Spring Harbor Symp. Quant. Biol.* **27**, 303–310.

Holland, J. J. and Kiehn, E. D. (1970). *Science* **167**, 202–205.

Holmes, K. V. and Choppin, P. W. (1966). *J. Exp. Med.* **124**, 501–520.

Holmes, K. V., Klenk, H.-D., and Choppin, P. W. (1969). *Proc. Soc. Exptl. Biol. Med.* **131**, 651–657.

Horne, R. W. and Waterson, A. P. (1960). *J. Mol. Biol.* **2**, 75–77.

Horne, R. W., Waterson, A. P., Wildy, P., and Farnham, A. E. (1960). *Virology* **11**, 79–98.

Hosaka, Y. (1968). *Virology* **35**, 445–457.

Hosaka, Y. and Shimizu, K. (1968). *J. Mol. Biol.* **35**, 369–373.

Hosaka, Y., Kitano, H., and Ikeguchi, S. (1966). *Virology* **29**, 205–221.

Howe, C., Morgan, C., De Vaux St. Cyr, C., Hsu, K. C., and Rose, H. M. (1967). *J. Virol.* **1**, 215–237.

Hoyle, L. (1952). *J. Hyg.* **50**, 229–245.

Hoyle, L., Horne, R. W., and Waterson, A. P. (1961). *Virology* **13**, 448–459.

Iwai, Y., Iwai, M., Okumoto, M., Hosokawa, Y., and Asai, T. (1966). *Biken J.* **9**, 241–248.

Joncas, J., Berthiaume, L., and Pavlanis, V. (1969). *Virology* **38**, 493–496.

Joss, A., Gandhi, S. S., Hay, A. J., and Burke, D. C. (1969). *J. Virol.* **4**, 816–822.

Kates, M., Allison, A. C., Tyrrell, D. A. J., and James, A. T. (1962). *Cold. Spring Harbor Symp. Quant. Biol.* **27**, 293–301.

Kilbourne, E. D. (1963). *Progr. Med. Virol.* **5**, 79–126.

Kilbourne, E. D. and Murphy, J. S. (1960). *J. Exp. Med.* **111**, 387–406.

Kilbourne, E. D., Laver, W. G., Schulman, J. L., and Webster, R. G. (1968). *J. Virol.* **2**, 281–288.

Kingsbury, D. W. (1966). *J. Mol. Biol.* **18**, 195–203.

Kingsbury, D. W. and Darlington, R. W. (1968). *J. Virol.* **2**, 248–255.

Kingsbury, D. W. and Webster, R. G. (1969). *J. Virol.* **4**, 219–225.

Klenk, H.-D. and Choppin, P. W. (1969a). *Virology* **37**, 155–157.

Klenk, H.-D. and Choppin, P. W. (1969b). *Virology* **38**, 255–268.

Klenk, H.-D. and Choppin, P. W. (1970a). *Virology* **40**, 939–947.

Klenk, H.-D. and Choppin, P. W. (1970b). *Proc. Nat. Acad. Sci. U. S.* **66**, 57–64.

Klenk, H.-D. and Choppin, P. W. (1971). In preparation.

Klenk, H.-D., Caliguiri, L. A., and Choppin, P. W. (1970). *Virology* **42**, 473–481.

Laver, W. G. (1964). *J. Mol. Biol.* **9**, 109–124.

Laver, W. G. and Valentine, R. C. (1969). *Virology* **38**, 105–119.

Laver, W. G. and Webster, R. G. (1966). *Virology* **30**, 104–115.

Laver, W. G. and Webster, R. G. (1968). *Virology* **34**, 193–202.

Lee, L. T., Howe, C., Meyer, K., and Choi, H. U. (1969). *J. Immunol.* **102**, 1144–1155.

Lief, F. S. and Henle, W. (1956). *Virology* **2**, 753–771.

Meier-Ewert, H., Gibbs, A. J., and Dimmock, N. J. (1970). *J. Gen. Virol.* **6**, 409–419.

Mills, K. C. and Dochez, A. R. (1944). *Proc. Soc. Expt. Biol. Med.* **57**, 140–143.

Moore, D. H., Davies, M. C., Levine, S., and Englert, M. E. (1962). *Virology* **17**, 470–479.

Morgan, C., Hsu, K. C., and Rose, H. M. (1962). *J. Exp. Med.* **116**, 553–564.

Mountcastle, W. E., Compans, R. W., Caliguiri, L. A., and Choppin, P. W. (1970). *J. Virol.* **6**, 677–684.

Mountcastle, W. E., Compans, R. W., and Choppin, P. W. (1971). *J. Virol.*, **7**, 47–52.

Murphy, J. S. and Bang, F. B. (1952). *J. Exp. Med.* **95**, 259–268.

Nakai, T., Shand, F. L., and Howatson, A. F. (1969). *Virology* **38**, 50–67.

Pons, M. W. (1967). *Virology* **31**, 523–531.

Pons, M. W. and Hirst, G. K. (1968). *Virology* **34**, 385–388.

Pons, M. W., Schulze, I. T., and Hirst, G. K. (1969). *Virology* **39**, 250–259.

Prose, P. H., Balk, S. D., Liebhaber, H., and Krugman, S. (1965). *J. Exp. Med.* **122**, 1151–1160.

Rott, R. and Schäfer, W. (1964). *In* "Cellular Biology of Myxovirus Infections" (G. E. W. Wolstenholme and J. Knight, eds.), pp. 27–42. Little Brown, Boston, Massachusetts.

Scholtissek, C. (1970). *In* "The Biology of Large RNA Viruses" (R. D. Barry and B. W. J. Mahy eds.), pp. 438–454. Academic Press, New York.

Schulman, J. L. and Kilbourne, E. D. (1969). *Proc. Nat. Acad. Sci. U. S.* **63**, 326–333.

Schulze, I. T. (1970). *Virology* **42**, 890–904.

Seto, J. T. and Rott, R. (1966). *Virology* **30**, 731–737.

Simpson, R. W. and Hauser, R. E. (1966). *Virology* **30**, 684–697.

Sjöstrand, F. S. (1968). *In* "The Membranes" (A. J. Dalton, and F. Haguenau, eds.), pp. 151–210. Academic Press, New York.

Sokol, F. and Szurman, J. (1959). *Acta Virol.* **3**, 175–180.

Sokol, F., Schlumberger, H. D., Wiktor, T. J., and Koprowski, H. (1969). *Virology* **35**, 651–665.

Tiffany, J. M. and Blough, H. A. (1969). *Science* **163**, 573–574.

von Magnus, P. (1954). *Advan. Virus Res.* **2**, 59–79.

Wagner, R. R., Schnaitman, T. A., Snyder, R. M., and Schnaitman, C. A. (1969). *J. Virol.* **3**, 395–403.

Webster, R. G. (1970). *Virology* **40**, 643–654.

Webster, R. G. and Darlington, R. W. (1969). *J. Virol.* **4**, 182–187.

Webster, R. G. and Laver, W. G. (1967). *J. Immunol.* **99**, 49–55.

Wecker, E. (1957). *Z. Naturforsch.* **126**, 208–210.

Yunis, E. J. and Donnelly, W. H. (1969). *Virology* **39**, 351–357.

Zakstelskaya, L. Y., Almeida, J. D., and Bradstreet, C. M. P. (1967). *Acta Virol.* **11**, 420–423.

CHAPTER 14 **A Plant Virus with Properties of a Free Ribonucleic Acid: Potato Spindle Tuber Virus**

T. O. DIENER

I. Introduction

During the last 14 years, it has become abundantly clear that nucleic acids isolated from virions contain all of the genetic information necessary to induce in a susceptible host not only replication of viral nucleic acid and of the other constituents of the virion, but also to induce the pathological condition characteristic of the particular virus disease. One consequence of this finding is the recognition that formation of virions is not a *conditio sine qua non* of virus infection; and that the virion, as far as the pathological condition of the host is concerned, is important only in the transmission of this condition from one host to another and, in some cases, in the initial process of infection. Pathological conditions that are initiated and transmitted by free nucleic acid are thus theoretically possible and, if such conditions could be demonstrated to occur naturally, they would necessitate a broadening of our usual concepts of virus and virus disease.

We must, however, immediately recognize that such pathogens would be at a serious disadvantage during the processes of transmission and initial infection, as compared with conventional, virion-producing viruses, primarily because known viral nucleic acids are readily inactivated and hydrolyzed by the ever-present nucleases. This expectation is, indeed, borne out by the observation that inoculation of a susceptible host with viral nucleic acid isolated from virions is generally a much less efficient process than inoculation with the complete virion. Evidently, the capacity of a viral pathogen to protect its vital genetic information by encapsulation in a protein coat of varying complexity has given such pathogens enhanced survival value in the process of evolution and explains why virion formation is such a general feature of virus infection.

The disadvantages of viral pathogens that lack the capacity of virion formation are well illustrated by the few cases that are known to exist in nature or that can be produced in the laboratory. Thus, forms of tobacco necrosis virus are known that multiply and occur in infected leaves as nucleic acid (Kassanis and Welkie, 1963). Similar forms are known of tobacco rattle virus (Sänger and Brandenburg, 1961; Cadman, 1962); and mutants of tobacco mosaic virus that lack the capacity to form virions have been produced by nitrous acid treatment (Siegel *et al.*, 1962). Although these pathogens differ from one another in some properties, they are all difficult to transmit, are unstable in extracts, and have low specific infectivities as compared with the stable forms of the respective viruses. Clearly, these defective virus forms are either laboratory curiosities or play only a very minor role in the natural epidemiology of virus diseases.

One must, however, keep in mind that the nucleic acids of the viruses mentioned are single-stranded RNA's; and one cannot exclude the possibility that viral pathogens that have their genetic information encoded in a more stable form of nucleic acid might be less disadvantaged without a capsid than are the above viruses. It is, for example, well known that double-stranded RNA is much more resistant to attack by ribonuclease (at relatively high ionic strength) than is single-stranded RNA, and it is conceivable that encapsulation of such a nucleic acid in a protein coat might not give the pathogen sufficient additional protection for the protein coat to be retained during evolution. This might particularly apply to nucleic acids in a circular, supercoiled configuration.

In the following, I wish to present evidence that an economically important viral pathogen of plants, potato spindle tuber virus (PSTV), multiplies and is naturally transmitted as a free nucleic acid. As will be seen, the existence of an extremely labile virion cannot be conclusively ruled out; but, if such an entity exists, it can, in the light of our evidence, have only very limited biological significance. In contrast to the defective viruses mentioned, PSTV nucleic acid is relatively stable in extracts, is easily transmitted by mechanical means, and has an extremely high specific infectivity.

II. Previous Studies on Potato Spindle Tuber Disease

A. EARLY WORK

The potato spindle tuber disease was first described by Schultz and Folsom (1923) and by Gilbert (1923). It was considered to be a disease closely allied to mosaic, leafroll, and other similar "degeneration" diseases of potato. Early work indicated that the disease could be transmitted by aphids and that it was readily transmitted to healthy plants under field conditions (Schultz and Folsom, 1923). Goss (1926) reported that the disease could be mechanically transmitted by cutting knives and by contact between healthy and infected seed (tuber) pieces. Since no microorganisms could be found as etiological factors by the then accepted criteria, the disease was considered to be of viral causation. Purification of the causative agent was apparently not attempted, even after suitable procedures had become available. Undoubtedly, the difficulty of detecting spindle tuber in the only host known at that time, potato, and the need of trained observers to recognize the variable foliage and tuber symptoms, contributed to neglect of the disease.

B. Bioassay on Tomato

Raymer and O'Brien (1962) were the first to transmit potato spindle tuber disease to a host other than potato. They reported that PSTV incited in tomato (*Lycopersicon esculentum* Mill., cv. Rutgers) distinctive systemic symptoms that were unlike those produced by any other known potato virus. The symptoms closely resembled, however, those reported by McClean (1931) from South Africa for the tomato bunchy-top virus.

Although a systemic indicator plant is not as useful as a local lesion host, systematic work with PSTV became possible by using tomato as a bioassay plant. All work reported here is based on bioassay of PSTV in tomato. An arbitrarily constructed "infectivity index," which is based on the dilution of the inoculum, the time required for symptom appearance in tomato plants, and the number of plants infected, proved to be of great value in estimating relative virus concentration (Raymer and Diener, 1969).

C. Attempts to Characterize PSTV

Subsequent to the discovery of tomato as an indicator host of PSTV, several workers have used this plant in attempts to purify and characterize PSTV.

Hunter (1964) studied the effects of various agents on the efficiency of virus extraction. He found that buffers of high ionic strength were more effective than buffers of low ionic strength for extraction of PSTV from infected tissue, and that addition of butanol or chloroform to the extraction medium had no deleterious effect on infectivity, but improved the clarification of infectious extracts. He reported that preparations from healthy and PSTV-infected tissue produced identical zones after rate-zonal centrifugation in sucrose density gradients.

Benson *et al.* (1964) reported that concentration of extracts from PSTV-infected tissue by filtration through a graded series of filters resulted in infectious preparations that contained spherical particles with diameters of 24–27.5 nm. Singh *et al.* (1966) reported further work on the purification of spherical particles, 25 nm in diameter, that were believed to be PSTV. However, more recent work by Singh and Bagnall (1968) confirms our report (Diener and Raymer, 1967, 1969) of an infectious nucleic acid in PSTV-infected tissue. Also, in our tests, the buffer system used by Benson *et al.* (1964) did not extract an appreciable amount of infectious material from PSTV-infected tissue. These recent findings, therefore, place in question the identity of the spherical particles.

Allington *et al.* (1964) and Ball *et al.* (1964) claimed that potato spindle tuber in Nebraska was caused by a strain of potato virus X (PVX). These investigators used a different source of PSTV, one from Nebraska. In our source of PSTV, we have been unable to find any evidence of PVX either by plant indicator tests, electron microscopy, serology, or extraction and sedimentation studies. Bagnall (1967) also found no evidence of PVX in our source of PSTV. The results of Allington *et al.* (1964) and Ball *et al.* (1964) could be explained by the assumption that PSTV and PVX were present as a mixture in their original source material and that they were not separated in subsequent transfers.

Bagnall (1967) reported production of antisera that reacted with a component in sap from PSTV-infected tomato and potato plants. No specific cross-reaction was obtained between PSTV antiserum and PVX-containing sap, or between PVX antiserum and PSTV-containing sap. Recently, however, Singh and Bagnall (1968) reported that this antigen is not a virus-specific component, but is actually a plant component, the quantity of which is much increased in PSTV-infected plants.

III. Properties of Crude Extracts from PSTV-Infected Tissue

A. Effect of pH and Phosphate Concentration

Preliminary results with either potato or tomato tissue indicated that infectivity levels of extracts were higher when basic phosphate buffers were used than when acidic phosphate buffers were used for extraction of the tissue. Low-speed centrifugation of extracts showed that with phosphate buffers of low ionic strength, most of the infectious material sedimented in 15 minutes at 6500 rpm and that this pelleted material could be extracted by resuspension of the pellets in phosphate buffer (Table I). Extraction of tissue with phosphate buffer of higher ionic strength (0.05–0.5 M) resulted in low-speed supernatants that contained a higher proportion of the total infectious material present in the extract (Table I). Extraction with 0.5 M K_2HPO_4 markedly increased the total amount of infectious material extracted, as compared with 0.005 M K_2HPO_4; but even with this high molarity buffer, an appreciable amount of infectious material pelleted during low-speed centrifugation (Table I). Clarified extracts prepared with 0.5 M

TABLE I

COMPARISON OF LOW AND HIGH IONIC STRENGTH BUFFERS FOR EXTRACTION OF PSTV

Extraction medium	Infectivity index[a]	
	Low-speed supernatant	Pellet[b]
0.005 M K$_2$HPO$_4$ + 1% TGA	5	30
0.005 M K$_2$HPO$_4$ + 1% TGA + chloroform– n-butanol[c]	0[d]	72
0.005 M K$_2$HPO$_4$ + 0.1 M ascorbic acid, pH 6.8[e]	7	28
0.5 M K$_2$HPO$_4$ + 1% TGA	48	33
0.5 M K$_2$HPO$_4$ + 1% TGA + chloroform– n-butanol[c]	83[d]	NT

[a] Assayed at dilutions of 10^{-1} to 10^{-4} in 0.005 M K$_2$HPO$_4$.

[b] Low-speed pellet resuspended in original volume of 0.5 M K$_2$HPO$_4$, homogenized for 3 minutes, centrifuged 15 minutes at 6500 rpm, supernatant diluted and assayed. NT = not tested.

[c] Mixture of chloroform and n-butanol: 1:1 (v/v). TGA, thioglycolic acid.

[d] Aqueous phase.

[e] Extraction medium used by Benson et al. (1964).

K$_2$HPO$_4$ alone were invariably brown. Addition of thioglycolic acid (1 ml per 200 ml of 0.5 M K$_2$HPO$_4$) reduced the pH from 9.1 to 7.8. The presence of thioglycolic acid prevented much of the browning reaction, but had no appreciable effect on the infectivity of the extracts.

B. EFFECT OF ORGANIC SOLVENTS

Preliminary experiments had shown that chloroform was useful for clarification of PSTV extracts and, as shown in Table I, the aqueous phases of extracts prepared in the presence of chloroform and butanol had a higher level of infectivity than extracts prepared identically but without organic solvents. Evidently, the infectious agent(s) of PSTV is insensitive to treatment with chloroform and butanol. Infectivity of extracts prepared with 0.5 M K$_2$HPO$_4$ in the presence of chloroform and butanol, followed by low-speed centrifugation, retained considerable infectivity after prolonged storage at 4°C (for as long as 3 months).

C. SEDIMENTATION PROPERTIES OF THE INFECTIOUS AGENT(S)

1. *High-Speed Centrifugation*

Attempts to purify the presumed virus particles by high-speed centrifugation led to unexpected results. As shown in Table II, when clarified extracts from PSTV-infected tissue were subjected to high-speed centrifugation (1 hour at 40,000 rpm, Spinco* model L, 40 rotor), most of the infectious material remained in the supernatant solution. To eliminate the possibility that the high salt content of the extract was responsible for the lack of sedimentation, portions of extracts were dialyzed against 0.005 M K_2HPO_4 prior to high-speed centrifugation. However, distribution of infectivity between high-speed supernatant and high-speed pellet was not affected by

TABLE II

DISTRIBUTION OF PSTV INFECTIVITY BEFORE AND AFTER HIGH-SPEED CENTRIFUGATION[a]

Expt. no.	Treatment[b] of aqueous phase	Infectivity[c]						
		High-speed supernatant					High-speed pellet[d]	
		10^{-1}	10^{-2}	10^{-3}	2×10^{-4}	10^{-4}	10^{-1}	10^{-2}
1[e]	Dialyzed 2 hours vs. 0.005 M K_2HPO_4	3/3	3/3	2/3	NT	0/3	2/3	0/3
2	None (pH 8.2)	3/3	3/3	2/3	1/3	NT	1/3	0/3
2	Adjusted to pH 7.5	3/3	2/3	2/3	1/3	NT	2/3	0/3
2	10^{-4} M Mg^{2+}, pH 8.2	3/3	2/3	2/3	2/3	NT	1/3	0/3
2	10^{-3} M Mg^{2+}, pH 8.2	3/3	2/3	1/3	2/3	NT	1/3	0/3

[a] Centrifuged 1 hour at 40,000 rpm, No. 40 rotor, Spinco Model L ultracentrifuge.

[b] Tissue extracted with 0.5 M K_2HPO_4, chloroform, and n-butanol (1:2:1:1, w:v:v:v). Phases separated by centrifugation for 15 minutes at 6000–6500 rpm; aqueous phase used for tests.

[c] Number of plants infected/number of plants inoculated. NT, not tested. All preparations were diluted for assay in 0.005 M K_2HPO_4.

[d] Resuspended in original volume of 0.5 M K_2HPO_4. No infectivity at dilutions of 10^{-3}, 2×10^{-4}, or 10^{-4}.

[e] The aqueous phase of experiment 1 yielded the following results on bioassay: 3/3 plants infected at a dilution of 10^{-1}, 2/3 at 10^{-2}, 2/3 at 10^{-3}, and 2/3 at 10^{-4}.

* Mention of specific equipment, trade products, or a commercial company does not constitute its endorsement by the U. S. Government over similar products or companies not named.

this treatment (Table II). Adjustment of the pH to 7.5 or addition of Mg^{2+} ions also did not cause more of the infectious material to pellet during high-speed centrifugation. Additional tests involving centrifugation for up to 4 hours at 40,000 rpm did not increase the proportion of infectious material recoverable from the high-speed pellet. Infectivity of high-speed supernatants was as stable as that of clarified extracts.

2. Rate-Zonal Density-Gradient Centrifugation

To arrive at a more accurate determination of the sedimentation properties of PSTV, numerous extracts were subjected to rate-zonal density-gradient centrifugation.

Extracts, clarified with chloroform and butanol, were dialyzed against 0.005 M K_2HPO_4 and were layered onto sucrose density gradients prepared with the same buffer. After centrifugation for 2–4 hours at 24,000 rpm (SW 25.1 rotor), the tubes were fractionated, and individual fractions assayed for infectivity. Infectivity was found only near the top of the tubes. Consequently, in further tests, centrifugation was extended to 16 hours. Results of experiments with high-speed supernatants and with resuspended high-speed pellets are shown in Table III.

Irrespective of whether or not organic solvents were used during the extraction of the tissue, infectivity of high-speed supernatants was confined to the upper and middle portions of the gradient columns. Distribution of infectivity from resuspended high-speed pellets was similar; but, as expected, levels of infectivity were lower than in the high-speed supernatants. The high-speed pellets from extracts preparared without organic solvents contained considerable amounts of green material. This material was large enough to sediment to the bottom of the tubes during centrifugation. Considerable infectivity was associated with these pelleted fractions (Table III).

The slow rate of sedimentation of the infectious material made it unlikely that the extracted infectious agent was a conventional virion. It appeared more likely that this material was free nucleic acid. The question thus arose whether the extraction procedures used had led to disruption of viral nucleoprotein particles present *in situ*, with the concomitant release of infectious nucleic acid. Efforts were, therefore, made to alter the extraction procedure in such a way as to maintain the integrity of these hypothetical virions.

It was conceivable that PSTV virions were extremely sensitive to exposure to salt, and that the failure to find infectious material sedimenting in the range of viral nucleoproteins was due to their disintegration in 0.5 M

TABLE III

DISTRIBUTION OF INFECTIVITY IN CENTRIFUGED SUCROSE DENSITY GRADIENTS[a] OF PSTV
EXTRACTS PREPARED WITH AND WITHOUT ORGANIC SOLVENTS

Fraction[b] no.	Expressed sap[d]	Infectivity[c]			
		Extraction			
		Without organic solvents[e]		With organic solvents[f]	
		HS[g]	HP[h]	HS	HP
1	0/3	0/3	0/3	0/3	0/3
2	0/3	3/3	2/3	0/3	1/3
3	0/3	3/3	0/3	2/3	1/3
4	0/3	2/3	0/3	3/3	0/3
5	0/3	3/3	0/3	0/3	0/3
6	0/3	2/3	0/3	0/3	0/3
7	0/3	2/3	1/3	3/3	0/3
8	0/3	0/3	0/3	1/3	0/3
9	0/3	0/3	0/3	0/3	0/3
10	0/3	0/3	0/3	0/3	0/3
11	0/3	0/3	0/3	0/3	0/3
12	1/3	0/3	0/3	0/3	0/3
Resuspended pellet	3/3	0/3	3/3	0/3	0/3

[a] Linear gradients, 0.2–0.7 M sucrose in 0.005 M K_2HPO_4, centrifugation at 24,000 rpm in Spinco L ultracentrifuge, SW 25.1 rotor.

[b] Consecutive 2-ml fractions. Fraction 1 is from top, fraction 12 from bottom of tube. All fractions diluted 1:4 with 0.005 M K_2HPO_4 for inoculation. Pellets resuspended in 2 ml of same buffer for inoculation.

[c] Number of plants with symptoms/number of plants inoculated.

[d] No buffer, solvent, clarification, or high-speed centrifugation used prior to density-gradient centrifugation. Density-gradient centrifugation for 3 hours at 24,000 rpm.

[e] Tissue extracted with 0.5 M K_2HPO_4 (2 ml/g tissue), extract clarified at 12,000 rpm for 15 minutes, supernatant centrifuged 1 hour at 40,000 rpm. Density-gradient centrifugation for 16 hours at 24,000 rpm.

[f] Tissue extracted with 0.5 M K_2HPO_4 + chloroform + n-butanol (2:1:1 ml/g of tissue), centrifuged for 15 minutes at 6500 rpm. Aqueous phase centrifuged for 1 hour at 40,000 rpm. Density-gradient centrifugation for 16 hours at 24,000 rpm.

[g] HS, high-speed supernatant.

[h] HP, high-speed pellet resuspended in original volume of 0.005 M K_2HPO_4.

K_2HPO_4. Consequently, extracts were prepared with phosphate buffers of lower ionic strength, and the distribution of infectivity was studied by density-gradient centrifugation. Table IV summarizes typical results. Similar results were obtained when extracts prepared with 0.01 or 0.005 M phosphate buffer were analyzed.

TABLE IV

DISTRIBUTION OF INFECTIVITY IN CENTRIFUGED SUCROSE DENSITY GRADIENTS[a] OF PSTV EXTRACTS PREPARED WITH LOW IONIC STRENGTH PHOSPHATE BUFFERS[b]

Fraction[e] number	Infectivity index				
	Extract 1		Extract 2		
			Centrifugation for 2.5 hours[e]		
	Centrifugation for 16 hours[d]	Centrifugation for 2.5 hours[e]	Immediately	After 1 day at 4°C	After 7 days at 4°C
1	14	18	17	18	14
2	20	12	19	11	13
3	21	17	19	12	3
4	16	10	16	12	1
5	14	4	15	13	1
6	12	11	14	15	6
7	9	5	16	11	0
8	8	0	10	4	0
9	7	0	12	2	0
10	0	0	16	2	0
11	0	0	15	0	0
12	0	0	14	2	0
13	0	0	10	0	0
Resuspended pellet[f]	6	17	15	13	0

[a] Linear gradients, 0.2–0.8 M sucrose in 0.02 M phosphate buffer, pH 7, centrifugation at 24,000 rpm in Spinco Model L Centrifuge, SW 25.1 rotor.

[b] Five grams fresh PSTV-infected tissue was extracted at 4°C with 20 ml of 0.05 M K_2HPO_4 (mortar and pestle); the extract was adjusted with NaOH to pH 8.3, and was then centrifuged for 15 minutes at 5000 g. The supernatant was used.

[e] Consecutive 2-ml fractions. Fraction 1 is from the top; fraction 13 is from the bottom of the tube. All fractions inoculated undiluted.

[d] Tobacco mosaic virus RNA added as a marker was located in fraction No. 10.

[e] Southern bean mosaic virus ($s_{20,W} = 115$) added as a marker was located in fraction Nos. 5–6.

[f] Pellets resuspended in 2 ml of 0.02 M phosphate buffer, pH 7.

Centrifugation of extracts for 16 hours disclosed that, in spite of the lower ionic strength used, much infectious material sedimented at very low rates. Note that in this experiment, tobacco mosaic virus RNA, added as a marker, was located in fraction No. 10. Centrifugation for 2.5 hours showed that although the top of the gradient contained the highest level of infectivity, considerable amounts of infectious material had sedimented as far down as the middle of the tube, and the resuspended pellet contained a relatively high level of infectivity. A study of the decay of infectivity upon storage (Table IV, last 3 columns) revealed that the infectious material that sedimented with a rate comparable to that of viral nucleoproteins did not possess greater stability than the very slowly sedimenting material. On the contrary, the slowest sedimenting material (fraction No. 1) appeared to be more resistant to being inactivated than faster sedimenting infectious material. It must be noted that infectivity distribution in individual extracts varied considerably and was dependent on the level of infectivity in the tissue extracted. An example of this variation is illustrated in Table IV (centrifugation for 2.5 hours, extract 1, and immediate centrifugation, extract 2). With low levels of infectivity, infectivity could only be detected in the upper fractions of a density-gradient tube; whereas with high levels of infectivity, often all fractions from a density gradient tube were infectious. Nevertheless, a tendency for infectivity levels to be highest in the uppermost fractions and lowest in the lower fractions of density gradients was usually evident. Irrespective of the extraction medium used, infectivity distributions were always heterogeneous; and no indication of the presence of a homogeneous, infectious particle with sedimentation properties similar to those of known viral nucleoproteins could be obtained.

IV. Nature and Subcellular Location of PSTV *in Situ*

In view of the unusual sedimentation properties of infectious material from PSTV-infected tissue, the question arose whether, *in situ*, PSTV might be present as a conventional virion which, however, upon extraction with phosphate buffer, rapidly disintegrated. Attempts were, therefore, made to study the properties of PSTV *in situ* and to determine its location within the cell.

A. Infectivity Distribution in Expressed Sap

One approach consisted in simply expressing sap from PSTV-infected tissue and to directly analyze the distribution of infectious material by density-gradient centrifugation. As shown in Table III, the sedimentation properties of the infectious material in expressed sap were radically different from those in phosphate buffer extracts. After centrifugation for only 3 hours, all infectious material was present in the pellet and the lowermost fraction of the gradient tubes. Other experiments revealed that the infectious material present in expressed sap could be completely pelleted by low-speed centrifugation (15 minutes at 5000 g).

B. Subcellular Location of PSTV

Since in expressed sap, PSTV sedimented at a very rapid rate, it had to be associated either with one of the larger cell organelles, such as chloroplasts or nuclei, or it was adsorbed to cellular debris. To decide between these possibilities, two methods were used to separate PSTV-infected tissue into various subcellular fractions. The method of Bandurski and Maheshwari (1962) was used to prepare nuclei and chloroplasts from PSTV-infected leaves. Bioassay of the various fractions showed that most infectivity remained with the tissue debris and with the pellets resulting from centrifugation of the tissue extract for 1 minute at 94 g (tissue debris) or 15 minutes at 94 g (nuclei and some chloroplasts). The chloroplast fraction (pellet after centrifugation for 30 minutes at 470 g) contained little infectivity. The soluble fraction (supernatant solution after centrifugation for 30 minutes at 470 g) was almost devoid of infectivity.

With the other method used (Kühl, 1964), infectivity was more efficiently released from the tissue debris; although with this method also, appreciable infectivity was present in the tissue debris (Table V). Remaining infectivity was almost exclusively associated with the nuclear fraction. The supernatant solution, after centrifugation for 10 minutes at 350 g (soluble, ribosomal, and mitochondrial fractions), was devoid of infectivity. Chloroplasts and nuclei were separated by the density-gradient centrifugation method of Tewari and Wildman (1966). The resulting green bands of chloroplasts were essentially noninfectious (Table V); whereas the resuspended nuclear pellet was highly infectious. Examination of the resuspended nuclear pellets by phase-contrast microscopy revealed the presence of many normal appearing nuclei, as well as of a small number of chlorophyll-bearing particles (ratio chloroplasts to nuclei was ca. 1:20) and of abundant amounts of crystalline material. Treatment of the nuclear fraction with Triton X-100 (Spencer and Wildman, 1964), followed by sedimentation and resuspension of the nuclei, resulted in removal of the remaining chloroplasts, yet had little effect on the infectivity of the nuclear fraction (Table V).

TABLE V

DISTRIBUTION OF PSTV INFECTIVITY IN SUBCELLULAR FRACTIONS[a]

Fraction	Infectivity index[b]	
	Experiment 1	Experiment 2
Tissue debris (pulp in cloth)	71	45
Supernatant after 10 minutes at 350 g (soluble, ribosomal, and mitochondrial fractions)	0	0
Crude nuclear fraction (nuclei and chloroplasts)	103	77
Chloroplasts[c]	0	6
Purified nuclear fraction[c]	106	60
Nuclear fraction after treatment with Triton X-100[d]	90	NT

[a] Prepared by the method of Kühl (1964).

[b] Preparations assayed undiluted and diluted 10^{-1}, 10^{-2}, 10^{-3}, and 10^{-4}. NT, Not Tested.

[c] Prepared by density-gradient centrifugation (Tewari and Wildman, 1966).

[d] Purified nuclei treated with 0.1% Triton X-100 for 1 hour at 0°C, followed by pelleting of the nuclei by centrifugation and resuspension in Kühl's medium.

Kühl (1964) determined that only ca. one-half of the nuclei present in a leaf were extracted with his procedure, and that the rest of the nuclei remained with the tissue debris. Thus, it is possible that the relatively large amount of infectivity in the tissue debris was associated with nonliberated nuclei.

The question arose whether the infectious material was located within the nuclei or whether it became adsorbed to the nuclei during extraction. The following observations tend to strengthen the former possibility. When crude nuclei were diluted for bioassay in the medium devised by Kühl (1964) to maintain integrity of nuclei, little or no infectivity could be demonstrated; whereas when the same preparations were diluted for bioassay with 0.02 M phosphate buffer, pH 7, infectivity increased conspicuously (Table VI).

The low level of infectivity in Kühl's medium was not entirely due to an inhibitory effect of this medium on the infectivity of PSTV. This was shown by the following experiments. Partially purified PSTV nucleic acid (see below) was suspended in Kühl's medium and diluted for bioassay in the same medium. Another equal portion of PSTV nucleic acid was suspended in, and diluted for bioassay with, phosphate buffer. PSTV nucleic acid

TABLE VI

INFECTIVITY OF NUCLEAR FRACTIONS IN KÜHL'S MEDIUM[a] AND EXTRACTION OF NUCLEI
WITH PHOSPHATE BUFFER

Fraction	Infectivity index[b]
Crude nuclei	
Diluted in Kühl's medium	8
Diluted in 0.02 M phosphate buffer, pH 7	71
PSTV RNA added to tissue homogenate in Kühl's medium and diluted with Kühl's medium	92
PSTV RNA in 0.02 M phosphate, pH 7, and diluted with same buffer	110
Purified nuclei[c] diluted with Kühl's medium	10
Extracted with 0.05 M K_2HPO_4[d]	
1st extract	62
2nd extract	35
3rd extract	24
4th extract	19

[a] Kühl (1964).

[b] Preparations assayed undiluted and diluted 10^{-1}, 10^{-2}, 10^{-3}, and 10^{-4}.

[c] Nuclei prepared by density-gradient centrifugation (Tewari and Wildman, 1966).

[d] Nuclei pelleted and extracted for 30 minutes at 4°C with 0.05 M K_2HPO_4, followed by pelleting and bioassay of supernatant.

suspended in Kühl's medium was nearly as infectious as PSTV nucleic acid suspended in phosphate buffer. Similar results were obtained when PSTV nucleic acid was added to crude nuclear preparations in Kühl's medium. Here again, infectivity of the preparations suspended and diluted in Kühl's medium and those in phosphate buffer were nearly equal (Table VI).

Further experiments showed that infectious material could be extracted from nuclei by treatment with phosphate buffer. For this purpose, purified nuclei (that expressed little or no infectivity when diluted for bioassay in Kühl's medium) were pelleted by low-speed centrifugation and were then suspended in 0.05 M K_2HPO_4. After 30 minutes storage at 4°C (with shaking), the nuclei were pelleted and the infectivity level of the supernatant solution was determined.

The pellets were then again suspended in 0.05 M K_2HPO_4, and the process was repeated. As shown in Table VI, the extracts so derived were highly infectious, and consecutive extractions of the same nuclei yielded lower and lower levels of infectivity.

Density-gradient centrifugation of the pooled extracts disclosed infectivity distributions indistinguishable from those obtained with crude tissue extracts prepared with phosphate buffer.

These results indicate that PSTV is probably contained within the nuclei of infected cells,* and that infectious material can be released therefrom by the action of phosphate buffer.

C. *In Situ* SENSITIVITY OF PSTV TO NUCLEASES

Although in extracts prepared from PSTV-infected tissue, no evidence could be found for the presence of a conventional viral nucleoprotein, such an entity might exist *in situ*, presumably in nuclei of infected cells. Evidently, in the light of our experiments, such a structure would have to be unusually labile. Nevertheless, efforts were made to demonstrate an infectious agent *in situ* that might be less susceptible to inactivation by nucleases than was the extractable infectious agent (see below). For this purpose, the technique of vacuum infiltration of infected leaves with solutions containing nucleases was chosen. The protocol for these experiments is detailed in Table VII.

To determine whether conventional virions would be affected by this procedure, control experiments were conducted with tomato leaves infected with tobacco mosaic virus (TMV). Table VIII shows that, as expected, vacuum infiltration of infected leaves with ribonuclease (RNase) had no measurable effect on the infectivity of TMV subsequently extracted from such leaves.

Similar experiments made with PSTV-infected tomato leaves yielded the results shown in Table IX.

Incubation of leaves infiltrated with RNase for 1 hour at 25°C had no effect on the infectivity of subsequently prepared extracts. Incubation of RNase-infiltrated leaves for 22–25 hour, however, led to a drastic reduction of infectivity. This reduction of infectivity, when based on dilution end point (and number of plants infected), was estimated to correspond to an inactivation of 95–100% of the originally present units of infection.

Infiltration of leaves with deoxyribonuclease (DNase) had little effect on the infectivity of subsequently prepared extracts (Table IX.)

---

* It is interesting to note that comparison of sections prepared from healthy and PSTV-infected leaves and petioles revealed a marked hypertrophy of nuclei in the infected tissue. Also, protoplasmic streaming was found to be consistently more active in cells from infected than in those from healthy tissue (J. F. Worley, personal communication).

TABLE VII

PROTOCOL FOR NUCLEASE INFILTRATION EXPERIMENTS

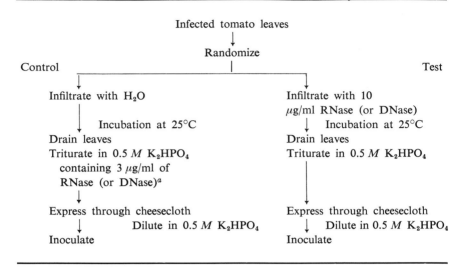

[a] Amount of nuclease added based on assumption of equilibration of nuclease within leaves of test sample.

TABLE VIII

EFFECT OF INFILTRATION[a] OF TMV-INFECTED TOMATO LEAVES WITH RIBONUCLEASE ON INFECTIVITY

Sample	Infectivity[b] at dilution				
	10^{-2}	10^{-3}	10^{-4}	10^{-5}	10^{-6}
RNase-infiltrated	154	89	49	8.6	1.3
Control, RNase added	166	118	40	5.6	0.5
Control, no RNase	205	147	62	10.5	1.4
Control, RNase added	205	145	44	4.9	0.4

[a] See Table VII for procedure. Incubation for 24 hours at 25°C.

[b] Average number of lesions per half-leaf on eight primary Pinto bean leaves. Figures in lines 1 and 2 and figures in lines 3 and 4 are each values from opposite half-leaves.

TABLE IX

EFFECT OF INFILTRATION[a] OF PSTV-INFECTED TOMATO LEAVES WITH RIBONUCLEASE OR DEOXYRIBONUCLEASE ON INFECTIVITY

Experiment number	Treatment	Time of incubation (hours at 25°C)	Infectivity index[b]
1	RNase-infiltrated	1	62
	Water-infiltrated[c]	1	68
2	RNase-infiltrated	22	0
	Water-infiltrated[c]	22	98
3	RNase-infiltrated[d]	25	24
	Water-infiltrated[d]	25	87
4	RNase-infiltrated	24	4
	Water-infiltrated	24	55
5	DNase-infiltrated	22	63
	Water-infiltrated	22	76

[a] See Table VII for procedure.
[b] Preparations assayed undiluted and at dilutions of 10^{-1}, 10^{-2}, 10^{-3}, and 10^{-4}.
[c] No RNase added to control.
[d] Bentonite added to grinding buffer.

V. Properties of PSTV Nucleic Acid

Since the sedimentation properties of some of the infectious material extractable from PSTV-infected tissue were compatible with those of free nucleic acid, treatments customarily used for nucleic acid extraction and purification were applied to PSTV extracts and PSTV-infected tissue.

A. EFFECT OF TREATMENT WITH PHENOL

PSTV-infected tissue was homogenized in 0.5 M K_2HPO_4 alone or together with chloroform and butanol. Portions of the clarified extracts were then treated three times in succession with equal volumes of phenol. Residual phenol was removed from the aqueous phases by repeated ether extraction. After appropriate dilution, the preparations were bioassayed. In other experiments, tissue was extracted directly with a mixture of 0.5 M K_2HPO_4 and phenol. After centrifugation and removal of residual phenol, the

TABLE X

EFFECT OF PHENOL TREATMENT ON INFECTIVITY OF PSTV

Method of extraction	Fraction assayed	Infectivity[a]							
		Extract, diluted				Phenol-treated extract, diluted			
		10^{-1}	10^{-2}	10^{-3}	10^{-4}	10^{-1}	10^{-2}	10^{-3}	10^{-4}
0.5 M K$_2$HPO$_4$	Clarified extract	NT	3/3	1/3	0/3	NT	3/3	1/3	0/3
	High-speed supernatant	NT	3/3	2/3	0/3	NT	3/3	3/3	1/3
	High-speed pellet	NT	0/3	0/3	0/3	NT	0/3	0/3	0/3
0.5 M K$_2$HPO$_4$ + CHCl$_3$ + BuOH	Clarified extract	NT	3/3	2/3	0/3	NT	3/3	1/3	0/3
	High-speed supernatant	NT	3/3	2/3	0/3	NT	3/3	2/3	0/3
	High-speed pellet	NT	3/3	0/3	0/3	NT	1/3	0/3	0/3
0.5 M K$_2$HPO$_4$	Aqueous phase[b]	NT	NT	NT	NT	3/3	3/3	1/3	0/3
0.5 M K$_2$HPO$_4$ + CHCl$_3$ + BuOH	Clarified extract	3/3	3/3	2/3	0/3	3/3	2/3	1/2	0/3

[a] Number of plants with symptoms/number of plants inoculated and surviving. Inocula diluted in 0.005 M K$_2$HPO$_4$. NT, not tested.

[b] Tissue extracted directly with 0.5 M K$_2$HPO$_4$ and phenol.

aqueous phases were bioassayed. As shown in Table X, treatment of extracts with phenol did not lead to an appreciable change in the infectivity of the preparations.

Infectious material could readily be concentrated by ethanol precipitation and resuspension in a smaller volume of buffer. Infectious material could be concentrated either from clarified extracts, from high-speed supernatants, or from phenol-treated extracts. The most highly concentrated preparations were obtained by ethanol precipitation of phenol-treated extracts. Many concentrates prepared in this manner had dilution end points of 10^{-6} or 10^{-7}.

B. SEDIMENTATION PROPERTIES

1. Rate-Zonal Density-Gradient Centrifugation

To investigate the effects of treatment with phenol on the sedimentation properties of PSTV, extracts were subjected to density-gradient centrifugation before and after treatment with phenol. As the sedimentation behavior of single-stranded nucleic acids is known to depend on the ionic strength of the suspending medium (Bishop, 1966; Billeter et al., 1966), analyses were performed in gradients with both low and high ionic strength. An extract from PSTV-infected tissue (0.5 M K_2HPO_4–chloroform–butanol) was subjected to high-speed centrifugation. The supernatant solution was then divided into two portions. One portion was dialyzed vs. 0.003 M K_2HPO_4 and was analyzed in a sucrose gradient containing 0.005 M K_2HPO_4 A sample of the other portion was directly analyzed in a sucrose gradient containing 0.5 M K_2HPO_4. Another sample was concentrated by ethanol precipitation, resuspended in 0.5 M K_2HPO_4, and analyzed in a gradient containing 0.5 M K_2HPO_4. A sample of the concentrated extract was treated with phenol and, after removal of residual phenol, the aqueous phase was analyzed in a gradient containing 0.5 M K_2HPO_4. After centrifugation, the four tubes were fractionated by collecting consecutive 2-ml fractions from each tube. All fractions were assayed for infectivity after dilution with 0.005 M K_2HPO_4. Table XI shows the results.

Although centrifugation of the supernatant solution in 0.5M K_2HPO_4 resulted in a somewhat wider distribution of infectivity within the gradient than in 0.005 M K_2HPO_4, most of the infectivity in either case was contained in the top four to five fractions of the gradient. In the tubes containing the concentrate, infectivity spread was much wider than in the tubes containing the supernatant solution, irrespective of whether analysis was performed at low or high ionic strength. Treatment of concentrates with

TABLE XI

EFFECT OF IONIC STRENGTH AND OF TREATMENT WITH PHENOL ON THE SEDIMENTATION PROPERTIES OF PSTV IN SUCROSE DENSITY GRADIENTS

Sample assayed[a]	Molar conc. of K_2HPO_4 in gradient	Infectivity[b] of consecutive 2-ml fractions from centrifuged gradient tubes[c]												P[d]
		1	2	3	4	5	6	7	8	9	10	11	12	
High-speed supernatant	0.005[e]	0/3	1/3	2/3	1/3	1/3	0/3	0/3	0/3	0/3	0/3	0/3	0/3	0/3
High-speed supernatant	0.5	3/3	3/3	3/3	2/3	2/3	1/3	1/3	1/3	0/3	0/3	0/3	0/3	1/3
Concentrate	0.005	3/3	3/3	3/3	2/3	2/3	2/3	1/3	1/3	1/3	1/3	1/3	0/3	0/3
Concentrate	0.5	3/3	3/3	3/3	3/3	3/3	0/3	3/3	3/3	3/3	1/3	2/3	1/3	3/3
Phenol-treated concentrate	0.005	0/3	1/3	3/3	3/3	3/3	2/3	1/3	3/3	0/3	1/3	1/3	1/3	0/3
	0.5	3/3	3/3	2/3	1/3	2/3	1/3	1/3	1/3	1/3	1/3	0/3	0/3	1/3

[a] PSTV-infected tissue was extracted with 0.5 M K_2HPO_4, chloroform, and butanol; the aqueous phase was used for the experiment.

[b] Number of plants with symptoms/number of plants inoculated.

[c] Fraction 1 is from top of gradient, fraction 12 from bottom of gradient. Fractions diluted 1/10 in 0.005 M K_2HPO_4.

[d] P, Pellet resuspended in 2 ml of 0.005 M K_2HPO_4.

[e] Sample dialyzed vs. 0.003 M K_2HPO_4 before analysis. Low infectivity was presumably due to losses incurred during dialysis.

phenol resulted in infectivity distributions similar to those found with concentrates that had not been treated with phenol, i.e., most of the infectivity was contained in the upper and middle fractions of the gradients.

Absorbance profiles of centrifuged sucrose gradients that contained clarified extracts (or high-speed supernatants) from healthy or PSTV-infected tissue did not disclose qualitative differences or well-defined components. Better absorbance profiles were achieved with ethanol concentrated preparations, and since such preparations had very high infectivity dilution end points (up to 10^{-7}), numerous correlations between location of infectivity within a gradient and the absorbance profile of the gradient were attempted.

As an example, Fig. 1 shows the absorbance profiles of ethanol concentrates prepared in identical fashion from equal amounts of healthy and PSTV-infected tissue and infectivity indexes of successive 2-ml fractions collected from the latter gradient. Evidently, no component with sedimentation properties identical with the infectious entity and present only in the extract from infected leaves could be found. Disregarding minor variations, the two absorbance profiles were identical.

Since PSTV had been shown to be insensitive to treatment with DNase, several extracts from PSTV-infected tissue were analyzed by density-gradient centrifugation after incubation with DNase. Figure 2 illustrates the results of one experiment. One portion of the extract, whose absorption profile is shown in Fig. 1(b), was incubated with DNase, another portion with RNase, and still another portion with both enzymes.

After incubation, the samples were concentrated by ethanol precipitation and were then analyzed. As shown in Fig. 2(a), the RNase-treated sample contained, aside from small material, only one component. The DNase-treated sample [Fig. 2(b)] contained small material, one major, and two minor components, whereas the sample treated with both enzymes contained only small material at the top of the gradient [Fig. 2(c)].

Infectivity indexes for consecutive 2-ml fractions taken from the tube containing the extract treated with DNase and from the tube with the nonincubated extract are shown in Fig. 3. It is evident that the major portion of the infectious particles sedimented at a slower rate after treatment with DNase than before. This observation was confirmed in a number of additional experiments.

Figure 4 shows the absorbance profile and infectivity indexes of a phenol-treated concentrate prepared from PSTV-infected tissue. Similar extracts from healthy leaves had essentially identical absorption profiles. Phenol-treated extracts had UV spectra typical of nucleic acids with maxima at 258 mμ, and maximum:minimum ratios of 2.2–2.4. Incubation of these

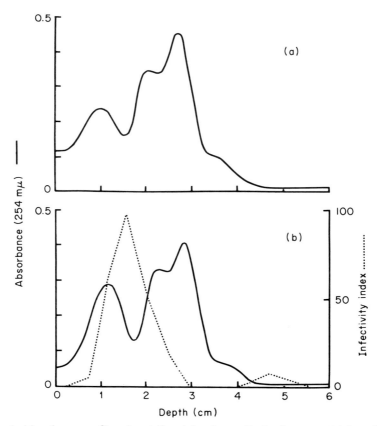

FIG. 1. Absorbance profiles of centrifuged density-gradient columns containing ethanol-concentrated extracts from healthy and PSTV-infected tissue, and infectivity distribution in the gradient containing the extract from infected tissue. (a) Extract from healthy tissue; 1 ml of a solution with 3.5 OD_{260} units/ml was layered onto the gradient. (b) Extract from infected tissue; 1 ml of a solution with 3.5 OD_{260} units/ml was layered onto the gradient. Solid line, absorbance profile; broken line, infectivity indexes of consecutive 2-ml fractions collected from gradient. Centrifugation for 16 hours at 24,000 rpm (SW 25.1 rotor, Spinco model L centrifuge), 0.2–0.8 M linear sucrose gradients in 0.005 M K_2HPO_4.

extracts with RNase or DNase, followed by density-gradient centrifugation, disclosed that peaks a, c, and d were RNA and that peak b was DNA (Fig. 4). Comparison of these profiles with those reported for nucleic acid preparations from plant tissues (Ralph and Bellamy, 1964) made it evident that peak a corresponded to transfer RNA (tRNA), and peaks c and d to the two species of ribosomal RNA (rRNA). The major portion of PSTV sedimented in sucrose gradients with approximately 10 S, based on rRNA's (17 and 27 S) used as markers.

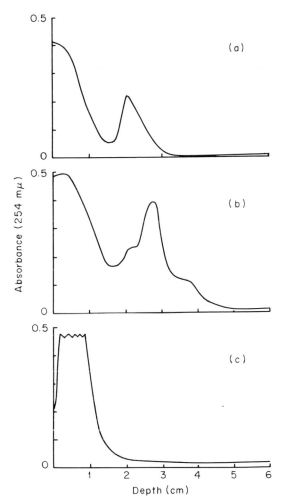

FIG. 2. Absorbance profiles of centrifuged density-gradient columns containing the extract from PSTV-infected tissue shown in Fig. 1(b) after treatment with nucleases for 1 hour at 25°C. (a) Extract treated with 1 μg/ml of RNase. (b) Extract treated with 1 μg/ml of DNase (incubation medium contained 10^{-2} M MgCl$_2$). (c) Extract treated with 1 μg/ml each of RNase and DNase (incubation medium contained 10^{-2} M MgCl$_2$). Centrifugation for 16 hour at 24,000 rpm (SW 25.1 rotor, Spinco model L centrifuge), 0.2–0.8 M linear sucrose gradients in 0.02 M phosphate buffer, pH 7.

2. Equilibrium Centrifugation

Preliminary experiments had shown that exposure of PSTV extracts to cesium chloride concentrations as high as 9 molal for 3 days at 4°C did not lead to an appreciable reduction in the infectivity of the preparations.

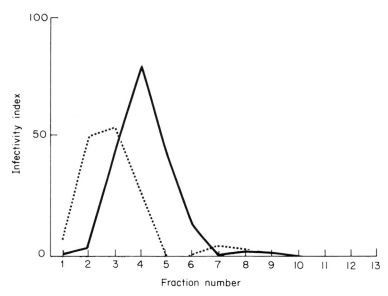

FIG. 3. Infectivity indexes of consecutive 2-ml fractions taken from gradients containing the extract illustrated in Fig. 1(b) before (solid line) and after (broken line) treatment with DNase. Fraction No. 1 is from top of gradient. All fractions were assayed at dilutions of 10^{-1}, 10^{-2}, 10^{-3}, and 10^{-4}. Centrifugation for 16 hours at 24,000 rpm (25.1 rotor, Spinco model L centrifuge), 0.2–0.8 M linear sucrose gradients in 0.02 M phosphate buffer, pH 7.

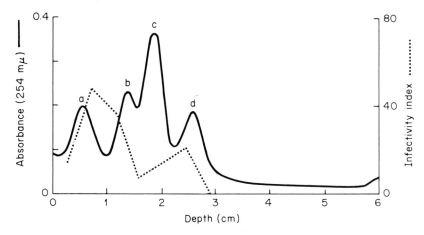

FIG. 4. Absorbance profile and infectivity indexes of a centrifuged density-gradient column containing a phenol-treated concentrate from PSTV-infected leaves. One milliliter of a solution with 2.5 OD units/ml was layered onto the gradient. a, tRNA; b, DNA; c and d, rRNA's. Centrifugation for 16 hours at 24,000 rpm (25.1 rotor, Spinco model L centrifuge), 0.2–0.8 M linear sucrose gradients in 0.02 M phosphate buffer, pH 7.

Concentrates from PSTV-infected tissue (not treated with phenol) were brought to a density of 1.70 gm/cm³ with cesium chloride or to a density of 1.617 gm/cm³ with cesium sulfate. After centrifugation, the tubes were fractionated and consecutive 0.5-ml fractions were collected, diluted, and bioassayed.

Figure 5 shows the results. The cesium chloride gradient contained, as expected, only one UV-absorbing peak, namely the DNA present in the

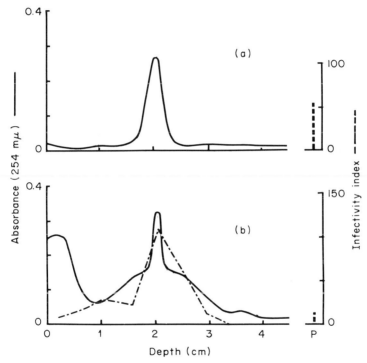

FIG. 5. Absorbance profiles and infectivity distributions in cesium chloride (a) and cesium sulfate (b) gradients that contained PSTV concentrates (not treated with phenol) and were centrifuged for 72 hours at 35,000 rpm (SW 39 rotor; Spinco model L-2 centrifuge; temperature 4°C). P, resuspended pellet.

sample, whereas all RNA had pelleted. All infectivity had similarly pelleted during centrifugation [Fig. 5(a)]. The cesium sulfate gradient contained one major UV-absorbing peak with pronounced shoulders on either side of this peak. Infectivity roughly coincided with the major peak, but a small amount of infectivity was also recovered from the pellet [Fig. 5(b)]. These observations are in accord with the view that the infectious material is RNA, not DNA.

C. Sensitivity to Nucleases

1. *Endonucleases*

Aliquots of a phenol-treated concentrate suspended in 0.005 M K_2HPO_4 (unadjusted, pH 8.2) were incubated in the presence of RNase or DNase. A control aliquot was incubated in the absence of nucleases. After incubation, the preparations were rapidly cooled to 0°C, diluted, and bioassayed. Other aliquots, to which equal amounts of RNase or DNase had been added, but which were not incubated at 25°C, were similarly diluted and bioassayed.

Table XII shows that no loss of infectivity occurred when the extract was

TABLE XII

Sensitivity of Phenol-Treated Concentrates of PSTV in 0.005 M K_2HPO_4 to Treatment with Nucleases

Enzymic treatment	Incubation at 25°C (hours)	Infectivity index[a]
None	0	150
None	1	183
1 μg/ml RNase	0	12
1 μg/ml RNase	1	0
1 μg/ml DNase[b]	0	156
1 μg/ml DNase[b]	1	171

[a] All preparations were assayed at dilutions of 10^{-1}, 10^{-2}, 10^{-3}, 10^{-4}, and 10^{-5}, made in 0.005 M K_2HPO_4.

[b] Incubation medium contained 10^{-2} M $MgCl_2$.

incubated in the absence of nucleases. Addition of RNase to the extract led to a drastic reduction in infectivity even without incubation. After incubation with RNase for 1 hour at 25°C, no infectivity remained. Treatment with DNase, on the other hand, had no effect on the infectivity of PSTV. Analyses of the untreated and DNase-treated extracts by density gradient centrifugation disclosed that the DNA peak present in the untreated extract had completely disappeared after incubation with DNase.

In another experiment, the question was asked whether quantitative differences existed in the RNase sensitivity of phenol-treated extracts as compared with those not treated with phenol. In this case, a high-speed supernatant was used, and one portion of it was treated with phenol. Both

TABLE XIII

SENSITIVITY TO RNASE OF CLARIFIED EXTRACTS AND OF PHENOL-TREATED EXTRACTS FROM PSTV-INFECTED TISSUES[a]

| Enzymic treatment | Incubation at 25°C (hours) | Infectivity[b] | | | | | | | |
| | | Extract, diluted | | | | Phenol-treated extract, diluted | | | |
		None	10^{-1}	10^{-2}	10^{-3}	None	10^{-1}	10^{-2}	10^{-3}
None	0	3/3	3/3	3/3	3/3	3/3	3/3	2/3	2/3
None	2	3/3	3/3	3/3	2/3	3/3	3/3	3/3	2/3
0.1 μg/ml RNase	0	3/3	3/3	3/3	0/3	3/3	3/3	3/3	2/3
0.1 μg/ml RNase	2	3/3	3/3	0/3	0/3	3/3	0/3	0/3	0/3
1 μg/ml RNase	0	3/3	0/3	0/3	0/3	3/3	0/3	0/3	0/3
1 μg/ml RNase	2	3/3	0/3	0/3	0/3	3/3	0/3	0/3	0/3

[a] PSTV-infected tissue was extracted with 0.5 M K_2HPO_4, chloroform, and butanol; the aqueous phase was centrifuged for 1 hour at 40,000 rpm. The resulting supernatant solution was used for the experiment.

[b] Number of plants with symptoms/number of plants inoculated. Dilutions were made with 0.005 M K_2HPO_4.

preparations were suspended in 0.5 M K$_2$HPO$_4$. Two levels of RNase concentration were used. After incubation, the preparations were diluted with 0.005 M K$_2$HPO$_4$ and infectivities of extracts not incubated or incubated for 2 hours at 25°C were compared. Table XIII shows that no appreciable difference was found in the sensitivity of the two extracts to RNase. It can be seen, however, that some infectivity survived all RNase treatments in the undiluted incubation mixtures.

The question thus arose whether PSTV was protected from RNase attack when suspended in media of high ionic strength. In one experiment, RNase-induced loss of infectivity in 0.15 M NaCl-0.015 M sodium citrate, pH 7, (SSC) was compared with that in 0.005 M K$_2$HPO$_4$ (Table XIV) and in

TABLE XIV

EFFECT OF IONIC STRENGTH ON RNASE SENSITIVITY OF PHENOL-TREATED CONCENTRATES OF PSTV[a]

Suspending medium	RNase concentration (μg/ml)	Infectivity index[b]	
		Not incubated	Incubated 1 hour at 25°C
0.005 M K$_2$HPO$_4$	0	NT	185
	0.01	187	16
	0.1	11	0
SSC[c]	0	NT	216
	0.01	193	121
	0.1	192	12

[a] PSTV-infected tissue was extracted with 0.5 M K$_2$HPO$_4$, chloroform, and butanol; the aqueous phase was separated, treated with phenol, and concentrated by ethanol precipitation. The pellets were resuspended in the respective buffer, and used for the experiment.

[b] All preparations were assayed at dilutions of 10^{-1}, 10^{-2}, 10^{-3}, 10^{-4}, and 10^{-5} made with 0.005 M K$_2$HPO$_4$. NT, not tested.

[c] SSC = 0.15 M NaCl–0.015 M sodium citrate, pH 7.

another experiment, RNase sensitivities in neutral phosphate buffers of low and high ionic strengths were determined (Table XV).

It can be seen that, in SSC, infectivity partially survived the RNase treatment and that a relatively large proportion of infectivity survived the RNase treatment, when the samples were incubated in 1 M phosphate buffer, pH 7.

TABLE XV

EFFECT OF IONIC STRENGTH OF NEUTRAL PHOSPHATE BUFFERS ON RNASE SENSITIVITY OF
PSTV CONCENTRATE[a]

RNase concentration (μg/ml)[b]	Diluent, phosphate buffer, pH 7 (M)	Infectivity index[c] of sample incubated in	
		0.005 M phosphate, pH 7	1.0 M phosphate, pH 7
0	0.005	78	69
0.1	0.005	0	69
0.5	0.005	0	53
1.0	0.005	0	22
0	1.0	NT	35
0.1	1.0	NT	55
0.5	1.0	NT	42
1.0	1.0	NT	33

[a] A concentrate of PSTV was dialyzed vs. 0.005 or 1.0 M phosphate buffer, pH 7, and used for the experiment.

[b] Incubation 1 hour at 25°C.

[c] All preparations were assayed undiluted and at dilutions of 10^{-1}, 10^{-2}, 10^{-3}, 10^{-4}, and 10^{-5}. NT, not tested.

It is known, however, that single-stranded RNA's are also more stable to treatment with RNase in media of high ionic strength (Billeter *et al.*, 1966). To investigate the effect of high ionic strength on the inactivation by RNase of a single-stranded viral RNA, experiments were made with southern bean mosaic virus RNA (Diener, 1965). SBMV RNA was precipitated with ethanol and dissolved in phenol-treated concentrates prepared from PSTV-infected tissue that had previously been dialyzed vs. 0.005 M or 1.0 M K_2HPO_4. SBMV RNA concentration was 0.1 mg/ml in each preparation. Thus, the total nucleic acid concentration was equal to or higher than that in solutions used for the investigation of RNase sensitivity of PSTV (Tables XII to XV), and equal or higher competition of nonviral RNA for the enzyme was present. RNase was then added to aliquots of both preparations. These samples, together with RNase-free controls, were incubated for 1 hour at 25°C. After dilution in the appropriate buffer, the preparations were assayed for infectivity.

As shown in Table XVI, incubation with RNase completely inactivated

TABLE XVI

INACTIVATION OF SBMV RNA BY RNASE IN EXTRACT FROM PSTV INFECTED TISSUE AT
LOW AND HIGH IONIC STRENGTH[a]

Molar concentration of K₂HPO₄ in incubation medium	Dilution of inoculum[b]	Infectivity[c] of SBMV RNA	
		Control[d]	RNase-treated[e]
0.005	10^{-1}	39	0
	10^{-2}	23	0
	10^{-3}	3	0
1.0	10^{-1}	14	0
	10^{-2}	6	0
	10^{-3}	2	0

[a] SBMV RNA was dissolved in a concentrate from PSTV-infected tissue, previously dialyzed vs. 0.005 or 1.0 M K₂HPO₄.

[b] Preparations in 0.005 M K₂HPO₄ were diluted with 0.005 M K₂HPO₄, those in 1.0 M K₂HPO₄ were diluted with 1.0 M K₂HPO₄.

[c] Average number of lesions per half-leaf on 10–16 primary Pinto bean leaves.

[d] Controls were treated identically but received no RNase.

[e] RNase concentration, 1 μg/ml; incubation 1 hour at 25°C.

SBMV RNA under the conditions used, irrespective of whether the RNA was dissolved in a medium of low or high ionic strength. Thus, the resistance to RNase action of PSTV suspended in 1 M phosphate buffer could not be explained solely by inhibition of RNase activity or by increased resistance of single-stranded RNA under conditions of high ionic strength.

2. Exonucleases

To determine whether PSTV was sensitive to inactivation by exonucleases, phenol-treated preparations of PSTV RNA were incubated with snake venom phosphodiesterase. As an internal control, TMV RNA was added to the preparations. Since tomato is a host for both TMV and PSTV, and since symptoms are readily distinguishable, preparations could be assayed for TMV and PSTV on one set of plants. TMV symptoms appeared within 1 week after inoculation, those of PSTV after 2–3 weeks. Presence or absence of TMV was verified by transfer to Pinto bean plants, a local lesion host of TMV. In all cases, extracts from tomato plants without TMV symptoms incited no lesions on Pinto, and extracts from tomato plants with TMV symptoms incited abundant lesions on Pinto.

TABLE XVII

EFFECT OF INCUBATION WITH SNAKE VENOM PHOSPHODIESTERASE AND ALKALINE PHOSPHATASE ON THE INFECTIVITY OF PSTV RNA AND TMV RNA

Experiment number	Treatment[a]	Infectivity index[b]	
		PSTV RNA	TMV RNA
1	Control, no incubation	136	66
	Control, 1 hour at 25°C	164	69
	Phosphodiesterase added after incubation	156	60
	One hour at 25°C with phosphodiesterase	165	0
2	Control, 1 hour at 25°C	138	240
	One hour at 25°C with alkaline phosphatase	144	186
	One hour at 25°C with phosphodiesterase	185	0
	One hour at 25°C with alkaline phosphatase and phosphodiesterase	185	0

[a] Phenol-treated preparations of PSTV RNA were mixed with TMV RNA and dialyzed against 0.02 M glycine–NaOH, 3 mM MgCl$_2$, pH 9.0. Final concentration of snake venom phosphodiesterase (Worthington, *Crotalus adamanteus*), 166 μg/ml; of alkaline phosphatase (Worthington, *E. coli*, code BAPF), 33 μg/ml.

[b] Assayed at dilutions of 10^{-1}, 10^{-2}, 10^{-3}, and 10^{-4}. In experiment 1, TMV RNA and PSTV RNA were mixed and were assayed on one set of tomato plants; in experiment 2, PSTV RNA and TMV RNA were treated and assayed separately.

As shown in Table XVII (experiment No. 1), incubation with snake venom phosphodiesterase completely inactivated TMV RNA, yet had no effect on the infectivity of PSTV RNA. These results indicate that PSTV RNA is either circular or is "masked" at the 3'-terminus in such a fashion that the enzyme cannot attack the terminal nucleotide. It is well known that nucleic acids phosphorylated at the 3'-terminus (the 5'-linked end) are resistant to snake venom phosphodiesterase (Singer and Fraenkel-Conrat, 1963). Consequently, in further experiments, PSTV RNA was incubated with a mixture of snake venom phosphodiesterase and alkaline phosphatase (Table XVII, experiment No. 2). Enzymic activity of the alkaline phosphatase used was verified by colorimetric assay of the enzyme. Evidently, PSTV RNA resisted attack by a combination of exonuclease and alkaline phosphatase. Thus, resistance of PSTV RNA to snake venom phosphodiesterase could not be explained by a phosphorylated 3'-terminus.

It was of interest to determine whether PSTV RNA was susceptible to inactivation by exonucleases that attack the nucleic acid from the 5'-terminus. As shown in Table XVIII, incubation of PSTV RNA with bovine spleen phosphodiesterase alone, or in combination with alkaline phosphatase had no effect on the infectivity of PSTV RNA.

TABLE XVIII

Effect of Incubation with Bovine Spleen Phosphodiesterase on the Infectivity of PSTV RNA

Treatment[a]	Infectivity index[b]
Control, 1 hour at 25°C	124
Phosphodiesterase added after incubation	99
One hour at 25°C with phosphodiesterase	84
One hour at 25°C with phosphodiesterase and alkaline phosphatase	94

[a] Bovine spleen phosphodiesterase (Schwarz) and alkaline phosphatase (Worthington, E. coli, code BAPF) were added to portions of a phenol-treated preparation of PSTV RNA in 0.02 M phosphate buffer, pH 8.0. Final concentration of bovine spleen phosphodiesterase, 0.33 units/ml; of alkaline phosphatase, 166 μg/ml.

[b] Assayed at dilutions of 10^{-1}, 10^{-2}, 10^{-3}, and 10^{-4}.

D. Chromatographic Properties of PSTV RNA

Since PSTV RNA could evidently not be separated from host RNA by rate-zonal density-gradient centrifugation, several chromatographic procedures were evaluated.

1. *Sephadex Gel Filtration*

Phenol-treated preparations of PSTV RNA were dialyzed against 1 M NaCl–0.01 M tris-HCl, pH 7.8 buffer, and were loaded onto Sephadex G-100 columns (Watson and Ralph, 1967). Nucleic acids were eluted with the same buffer. Infectivity assays of the collected fractions showed that under these conditions PSTV RNA was excluded from Sephadex and could readily be separated from tRNA, 5 S RNA, and DNA fragments (with DNase-treated preparations).

2. *Chromatography on Methylated Serum Albumin (MAK)*

MAK columns were prepared as described by Mandell and Hershey (1960). One to two milligrams of nucleic acids dissolved in 0.4 M saline were applied to the columns. Nucleic acids were eluted at room temperature with linear saline gradients. The eluent was continuously monitored at 254 nm, and fractions were collected in an automatic fraction collector. The nucleic acids in pooled fractions were precipitated with ethanol; the pellets were resuspended in phosphate buffer and dialyzed vs. 0.02 M phosphate buffer, pH 7.

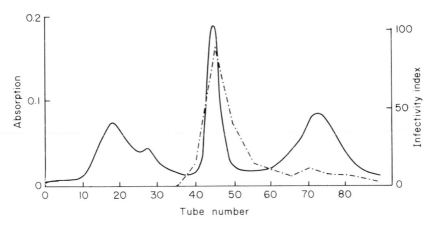

Fig. 6. Elution pattern from a MAK column of a phenol-treated PSTV concentrate. Solid line, absorbance profile; broken line, infectivity distribution. One milligram of nucleic acid (in 0.6 ml of 0.4 M saline-phosphate buffer) was applied to the column. Flow rate, 0.8 ml/minute; linear gradient 0.4–1.2 M saline-phosphate.

A typical experiment is detailed in Fig. 6. No unusual UV-absorbing components were found, and when identically prepared extracts from healthy leaves were fractionated on MAK columns, elution patterns were found to be similar to those of extracts from PSTV-infected tissue.

Bioassays of consecutive fractions eluted from MAK columns showed that infectivity eluted primarily at a saline-phosphate concentration of 0.7 M in those fractions that contained the host DNA or immediately followed DNA (Fig. 6).

3. *Chromatography on Cellulose*

The sensitivity of PSTV RNA to ribonuclease treatment in media of low ionic strength and its partial resistance in media of high ionic strength is in accordance with the hypothesis that PSTV RNA might be double-stranded. Thus, the usefulness of methods developed for the separation of double-stranded from single-stranded RNA was investigated.

Cellulose columns were prepared as described by Franklin (1966). Two to ten milligrams of nucleic acid in STE buffer (0.1 M NaCl, 0.001 M EDTA, and 0.05 M tris, pH 6.9) to which ethanol had been added to a final concentration of 35% (v/v), were applied to the columns. Elution was stepwise, first with STE-35% ethanol, then with STE-15% ethanol, and finally with STE alone. Fractions were collected in a fraction collector and analyzed at 260 nm.

As expected, most of the nucleic acid eluted with the STE-35% ethanol and STE-15% ethanol buffers; only 1–2% of the nucleic acid eluted in STE

TABLE XIX

Nucleic Acid Content and Infectivity of Fractions Eluted from Cellulose Columns Loaded with PSTV Extracts[a]

Eluting buffer	PSTV extract		DNase-treated PSTV extract[b]	
	NA content (mg)	Infectivity index[c]	NA content (mg)	Infectivity index[c]
STE-35% ethanol	2.25	37	1.76	0
STE-15% ethanol	8.01	57	8.92	50
STE	0.15	22	0.18	47

[a] Column preparation and elution schedule according to Franklin (1966). STE, NaCl-tris-EDTA buffer.

[b] Ten μg/ml of DNase, 10^{-2} M MgCl$_2$, incubation for 1 hour at 25°C.

[c] All preparations were assayed at dilutions of 10^{-1}, 10^{-2}, 10^{-3}, 10^{-4}, and 10^{-5} made in 0.005 M K$_2$HPO$_4$.

alone (Table XIX). Elution patterns of extracts from healthy leaves were similar, except that somewhat less nucleic acid eluted in STE alone. Since only double-stranded RNA was reported to be eluted in STE alone (Franklin, 1966),* thermal denaturation properties of these fractions were determined. However, denaturation occurred gradually over a wide range of

* Although the author showed a small amount of nucleic acid from uninfected *Escherichia coli* eluting in STE alone (Fig. 1a, Franklin, 1966).

temperatures, when either nucleic acid fractions eluted in STE from healthy or from PSTV-infected leaves were heated. Thus, no evidence for the presence of double-stranded RNA could be found.

Pooled fractions were assayed for infectivity. As shown in Table XIX, infectivity was found in both STE-ethanol fractions as well as in STE alone. Thus, no satisfactory separation of infectivity from host nucleic acid was achieved.

DNA does not elute as well as RNA from cellulose columns (Franklin,

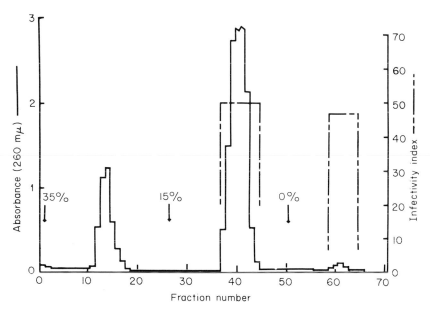

FIG. 7. Absorbance profile and infectivity distribution in eluate from a cellulose column of a phenol-and DNase-treated PSTV concentrate. Two milligrams of nucleic acid (in 0.92 ml of STE-35% ethanol) was applied to the column. Arrows indicate change of eluting buffer. 35%, STE-35% ethanol buffer; 15%, STE-15% ethanol buffer; 0%, STE buffer.

1966). Thus, it was conceivable that the unsatisfactory fractionation of PSTV infectivity could have been caused by association of the infectious entity with DNA; but treatment of PSTV extracts with DNase prior to cellulose column chromatography affected the elution pattern only slightly. Infectivity distribution, on the other hand, was altered in that no infectivity was now associated with the STE-35% ethanol fraction, and that a larger proportion was associated with the STE fraction (Fig. 7 and Table XIX). Thermal denaturation of nucleic acid in the STE fraction was again typical of that of single-stranded RNA and analysis of this nucleic acid in sucrose density gradients revealed only low molecular weight RNA. Infectivity did not correlate with the bulk of the nucleic acid present in the STE fraction.

4. *Hydroxyapatite Chromatography*

Chromatography on hydroxyapatite (Bernardi, 1965) has been shown to be a powerful tool for the fractionation of nucleic acids, and particularly for the separation of single-stranded from double-stranded RNA (Bockstahler, 1967).

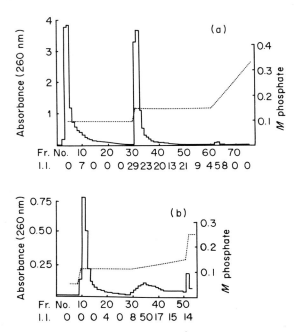

FIG. 8. Absorbance profiles and infectivity distributions of eluates from hydroxyapatite columns. (a) Elution profile of a phenol- and DNase-treated PSTV preparation. (b) Elution profile of the pooled, reconcentrated, and dialyzed 0.15 M phosphate buffer eluate of (a). Solid lines, absorbance profiles; broken lines, molarity of phosphate buffer; FR. NO., fraction number; I.I., infectivity indexes of individual fractions.

Hydroxyapatite (Bio-Gel HT, Calbiochem) columns (1 × 10 cm) were equilibrated with 0.01 M phosphate buffer (primary sodium and secondary potassium phosphates), pH 6.7. The columns were loaded with phenol-treated and DNase-digested preparations of PSTV RNA suspended in 0.01 M phosphate buffer, pH 6.7. Elution was carried out with phosphate buffers of increasing molarity.

As expected, most of the nucleic acid eluted in 0.10 and 0.15 M phosphate buffers; only traces of nucleic acid eluted at higher molarities [Fig. 8(a)]. Only traces of infectious material eluted in 0.10 M buffer; most of the infectious material eluted in 0.15 M buffer. Elution of infectious material was, however, not complete at 0.15 M; small amounts of it continued to elute upon prolonged washing of the column with 0.15 M buffer and with buffers of higher molarity. Hydroxyapatite chromatography of nucleic acid extracts, prepared in identical fashion from healthy plants, gave identical elution patterns. Obviously, the 0.15 M fraction, although containing the bulk of infectivity, was still heavily contaminated with host nucleic acid. To determine whether this host nucleic acid could be separated from the infectious nucleic acid by a different elution protocol, the fraction that had eluted in 0.15 M buffer was reconcentrated, dialyzed against 0.01 M phosphate buffer, and loaded onto a new column of hydroxyapatite. This time, elution was commenced with 0.11 M buffer, followed by a linear gradient of buffer from 0.11 to 0.15 M. As shown in Fig. 8(b), much of the nucleic acid eluted in 0.11 M buffer; yet only traces of infectivity were associated with this fraction. The bulk of infectious material eluted during gradient elution. The highest level of infectivity, as well as the highest nucleic acid concentration, was contained in a fraction that eluted in 0.125 M buffer; and more infectious material could be recovered by elution with higher molarity phosphate buffer [Fig. 8(b)].

Parallel chromatography with extracts from healthy plants resulted in identical elution profiles. Thus, most of the nucleic acid that eluted in the gradient from 0.11–0.15 M buffer was host nucleic acid. This conclusion was confirmed by analysis of the 0.11–0.15 M gradient fraction by rate-zonal density-gradient centrifugation. Most of the nucleic acid remained at the top of the gradient; whereas the infectious material sedimented to its usual position in the gradient (ca. 10 S). No optical density peak was evident at this position.

VI. Attempted Purification of PSTV RNA

In the course of these studies, it became increasingly evident that, in spite of the high infectivity dilution end points, PSTV RNA was only a minor constituent of purified nucleic acid preparations from infected tissue and that, for the isolation of microgram quantities of PSTV RNA, kilograms of infected leaves would have to be processed.

A. PURIFICATION BY CHROMATOGRAPHY ON MAK COLUMNS

In one experiment, 3 kg of PSTV-infected tomato leaves were extracted in batches of 100 gm. After ethanol precipitation and treatment with phenol, each batch was chromatographed on MAK columns (60 runs). The peaks that contained host DNA were collected from each column, pooled, and then incubated with DNase. DNA fragments were separated by a second cycle of MAK chromatography (20 runs). Infectivity eluted, as before, at a saline phosphate concentration of 0.7 M; however, no optical density peaks were discernible in these fractions. The infectious fractions were again pooled, reduced in volume, dialyzed, and subjected to final chromatography on a single MAK column. A small optical density peak (0.06 OD_{260}) that coincided with infectivity eluted at 0.7 M saline-phosphate. Since the infectivity dilution end point of this fraction was 10^{-4}, 2.4×10^{-4} $\mu g/ml$ of the RNA were sufficient to infect tomato plants. Analysis by density gradient and analytical centrifugation disclosed, however, that the RNA was heterogeneous and sedimented much slower than the infectious material. No optical density peak was discernible at the position in the gradient where most of the infectious material was located. Since area comparisons showed that only about 1% of the RNA present in the preparation sedimented into the region where infectivity was demonstrable, it follows that PSTV RNA at a concentration of not more than 2.4×10^{-6} $\mu g/ml$ was sufficient to infect tomato plants.

B. PURIFICATION BY CELLULOSE CHROMATOGRAPHY

In another experiment, 5 kg of PSTV-infected tomato leaves were extracted and the extracts processed as previously described. The resulting nucleic acid preparation was treated with DNase and portions of it were chromatographed on cellulose columns (Franklin, 1966). The fractions that eluted in ethanol-free buffer were pooled, the nucleic acid was concentrated by ethanol precipitation, and was analyzed by density gradient and analytical centrifugation. Results were similar to those achieved with MAK chromatography in that no correlation existed between the sedimentation properties of the small amount of heterogeneous RNA discernible by UV spectrophotometry and the infectious nucleic acid. Quantitative determinations indicated that not more than 5.10^{-6} $\mu g/ml$ of PSTV RNA was sufficient to infect tomato plants.

Thermal denaturation of the RNA that eluted in ethanol-free buffer was typical of that of single-stranded RNA.

C. PURIFICATION BY HYDROXYAPATITE CHROMATOGRAPHY

In further experiments, the feasibility of hydroxyapatite chromatography for the separation of PSTV RNA from host RNA was investigated. Kilogram amounts of PSTV-infected tissue were extracted as before in batches of 100 gm. After treatment with phenol and concentration by ethanol precipitation, polysaccharides were removed from the extracts by the method of Ralph and Bellamy (1964). The extracts were incubated with DNase; and low molecular weight RNA, as well as DNA fragments, were then removed by Sephadex gel filtration. RNA that was excluded from Sephadex was loaded onto hydroxyapatite columns. The elution protocol illustrated in Fig. 8(b) was followed. Most of the RNA eluted in 0.11 M phosphate buffer. Bioassays showed that, as before, this RNA was devoid of infectivity. Elution was then continued with a linear gradient from 0.11 to 0.15 M buffer. As expected, most infectivity eluted in this fraction. Infectious fractions were pooled from several columns, concentrated, and were again chromatographed on hydroxyapatite. After a further cycle of hydroxyapatite chromatography, the final preparation was bioassayed and analyzed by density-gradient and analytical centrifugation. With this procedure, coincidence between optical density and infectivity in sucrose gradients was better than that achieved earlier; but preparations from healthy plants that were processed identically contained RNA in similar quantities and with equal sedimentation properties as those from infected plants. Thus, again, most of the RNA in the final preparation was host RNA, not PSTV RNA. Thermal denaturation of this RNA was typical of that of single-stranded RNA.

D. PURIFICATION BY EXONUCLEASE TREATMENT AND HYDROXYAPATITE CHROMATOGRAPHY

Recently, nucleic acid extracts treated with exonuclease were subjected to hydroxyapatite chromatography. Extracts from healthy and PSTV-infected plants were processed as described above; except that, after treatment with DNase, the extracts were dialyzed against 0.02 M glycine–NaOH buffer, 3 mM Mg^{2+}, pH 9. They were then extensively incubated with a combination of snake venom phosphodiesterase and alkaline phosphatase. Bioassays showed that this treatment reduced the infectivity level of the extracts only slightly.

RNA and DNA fragments were removed by Sephadex gel filtration. RNA excluded from Sephadex was chromatographed on hydroxyapatite, and the fractions that eluted in 0.11–0.15 M phosphate buffer were concentrated, rechromatographed, and finally analyzed.

Density-gradient centrifugation revealed the presence of an apparently homogeneous RNA that sedimented at ca. 10 S, and that coincided with the distribution of infectivity in the gradients. Preliminary electron microscopy (Theodor Koller, personal communication) indicated that this RNA was probably circular. Exonuclease-resistant RNA could, however, be isolated from both healthy and PSTV-infected tissue.

Work is now in progress to elucidate the properties of exonuclease-resistant RNA and to determine its relation to PSTV RNA.

VII. Discussion and Conclusions

The spindle tuber disease of potato is characterized by symptoms and natural means of transmission that are basically similar to those of many other virus diseases of plants. Crude extracts from spindle tuber diseased tissue, however, are radically different from similar extracts prepared from tissues infected with other known plant viruses in that (a) the presence of virions cannot readily be demonstrated, and (b) such extracts contain slowly sedimenting, nuclease-sensitive, infectious material. These observations immediately pose two distinct problems: (a) what is the nature of PSTV *in situ*, and (b) what is the nature of the extractable, infectious material?

Our studies showed that, irrespective of ionic strength, phosphate buffer extracts from PSTV-infected tissue contained infectious material that sedimented at a rate of approximately 10 S (Fig. 1). This infectious material was shown to be sensitive to RNase, but not to DNase (Table XII); and for the following reasons, it is considered to be free RNA: (1) incubation with RNase destroyed its infectivity, (2) treatment with phenol did not alter its sedimentation properties, and (3) its slow rate of sedimentation appears incompatible with a virionlike particle. The RNA enclosed in such a structure would have to be so small that, in light of present coding concepts, it could not contain the necessary genetic information for its own independent replication as well as for the specification of a coat protein.

Although highest levels of infectivity were usually associated with the 10 S material, some infectious material always sedimented at faster rates, and the fastest sedimenting material had S values well above 200. Particularly, extracts prepared with low molarity phosphate buffers contained considerable amounts of infectious material that sedimented in the range of viral nucleoproteins (virions) (Table IV). Although distribution of this faster-sedimenting material was always most polydisperse, and differed from one extract to the next, it was nevertheless conceivable that this material was composed of viral nucleoprotein particles. If one assumes that the free infectious nucleic acid that was present in all extracts was derived from virions that had been degraded during extraction, the observed polydispersity of the infectious material could be explained by the presence of complete virions as well as of virions in various stages of degradation. In this case, one would expect the faster-sedimenting infectious material to be more stable during storage and more resistant to attack by RNase than the liberated nucleic acid. This was, however, not the case. Incubation of extracts with RNase led to inactivation of all infectivity of the extracts, including that associated with the fast sedimenting material (Tables XII to XV). Similarly, infectivity associated with fast-sedimenting material decayed upon storage at least as fast as, if not faster than, that associated with free RNA (Table IV). The contention that the faster-sedimenting infectious material is composed of viral nucleoprotein particles is further weakened by the finding that treatment of crude extracts with phenol did not materially change the sedimentation properties of the infectious material in the extracts (Table XI). Also, low molarity phosphate buffer extracts that were treated with phenol contained similar amounts of infectious material sedimenting in the range of viral nucleoproteins, as did untreated extracts (Diener, unpublished). Thus, if the faster sedimenting material were composed of complete and/or partially degraded virions, these virions would have unusual properties. On the one hand, their protein coats would have to be loose enough to allow access to RNase, yet, on the other hand, these structures would have to be resistant to treatment with phenol. No nucleoproteins with these properties are known; and it is, therefore, most unlikely that the faster sedimenting, infectious material is composed of virions.

One might, however, speculate that PSTV virions do exist *in situ*, but that they are so fragile that even extraction of tissue with low molarity phosphate buffer leads to their complete degradation with release of infectious nucleic acid. Although the absence of such labile structures cannot conclusively be demonstrated, infiltration of nucleases into infected leaves showed that, *in situ*, PSTV was sensitive to attack by RNase (Table IX). Furthermore, electron microscopy of thin sections prepared from PSTV-infected and healthy tissue failed to disclose the presence of virionlike particles in cells of infected plants (R. H. Lawson, personal communication) and systematic analyses of proteins in PSTV-infected plants gave no evidence for the production of viral coat proteins under conditions where coat protein of a defective strain of TMV could readily be demonstrated (M. Zaitlin and V. Hariharasubramanian, personal communication). Last, in spite of a number of attempts, no virus-specific antisera could be produced from PSTV-infected tissues (Singh and Bagnall, 1968; W. B. Raymer, personal communication).

In light of these observations, it appears unlikely that virions exist in PSTV-infected tissue. However, even if such structures should exist, our findings show that they must be extremely fragile and that they can have, at most, limited significance in the natural transmission of the disease agent.

We now turn our attention to the second problem. What is the nature of the extractable, infectious material? As shown, the slowest sedimenting infectious material almost certainly is free RNA, but what is the chemical composition of the faster sedimenting material?

It is difficult to believe that all of this infectious material is free RNA; yet, in light of the observation that treatment of infectious extracts with phenol had no significant effect on the sedimentation properties of the faster sedimenting infectious material, some alternative explanations are even less acceptable. Thus, attachment of PSTV RNA to host protein during extraction appears to be ruled out; and in view of the strong dissociating properties of phenol, attachment to other cell constituents also appears unlikely. Indeed, infectivity distributions were similar whether crude extracts, extracts from purified nuclei, or partially purified preparations of PSTV RNA were analyzed.

Some experimental results, however, indicate that PSTV RNA may be associated *in situ* with host DNA. Thus, when chromatin was prepared from PSTV-infected tissue by the method of Bekhor *et al.* (1969), much of the infectivity was found associated with the isolated chromatin (Diener, unpublished). Also, the observation of a reduced sedimentation rate of PSTV RNA after treatment with DNase (Fig. 3) could be caused by association of PSTV RNA with DNA. This explanation cannot, however, account for all of the polydispersity observed since, after DNase treatment, infectivity was still distributed in polydisperse fashion.

PSTV RNA, on the other hand, may have an unusual propensity to aggregate. Indeed, preliminary electron microscopy revealed the presence of numerous large aggregates of nucleic acid and of relatively few single molecules (Theodor Koller, personal communication). Aggregation occurring during preparation of the samples for electron microscopy can, however, not be ruled out at present.

PSTV RNA evidently is more stable in crude extracts than are known single-stranded viral RNA's. This is indicated by the fact that PSTV RNA can readily be extracted in infectious form from tissue without precautions against hydrolysis by RNase. This is in sharp contrast to the difficulties encountered in the isolation of infectious RNA from defective viruses, which is only successful if adequate precautions are taken to eliminate RNase activity.

Another unusual property of PSTV RNA is its high specific infectivity. Analysis of infectious fractions eluted from MAK columns (Fig. 6) and from cellulose columns (Table XIX and Fig. 7) indicated that PSTV RNA is infectious at very low concentrations (10^{-5} to 10^{-6} μg/ml).

It appears unlikely that the 10 S material could be single-stranded RNA similar to that found with most plant viruses. First, if the formula $M = 1100 \times S^{2,2}$ (Gierer, 1958), relating sedimentation constant (S) and molecular weight (M) of single-stranded RNA's is applied, a value of 1 to 2×10^5 daltons results. It is doubtful that a nucleic acid of this size could contain the genetic information required for independent replication. Second, the relative resistance of PSTV RNA to treatment with RNase in media of high ionic strength is contrary to known properties of single-stranded RNA's.

Sedimentation properties, elution from MAK columns, and nuclease sensitivity data are, however, compatible with the hypothesis that PSTV RNA is a nucleic acid of double-helical structure, at least one strand of which is composed of RNA. The partial loss of infectivity after incubation of PSTV RNA with RNase in media of high ionic strength is comparable to that reported for polio virus-induced double-stranded RNA (Bishop and Koch, 1967). The relative stability and high specific infectivity of PSTV RNA may be a consequence of the inherently greater stability of double-stranded, as compared with single-stranded RNA. The specific infectivity of double-stranded poliovirus RNA, for example, was reported to be 30-fold greater than that of single-stranded polio virus RNA (Bishop and Koch, 1967).

However, other properties of PSTV RNA do not appear to be compatible with a double-stranded structure. The elution pattern of PSTV RNA from cellulose columns (Fig. 7) is not typical of that of double-stranded RNA's (Franklin, 1966), and from hydroxyapatite (Fig. 8) PSTV RNA eluted at a much lower phosphate buffer concentration (0.125 M) than has been determined with double-stranded RNA's (0.18–0.20 M) (Bockstahler, 1967; Pinck et al., 1968). Also, infectious nucleic acid preparations had melting characteristics typical of those of single-stranded and not of double-stranded RNA's.

The resistance of PSTV RNA toward attack by exonucleases (Tables XVII and XVIII) indicates that the RNA is either a circular molecule or is "masked" at either terminus in such a manner that exonuclease cannot attack the terminal nucleotide. In the latter case, this "masking" cannot be due to phosphorylation of the terminal nucleotides, since incubation of the RNA with alkaline phosphatase did not lead to loss of resistance to exonucleases.

Evidently, elucidation of the structure of PSTV RNA requires its isolation in pure form. This isolation is difficult because PSTV RNA, even in the most highly purified preparations achieved so far, constituted only a small fraction of the total nucleic acid present; more discriminatory techniques of nucleic acid separation than the ones evaluated so far appear necessary for achieving this goal.

Thus, explanation of the unusual properties of PSTV RNA in terms of molecular structure is a subject for future investigation. The present studies appear to clearly demonstrate, however, that the potato spindle tuber disease is not transmitted by conventional virions, and that virions, if they exist at all, play at most a minor role in the replication of PSTV and in its dissemination. *In situ*, the infectious nucleic acid appears to be localized in the nuclei of infected cells. In extracts prepared under conditions that favor retention of nuclear integrity, the infectious nucleic acid remains associated with nuclei; but it is slowly released from nuclei by the action of phosphate buffers. Release of infectious nucleic acid is accelerated when nuclei are suspended in high molarity phosphate buffer; and, particularly, when organic solvents, such as chloroform or butanol, are used concurrently. However, even with the most efficient extraction medium found, considerable infectivity remains associated with nuclei (or with the tissue debris in crude extracts).

One wonders whether PSTV is the only representative of this type of viral pathogen. This appears unlikely and, indeed, one other virus disease of plants, exocortis disease of citrus, appears to be caused by a viral pathogen with properties similar to those of PSTV (Semancik and Weathers, 1968). In addition, some reported properties of a disease of sheep, scrapie (Alper *et al.*, 1967), bear certain resemblances with those of PSTV; and it is possible that other pathological conditions whose causes are obscure, may be incited by similar viral agents. The ease with which PSTV is transmitted mechanically may not be a general characteristic of this group of viral pathogens, and elucidation of the nature of the causal agents may thus be more difficult with other representatives.

Finally, one wonders how viral pathogens of this type originated. One might speculate that they represent defective forms of viruses whose nucleic acids were originally enclosed by protein coats and that these protein coats, by virtue of the great stability of the viral genome, did not impart sufficient additional protection to the genome to be retained during evolution.

Alternatively, PSTV-like pathogens may be aberrant forms of host RNA that have become capable of self-replication. This might occur, for example, if a messenger RNA that is normally transcribed on a DNA template and that codes for a DNA-dependent RNA polymerase undergoes mutation in such a fashion that the enzyme translated from the mutated messenger RNA now recognizes RNA instead of DNA as a template. The new enzyme would thus be an RNA-dependent RNA polymerase or replicase, and the mutated messenger RNA would be self-replicating. In this view, PSTV-like pathogens might be considered to be primitive viruses that have not yet acquired the genetic sophistication required to code for a self-assembling coat protein and to induce their host to synthesize this protein.

VIII. Summary

Phosphate buffer extracts from tissue infected with potato spindle tuber virus (PSTV) contain infectious RNA that sediments with a rate of ca. 10 S, as well as faster-sedimenting infectious material. Treatment of buffer extracts with phenol neither changes the sedimentation properties nor the relative infectivity of the infectious material. The infectious material is located within the nuclei of infected cells, and is liberated therefrom by the action of phosphate buffer. In extracts, as well as *in situ*, the infectious material is sensitive to treatment with ribonuclease. Although some infectious material sediments within the range of viral nucleoproteins, several properties of this material indicate that it is not composed of conventional virions. Virions, if they exist at all, must be extremely fragile structures and cannot play a significant role in the transmission of the disease.

The infectious 10 S material is almost certainly free RNA; but the chemical composition of the faster sedimenting infectious material is less certain. The latter, polydisperse material, may be aggregated PSTV RNA or, alternatively, PSTV RNA associated with various amounts of host DNA. PSTV RNA is remarkably stable in extracts, is readily transmitted mechanically, and has a very high specific infectivity. Its resistance to attack by exonucleases indicates that PSTV RNA may be circular.

ACKNOWLEDGEMENT

I wish to thank Dr. W. B. Raymer for bringing PSTV to my attention and for the many valuable contributions he made during the earlier stages of this investigation.

REFERENCES

Allington, W. B., Ball, E. M., and Galvez, G. E. (1964). *Plant Disease Reptr* **48**, 597–598.

Alper, T., Cramp, W. A., Haig, D. A., and Clarke, M. C. (1967). *Nature (London)* **214**, 764–766.

Bagnall, R. H. (1967). *Phytopathology* **57**, 533–534.

Ball, E. M., Allington, W. B., and Galvez, G. E. (1964). *Phytopathology* **54**, 887 (Abstr.).

Bandurski, R. S. and Maheshwari, S. C. (1962). *Plant Physiol.* **37**, 556–560.

Bekhor, I., Kung, G. M., and Bonner, T. (1969). *J. Mol. Biol.* **39**, 351–364.

Benson, A. P., Salama, F. M., and Singh, R. P. (1964). *Amer. Potato J.* **41**, 293 (Abstr.).

Bernardi, G. (1965). *Nature (London)* **206**, 779–783.

Billeter, M. A., Weissmann, C., and Warner, R. C. (1966). *J. Mol. Biol.* **17**, 145–173.

Bishop, D. H. L. (1966). *Biochem. J.* **100**, 321–329.

Bishop, J. M. and Koch, G. (1967). *J. Biol. Chem.* **242**, 1736–1743.

Bockstahler, L. E. (1967). *Mol. Gen. Genet.* **100**, 337–348.

Cadman, C. H. (1962). *Nature (London)* **193**, 49–52.

Diener, T. O. (1965). *Virology* **27**, 425–428.

Diener, T. O. and Raymer, W. B. (1967). *Science* **158**, 378–381.

Diener, T. O. and Raymer, W. B. (1969). *Virology* **37**, 351–366.

Franklin, R. M. (1966). *Proc. Nat. Acad. Sci. U. S.* **55**, 1504–1511.

Gierer, A. (1958). *Z. Naturforsch.* **13B**, 477–484.

Gilbert, A. H. (1923). *Vermont Agr. Ext. Serv. Circ.* **28**, 4 pp.

Goss, R. W. (1926). *Phytopathology* **16**, 299–303.

Hunter, J. E. (1964). Studies on potato spindle tuber virus. Ph. D. Dissertation, Univ. of New Hampshire, Durham, New Hampshire.

Kassanis, B. and Welkie, G. W. (1963). *Virology* **21**, 540–550.

Kühl, L. (1964). *Z. Naturforsch.* **19B**, 525–532.

McClean, A. P. D. (1931). *Union So. Africa Dept. Agr. Sci. Bull.* **100**, 36 pp.

Mandell, J. D. and Hershey, A. D. (1960). *Anal. Biochem.* **1**, 66–77.

Pinck, L., Hirth, L., and Bernardi, G. (1968). *Biochem. Biophys. Res. Commun.* **31**, 481–487.

Ralph, R. K. and Bellamy, A. R. (1964). *Biochim. Biophys. Acta* **87**, 9–16.

Raymer, W. B. and Diener, T. O. (1969). *Virology* **37**, 343–350.

Raymer, W. B. and O'Brien, M. J. (1962). *Amer. Potato J.* **39**, 401–408.

Sänger, H. L. and Brandenburg, E. (1961). *Naturwissenschaften* **48**, 391.

Schultz, E. S. and Folsom, D. (1923). *J. Agr. Res.* **25**, 43–117.

Semancik, J. S. and Weathers, L. G. (1968). *Virology* **36**, 326–328.

Siegel, A., Zaitlin, M., and Sehgal, O. P. (1962). *Proc. Nat. Acad. Sci. U. S.* **48**, 1845–1851.

Singer, B. and Fraenkel-Conrat, H. (1963). *Biochim. Biophys. Acta* **72**, 534–543.

Singh, R. P. and Bagnall, R. H. (1968). *Phytopathology* **58**, 696–699.

Singh, R. P., Benson, A. P., and Salama, F. M. (1966). *Phytopathology* **56**, 901–902 (Abstr.).

Spencer, D. and Wildman, S. G. (1964). *Biochemistry* **3**, 954–959.

Tewari, K. K. and Wildman, S. G. (1966). *Science* **153**, 1269–1271.

Watson, J. D. and Ralph, R. K. (1967). *J. Mol. Biol.* **26**, 541–544.

CHAPTER 15 **The Viruses Causing the
Polyhedroses and Granuloses
of Insects**

KENNETH M. SMITH

I. Introduction

In spite of the fact that the intensive study of insect viruses is less than 20 years old, this branch of virology now compares favorably with other branches in its interest and significance.

The insect viruses possess the same wide range of size and shape as the plant viruses and the viruses affecting the higher animals; both DNA and RNA types occur.

Although many insect viruses occur freely in the tissues of the host, there are three large groups which differ sharply from all other viruses. In these groups the virus particles are occluded in protein crystals, called, respectively, "polyhedra" and "capsules" and cause what are known sometimes as "inclusion body" diseases.

This chapter is concerned only with these groups and they are divided into the *nuclear polyhedroses*, the *cytoplasmic polyhedroses*, and the *granuloses*.

In the nuclear polyhedroses the virus particles are rod-shaped and contain DNA; as the name implies, replication is confined to the cell nucleus. In the cytoplasmic polyhedroses the virus particles are near spherical (icosahedra), occur only in the cytoplasm, and as so far ascertained, contain RNA. In both types of polyhedroses the crystals, or polyhedra, contain many hundreds of virus particles.

In contrast, however, the occlusion bodies of the granuloses are extremely minute crystals (capsules) and are just visible with the oil immersion lens of the optical microscope. The virus particles are rod-shaped and, as a rule, occur singly in each capsule. They may be found in both nucleus and cytoplasm and appear to contain DNA.

The nuclear polyhedroses are the oldest known insect virus diseases and a disorder of the silkworm, now recognized to be of this type, is referred to in a poem by Vida, which was published in 1527. Because of its association with the silkworm, an insect of economic importance, a great deal of attention has been paid to this disease.

The discovery of the cytoplasmic polyhedroses is quite recent and the existence of this separate and distinct type of polyhedrosis virus was first demonstrated by Smith and Wyckoff (1950) in the larvae of two species of tiger moths *Arctia caja* Linn. and *A. villica* Linn.

Although the majority of insect viruses attack the larval stages, there are some, such as that causing bee paralysis, which attack the adult insect only.

The three groups of viruses discussed in this chapter are all primarily associated with the larval stages, although the adult insects are liable to infection under certain circumstances.

The polyhedroses affect mainly the larvae of Lepidoptera and Hymenoptera (sawflies), though there is at least one virus of the nuclear type which attacks dipterous larvae. Both types of polyhedroses are also associated with certain species of Neuroptera.

So far as is known at present the granuloses have only been recorded as attacking the larvae of Lepidoptera.

II. The Virus Particles (Virions)

A. MORPHOLOGY

1. *Size, Shape, and Ultrastructure*

As has already been stated briefly in the introduction, the virus particles of the nuclear polyhedroses and of the granuloses are rod-shaped, whereas those of the cytoplasmic polyhedroses are icosahedral.

The following list of the various structures which are concerned in these diseases is given to avoid confusion during the later discussion.

Nuclear Polyhedroses. The virus rod, the intimate membrane, the outer membrane, the occluding polyhedral crystal, and the membrane surrounding the polyhedral crystal.

Cytoplasmic Polyhedroses. The icosahedral virus particle; this consists of the capsid, containing an electron-dense core. With this type of virus no intimate or outer membranes have been described. The occluding crystal may or may not be polyhedral. There is no membrane surrounding it.

Granuloses. A state of affairs, rather similar to that which exists in the nuclear polyhedroses, is found. There is a rod-shaped virus particle, an inner and outer membrane, and an inclusion body which is very small and, as a rule, contains only one or more particles.

The actual dimensions of the nuclear polyhedrosis virus rods may vary considerably, from 20–50 mμ in diameter to between 200 and 400 mμ in length.

When a purified suspension of the virus from a nuclear polyhedrosis of the silkworm, *Bombyx mori* L., is treated with weak alkali, the nucleoprotein content of the rod is liberated from the occluding membranes. Although there is little information on its ultrastructure, at very high magnification it has the appearance of a widely spaced helix sometimes with a terminal protrusion.

By prolonged treatment with sodium carbonate it is possible to dissolve away the virus protein and to liberate fibrillar or rod-shaped structures which are presumably the DNA cores of the virus particle (Fig. 1). This has been demonstrated by Ponsen (1965) with the nuclear polyhedroses of three species of Lepidoptera, *Barathra brassicae* (L.), *Adoxophyes reticulana* (Hb.), and *Orgyia antiqua* (L.). It is possible that the protrusion, frequently observed at one end of the nucleoprotein rod, may be part of the DNA core as suggested by Krieg (1961). According to Bergold (1963a), the central core consists of eight subunits.

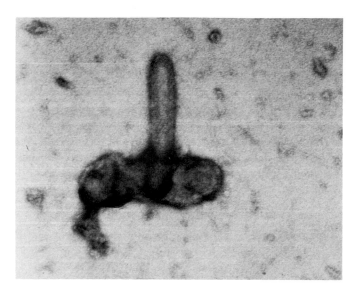

FIG. 1. A single virus particle from a nuclear polyhedrosis of *Orgyia antiqua* showing the shed outer membrane and the DNA core in the virus particle. × 120,000. [Ponsen, M. B. (1965). *Netherland J. Plant Pathol.* **71**, 54–56.]

As we have already mentioned, the appearance of the virus particles in the cytoplasmic polyhedroses is very different from that of nuclear polyhedrosis viruses. In all the cases examined the particles are of the same general shape, icosahedra with a number of projections. They have been observed in the silkworm (Hosaka and Aizawa, 1964), in *Orgyia leucostigma* (Bird, 1965), in the noctuid *Autographa brassicae* Riley (Smith, 1967), and in *Danaus plexippus* (Arnott *et al.*, 1968).

The cytoplasmic viruses from the silkworm *Bombyx mori* and *Danaus plexippus*, the monarch butterfly, are dealt with here as being typical of the group.

As described by Hosaka and Aizawa (1964) the silkworm virus particle consists of two concentric icosahedral shells. The maximum diameter of the outer icosahedral shell is 69 mμ and the diameter of the inner shell is approximately 45 mμ. Each shell has twelve subunits localized at twelve vertices of the icosahedron. Each subunit of the outer shell is a hollow pentagonal disc of about 200 Å in the outer, and about 50 Å in the inner diameters. It consists of five smaller units at its five corners and has a protruding segmented hollow projection. The subunit of the inner shell is connected, at the same symmetrical axis, with the subunit of the outer shell by a tubular structure about 100 Å in length and about 150 Å in the outer,

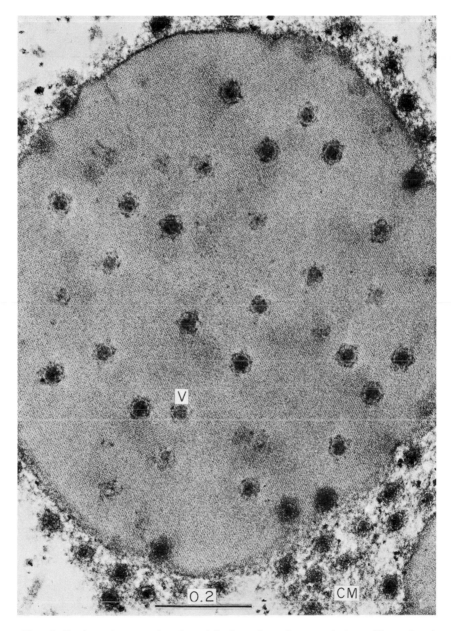

FIG. 2. Section of a cytoplasmic polyhedron from *Danaus plexippus* illustrating its crystalline nature and the distribution of the occluded virus particles. The virus particles (V) show a dense core and an external capsid. [Arnott, H. J., Smith, K. M., and Fullilove, S. L. (1968). *J. Ultrastruct. Res.* **24**, 479–507.]

and about 50 Å in the inner diameters. The region within the inner shell appears to correspond to the core of the virus particle.

In the cytoplasmic polyhedrosis of *Danaus plexippus*, the monarch butterfly, the virus particles exhibit different appearances in various circumstances. They will be described first as they appear in sectioned polyhedra and then in purified suspension. With satisfactory resolution, it is possible to see a central electron-opaque core surrounded by a less dense capsid (Fig. 2). The core which presumably contains the nucleic acid, usually appears to be hexagonal but sometimes may be pentagonal, round, or irregular. The core measures about 350 Å in its longest dimension; it is approximately 1/2 to 3/5 the diameter of the particle itself which may measure as much as 670 Å (point-to-point). With proper printing of the negative, a series of fibrils about 20 Å in diameter can be resolved within the core; the packing of these fibrils is difficult to resolve. Within the polyhedra, the capsid consists of an electron-opaque layer about 25 Å thick separated from the core by a less opaque layer about 25 Å thick (Fig. 2).

This differentiation into core and capsid can be seen in most polyhedra. However, an additional feature of the capsid, the projections previously mentioned, can only be seen under favorable circumstances. Often six of these projections, forming a six-pointed star, can be seen; they measure about 170 Å in length.

In many polyhedra, a second feature in the virus particle can be observed; namely the presence of a tail extending from a vertex of the hexagon (Fig. 3). The length of the tail is almost that of the diameter of the virus particle itself and is much longer than the regular projections. The nature of the tail is problematical, but it has also been seen in PTA negatively stained particles (Fig. 4). The tails do not show an exact orientation preference with respect to the lattice of the crystalline polyhedron, but often they extend toward the nearest edge of the polyhedron.

What may be called "false tails" also occur; these are several times the length of the virus particle, are quite narrow, and always extend from the virus particle toward the periphery of the polyhedron. Apparently these represent the incomplete fusing of the crystal as the virus particle becomes incorporated and produces a defect in the crystal lattice.

Virus particles, released from the polyhedra and stained with PTA, exhibit a central cavity apparently outlining the polygonal nucleoprotein tpace. The double nature of the capsid can frequently be seen and is like shat reported by Hosaka and Aizawa (1964). In their general appearance the virus particles of *Danaus* are quite similar to those described for the silkworm. There are twelve projections apparently falling on the twelve

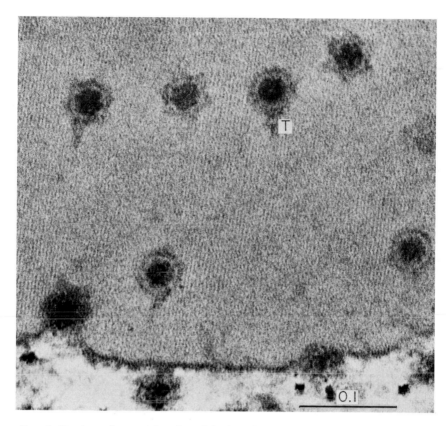

FIG. 3. Portion of a cytoplasmic polyhedron from *Danaus plexippus* illustrating the crystalline lattice associated with several virus particles. Two particles clearly show a tail (T); a third shows a portion of a tail. [Arnott, H. J., Smith, K. M., and Fullilove, S. L. (1968). *J. Ultrastruct. Res.* **24**, 479–507.]

vertices of the icosahedron which are found on the fivefold axes. These projections, however, are not clearly in four parts although it is possible, on occasion, to make out two components. It should be remembered, however, that the four-component system of *Bombyx mori* was established by measurements of the lengths of about 125 projections which seemed to fall in four groups with mean differences of about 75 Å.

Previous reports (Bergold and Suter, 1959; Bird, 1966; Hosaka and Aizawa, 1964) have not described a "tail" associated with this kind of virus, but, as already mentioned, what appears to be a true tail has been found both by negative staining and by section techniques (Arnott *et al.*, 1968). Table I gives the characteristics of some cytoplasmic polyhedroses.

So far as the morphology of the virus particle in the granuloses is concerned, there is a close resemblance to the rod-shaped particle of the nuclear polyhedroses. The rod is short, rather thick, and slightly curved. The dimensions of these short rods are available for several granuloses. For twenty-five naked rods from the armyworm *Pseudoletia unipuncta* Haworth, the width varied from 57.8 to 73.5 mμ with a mean of 62.0 mμ, and the lengths ranged from 367.5 to 441.0 mμ with a mean of 411.6 mμ. The average size of ten virus particles from the larva of the codling moth, *Carpocapsa pomonella* L. was 313.5 mμ in length, and 50.7 mμ in width (Tanada 1959, 1964). In a granulosis of *Plodia interpunctella* (Hbn.), the Indian meal moth, the virus rods measure from 210 to 260 mμ in length and from 30 to 33 mμ in diameter (Arnott and Smith, 1968a). Huger (1963) summarizes the average measurements of granulosis viruses as a whole as having a width ranging from 36 to 80 mμ and a length from 245 to 411 mμ.

Not very much is known about the ultrastructure of the granulosis virus particle; Smith and Hills (1960), by means of negative staining, detected a closely packed helical structure of the intimate membrane of isolated granulosis virus rods which could not be resolved on empty membranes. So far as the granulosis of *Plodia interpunctella* is concerned, thin sections and negative staining with phosphotungstic acid and uranyl acetate, after

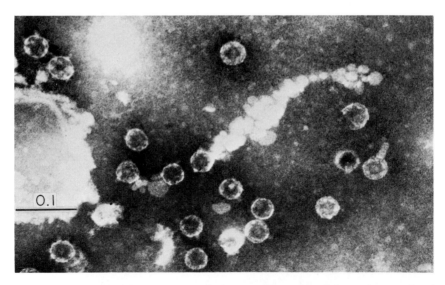

FIG. 4. Negative stain preparation of virus particles released from the cytoplasmic polyhedra from *Danaus plexippus*; note pentagonal shape of core and the tail on the right-hand particle. [Arnott, H. J., Smith, K. M., and Fullilove, S. L. (1968). *J. Ultrastruct. Res.* **24**, 479–507.]

TABLE I

CHARACTERISTICS OF CYTOPLASMIC POLYHEDROSES

Insect species:	Danaus	Bombyx	Orgyia	Choristoneura	Autographa
Author(s):	Present	Hosaka and Aiyawa (1964)	Bird (1965)	Bird (1966)	Smith (1967); Arnott and Smith (unpublished)
Shape of virus:	Icosahedral	Icosahedral	Icosahedral	Icosahedral	Icosahedral
Diameter of virus particle:	670 Å	690 Å	680 Å	700 Å	ca. 690 Å
Projection present:	+	+	+		+
Length of projection:	ca. 170 Å	75–300 Å			
Width of projection:	ca. 170 Å	162 Å			ca. 100 Å
Number of projections:	12	12	12		12
Double capsid:	+	+	+		+
Tail:	+	—	—	—	—
Shape of inner core:	Polygonal	Polygonal			Polygonal
Polyhedron shape:	Round or cubic	Tetragonal or polygonal[a]	Polygonal	Polygonal	Round
Size:	to 5 μ	3 μ		1.2 μ	up to 4.7 μ

[a] Personal communication.

[b] From Arnott et al. (1968).

$NaCO_3$ treatment to remove the capsule, have so far failed to give much information on the ultrastructure of the virus rod. Some micrographs do, however, suggest a helical structure similar to that described by Smith and Hills (1960). Both transverse sections and negatively stained preparations reveal some hollow rods, possibly suggesting the existence of a DNA core similar to that described for the virus rods of the nuclear polyhedroses (Arnott and Smith, 1968a).

B. THE VIRUS MEMBRANES

1. Intimate (Inner) Membrane

The virus rods of both the nuclear polyhedroses and the granuloses are enclosed in two membranes, the *intimate or inner* membrane which contains the virus nucleoprotein, and the *outer membrane* which contains both of these. According to Bergold (1963a), in the case of the nuclear polyhedroses there is a space of about 60 Å between the intimate membrane and the virus nucleoprotein. This is in contrast with what is found in sodium carbonate-treated preparations, where the inner membranes are closely attached to the virus rods (Ponsen *et al.*, 1965).

In the granulosis of *Plodia interpunctella*, the intimate membrane appears to develop *after* the formation of the outer membrane, as the virus rods do not have an intimate membrane at the time they are inserted into their outer membranes. Subsequent to this, the material between the outer membrane and the virus rod appears to become more concentrated and eventually forms the intimate membrane. This is clearly not a membrane in the sense of a unit membrane, but rather it appears to represent the condensation of some substance. Whether all the material forming the intimate membrane is present within the "loose" outer membrane at the time of insertion is not known; it may possibly have its origin in materials contained within the endoplasmic reticulum (Arnott and Smith, 1968a).

In the cytoplasmic polyhedroses no membranes have been observed surrounding the virus particle.

2. Outer Membrane

There seems little doubt that the outer membrane is a true unit membrane. In the nuclear polyhedroses it may include a varying number of virus rods, each of which has its own intimate membrane. Bergold (1963a) records up to nineteen rods enclosed in one outer membrane in *L. monacha*, but this is unusual. He also points out that these bundles of rods are always arranged in the closest possible packing forming, triangles, squares, pentagons, hexagons, etc. This suggests that the bundles of rods must be suspended in a fairly liquid medium which permits the shaping of symmetrical patterns.

According to Ponsen *et al.* (1965) the outer membranes have a central, bimolecular leaflet of lipids with a width of about 5 mμ bounded on either side by carbohydrates and proteins. The much less dense layer which can be clearly seen after staining with potassium permanganate is the lipid layer of the outer membranes.

The formation of the outer membrane of the virus particle in a granulosis has been studied in some detail in a disease of this type affecting the larvae of *Plodia interpunctella*, the Indian meal moth. Previous studies (Bergold, 1963b; Bird, 1963; Huger and Krieg, 1961; Hughes, 1952) of granulosis viruses have not disclosed the ordered arrays of naked virus particles associated with the endoplasmic reticulum (ER). These arrays are a common feature of the granulosis of *P. interpunctella* and appear to be important in the mechanism by which the naked rods receive their outer membranes. The rods are found both in precise arrays and in less ordered situations in very intimate association with the ER (Fig. 5). The ordered arrays come about almost certainly because of the characteristics of the virus rods and hexagonal close packing can commonly be seen. The reason why such large numbers of rods accumulate is not clear, but perhaps they represent a stage in the replication cycle during which a sequence of replication has proceeded more rapidly than another and therefore causes a build-up of the naked rods; it may be that capsule protein is not yet present in sufficient quantity for capsule formation to occur. It is clear that naked rods are present in the infected cells: these rods have the same dimensions as do those which are contained within the complete capsules; naked rods and complete capsules can often be found in the same cell. The process of capsule assembly, analogous to the formation of the polyhedra in the polyhedroses, begins with the formation of the outer membrane around the naked virus rod. Although the outer membrane is not a part of the capsule it represents the stratum upon which the capsule protein will be deposited. The formation of the outer membrane occurs by a process by which the naked virus rod is inserted into the membrane which is supplied by the ER (Fig. 5). All the details of this are not understood, but smooth ER is very commonly associated with the virogenic stroma both in large masses and in individual branched strands. Other membranes are not common in these cells and the ER seems the most probable source of the outer membrane. When the outer membrane is first joined round the rod it fits rather loosely and would

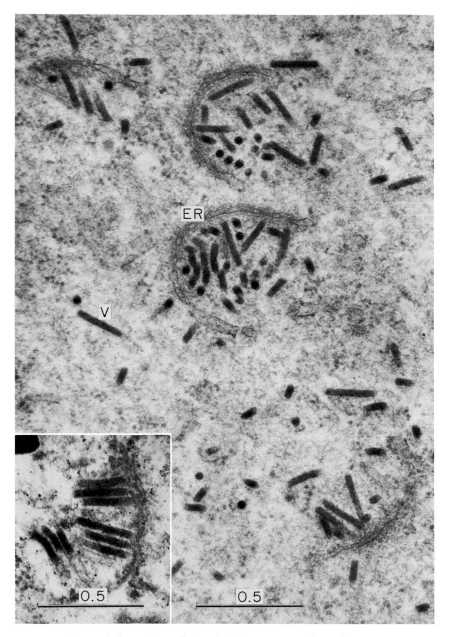

FIG. 5. Virus rods from a granulosis of *Plodia interpunctella* in close association with smooth endoplasmic reticulum from which the outer membrane appears to be derived. *Inset*. Virus rods apparently in the process of acquiring their outer membrane. [Arnott, H. J. and Smith, K. M. (1968a). *J. Ultrastruct. Res.* **21**, 251–268.]

contain a volume much greater than that of the virus rod. This membrane has the structure and dimensions of a unit membrane and in this sense is similar to other membranes associated with these cells. As the intimate membrane begins to develop the nature of the outer membrane changes. This occurs by a shrinkage of the membrane profile which not only makes it smaller but at the same time brings it into much greater conformity with the outline of the virus rod (Arnott and Smith, 1968a). According to Bergold and Wellington (1954), the amino acid composition of the virus membranes and that of the viruses is quite different, the virus membranes containing more aspartic acid and much less arginine than the virus.

C. CHEMICAL COMPOSITION

There is now no doubt that the virus rods of the nuclear polyhedroses contain DNA and no RNA (Wyatt, 1952a,b; Bergold, 1959). As analysis of purified virus particles suggested 13% DNA in *Bombyx mori* and 16% in *Porthetria dispar* (Bergold, 1947; Bergold and Pister, 1948). A reinvestigation of very highly purified *B. mori* particles by Bergold and Wellington (1954) revealed the percentage of DNA to be about 7.9% and 0.915% phosphorus, of which, however, only about 87% is found in the DNA.

Krieg (1956) has studied the nucleic acid of a nuclear polyhedrosis virus from *Aporia crataegi* (Linn.) in which he found 9% DNA but no RNA. Wellington (1954) determined the amino acid content of several nuclear polyhedrosis viruses. She found that whereas the seven viruses analyzed all had a similar pattern of amino acid composition, they differed markedly from the pattern of their respective inclusion body proteins. Kawase (1964) determined the amino acid composition of some polyhedral proteins and viruses of the polyhedroses affecting *B. mori* by means of the Beckman-Spinco Model 120 amino acid analyzer.

Three viruses, two cytoplasmic and one nuclear, were examined together with their polyhedral proteins and they showed considerable similarity in the pattern of their amino acid composition. The relative proportion of arginine in the viruses was similar to that of the polyhedral proteins except that in the nuclear type there was more arginine in the inclusion body protein than in the virus. The viruses, however, had less tyrosine than the polyhedra, just as reported by Wellington (1954). Furthermore, the viruses had much more threonine, proline, and glycine, but less cystine than the polyhedral protein. J. D. Smith and Wyatt (1951) and Wyatt (1952b) have investigated the bases of several insect viruses. In the nuclear polyhedrosis of *Lymantria dispar*, the gypsy moth, the DNA was shown to contain the purines, adenine and guanine, and the pyrimidines, cytosine and thymine; 5-methylcytosine could not be detected. Chromatograms of the gypsy moth virus hydrolyzed whole, without isolation of its nucleic acid, showed no uracil, thus confirming the absence of RNA from this virus.

As regards the nucleic acid content of the cytoplasmic polyhedrosis viruses, Xeros (1962) has carried out chromatographic and electrophoretic analysis of such viruses from *Laothoe populi* Linn. This showed that they contained 0.95% RNA and no DNA. Chromatographic analysis of similar viruses from five further species showed the presence of similar quantities of RNA and the absence of DNA. The cytoplasmic polyhedrosis virus consists of nucleic acid, protein, and a small amount of lipid (Kawase, 1967). Recent studies on the RNA of the virus from a cytoplasmic polyhedrosis of the silkworm have been carried out by Miura *et al.* (1968). The RNA was extracted by phenol treatment of the virus and appeared as threads when precipitated in alcohol; these consisted of two components with different sedimentation constants. The extracted RNA seemed to split into segments at a preferential breaking point, and was soluble in concentrated salt solution, thus differing from single-stranded, high molecular weight RNA. The base composition of the RNA was complementary in the ratios of adenosine to uridine and guanosine to cytosine. It contained 43% guanosine plus cytosine. Based on these and other factors the authors conclude that this RNA is double-stranded and has regular base pairings of guanosine/cytosine.

The RNA extracted from the virus of the silkworm cytoplasmic polyhedrosis shows a comparatively high degree of infectivity (Kawase and Miyajima, 1968).

Wyatt (1952a,b) has studied the DNA content of granulosis viruses in *Choristoneura murinana* and *C. fumiferana*. He found that they contain the purines, adenine and guanine, and the pyrimidines, cytosine and thymine, but no methylcytosine or uracil. Although the two viruses come from closely related hosts they differ in their nucleic acid composition.

D. Virus Assembly

Although there is a good deal of information on the patterns and changes of nucleic acid synthetic activity during the course of nuclear polyhedrosis development, knowledge of the morphology of virus replication, i.e., the assembly of the virus particle itself with which this chapter is concerned, is somewhat meager.

The development of the nuclear polyhedrosis virus in the cell nuclei of the silkworm, *B. mori*, at various stages of infection has been studied by Smith and Xeros (1953). The study arose out of the discovery of a peculiar structure called the "nuclear net." What was presumably the nuclear net has been studied further by Xeros (1955, 1956) under the name of a "virogenic stroma." The proteinaceous virogenic stromata form *de novo* in the nuclear sap of infected cells; they become increasingly proteinaceous and Feulgen-positive as they grow and develop. Morphologically they are networks, and virus rods differentiate within vesicles in their cords. According to Xeros these begin as fine rodlets about 60×1200 Å in size and increase *in situ* to their final size of 280×2800 Å. They are then set free from their vesicles into the pores of the net by disruption of the surrounding cord material and may ultimately reach the ring zone between the centrally placed virogenic mass and the nuclear membrane. The freed virus rods then become enveloped by independently formed membranes. A similar state of affairs in a nuclear polyhedrosis affecting the larvae of *Aglais urticae*, the small tortoise shell butterfly has been described by Harrap and Robertson (1968). Initially, the nuclear content of the fat-body cells appeared as a mesh or network of strands from which naked virus particles were produced. The individual virus particles then acquired a membrane in the spaces within this mesh, a clear space being visible between the membrane and the enclosed virus particle.

In a study of the nuclear polyhedrosis of the silkworm, *B. mori*, the supernatant of infected hemolymph was subjected to differential, and cesium chloride density gradient centrifugation. This was highly infectious but contained no virus rods, the size of the particles found appeared to be about $20 \, m\mu$ in diameter (Aizawa, 1967). There does not seem to be any information on the morphology and appearance of these particles.

The nature of the earliest replication process of the virus in the cytoplasmic polyhedroses is at present unknown but the following account of the replication of the virus in a cytoplasmic polyhedrosis of *Danaus plexippus*, the monarch butterfly, is taken from some recent work by Arnott *et al.* (1968). At the earliest stages of the disease, 48 hours after infection, polyhedra were already present in the cells of the midgut. In these cells a virogenic stroma was found associated with the polyhedra; within this stroma only a very few complete virus particles were found and these were usually near the edge of the stroma. However, many empty or partly filled particles were present in the stroma, these apparently represent stages in the replication of the virus. Particles consisting only of a naked capsid, a capsid containing a small central core, and capsids with a large central core were observed. All these were associated with fibrils which extended from their

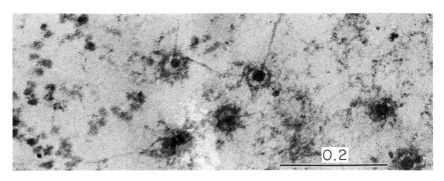

Fig. 6. Larvae of *Danaus plexippus* fixed 48 hours after infection with a cytoplasmic polyhedrosis. Many virus particles have one or more long filaments that appear to penetrate the capsid and interconnect with the dense core. [Arnott, H. J., Smith, K. M., and Fullilove, S. L. (1968). *J. Ultrastruct. Res.* **24**, 479–507.]

surface for some distance into the surrounding stroma. The capsid often appeared to be irregular in thickness. In many the fibrils appear to pass through the capsid and to attach to the central core (Fig. 6). The material surrounding these particles is in general appearance exactly the same as the crystallogenic matrix (see Section III,A). Whether one is derived from the other or whether they are independent structural elements is not clear at the present time. It is clear, however, that the crystallogenic matrix has many complete particles in it while the virogenic stroma has few complete particles and many incomplete particles present.

It may therefore be suggested that these incomplete particles represent an early stage in the development of the virus. They do, in fact, have a precedent in the studies of the development of *Tipula* and *Sericesthis* iridescent viruses (Bellett, 1965a,b; Smith, 1958; Xeros, 1964) in which the capsid is first produced and is subsequently filled with the nucleic acid.

A tentative drawing of the replication cycle and assembly of a cytoplasmic polyhedrosis virus is shown in Fig. 7.

Not much is known of the method of assembly of the rod-shaped virus particles of the granuloses. In the case of the granulosis of *Plodia interpunctella*, the naked virus particles consisting of the protein coat and DNA are assembled in the virogenic stroma and can be seen scattered throughout the cytoplasm. The occurrence, in both transverse and longitudinal section, of empty rods suggests the existence of an outer protein membrane which contains the DNA. A diagrammatic representation of the sequence of virus assembly in the cells of *Plodia interpunctella* is given in Fig. 8 (Arnott and Smith, 1968a).

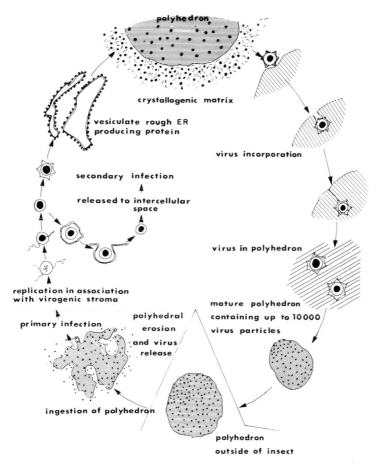

FIG. 7. Tentative drawing of the replicative cycle of a cytoplasmic polyhedrosis virus in the larva of the monarch butterfly. [Arnott, H. J., Smith, K. M., and Fullilove, S. L. (1968). *J. Ultrastruct. Res.* **24**, 479–507.]

III. The Inclusion Bodies

A. MORPHOLOGY, MEMBRANE, AND ULTRASTRUCTURE

The nuclear polyhedra vary considerably, both in size and shape. The variation occurs both in different insects and in the same insect, but not as a rule in the same cell where the polyhedra tend to be the same size. Bergold (1963a) points out that in the silkworm, *Bombyx mori*, the prevailing types of polyhedra are dodecahedra, whereas those of *Lymantria monacha* consist

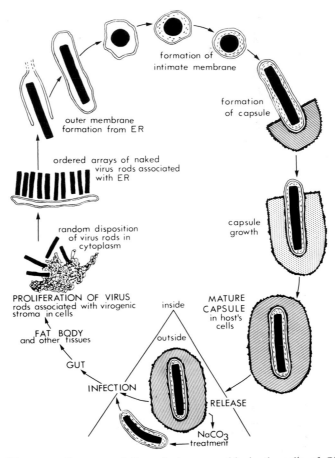

Fig. 8. Diagrammatic representation of virus assembly in the cells of *Plodia inter-punctella*. [Arnott, H. J., and Smith, K. M. (1968). *J. Ultrastruct. Res.* **21**, 251–268.]

mostly of tetrahedra. The polyhedra from *Porthetria dispar* are irregular in shape. The diameter of the polyhedra varies from 0.5 to 15 μ, according to the species. In the case of a nuclear polyhedrosis of *Barathra brassicae*, the diameters varied from 0.8 to 2.7 μ with an average of 1.8 μ (Ponsen and De Jong, 1964).

In the nuclear polyhedroses of some species the polyhedra have an extremely characteristic appearance. In the larvae of the scarlet tiger moth, *Panaxia dominula*, for example, the polyhedra are rectangular (Smith, 1955). Similarly in *Tipula paludosa*, the polyhedra are crescent-shaped, rather like a segment of an orange. There has been some controversy as to whether there is actually a membrane surrounding the polyhedral crystal and the suggestion has been made that there is only a hardened outer layer of the crystal. This is largely because it is difficult to see a membrane in sections of the polyhedra. However, the fact remains that there is a membrane of some kind. When the polyhedra are dissolved in weak alkali, there is left a structure, closely following the outline of the polyhedral crystal, which contains the virus rods. This is in strong contrast to what happens when the cytoplasmic polyhedra are similarly treated.

In a study of some nuclear polyhedroses of lepidopterous larvae, Ponsen *et al.* (1965) state that the polyhedral membranes appear as very thin dense bands closely surrounding the polyhedra and suggest that they consist of protein.

Examination of ultrathin sections of nuclear polyhedra at very high magnification under the electron microscope reveals regular dot and line patterns. Bergold (1963a) has made a study of the crystalline lattice in the polyhedra of various species by means of X-rays and electron microscopy of thin sections stated to be only 100 Å thick and his conclusions are given here. The crystalline lattice has a very high degree of regularity without dislocations, the protein molecules being spheres with a diameter of about 60 Å (Hall, 1960), but this diameter may vary from 65 to 90 Å according to the insect species from which the polyhedra have been obtained. Between the rows of molecules angles of 90° and 120° could readily be observed. All the observed dot-and-line patterns can be explained with the aid of light and X-ray micrographs of molecule models arranged in a cubic system, but cut at different angles.

The cytoplasmic polyhedra in the different diseases vary greatly in shape, in some cases they tend to be very large as in the larvae of *Bombyx mori L.*, *Ourapteryx sambucaria L.*, and *Estigmene acrea*; when large the polyhedra in the latter insect sometimes lose their many-sided character and appear almost spherical.

In a cytoplasmic polyhedrosis of *Danaus plexippus*, the crystalline polyhedra are oval in sectional view; presumably they are approximately spherical in shape. The polyhedra vary in size from quite small bodies to bodies as much as 5 μ in diameter. The margin of the polyhedra is usually slightly more electron opaque than the interior which is more or less uniform. Examination of the polyhedra at high magnification allows the resolution of a lattice indicating their crystalline nature. Two general lattice patterns were commonly observed; the first a dot pattern, the second a line pattern. The center-to-center spacing between the rows of dots or the lines forming the lattice patterns in the polyhedra has a mean distance of 41 Å (range 38–44 Å). The dots or lines forming the patterns are approximately 20–25 Å in diameter or width and the dot pattern forming the lattice can be clearly seen, in which the vertical and horizontal lattice pattern is formed by dots aligned into rows. In this case the vertical and horizontal rows are arranged at an angle of 86° with respect to each other. At some points the dots appear to merge along a single row and there is a certain amount of variation in the size and shape of the single dots.

The elements forming the line pattern appear to consist of fibrils which often have a zigzag shape. These elements are frequently continuous for a few hundred angstroms and from their zigzag shape might be interpreted as a side view of a helix; they are approximately 20–25 Å in width.

The lattice planes are continuous across the entire polyhedron, precise alignment from one edge to the other being characteristic. However, while the overall lattice structure is not affected, the lattice shows slight alteration at points where it passes near the virus particles. These modifications consist of the bending of the closest lattice planes toward the virus particle. While the deflection is only a matter of 20–30 Å, these regular deformations can be clearly seen (Arnott et al., 1968).

When the cytoplasmic polyhedra are subjected to treatment with alkali, the reaction is very different from that shown by the nuclear polyhedra in similar circumstances. Instead of dissolving completely, as in the latter case, leaving a membrane behind filled with rod-shaped virus particles, the cytoplasmic polyhedra dissolve only partially leaving a matrix pitted with holes in which the near-spherical virus particles had been. There is no surrounding membrane. Another point of difference between the two types of polyhedra is that the cytoplasmic polyhedra readily take up stains such as giemsa's solution or methylene blue whereas, unless overheated in fixation, the nuclear polyhedra do not stain. This affords an easy method of differentiation on the optical microscope.

In the granuloses, the normal capsule form does not seem to conform to any precise geometric pattern. The ovocylindrical shape sometimes appears angular or even hexagonal, but at other times the contours are rounded. Despite the difficulty in assigning a specific shape, capsules are relatively uniform in size. Although "monsters" of various sorts (see Section III, D) are found in association with otherwise normal appearing capsules in the granulosis of *Plodia interpunctella*, there is a general size category. Since the capsule is composed of a protein crystal, the forces that control the size and shape although unknown are of considerable interest.

A general average of the sizes of capsules is given by Huger (1963) as ranging from 300 to 511 mμ in length and 119 to 350 mμ in width.

There is no membrane around the crystal in the sense of a unit membrane, but there is a distinct zone at the surface of the crystal in the shape of a loosely attached particulate substance of increased electron density (Fig. 9). It seems likely that this surface represents the zone where the molecular organization of the complex protein molecules takes place, but how it takes place or why it stops at a uniform size is not known.

The arrangement of the crystal lattice with reference to the longitudinal axis of the virus rod appears to be random; any combination of lattice arrangement and virus orientation is apparently possible (Arnott and Smith, 1968a).

According to Bergold (1959) the molecules are in a cubic arrangement. In capsules of *Choristoneura murinana* they have dimensions of $57 \times 57 \times 229$ Å: the sides of the unit all measure 57 Å and the angle (Y) is $120°$.

A recent paper by Summers and Arnott (1969) on a nuclear polyhedrosis

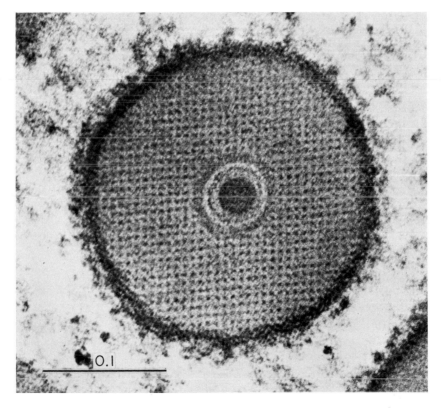

FIG. 9. Transverse section through a virus rod in its capsule from a granulosis of *P. interpunctella*; note the concentric position of the virus rod relative to its membranes, the crystalline lattice, and the loosely attached particulate substance around the exterior of the crystal. [Arnott, H. J. and Smith, K. M. (1968a). *J. Ultrastruct. Res.* **21**, 251–268.]

and granulosis of the cabbage looper, *Trichoplusia ni*, describes large quantities of fibrous material which seems to be intimately associated with the formation of the inclusion bodies in both types of disease. They suggest that crystal formation may occur by a series of associated events involving "membranelike" profiles, fibrous material, and polyhedral formation. They conclude that "it is interesting to note that these crystals are characteristic of insect viruses with striking morphological and nucleic acid differences, and it has been repeatedly demonstrated that the inclusion body is an intimate part of their replication cycles. It remains to be seen whether the inclusion body is an absolute necessity for virus replication and survival and/or to what degree this process of crystal occlusion adds to the viability of the particular virus."

B. CHEMICAL COMPOSITION

Bolle (1874, 1894) was the first to analyze nuclear polyhedra; he found them to consist of protein and to contain no lipids. It must be remembered that any analysis of the whole polyhedra is an analysis of both polyhedral protein and the virus particles, the virus being about 5% of the whole polyhedral crystal. Chemical analyses together with sedimentation and diffusion measurements have revealed that the nuclear polyhedra of *Porthetria dispar* and *Bombyx mori* consist of 95% protein with molecular weights of 276,000 and 378,000, respectively (Morgan *et al.*, 1955).

The nitrogen content of polyhedra and the purified polyhedron proteins from different hosts do not vary much and is about 14–15%. However, the phosphorus content of polyhedra varies with the preparation of purified polyhedron protein (Bergold, 1963a).

Faulkner (1962) found that the nuclear polyhedra of *B. mori* contained RNA. Aizawa and Iida (1963) state that those nuclear polyhedra always contain not only RNA, but DNA as well. The RNA content seemed to vary with the silkworm strains under study. Studies on the nuclear polyhedra from the cabbage looper, *Trichoplusia ni* indicate a ratio of DNA to RNA of 7:5 (Faust and Estes, 1965). Investigations of metal ion content of polyhedral inclusion bodies from *B. mori* revealed the presence of only iron and magnesium (Holoway and Bergold, 1953, 1955). Analysis of intact and whole nuclear inclusion bodies from the corn earworm, *Heliothis zea*, showed a silicon content of 0.12% of the weight of the whole body. The presence of silicon in the polyhedral protein and its absence from the virus rod was demonstrated (Estes and Faust, 1966).

The amino acid composition of two types of cytoplasmic polyhedra, hexahedra and icosahedra, from the silkworm was determined by means of the Beckman-Spinco Model 120 amino acid analyzer (Kawase, 1964). The cytoplasmic polyhedra were found to contain more leucine and iso-leucine, but less tyrosine and phenylalanine than the nuclear polyhedra. Cytoplasmic polyhedra vary in size according to their position in the midgut. The amino acid composition of the polyhedra formed in the caudal portion of the midgut shows much more similarity in pattern to that of the poly-hedra formed in starved larvae than to the amino acid composition of poly-hedra formed in the cephalic portion of the midgut.

There is not much information on the chemical composition of the gran-ulosis capsules. Wellington (1951, 1954) has studied the amino acids of the virus and the inclusion body protein from a granulosis of C. murinana (Hb.). She found the following amino acids in the acid hydrolyzates of both virus and capsular protein: cysteic acid, aspartic acid, glutamic acid, serine, threonine, alanine, tyrosine, methionine, histidine, lysine, arginine, proline, valine, leucine, isoleucine, phenylalanine, and glycine. Tryptophan was determined in the unhydrolyzed samples.

C. Number of Virions Enclosed

It is clear from the examination of thin sections of both nuclear and cyto-plasmic polyhedra that the number of occluded virions is very large. Esti-mates of their number in the former type of disease do not seem to have been made but some attempt has been made to arrive at a figure for the cytoplasmic polyhedroses. Using a section thickness of 1000 Å, or less, and assuming the polyhedra are approximately spherical, it can be estimated that as many as 10,000 or more virus particles are present in a single poly-hedron. They appear to be distributed within the polyhedra in a random fashion; however, particles are never found side by side but rather they are separated by a space, usually a minimum 1000 Å or more but on rare occasions as little as 500 Å apart (Arnott et al., 1968).

In the nuclear polyhedra the distribution of the virus rods appears to be haphazard, with the possible exception of the nuclear polyhedrosis of the fly, Tipula paludosa, in which there may be some arrangement of the par-ticles. The inclusion bodies of the granuloses differ markedly from those of the polyhedroses by their minute size and by the occlusion, on the average, of only one virus particle. The occurrence of more than one particle per capsule is discussed in the next section.

It is a rather remarkable fact that no cell organelles or components, other than virus particles, have ever been found inside the inclusion bodies.

D. Mutations and Abnormalities

Gershenson (1959, 1960) considers that it is the virus that controls the shape of the polyhedra rather than the host cell and he isolated a strain of nuclear polyhedrosis virus from *Antheraea pernyi* G-M. which induced the formation of a hexagonal polyhedron instead of the more usual tetragon-tritetrahedral shape. He also found that the new shape of polyhedron "breeds true" in the sense that infection with the new type produces only similarly shaped polyhedra.

A similar phenomenon has been observed in a granulosis of *Choristoneura fumiferana* (Clemens.). In this instance the aberrant capsules were cube-shaped instead of the normal oval type. Propagation of these aberrant forms proved that they retained their cubic shape (Stairs, 1964).

It is probable that this type of mutation is relatively common in the

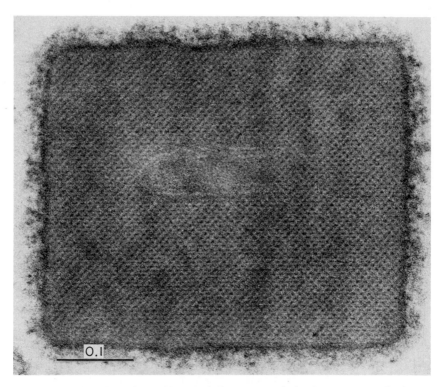

Fig. 10. Section through a cubic capsule from a granulosis of *P. interpunctella* showing the crystalline lattice, the virus rod and its membrane, and the outer layer of particulate matter. Scale equals 1 μ or portion thereof as indicated. [Arnott, H. J. and Smith, K. M. (1968b). *J. Ultrastruct. Res.* **22**, 136–158.]

inclusion-body diseases. Figure 10 illustrates a cubic capsule from a granulosis of *Plodia interpunctella* (Hbn.), showing the crystalline lattice. These cubic capsules were organized into orderly arrays within the cell, in contrast to the haphazard assembly of the normal capsules.

Inoculation of *Plodia* larvae with the cubic capsule gave rise only to the cubic type (Arnott and Smith, 1968b).

Similarly with the cytoplasmic polyhedrosis of *Danaus plexippus*, a cubic polyhedron developed which also produced only cubic types when transmitted to fresh larvae (Arnott and Smith, unpublished).

The most extreme aberrant forms and abnormalities have been observed in the granulosis affecting *Plodia interpunctella*. Giant capsules exhibiting more or less normal morphology but a severalfold increase in volume were found associated with both the normal and cubic capsule forms. The following aberrant capsule types were found irregularly but with enough frequency to be characterized: elongate capsules are occasionally bent and consist of a long crystalline structure with a central channel; compound capsules also have a central channel but are made up of several component crystals; multiparticulate capsules are found with from two to nine rods in a single capsule; agglomerated capsules consist of large irregular aggregations of crystalline capsular material but with only an occasional virus rod occluded in the structure. It may be suggested that aberrant capsules can arise at almost any stage in development from a malfunction of the replication process. Depending on the stage at the time of this malfunction, a characteristic aberrant capsule will result. The factors controlling the malfunction are not known. However, it is possible, on the basis of the distribution of such capsules, that certain cells may provide the environment in which malfunction, and hence aberrant capsules, can occur. It is also possible that viral mutations are involved, as they surely must be in the origin of the cubic capsule (Arnott and Smith, 1968b).

Large numbers of viruslike threads or rods, frequently branched, sometimes at right angles, and forming bizarre shapes are found in the cells of lepidopterous larvae, especially the noctuidae, when they are infected with a granulosis (Schmidt and Philips, 1958; Steinhaus and Marsh, 1960; Smith *et al.*, 1964; Smith and Brown, 1965). These rods are found in association with capsules which contain the occluded virus rod. It was suggested by Smith *et al.* (1964) and by Smith and Brown (1965) that these threads were produced by the virus rod which had escaped from the crystalline capsule, the virus rod then proceeding to elongate and branch. The evidence for this conclusion seemed fairly good at the time. However, bizarre shapes assumed by the capsules in a granulosis of *Plodia interpunctella* suggested an alternative hypothesis. When present, the rods are smaller in diametre than the virus rods, and possess in common with the capsules an epicapsular layer. The branched rods were found attached to capsulelike bodies also possessing an epicapsular layer; at points of junction this layer was continuous from the rods to the capsulelike bodies. The alternative hypothesis is, then, that the rods may represent a peculiar arrangement of the capsular protein rather than an extension of the virus rod itself. (Arnott and Smith, 1969).

IV. Serological Relationships

This is a fruitful field of study which has been much neglected in insect virology, and it should yield information which would be useful in many ways. The number of apparently distinct viruses would almost certainly be reduced, and light would probably be thrown onto the question as to how far the large numbers of cytoplasmic polyhedrosis viruses are one and the same.

Krywienczyk *et al.* (1958) have shown that viruses from the nuclear polyhedroses and the granuloses belong to two serologically distinct groups.

In a further paper (Krywienczyk and Bergold, 1960a) the serological relationships of seventeen insect viruses from four continents were investigated by the complement fixation technique. As would be expected, the results support the grouping of the viruses into polyhedroses and granuloses, although one granulosis virus from *Recurvaria milleri* Busck gave detectable cross-reactions with six polyhedrosis viruses. The two polyhedrosis viruses from the Hymenoptera gave strong cross-reactions only within their own group and are probably closely related, if not identical.

Krywienczyk and Bergold (1960b) suggest that serological studies of intact insect viruses, so far reported, are, in reality, studies of the antigenic structure of the virus membranes, and not of the virus nucleoprotein. The same workers (Krywienczyk and Bergold, 1960c) have studied the serological relationships between the inclusion body proteins of Lepidoptera and Hymenoptera. They find that the proteins belong to three serologically distinct groups. Two of these groups, the nuclear polyhedron proteins and the granulosis proteins from Lepidoptera, are the same as those established for the corresponding viruses. The granulosis protein from *R. milleri*, however, showed, like its virus, a higher cross-reactivity with polyhedron proteins than with the other granulosis proteins. The third group is represented by the nuclear polyhedron proteins of the sawflies (Hymenoptera).

In serological studies of the cytoplasmic- and nuclear-polyhedrosis viruses of the silkworm, *B. mori*, it was found that two strains of the cytoplasmic polyhedrosis virus, which formed icosahedral and haxahedral polyhedra, respectively, were closely related. On the other hand, both strains were unrelated to the nuclear polyhedrosis virus of the same insect (Hukuhara and Hashimoto, 1966). An attempt has been made by Cunningham and Longworth (1968) to differentiate by serological methods some of the many cytoplasmic polyhedrosis viruses recorded. They extracted the virus from the cytoplasmic polyedra from *Aglais urticae, Nymphalis io, Vanessa cardui, Arctia caja, Porthetria dispar, Phalera bucephala,* and *Euproctis chrysorrhoea.* Antisera were prepared in guinea pigs to the virus particles extracted from three of these isolates, and all seven isolates were found to be serologically very closely related when compared by means of the complement fixation test. The results of this work suggest still further that there are not as many distinct cytoplasmic polyhedrosis viruses as was thought at first. The fact that many of these viruses are morphologically so similar seems to bear this out.

REFERENCES

Aizawa, K. (1967). *J. Sericult. Sci.* **36,** 327–331.

Aizawa, K. and Iida, S. (1963). *J. Insect Pathol.* **5,** 344–348.

Arnott, H. J. and Smith, K. M. (1968a). *J. Ultrastruct. Res.* **21,** 251–268.

Arnott, H. J. and Smith, K. M. (1968b). *J. Ultrastruct. Res.* **22,** 136–158.

Arnott, H. J. and Smith, K. M. (1969). *J. Invert. Pathol.* **13,** 345–350.

Arnott, H. J., Smith, K. M., and Fullilove, S. L. (1968). *J. Ultrastruct. Res.* **24,** 479–507.

Bellett, A. J. D. (1965a). *Virology,* **26,** 127–131.

Bellett, A. J. D. (1965b). *Virology,* **26,** 132–141.

Bergold, G. H. (1947). *Z. Naturforsch.* **2B,** 122–143.

Bergold, G. H. (1959). *Proc. 4th Int. Congr. Biochem. Vienna 1958,* Vol. 7, pp. 95–98. Pergamon, Oxford.

Bergold, G. H. (1963a). *In* "Insect Pathology" (E. A. Steinhaus, ed.), pp. 413–456. Academic Press, New York.

Bergold, G. H. (1963b). *J. Insect Pathol.* **5,** 111.

Bergold, G. H. and Pister, L. (1948). *Z. Naturforsch.* **3B,** 406–410.

Bergold, G. H. and Suter, J. (1959). *J. Insect Pathol.* **1,** 1.

Bergold, G. H. and Wellington, E. F. (1954). *J. Bacteriol.* **67,** 210–216.

Bird, F. T. (1963). *J. Insect Pathol.* **5,** 368.

Bird, F. T. (1965). *Can. J. Microbiol.* **11,** 497.

Bird, F. T. (1966). *Can. J. Microbiol.* **12,** 337.

Bolle, J. (1874). *Jahrb. der k.k. Seidenbau-Versuchsstation in Götz Verlag der k.k. Seidenbau-Versuchsstation Götz 1874,* p. 129.

Bolle, J. (1894). *Jahrb. der k.k. Seidenbau-Versuchsstation in Götz Verlag der k.k. Seidenbau-Versuchsstation Götz 1894,* p. 112.

Cunningham, J. C. and Longworth, J. F. (1968). *J. Invert. Pathol.* **11,** 196–202.

Estes, Z. E. and Faust, R. M. (1966). *J. Invert. Pathol.* **8**, 145–149.

Faulkner, P. (1962). *Virology* **16**, 479–484.

Faust, R. M. and Estes, Z. E. (1965). *J. Invert. Pathol.* **7**, 521–522.

Gershenson, S. M. (1959). *Dokl. Akad. Nauk. SSSR* **128**, 622–625.

Gershenson, S. M. (1960). *Problems Virol* (*U.S.S.R.*) **6**, 720–725 (English Transl.).

Hall, C. E. (1960). *J. Biophys. Biochem. Cytol.* **7**, 613–618.

Harrap, K. A. and Robertson, J. S. (1968). *J. Gen. Virol.* **3**, 221–225.

Holoway, D. F. and Bergold, G. H. (1953). *Science* **117**, 251–252.

Holoway, D. F. and Bergold, G. H. (1955). *Science* **122**, 1266–2167.

Hosaka, Y. and Aizawa, K. (1964). *J. Insect Pathol.* **6**, 53–77.

Huger, A. (1963). *In* "Insect Pathology" (E. A. Steinhaus, ed.), Vol. I p. 531–538. Academic Press, New York.

Huger, A. and Krieg, A. (1961). *J. Insect Pathol.* **3**, 183.

Hughes, K. M. (1952). *J. Bacteriol.* **64**, 375.

Hukuhara, T. and Hashimoto, Y. (1966). *J. Invert. Pathol.* **8**, 234–239.

Kawase, S. (1964). *J. Insect Pathol.* **6**, 156–163.

Kawase, S. (1967). *J. Invert. Pathol.* **9**, 136–138.

Kawase, S. and Miyajima, S. (1968). *J. Invert. Pathol.* **11**, 63–69.

Krieg, A. (1956). *Naturwissenschaften* **43**, 537.

Krieg, A. (1961). *Z. Naturforsch.* **16**, 115–117.

Krywienczyk, J. and Bergold, G. H. (1960a). *Virology* **10**, 308–315.

Krywienczyk, J. and Bergold, G. H. (1960b). *Virology* **10**, 549–550.

Krywienczyk, J. and Bergold, G. H. (1960c). *J. Immunol.* **84**, 404–408.

Krywienczyk, J., McGregor, D. R., and Bergold, G. H. (1958). *Virology* **5**, 476–480.

Miura, K., Fujii, I., Sakaki, T., Fuke, M., and Kawase, S. (1968). *J. Virol.* **2**, 1211–1222.

Morgan, C., Bergold, G. H., Moore, D. H. and Rose, H. M. (1955). *J. Biophys. Biochem. Cytol.* **1**, 187–190.

Ponsen, M. B. (1965). *Netherlands J. Plant Pathol.* **71**, 54–56.

Ponsen, M. B. and De Jong, D. J. (1964). *Entomophaga* **9**, 253–255.

Ponsen, M. B., Henstra, S., and Van der Scheer, C. (1965). *Netherlands J. Plant Pathol.* **71**, 20–24.

Schmidt, L. and Philips, G. (1958). *Fac. Agr. Forestry Inst. Entomol. Zagreb* [No. 1], 27 pp.

Smith, J. D. and Wyatt, G. R. (1951). *Biochem. J.* **49**, 144–148.

Smith, K. M. (1955). *Advan. Virus Res.* **3**, 199–220.

Smith, K. M. (1958). *Parasitology* **48**, 459–462.

Smith, K. M. (1967). "Insect Virology." Academic Press, New York.

Smith, K. M. and Brown, R. M., Jr. (1965). *Virology* **27**, 512–519.

Smith, K. M. and Hills, G. J. (1960). *Proc. 11th Int. Congr. Entomol.* Vol. 2, p. 823. Springer, Berlin.

Smith, K. M. and Wyckoff, R. W. G. (1950). *Nature* (*London*) **166**, 861.

Smith, K. M. and Xeros, N. (1953). *Nature* (*London*) **172**, 670–671.

Smith, K. M., Trontl, Z. M., and Frist, R. H. (1964). *Virology* **24**, 508–513.

Stairs, G. R. (1964). *Virology* **24**, 514.

Steinhaus, E. A. and Marsh, G. A. (1960). *J. Insect Pathol.* **2**, 115–117.

Summers, M. D. and Arnott, H. J. (1969). *J. Ultrastruct. Res.* **28**, 462–480.

Tanada, Y. (1959). *J. Insect Pathol.* **1**, 215–231.

Tanada, Y. (1964). *J. Insect Pathol.* **6**, 378–380.

Wellington, E. F. (1951). *Biochim. Biophys. Acta* **7**, 238–243.
Wellington, E. F. (1954). *Biochem. J.* **57**, 334–338.
Wyatt, G. R. (1952a). *Exp. Cell Res. Suppl.* **2**, 201–217.
Wyatt, G. R. (1952b). *J. Gen. Physiol.* **36**, 201–205.
Xeros, N. (1955). *Nature (London)* **175**, 588.
Xeros, N. (1956). *Nature (London)* **178**, 412–413.
Xeros, N. (1962). *Biochim. Biophys. Acta* **55**, 176–181.
Xeros, N. (1964). *J. Insect Pathol.* **6**, 261–283.

CHAPTER 16

Oncogenic Viruses: A Survey of Their Properties

A. F. HOWATSON

I. Introduction

The history of oncogenic viruses, defined as viruses capable of inducing neoplasms, either benign or malignant, and including leukemia, extends almost as far back as the first recognition of viruses as agents of disease. In 1907, Ciuffo reported successful transmission of human warts by inoculation of cell-free filtrates, and in the following year Ellermann and Bang (1908) demonstrated cell-free transmission of chicken leukemia. This was followed in 1911 by the experimental transmission by Rous (1911) of malignant sarcomas in chickens by cell-free filtrates. Some 20 years elapsed before the range of species shown to be susceptible to virus-induction of tumors was extended to rabbits by the work of Shope on rabbit fibromas and papillomas (Shope, 1932, 1933). A few years later, the studies of Bittner (1936) established the role of a viral agent in the induction of a typical mammalian cancer—mammary carcinoma of mice. Despite the irrefutable evidence for the oncogenic capacity of these viruses, the possibility that viruses play an important role in the etiology of cancer seldom received serious consideration. The general view of viruses as highly infectious, destructive agents could not be reconciled in many people's minds with the concept of an agent causing abnormal, uncontrolled cell proliferation. This climate of scepticism persisted for many years despite the efforts of a small band of "cancer-virus" enthusiasts among whom one of the most prominent was Charles Oberling who presented a strong case for a viral etiology of cancer in his book "The Riddle of Cancer" (Oberling, 1952).

The turning point can now clearly be recognized as the discovery by Gross in 1951 of a virus causing leukemia in mice, and his subsequent realization that the same preparations contained a second virus (now known as polyoma virus) that was capable of inducing a variety of different tumor types in rodents (Gross, 1961). Study of the properties of polyoma virus was greatly facilitated by the finding by Stewart et al. (1957) that it could readily be propagated in mouse cells in vitro.

Polyoma virus shows in a very striking fashion that a single viral species is capable of eliciting in cells of the same host either in vivo or in vitro one or other of two very different types of response—a lytic or productive response which results in viral replication with concomitant cell destruction, or a nonproductive response in which little or no infectious virus is produced but the growth control mechanism of the host cell is disturbed, resulting in abnormal cell proliferation (McCulloch et al., 1961).

The great impetus given to work on tumor viruses by these discoveries led to the isolation of further viruses capable of causing leukemia or solid tumors in numerous species of laboratory and domestic animals. Demonstration of the widespread occurrence of oncogenic viruses in animal species has dissipated the earlier scepticism concerning the importance of viruses in the etiology of cancer and has led to an intensified search for human oncogenic viruses.

Oncogenic viruses occur in both of the two major classes of viruses— DNA- and RNA-containing. The purpose of this article is to examine the properties of oncogenic viruses within each of these major classes, to compare them one with another and with nononcogenic members of the class and to discuss and compare virus–host cell relationships in the two classes. By comparisons of this sort, one might hope to obtain answers to questions such as: Are oncogenic viruses basically different from other viruses?: Do oncogenic viruses have properties in common and, in particular, is the mechanism of oncogenesis the same for all oncogenic viruses?

Oncogenic viruses are defined in terms of the effects they elicit in the host animal or host cell. However, classification of viruses in general is no longer based on host response but primarily on the properties of the virion. Thus a survey of the distribution of oncogenic viruses among the various families of viruses that have been established in this way should reveal whether they have basic properties in common other than ability to induce tumors. DNA and RNA viruses will first be considered separately.

II. Distribution of Oncogenic Viruses

A. DNA ONCOGENIC VIRUSES

Table I shows the principal families of DNA-containing viruses and the distribution of oncogenic viruses among them. It is immediately apparent that oncogenic viruses are widely distributed among the DNA virus groups. In their morphology and content of DNA they cover a wide spectrum, ranging from the large, DNA-rich poxviruses to the small polyoma and SV40 viruses and the even smaller parvoviruses, the oncogenicity of some members of which is suspected but not yet established. Further, it is noteworthy that, within each of the oncogenic-virus containing groups, there are viruses with similar physicochemical properties that have not been shown to be oncogenic.

The papova- and adenoviruses have been studied most extensively in

TABLE I

ANIMAL DNA VIRUSES

	Mol. wt. of DNA ($\times 10^6$)	Morphology and dimensions (nm)	Envelope ($+$ or $-$)	Symmetry of capsid	Oncogenic members
Poxvirus	160–200	Brick-shaped 300 × 230	$-$	Complex	Rabbit fibroma, myxoma, Yaba, molluscum contagiosum
Adenovirus	20–25	Spherical 70–80	$-$	Icosahedral	Adeno 12, 18, etc., in man. Simian and avian adenoviruses
Papovavirus	3–5	Spherical 45–55	$-$	Icosahedral	Polyoma, SV40, papilloma
Herpesvirus	60–80	Approx. spherical 100–150	$+$	Icosahedral	Lucke renal tumor Marek's disease Burkitt's lymphoma?
Parvovirus	2	Spherical 20	$-$	Icosahedral	Minute virus of mice?

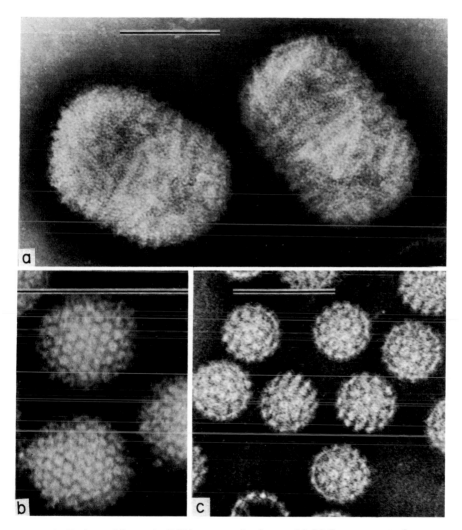

FIG. 1. Variety of forms in DNA oncogenic viruses. (a) Molluscum contagiosum, a poxvirus, magnification, × 250,000. (b) Adenovirus type 12, magnification, × 450,000. (c) Human papilloma (wart) virus, magnification, × 275,000. Bars represent 0.1 μ.

recent years but some of the oldest established oncogenic viruses belong to the pox group. The tumors produced in the natural hosts by these viruses are usually small but may be quite large as, for example, those induced in monkeys by the Yaba virus. Poxvirus-induced tumors are usually benign, but in at least one instance, rabbit fibroma virus, malignant tumors can be induced (Duran-Reynals, 1940). The poxviruses have a complex structure quite distinct from that of other DNA viruses (Fig. 1a). They differ also in that they replicate entirely in the cytoplasm and that infectious progeny virus are commonly present in large numbers in host cells and can readily be extracted from tumors.

The viruses in the remaining groups all possess well-defined capsids having icosahedral symmetry, but the DNA content, the diameter of the capsid, the number of capsomeres, and other structural features differ from one group to another (Fig. 1b,c). The herpes group is the only one in which the capsid is usually surrounded by an envelope. Until recently only one member of the herpes group—the Lucké virus—associated with tumors of the kidney in leopard frogs was regarded as oncogenic but within the last few years two new, probably oncogenic members have been discovered—the viruses associated with Marek's disease (neurolymphomatosis of chicks) and Burkitt's lymphoma.

B. RNA ONCOGENIC VIRUSES

Examination of a similar table of the distribution of oncogenic viruses among RNA animal viruses (Table II) shows that the situation is very different. With one possible exception (reovirus 3), oncogenic viruses are confined to a single rather homogeneous group. The name *thylaxovirus* was recommended for the group by workers in the field (Dalton *et al.*, 1966). However the more euphonious term *leukovirus* suggested by Fenner (1968) has recently been approved by the International Commission for the Nomenclature of Viruses and will be used here. The grouping is based on the properties of the virions and on their mode of development in infected cells. The leukovirus group contains a large number of different viruses capable of inducing leukemia and similar diseases in a variety of mammalian and avian species. It also includes several viruses capable of inducing solid tumors; the avian (Rous) and mouse sarcoma viruses (RSV and MSV) and the mouse mammary tumor virus (MTV). Apart from the last virus which is morphologically distinct, the members of this group differ only in minor details in their structure which is best depicted in thin sections (Fig. 2). The virions are of the C type designated by Bernhard (1960) and mature by a well-defined budding process at the cell surface or intracytoplasmic membrane (De Harven, 1968). Similar particles are present in large numbers in leukemic tissues of guinea pig and cat. Recently virions of the "C" type have been observed in a spleen cell line derived from a tumor-bearing viper, the first association of such particles with a cold-blooded vertebrate (Zeigel

TABLE II

ANIMAL RNA VIRUSES

Group	Mol. wt. of RNA ($\times 10^6$)	Morphology and dimensions (nm)	Envelope (+ or −)	Symmetry of capsid	Oncogenic members
Picornavirus	2–4	Spherical 20–30	−	Icosahedral	None known
Reovirus	10	Spherical 70–75	−	Icosahedral	Type 3?
Myxovirus	2–5	Approx. spherical 80–100	+	Helical	None known
Paramyxovirus	6–8	Pleomorphic 100–300	+	Helical	None known
Rhabdovirus	3–4	Bullet-shaped 70 × 170	+	Helical	None known
Leukovirus	10–12	Spherical 100–120	+	?	MLV, MSV, MTV, RSV, ALV, AMV, etc.

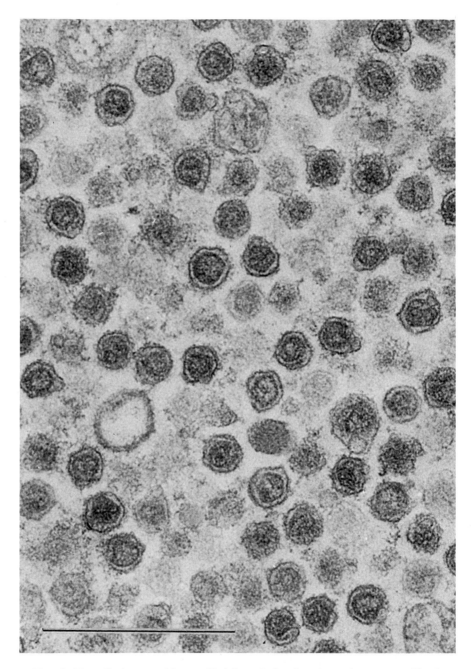

FIG. 2. Type C virus particles purified from leukemic mouse plasma, magnification, × 100,000. Bar represents 0.5 μ.

and Clark, 1969). The mouse mammary tumor virus (type B) differs in some respects from the other viruses. It is not associated with any form of leukemia; also the virion has an eccentric nucleoid and the envelope possesses more prominent surface projections than the type C virions. Because of these differences it should perhaps be assigned to a subgroup. A peculiarity of all the leukoviruses is the difficulty that has been experienced in obtaining clear-cut evidence of the structural configuration of the nucleocapsid. In all the other groups, as illustrated in Table II, the capsid is clearly either icosahedral or helical, but in the leukoviruses the configuration has yet to be determined. The leukoviruses are all ether sensitive and have a common buoyant density in sucrose of 1.15–1.18 gm/ml. Further evidence of similarity among different members of the group is shown by comparing the molecular weight and base composition of the viral RNA's (Table III).

TABLE III

RNA FROM LEUKOVIRUSES (70 S COMPONENT)[a]

Virus	Mol. wt. ($\times 10^6$)	Base composition (%)			
		A	G	C	U
AMV	12	25.3	28.7	23.0	23.0
RSV (RAV)	12	25.1	28.3	24.2	22.4
Rauscher	13	25.5	25.1	26.7	22.7
MTV	12	—	—	—	—

[a] Data from Robinson and Baluda (1965) (AMV); Robinson et al. (1965) RSV (RAV); and Duesberg and Robinson (1966) (Rauscher).

Another feature peculiar to this group of RNA viruses is a requirement for synthesis of DNA during the early stages of infection. Continuous transcription of DNA into RNA at later stages is necessary for virus production (Temin, 1964; Bader, 1964). It is apparent that, in contrast to the DNA viruses, the RNA oncogenic viruses show a rather strong similarity in their physical and chemical properties and also in their mode of development and release from cells.

III. Virus-Host Cell Interaction

The RNA oncogenic viruses form a homogeneous group and differ markedly in their structure and physicochemical properties from the DNA oncogenic viruses. Differences in host cell response to DNA and RNA tumor viruses are equally striking. DNA oncogenic viruses (with the possible exception of poxviruses which will be considered later) can have one or other of two very different effects on host cells: *either* productive infection in which the virus multiplies and the cell is destroyed (the normal cytolytic effect of most viruses) *or* an abortive type of infection which may lead *in vitro* to cell transformation with permanent alteration in the growth characteristics of the cells or *in vivo* to tumor cell conversion. If the response is of the second type, little or no infectious virus is produced. Transformation *in vitro* corresponds closely to the neoplastic change that occurs *in vivo* and lends itself more readily to study by quantitative virological techniques. Transformed cells are distinguished from normal cells by their morphology and altered growth habits, especially their loss of contact inhibition. This results in the formation from individual transformed cells in a monolayer of localized multiple cell layers or foci of altered cells. Enumeration of such foci forms the basis for a transformation assay. It is found that transformation by DNA viruses is a rare event, some 10^6–10^7 particles being required for each transforming event. The transformation, however, is brought about by a single virus particle. Viral DNA alone is capable of initiating transformation though it does so less efficiently than the complete virion.

Transformation by RNA viruses is a very different process. Cells transformed *in vitro* usually continue to produce virus in large quantities, and in

TABLE IV

VIRUS–HOST INTERACTION

DNA Oncogenic Viruses	RNA Oncogenic Viruses
Either: productive infection with cell lysis	Cell transformation with production of virus by transformed cells
Or: abortive infection which may lead to cell transformation with little or no virus production	Abortive infection and transformation with no virus production (RSV, MSV)
Transformation inefficient 10^6–10^7 particles/transformant	Transformation fairly efficient 10^2–10^3 particles/transformant
Not important natural cause of malignant tumors	Important natural pathogens in chicks and several mammalian species

diseased animals infectious virus is, in general, readily recoverable from blood or tumor tissue. Some exceptions, such as Rous virus-induced tumors in mammals which do not contain infectious virus, will be considered later. *In vitro* transformation by RNA viruses is relatively efficient, 10^2–10^3 particles per transformant being required. Transformation by viral RNA has not been demonstrated. The differences in the effects of DNA and RNA viruses on host cells are summarized in Table IV.

Under natural conditions, RNA oncogenic viruses are found in association with leukemia or solid tumors in chickens, mice, and cats, and appear to be an important cause of malignant disease in these animals. Similar C-type particles have been reported in leukemic tissues of several other species, including man. On the contrary, there is no evidence that DNA oncogenic viruses are important *natural* carcinogens. Induction of malignant tumors by these viruses requires rather artificial conditions, usually injection of large doses of virus into newborn rodents.

The benign tumors produced by papilloma and poxviruses seem to be an exception. Infectious virus is present in these tumors and is responsible for the spread of these tumors under natural conditions. In poxvirus infection there is an initial stimulus to cell proliferation which is usually superseded by cell damage and destruction associated with viral multiplication. The outcome of the opposing tendencies to cell proliferation and cell destruction depends on a number of factors including the type of virus and the host species and age. Usually the destructive tendencies prevail and either no tumor is formed or a tumor develops but eventually regresses. However, in at least one instance, rabbit fibroma virus, the outcome can be influenced toward malignant tumor formation by cotreatment of the host with X-rays or tar, or by selecting very young recipients (Andrewes and Ahlstrom, 1938; Clemmeson, 1939; Duran-Reynals, 1940). It is of interest that infectious virus could not be detected in fibrosarcomas induced by cotreatment with tar. Moreover, it has been shown that in papillomas of several species, including man (Williams *et al.*, 1961), virus is not detectable in cells of the basal layer where abnormal proliferation occurs, but appears only in cells that have differentiated and moved toward the surface of the lesion. Also when the tumors become malignant, as may happen in the rabbit, virus can no longer be detected in them. Thus it appears that in this tumor also, malignant transformation is associated with absence of infectious virus. Replication of DNA viruses, at least those that multiply in the nucleus, does not appear to be compatible with cell survival and cell proliferation. On the contrary, in the case of RNA oncogenic viruses a symbiotic relationship can be established, and cell division and viral production can coexist.

IV. Polyoma Virus, Simian Virus 40 (SV40), and Adenovirus

A. GENERAL PROPERTIES

These three oncogenic viruses have been the subject of intensive study in recent years. This is no doubt due to a large extent to their amenability to investigation by quantitative *in vitro* virological techniques, but they are of interest for other reasons too. Polyoma virus is remarkable for the wide variety of tumor types it can elicit in mice (Stewart *et al.*, 1958), and for the rapidity and consistency with which it induces tumors in baby hamsters (Axelrad *et al.*, 1960). SV40 achieved importance and notoriety because of its accidental presence in some of the early polio virus vaccines. The adenoviruses are the only viruses of human origin known to induce malignant tumors in animals.

B. VIRION PROPERTIES

Polyoma virus and SV40 closely resemble each other in most of their properties. They are both members of the papovavirus group which includes the K mouse virus, recently reported to be capable of transforming mouse cells *in vitro* (Takemoto and Fabisch, 1970), and also rabbit vacuolating virus. The other members of the group, the papilloma viruses, are structurally similar to those already mentioned but are appreciably larger (55 nm in diameter compared with 45 nm). The molecular weight of papilloma virus DNA is correspondingly greater ($5 \cdot 10^6$ vs. $3 \cdot 10^6$ daltons). These differences would seem to be sufficiently great to justify a division of the papovaviruses into two subgroups—the polyomalike viruses and the papilloma viruses.

The DNA extracted from polyoma and SV40 virions is double-stranded and is in the form of a closed loop which itself is twisted or supercoiled. Purified viral DNA has been shown to be capable of causing transformation as well as productive infection, the type of response depending on the host cell. DNA from papilloma viruses has a similar conformation. Study of the properties of papilloma viruses is hampered, however, by difficulties encountered in propagating the viruses *in vitro*.

The adenoviruses (31 human serotypes) are larger (70 nm in diameter) and more complex than the papovaviruses. The structure of the capsid is well established. It forms an icosahedral shell consisting of 252 capsomers; 12 are situated at the vertices and the remaining 240 complete the facets of the icosahedron. Attached to each of the 12 vertex capsomers is a fiber with a terminal knob, the whole unit being termed a penton since it is surrounded by five neighbors. The length of the penton fiber is different for different adenovirus types. The other capsomers are each surrounded by six neighbors and are referred to as hexons. The pentons and hexons are antigenically distinct. The capsid contains within it other structural proteins and the viral DNA. The latter is double-stranded, but unlike the DNA of papovavirues is a linear molecule. The molecular weight is $20\text{--}25 \cdot 10^6$ daltons. The DNA of adenoviruses has not been shown to be capable of causing cell transformation.

C. Transformation Properties

At the level of a single cell, transformation of the cell and multiplication of DNA viruses are mutually exclusive. However, this does not necessarily apply at the level of whole organisms or populations of cells. In the kidneys of newborn hamsters infected with polyomavirus, cells undergoing lysis associated with viral multiplication can be observed along side cells recognizable as transformed or tumor cells. Mouse embryo cell cultures infected with polyoma virus undergo considerable lysis before the emergence of a population of transformed cells. For *in vitro* work, however, it is possible to select cell lines which are overwhelmingly either permissive, i.e., allow the production of progeny virus, or nonpermissive, i.e., are susceptible to transformation but do not allow the full expression of the viral genome necessary for production of infectious virus. Thus polyoma virus is usually propagated in mouse cell lines, but in order to study transformation, rat or hamster cell lines are used. Likewise, SV40 can be propagated readily in green monkey cells but will transform cells of other species including human cells. Human adenoviruses replicate in human cells but transform rat embryo and hamster cells in culture.

Transformed cell lines and tumors induced by these DNA viruses are usually completely free of infectious virus. However, cells transformed by SV40 virus in some instances retain the capacity to produce infectious virus. This can be detected by growing the transformed cells in the presence of cells of a permissive line (Gerber, 1966). The use of recently developed techniques for cell fusion has resulted in improved recovery of infectious virus from SV40-transformed cells. Virus can also be recovered by treating transformed cells with agents such as ultraviolet light or mitomycin, well known as inducers of prophage in lysogenic bacteria (Gerber, 1964).

Infectious virus cannot be obtained, however, under the same conditions from cells transformed by polyoma or adenovirus. Nevertheless, there is strong evidence that the viral genome, or part of it, is present in these cells. Specific viral antigens of two types can be detected in transformed cells, T antigen present in the nucleus and a specific transplantation antigen (TSTA) at the cell surface. In addition, but only in adenovirus-transformed cells, viral capsid antigens can be detected in some instances. Further evidence for the presence and expression of the viral genome comes from the work of Green and others on homology between viral DNA and messenger RNA from transformed cells (Green, 1969). A surprisingly high percentage of the messenger RNA in transformed cells is viral coded, 2–5% in the case of the adenoviruses. DNA homologous with that of the virus can also be detected in transformed cells but the proportion of viral to cellular DNA is much smaller, of the order of one thousandth of one percent. Green estimates that this represents some 20–80 viral DNA equivalents per transformed cell. The viral DNA does not appear to be associated with specific chromosomes.

Estimates of the number of viral DNA equivalents in SV40-transformed cells give values similar to those for adenovirus. Corresponding estimates for polyoma virus-transformed cells give somewhat smaller values. The persistence of viral-specific messenger RNA and protein throughout many generations of transformed cells indicates that part or all of the viral genome is present in transformed cells and replicates with the cells. The state of the viral DNA in the transformed cell is of obvious interest. It has been suspected for some time that the viral DNA may be integrated into the cell genome in much the same way as prophage DNA in lysogenic bacteria. Evidence has recently been obtained (Sambrook et al., 1968) that this is indeed so in one line of transformed cells, 3T3 mouse cells transformed by SV40. The site or sites of insertion of the viral DNA are not yet known. It is of interest in this connection to note that transformation is not always stable but may be abortive, i.e., after a number of divisions the transformed cells revert to the normal state. Abortive transformation is thought to be due to failure of the viral DNA to become integrated with the cell DNA.

Human adenoviruses are not all equally oncogenic as measured by their capacity to induce tumors in hamsters. The 31 known types can be divided into three groups A, B, and C, according to their oncogenic potential. It would, of course, be of interest to know why some are more oncogenic than others. A possible clue was obtained by comparing the base compositions of DNA from different types of adenoviruses (Pina and Green, 1965). The percentage of guanine plus cytosine [(G + C)%] varied from values approaching those of animal cell DNA for the most oncogenic group to

TABLE V

PROPERTIES OF ADENOVIRUS TYPES[a]

Group	Members	Oncogenicity	(G + C)%	Mol. wt. ($\times 10^6$)
A	12, 18, 31	High	48–49	20–22
B	3, 7, 11, etc.	Intermediate	49–52	23
C	1, 2, 5, and 6	Low	57–59	23–25

[a] Data taken from Green (1969).

appreciably higher values in the least oncogenic group (Table V). It is noteworthy that the [(G + C)%] of the carcinogenic papovaviruses is also close to that of cell DNA. However, an exception to the rule for adenoviruses has recently been found in the oncogenic simian virus SA-7 which has a [(G + C)%] of 58–60 (Green, 1969).

It may be that the [(G + C)%] of the whole viral DNA is not really very significant if only a small portion of the adenovirus genome is involved in transformation. There is evidence for this in that the viral messenger RNA in tumor and transformed cells has lower G + C content than the viral DNA, suggesting that only genes with low G + C content are transcribed.

There are also small but possibly significant differences in the molecular weights of the DNA obtained from viruses in groups A, B, and C (Table V). Further indication of differences between the groups comes from studies on viral DNA homology. Between viruses of any one group the percentage of homology is high but there is little homology between viruses of different groups (Fig. 3). In line with this, the messenger RNA's synthesized by viruses in different groups are distinctly different; the T antigens are also

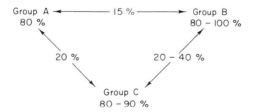

FIG. 3. DNA-DNA homology between and within the three groups of adenoviruses. Data taken from Green (1969).

different. Thus it would appear that the viral-coded information involved in transformation differs from one virus to another according to the group to which it belongs.

Further information on the extent to which the adenovirus genome is transcribed in transformed cells compared with productively infected cells has been obtained by Green and co-workers using hybridization techniques (Green, 1969). Adenovirus 2 contains DNA of molecular weight 23×10^6 daltons which has sufficient genetic information to code for an estimated 23 to 46 proteins. During productive infection, it was found that nearly all the genome (80–100%) is transcribed, representing 18 to 46 genes. Of these, 2 to 10 are transcribed in the early stages of productive infection. In adenovirus 12-transformed cells, however, only 1 to 5 viral genes are transcribed. In further experiments, Green and his colleagues detected four virus-specific antigenic components in adenovirus 12-transformed cells and suggested that these may represent the products of the genes transcribed in the transformed cells.

In similar experiments, Dulbecco and collaborators (Dulbecco, 1969) compared the viral-specific RNA synthesized in 3T3 cells transformed by SV40 with that synthesized in productively infected BSC-1 cells. In the latter cells about one-third of the viral genome is transcribed early in infection, but the whole viral genome is eventually transcribed. In the transformed cells only one-third of the viral genome is transcribed but the transcribed genes are not all the same as those that are transcribed early in infection. One-third of the SV40 genome corresponds to two or three genes, so that the number of genes involved in transformation is very small. Identification and determination of the functions of the viral genes concerned in transformation are clearly important objectives. A promising approach involving the use of temperature-sensitive mutants has been applied to polyoma virus (Eckhart, 1969; di Mayorca et al., 1969) but so far it has not been possible to define the relationship between viral genes and the new functions appearing in transformed cells. Another question concerns the basis for the restriction in transcription of viral genes in transformed cells. Incomplete transcription may be due in some instances to a defective viral genome but this cannot be generally true as the whole viral genome is known to be present in some transformed cells.

Evidence has recently been obtained by Cassingena and Tournier (1968) that there exists in cells transformed by SV40 a specific protein which appears to act as a repressor of viral DNA in permissive cells. These observations suggest a mechanism for "shutting off" viral genes that could be of great importance in understanding the nature of cell transformation.

Another aspect of the virus–host cell interaction that has received considerable attention is the induction of the synthesis of cellular DNA that occurs in both productively and abortively infected cells at an early stage of infection. It is surmised that induction of cell DNA synthesis may be related to transformation since both phenomena are associated with release of the cell from the controls restricting growth (Winocour, 1968).

Table VI summarizes the properties of the three DNA oncogenic viruses discussed in this section.

TABLE VI

PROPERTIES OF THREE DNA ONCOGENIC VIRUSES

	SV 40	Polyoma	Adeno
Tumor induction	Baby hamsters, rats	Baby hamsters, mice, rats	Baby hamsters, mice, rats
Transformation *in vitro*	Cells of many species	Hamster, rat cells	Hamster, rat cells
Recovery of infectious virus from tumor and transformed cells	+	−	−
Transformation by viral DNA	+	+	−
Conformation of viral DNA	Supercoiled, circular, ds[a]	Supercoiled circular, ds	Linear, ds
Stimulation of cell DNA synthesis	+	+	−
Antigens in tumor and transformed cells	T, TSTA	T, TSTA	T, TSTA virion

[a] ds, double-standed.

V. RNA Oncogenic Viruses

A. GENERAL INFORMATION

RNA oncogenic viruses differ in many ways from the DNA viruses we have been discussing. One of the most striking differences is the ability of cells transformed by RNA viruses to continue to produce infectious virus. Using selected strains of avian myeloblastosis virus, Beard *et al.* (1963) have obtained particle counts as high as 10^{12}/ml in plasma of infected chicks.

High viral titers are also regularly observed in plasma or tissue extracts of mice rendered leukemic by injection of some strains of mouse leukemia virus. The biological evidence of virus production is confirmed by electron microscopic evidence of virus particles budding from cytoplasmic membranes of leukemic cells.

However, the relationship between tumor formation and virus production is quite complex. This has been recognized for many years and was in fact a puzzling aspect of early work on transmission of Rous sarcomas. It was found that the yield of virus varied from one tumor to another and that some tumors contained no detectable virus at all (Rous and Murphy, 1914). Conversely, detection of oncogenic virus in an animal does not necessarily indicate that it is suffering from neoplastic disease. Many flocks of chickens are infected with avian leukosis virus but most birds remain symptomless throughout their life-span. In some high leukemic strains of mice such as AKR and C58, virus can be detected long before the animals become leukemic and even prior to birth. Some of the problems concerning the relationship of RNA viruses to the neoplastic cell have been clarified in recent years, especially by work in tissue culture systems, which will be discussed in the next section.

B. Avian Leukosis Virus Complex

Irregularities in the transformation of chick cells by Rous sarcoma virus (RSV) have been traced to the presence in the virus stocks or in the host cells of the closely related but distinct avian leukosis viruses (ALV). The latter viruses can affect transformation by RSV in two different ways (a) as interfering agents or (b) as helper viruses. The first action is exemplified by the resistance of cells infected with ALV (but apparently normal) to super-infection with RSV and thus to transformation by this virus. The second effect is shown by cells free from ALV which are transformed by RSV but are unable to replicate the virus (nonproducer cells). These cells, when superinfected with ALV, are able to produce both types of virus. The ALV provides some factor which promotes formation of viral envelope and helps the RSV particles to complete their maturation. The resultant RSV particles, which have the RSV genome but the envelope characteristic of the helper virus, are denoted by RSV (ALV). It was originally thought that the RSV strains that were unable to produce detectable virus without the aid of helper virus were genetically defective but it has since been shown that "nonproducing" cells do in fact release mature RSV particles. The particles are essentially noninfectious as measured by their ability to transform chick embryo cells, but show detectable biological activity in other cell lines. RSV produced without the aid of helper virus is denoted by RSV (0).

Though originally thought to have a very restricted host range, RSV has been shown to be capable of producing tumors not only in avian species other than the chicken but also in several mammalian species, including hamster, rat, and monkey. Mammalian tumors and transformed cells do not release infectious RSV even when helper virus is present. The presence of the viral genome can be demonstrated, however, by cocultivating transformed mammalian cells with chick fibroblasts or inoculating chicks with tumor or transformed cells. In both instances, the chick cells are transformed and release infectious RSV. The virus–host cell relationship in mammalian cell systems seems to be similar to that in chicken cells in that although transformed cells do not produce infectious virus, they may produce virions which can be detected by electron microscopy or by banding in sucrose density gradients (Valentine and Bader, 1968). The production of virus appears to be dependent on unidentified host cell factors as well as the strain of virus used.

The avian sarcoma-leukosis viruses have a common, group-specific (gs) antigen associated with the internal component or nucleoid of the virions. The antigen can be detected by complement fixation, gel diffusion, and fluorescent antibody techniques. The antigen is present in transformed and tumor cells, and is detectable even in cells that do not produce infectious virus such as RSV-induced mammalian tumor cells. A second antigen, detectable by its neutralizing activity, is associated with the viral envelopes, and is type specific. There is also some evidence for the existence of a tumor-specific transplantation antigen on the surface of RSV-transformed mammalian cells (Jonsson and Sjögren, 1965). This antigen seems to be different from the other two antigens.

C. MURINE LEUKEMIA AND SARCOMA VIRUSES

A close parallelism exists between the avian leukemia virus complex and the murine leukemia viruses. The resemblance was strengthened by the discovery of a murine equivalent of the Rous sarcoma virus, the murine sarcoma virus (MSV) (Harvey, 1964). MSV, which is structurally indistinguishable from MLV, produces solid tumors in mice and transforms mouse embryo cells *in vitro*. It also produces tumors in hamsters and rats. MSV, like RSV, is "defective," requiring a helper virus, MLV, to enable it to replicate (Hartley and Rowe, 1966). It differs from RSV in that it appears to be unable to transform cells (or at least form foci in mouse cell mono-layers) in the absence of helper virus. O'Connor and Fischinger (1968) distinguish between "defective" and "competent" forms of MSV present in preparations of the murine virus complex. The latter form of MSV is able to produce foci without superinfection with MLV. The competent virion is believed to consist of a temporary association of MLV and MSV as an effective single particle.

The murine sarcoma-leukemia viruses resemble their avian counterparts in having an internal group-specific (gs) antigen and also a type-specific antigen associated with the viral envelope. The gs antigen, however, is distinct from that of the avian complex. The murine viruses appear to be even more closely related to the feline leukemia virus which will be discussed next.

D. FELINE LEUKEMIA VIRUS

It was shown by Jarrett *et al.* (1964, 1969a) that leukemia (lymphosar-coma) in cats is associated with a virus having morphological and physical characteristics similar to those of the murine and avian leukemia viruses (Fig. 4a). The feline leukemic virus (FeLV) induces leukemia in kittens inoculated shortly after birth. It has also been shown to induce lympho-sarcoma in dogs (Rickard *et al.*, 1969). The virus replicates in cell cultures derived from feline embryos (Fig. 4b) and also in cells of canine, porcine, and human origin (Jarrett *et al.*, 1969b). A virus that may be the feline equivalent of the chicken and murine sarcoma viruses has been isolated from fibrosarcomas in a Siamese cat by Snyder and Theilen (1969). Both cell homogenates and cell-free filtrates obtained from the tumors were biologically active, producing fibrosarcomas when injected into newborn kittens.

It has been shown by Geering *et al.* (1968) that an immunological relation-ship exists between the feline and murine leukemia viruses, the gs antigens having a component in common. The relationship between the viruses was further emphasized by Fischinger and O'Connor (1969) who obtained by high-speed centrifugation of FeLV together with defective MSV an "inter-viral pellet" which could infect cat cells *in vitro* and produce foci of mor-phologically altered cells. The progeny virus was infectious for cat and dog cells but not for mouse cell cultures. However, the viral progeny from cat cells could be readapted for growth in mouse cells by forming an interviral

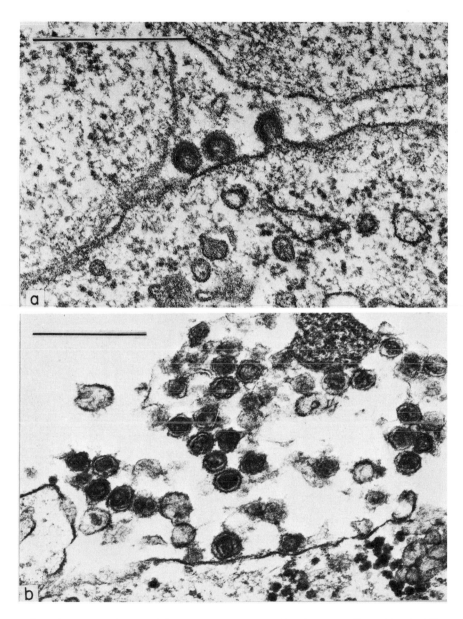

Fɪɢ. 4. (a) Field case of feline lymphosarcoma showing type C particles budding from surface of bone marrow cells, magnification × 80,000. (b) Normal feline embryo lung culture 29 days after infection with extract of leukemic tissue from field case of feline lymphosarcoma showing extracellular aggregate of type C particles, magnification × 60,000. Bars represent 0.5 μ. Micrographs courtesy of Dr. Helen M. Laird.

pellet with MLV. The ability of a viral genome, by acquiring a suitable envelope, to extend the range of host cells it can infect or transform even to cells of distantly related species is clearly of interest and importance in connection with the induction and spread of cancer.

E. TYPE C VIRUSES AND CANCER

The increasing evidence for the involvement of RNA type C viruses or viral genomes in the induction of leukemia and solid tumors in several animal species has led Huebner and Todaro (1969) to propose a hypothesis which provides a basis for a unifying concept of the cause of cancer. According to this hypothesis, cells of most or all vertebrates contain C-type virus genomes which are transmitted vertically, i.e., from parent to offspring, rather than by the usual (horizontal) mode of viral infection. The expression of the viral genes including the oncogenes responsible for transformation is under the control of the cellular gene control mechanism. Various factors such as mutations and genetic defects and agents such as radiation and chemical carcinogens can bring about a derepression of the viral genes. The outcome, depending on the host genotype and various modifying environmental factors, may be either virus production or cell transformation or both in the animals and/or in their cells grown in culture.

Activation of latent viruses is not a new concept. It has been proposed, for example, as a likely explanation for the appearance of infectious virus in lymphomas induced in mice of low leukemia incidence by X-radiation and chemical carcinogens. Huebner and Todaro (1969), however, have broadened the concept and have revived a unitary theory of cancer (cf. Oberling, 1952) in which derepressed RNA viruses of the C type are the prime determinants. They have even suggested that the oncogenic DNA viruses may function, at least in part, by derepressing C-type RNA genetic information which is indigenous to the host cells. In this connection, it is interesting to note that in our early work on polyoma virus isolated from a mammary tumor in a C_3H mouse (Howatson et al., 1960), the most prominent particles in tumors induced in Swiss mice were not polyoma virus or mammary tumor virus but type C virus which presumably was latent in the host animals and was activated by the experimental procedures. It is conceivable that the type C particles played an important role in the induction of the tumors.

Table VII lists viruses and combinations of viruses in the rapidly increasing leukovirus group, together with symbols used to denote the viral genome and envelope.

TABLE VII

RNA TUMOR VIRUSES (LEUKOVIRUSES)

RSV:	Rous sarcoma virus. Several strains, e.g., Bryan, Schmidt-Ruppin, Carr-Zilber.
RSV (o):	RSV particles of low infectivity released by "nonproducer" cells.
ALV:	Avian leukosis virus. Includes avian lymphomatosis and avian myeloblastosis virus (AMV); also known as Rous associated (helper) virus (RAV) and Rous inhibitory factor or interfering virus (RIF).
RSV (ALV):	RSV with envelope derived from ALV.
MLV:	Murine leukemia virus. Different strains; Gross, Moloney, etc.
MSV:	Murine sarcoma virus.
MSV (MLV):	MSV with envelope derived from MLV.
FeLV:	Feline leukemia virus.
MSV (FeLV):	MSV with envelope derived from FeLV.
MTV:	Murine mammary tumor virus.

VI. Possible Human Tumor Viruses

A. TYPE C VIRUS

The possibility that a C-type virus or viral genome exists in most or all animal species has already been referred to. The association of such a virus with human leukemia has been suspected for many years. Electron microscopic evidence suggestive of the existence of C-type particles in human leukemic tissue has been presented (Dmochowski et al., 1967). However, conclusive evidence for a role of C-type particles in the etiology of human neoplasia has been difficult to come by, although much effort has been expended in recent years in attempts to obtain such evidence. Investigations of human viruses are subject to the limitation that most of the work has to be done in vitro, and these conditions are not generally favorable for studies on C-type viruses.

It seems clear, however, that if type C viruses exist in human leukemic tissues they do not replicate as abundantly as do the corresponding viruses in chickens, mice, cats, and other animals. Nevertheless, a type C viral genome [or oncogene of Huebner and Todaro, 1969] may be present in human tissue and play a significant role in cell alterations leading to neoplasia without necessarily producing mature virions. Evidence for the functioning of viral genes may be obtainable by use of techniques for detection of viral-specific RNA and protein that have been developed recently with animal systems.

B. HERPESLIKE VIRUS

Herpeslike virions were discovered by Epstein *et al.* (1965) in cell cultures derived from neoplastic cells of patients with Burkitt's lymphoma, a disease prevalent among children in certain parts of Africa. It had already been suspected that the disease might be virus induced (Burkitt, 1963). This suggestion was based on the observation that high incidence of the disease was confined to areas of low altitude and high humidity where mosquitoes, which are known to transmit viruses, were abundant. The virus was found to be serologically distinct from previously identified herpes viruses. It is known as the Epstein-Barr virus (EBV) or herpeslike virus (HLV). Somewhat later the same, or a very similar virus, was detected in long-term cultures established from peripheral blood cells of American and other non-African patients suffering from leukemia and lymphoma (Ikawata and Grace, 1964). It was also found in cell cultures derived from patients with infectious mononucleosis and in a few cultures from apparently normal individuals. The EBV is not readily propagated *in vitro* and appears to be without biological effect when injected into newborn animals of various species. These negative characteristics have hampered purification of the virus and investigation of its properties. However, antibody to the virus can be detected by several immunological techniques including immunofluorescence complement fixation, immunodiffusion, and antibody coating observed by electron microscopy. These studies showed that high levels of antibody to EBV are present in sera of all Burkitt lymphoma patients. About 50% of control African children also showed detectable antibody. Further studies showed that a substantial proportion of normal individuals in various parts of the world have antibody to EBV or a very similar virus. The percentage of positive sera varies according to the detection techniques used and the age of the sample population. In general, the age distribution of EBV antibodies among children is similar to that found in other common viral infections. The significance of these observations was clarified especially by the work of Henle and Henle (1968) who showed that there is a correlation between high antibody titers and recent or past exposure to infectious mononucleosis. Patients suffering from infectious mononucleosis invariably had high antibody titers as determined by indirect immunodiffusion tests on smears of virus-infected Burkitt lymphoma cells. The evidence strongly suggests that EBV is the causative agent of infectious mononucleosis. The relation of the virus to Burkitt's lymphoma and to other forms of lymphoma and leukemia is not so clear. The high incidence of EBV antibodies in Burkitt lymphoma patients compared with controls suggests that it is not merely a passenger virus in the former, and may play a role in the etiology of this disease as well as infectious mononucleosis. It is interesting to note that the clinical features of early infectious mononucleosis may resemble those of acute leukemia and in fact infectious mononucleosis is sometimes regarded as an abortive form of leukemia. However, the epidemiological data do not support a relationship between EBV and human leukemia in general.

Recently, however, a third disease associated with high antibody titers to EBV has been found; this is nasopharyngeal carcinoma, a disease particularly prevalent among southern Chinese (Old et al., 1966). The reason for the apparent association of the virus with this type of malignancy is obscure. Whether and to what extent the virus is implicated in other human malignancies remains to be determined.

C. OTHER VIRUSES

1. Herpesvirus, Type 2

This virus is associated with genital herpesvirus infection and can be distinguished immunologically from type 1 (oral) herpesvirus. Data suggesting an association of type 2 herpesvirus and carcinoma of the cervix have been presented (Naib et al., 1969; Rawls et al., 1969). In the latter study, type 2 antibodies were found in 83% of patients with carcinoma of the cervix compared with 0–20% in the control groups tested.

2. Polyomalike Virus

A rare, and usually fatal disease of the central nervous system, termed progressive multifocal leukoencephalopathy (PML) is of interest in relation to cancer for two reasons. First, it is usually superimposed on some form of chronic neoplastic disease, especially chronic leukemia and Hodgkin's disease, and, second, it is almost certainly caused by a virus that is morphologically indistinguishable from polyoma virus (Zu Rhein and Chou, 1965; Howatson et al., 1965). The relationship of the virus to the neoplastic disease on which PML is commonly superimposed is not clear; but in view of the oncogenic potential of viruses in the polyoma subgroup the possibility that the virus plays a role in the initiation of the neoplastic disease cannot be ignored.

VII. Concluding Remarks

We now return to a consideration of the questions posed at the beginning of this article. First, is there any evidence that oncogenic viruses are fundamentally different from other viruses? Considering first the DNA viruses, the most striking impression is the "normality" of oncogenic members. Although there is great variability in the properties of different oncogenic viruses they do not differ in any obvious way from other nononcogenic viruses in the same group. Moreover, in a suitable cellular environment these oncogenic viruses can initiate a normal productive viral replication cycle leading to cell lysis. Transformation by these viruses is a rare event which seldom occurs under natural conditions. An obvious prerequisite for transformation is avoidance of cell destruction which in turn, in the case of DNA viruses, means avoidance of viral replication. Some transforming viruses may have a defective genome which enables them to initiate, but prevents them from completing, the normal viral growth cycle. For transformation to occur the virus must in addition be capable of permanently altering the growth properties of the cell. Such viruses are clearly abnormal. However, a defective viral genome is not essential for transformation, although incomplete expression of the genome is necessary. Since the outcome of infection is strongly dependent on the host cell it is clear that host cell factors are involved in determining the extent of expression of the viral genome. Thus the emphasis is not on any abnormality of the virus but on the factors in the cell-virus interaction that abort normal viral replication and make transformation possible.

In contrast to the DNA oncogenic viruses, the RNA oncogenic viruses form a homogeneous group with many structural, physical, and chemical properties in common. These include properties that distinguish them from other RNA viruses. It is not known which of these properties are related to their oncogenic capacity. The RNA oncogenic viruses differ from their DNA counterparts in that viral multiplication is compatible with cell survival, and transformed cells usually continue to release infectious virus. However, transformation can occur in the absence of virus formation and virus formation without transformation. The RNA oncogenic viruses appear to be of a special, perhaps unique, type—ubiquitous agents that may be present in some form (virogene, oncogene) in animals of many different species. The viral genetic information is expressed to varying extents according to the genetic constitution and age of the animal and its exposure to various environmental factors. Exactly how the expression of the viral genome is related to the transformation event remains to be elucidated.

The question whether the mechanism of transformation is the same for all oncogenic viruses was also raised. The structural and functional diversity of tumor viruses and the variety of possible types of interaction with host cells make it difficult to conceive of a single mechanism of viral carcinogenesis. However, there are some fundamental similarities between different systems, not only between similar groups such as the papova- and adenoviruses but also between DNA and RNA oncogenic viruses. For example, cells transformed by both types of virus show evidence of continued presence and functioning of the viral genome, even in the absence of viral replication, in the form of viral-specific products; also, virus can be rescued from nonproducing transformed cells by cocultivation with a susceptible cell line. It is not known, however, whether some form of integration of the viral genome with the cell genome can occur with RNA viruses.

In conclusion, although there are obvious differences in the physical and biological properties of oncogenic viruses and deficiencies in our knowledge of the mechanism of transformation and the transformed state, especially in the RNA virus systems, it remains a possibility that the crucial events in transformation may be the same or similar for all oncogenic viruses. If this proved to be so it might provide a basis for an understanding of the process of carcinogenesis in general.

VIII. Summary

The distribution of oncogenic viruses among the various families of viruses classified according to current criteria was examined. The wide distribution of DNA oncogenic viruses contrasts with a single homogeneous class of RNA oncogenic viruses. The different types of virus-host cell interaction encountered in DNA and RNA oncogenic virus infection were compared and contrasted. The virion properties and transformation characteristics of the three DNA viruses, polyoma, SV40, and adenovirus were discussed in detail. This was followed by a survey of the properties of RNA viruses causing leukemia and solid tumors in chickens, mice, and cats, and a brief discussion of the relevance of viruses of this type to cancer in general. Finally, there was a discussion of the present status of four suspect human tumor viruses.

ACKNOWLEDGMENTS

The research from this laboratory to which reference is made in this report was supported by the National Cancer Institute of Canada.

REFERENCES

Andrewes, C. H. and Ahlstrom, C. G. (1938). *J. Pathol. Biol.* **47**, 87–99.

Axelrad, A. A., McCulloch, E. A., Howatson, A. F., Ham, A. W., and Siminovitch, L. (1960). *J. Nat. Cancer Inst.* **24**, 1095–1111.

Bader, J. P. (1964). *Virology* **22**, 462–468.

Beard, J. W., Bonar, R. A., Heine, U., de Thé, G., and Beard, D. (1963). *In* "Viruses, Nucleic Acids, and Cancer," pp. 344–373. Williams & Wilkins, Baltimore, Maryland.

Bernhard, W. (1960). *Cancer Res.* **20**, 712–727.

Bittner, J. J. (1936). *Science* **84**, 162.

Burkitt, D. (1963). *In* "Viruses, Nucleic Acids, and Cancer," pp. 615–629. Williams & Wilkins, Baltimore, Maryland.

Cassingena, R. and Tournier, P. (1968). *C.R. Acad. Sci. (Paris) Ser.* **D 267**, 2251.

Ciuffo, G. (1907). *Giorn. Ital. Mal. Vener.* **48**, 12.

Clemmesen, J. (1939). *Amer. J. Cancer* **35**, 378–385.

Dalton, A. J., *et al.* (1966). *J. Nat. Cancer Inst.* **37**, 395–397.

De Harven, E. (1968). *In* "Experimental Leukemia" (M. Rich, ed.), pp. 97–129. Appleton-Century-Crofts, New York.

di Mayorca, G., Callender, J., Marin, G., and Giordano, R. (1969). *Virology* **38**, 126.

Dmochowski, L., Yumoto, T., Grey, C. E., Hales, R. L., Taylor, H. G., Freireich, J., Shullenberger, C. C., Shiveley, J. A. and Howe, C. D. (1967). *Cancer* **20**, 760–777.

Duesberg, P. H. and Robinson, W. S. (1966). *Proc. Nat. Acad. Sci. U.S.* **55**, 219.

Dulbecco, R. (1969). *Science* **166**, 962–968.

Duran-Reynals, F. (1940). *Yale J. Biol. Med.* **13**, 99–110.

Eckhart, W. (1969). *Virology* **38**, 120.

Ellermann, V. and Bang, O. (1908). *Zeutrabl. Bakteriol. Parasiteuk. Abt. I* **46**, 595.

Epstein, M. A., Henle, G., Achong, B. G., and Barr, Y. M. (1965). *J. Exp. Med.* **121**, 761.

Fenner, F. (1968). "The Biology of Animal Viruses," p. 26. Academic Press, New York.

Fischinger, P. J. and O'Connor, T. E. (1969). *Science* **165**, 714–716.

Geering, G., Hardy, W. D., Old, L. J., De Harven, E., and Brodey, R. S. (1968). *Virology* **36**, 678–707.

Gerber, P. (1964). *Science* **145**, 833.

Gerber, P. (1966). *Virology* **28**, 501–509.

Green, M. (1969). *In* "Proceedings of the Eighth Canadian Cancer Research Conference" (J. F. Morgan, ed.), Vol. 8, pp. 261–285. Macmillan (Pergamon), New York.

Gross, L. (1961). "Oncogenic Viruses." Macmillan (Pergamon), London.

Hartley, J. W. and Rowe, W. P. (1966). *Proc. Nat. Acad. Sci. U.S.* **55**, 780–786.

Harvey, J. J. (1964). *Nature (London)* **204**, 1104–1105.

Henle, W. and Henle, G. (1968). *Perspectives Virol.* **VI**, 105–124.

Howatson, A. F., McCulloch, E. A., Almeida, J. D., Siminovitch, L., Axelrad, A. A., and Ham, A. W. (1960). *J. Nat. Cancer Inst.* **24**, 1131–1151.

Howatson, A. F., Nagai, M., and Zu Rhein, G. M. (1965). *Can. Med. Assoc. J.* **93**, 1.

Huebner, R. J. and Todaro, G. J. (1969). *Proc. Nat. Acad. Sci. U.S.* **64**, 1087.

Ikawata, S. and Grace, J. T. (1964). *N.Y. State J. Med.* **64**, 2279.

Jarrett, W. F. H., Crawford, E. M., Martin, W. B., and Davie, F. (1964). *Nature (London)* **202**, 567.

Jarrett, O., Laird, H. M., and Hay, D. (1969a). *J. Small Animal. Pract.* **10**, 599–603.

Jarrett, O., Laird, H. M., and Hay, D. (1969b). *Nature (London)* **224**, 1208–1209.

Jonsson, N. and Sjögren, H. O. (1965). *J. Exp. Med.* **122**, 403.

McCulloch, E. A., Siminovitch, L., Ham, A. W., Axelrad, A. A., and Howatson, A. F. (1961). *Can. Cancer Conf.* **4**, 253.

Naib, Z. M., Nahmias, A. J., Josey, W. E., and Kramer, J. H. (1969). *Cancer* **23**, 940–945.

Oberling, C. (1952). "The Riddle of Cancer" (Translated by W. H. Woglom). Yale Univ. Press, New Haven, Connecticut.

O'Connor, T. E. and Fischinger, P. J. (1968). *Science* **159**, 325–329.

Old, L. J., Boyse, E. A., Oettgen, H. F., de Harven, E., Geering, G., Williamson, B., and Clifford, P. (1966). *Proc. Nat. Acad. Sci. U.S.* **56**, 1699.

Pina, M. and Green, M. (1965). *Proc. Nat. Acad. Sci. U.S.* **54**, 547–551.

Rawls, W. E., Tompkins, W. A. F., and Melnick, J. L. (1969). *Amer. J. Epidemiol.* **89**, 547–554.

Rickard, C. G., Post, J. E., Noronha, F., and Barr, L. M. (1969). *J. Nat. Cancer Inst.* **42**, 987.

Robinson, W. S. and Baluda, M. A. (1965). *Proc. Nat. Acad. Sci. U.S.* **54**, 1686.

Robinson, W. S., Pitkanen, A., and Rubin, H. (1965). *Proc. Nat. Acad. Sci. U.S.* **54**, 137.

Rous, P. (1911). *J. Exp. Med.* **13**, 397.

Rous, P. and Murphy, J. B. (1914). *J. Exp. Med.* **20**, 419–432.

Sambrook, J., Westphal, H., Srinivasan, P. R., and Dulbecco, R. (1968). *Proc. Nat. Acad. Sci. U.S.* **60**, 1288.

Shope, R. E. (1932). *J. Exp. Med.* **56**, 803–822.

Shope, R. E. (1933). *J. Exp. Med.* **56**, 607–624.

Snyder, S. P. and Theilen, G. H. (1969). *Nature (London)* **221**, 1074.

Stewart, S. E., Eddy, B. E., Gochenour, A. M., Borgese, N. G., and Grubbs, G. E. (1957). *Virology* **3**, 380–400.

Stewart, S. E., Eddy, B. E., and Borgese, N. (1958). *J. Nat. Cancer Inst.* **20**, 1223–1243.

Takemoto, K. K. and Fabisch, P. (1970). *Virology* **40**, 135–143.

Temin, H. M. (1964). *Virology* **23**, 486–494.

Valentine, A. F. and Bader, J. P. (1968). *J. Virol.* **2**, 224–237.

Williams, M. G., Howatson, A. F., and Almeida, J. D. (1961). *Nature (London)* **189**, 895–897.

Winocour, E. (1968). *Advan. Virus Res.* **14**, 177.

Zeigel, R. F. and Clark, H. F. (1969). *J. Nat. Cancer Inst.* **43**, 1097–1102.

Zu Rhein, G. M. and Chou, S.-M. (1965). *Science* **148**, 1477–1479.

Author Index

Numbers in italics refer to the pages on which the complete references are listed.

539

C

Cadman, C. H., 434, *478*
Cain, A. J., 38, *41*
Cairns, H. J. F., 179, *204*
Calberg-Bacq, C. M., 222, *251*
Caliguiri, L. A., 415, 420, 421, 422, 428, *430, 431, 432*
Calisher, C. H., 321, *333*
Callender, J., 524, *536*
Callis, J. J., 281, *300*
Campbell, R. N., 364, 267, *385*
Candler, E. L., 116, *131*
Canelo, E. S., 225, *251*
Canivet, M., 384, *386*
Carley, J. G., 322, *332*
Carmichael, L. E., 117, *130*, 157, *165*
Caro, L. G., 234, 235, *251*
Cartwright, B., 268, 280, *301*, 371, 375, *385*, 428, *430*
Cartwright, S. F., 263, 279, *301*
Casals, J., 307, 308, 309, 317, 318, 320, 329, 331, *332, 333*
Casey, M. J., 126, *131*
Caspar, D. L. D., 94, *102*, 156, *165*, 261, 287, 288, 291, *301, 303*, 336, 337, 349, 358, 359, *360*
Cassingena, R., 524, *536*
Casto, B. C., 45, 46, 48, 65, 72, *76*, 123, 125, *131*
Causey, O. R., 363, 375, *386*
Chambers, T. C., 369, *386*
Chambers, V. C., 93, 101, *102*, 363, *385*
Chandler, B., 232, 234, *251*
Chandler, R. L., 124, *130*
Chandra, S., 44, 45, 47, 49, *77, 79*
Chanessian, A., 364, *385*
Chaney, C., 47, 51, *77*
Chang, C., 47, 60, *76*
Chanock, R. M., 126, *131*, 257, 259, 266, *301, 305*
Chany, C., 46, *77*
Chardonnet, Y., 106, *133*
Charney, J., 264, 280, *301, 303*
Chen, M., 364, *386*
Chenaille, P., 84, 93, *102*
Cheong, L., 64, *77*
Cherny, N. E., 364, *386*

Cheville, N. F., 84, 93, 101, *102*
Cheyne, I. M., 420, 421, *431*
Chiu, R. M., 377, 382, 383, *385*
Choi, H. U., 416, *432*
Choppin, P. W., 410, 411, 412, 413, 414, 415, 416, 417, 419, 420, 421, 422, 423, 424, 427, 428, *430, 431, 432*
Chou, S., 82, 93, *104*, 533, *537*
Chu, C. M., 410, *430*
Churchill, A. E., 163, *165*
Chwat, J. C., 384, *386*
Ciuffo, G., 510, *536*
Clark, H. F., 517, *537*
Clark, I. A., 375, *385*
Clarke, M. C., 108, 113, 124, *130*, 476, *478*
Clarke, S. K. R., 126, *132*
Clayton, D. A., 87, *103*
Clemmer, D. I., 109, 110, 113, 116, *130*
Clemmesen, J., 519, *536*
Clifford, P., 533, *537*
Clothier, F. W., 261, 265, 269, *301*
Coetzee, J. N., 220, *251, 252*
Cohen, A., 15, *42*
Coleman, P. H., 318, 321, 327, *333*, 363, *386*
Colter, J. S., 261, 263, 264, 265, 266, 267, 268, 269, 272, 275, 283, 285, *301, 303, 304, 305*
Compans, R. W., 411, 412, 414, 415, 417, 419, 420, 421, 422, 423, 424, 425, 427, 428, *430, 431, 432*
Conti, M., 364, *385*
Cooper, J. E. K., 123, *130*
Cooper, P. D., 270, 281, *301, 305*, 324, *332*
Cornfeit, F., 122, 124, *130*
Correll, D. L., 220, *252*
Costa, A. S., 364, 382, 383, *385*
Cota-Robles, E. H., 238, *251*
Cote, J. R., 8, *42*, 49, 55, *78*
Coto, C., 142, *165, 168*
Cowan, K., 266, 280, *302*
Cowan, K. M., 277, 278, 279, 280, *301, 303*
Coyette, J., 222, *251*
Cramer, R., 101, *102*
Cramp, W. A., 476, *478*
Crawford, E. M., 85, 93, 97, *102*, 140, *166*, 235, *251*, 528, *537*
Crawford, L. B., 140, *166*

Polatnick, J., 271, 279, 280, *300, 304*
Poliakova, G. P., 364, *386*
Pollard, M., 123, *133*
Pomeroy, B. S., 110, *130*
Pomroy, B. C., 74, *77*
Pons, M. W., 405, *406,* 412, 413, 414, 415, 425, *430, 432*
Ponsen, M. B., 481, 482, 488, 489, 496, 497, *506*
Pootjes, C. F., 223, *252*
Popova, G. A., 347, *359*
Post, J. E., 528, *537*
Potter, C. W., 116, *133*
Prage, L., 106, 116, *133*
Pratt, D., 239, *252*
Preston, R. E., 44, 64, 72, *78*
Preuss, A., 239, *251*
Prevec, L., 371, 375, *385,* 397, 398, 399, 400, *406*
Prier, J. E., 124, *133*
Printz, P., 384, *386*
Prose, P. H., 417, 423, *432*
Prozesky, O. W., 220, *252*
Prunieras, M., 106, *133*
Pugh, W, E., 108, *133,* 315, *333*
Pyl, G., 281, *304*

R

Rabin, E. R., 157, 163, *166,* 321, *333*
Rabson, A. S., 60, *78*
Rada, B., 397, 402, *406*
Radloff, R., 86, *104*
Rafaiko, R., *77*
Rafajko, R. R., 122, *133*
Raimondo, L. M., 235, *252*
Ralph, R. K., 95, *103,* 454, 464, 471, *478*
Randall, C. C., 138, 140, *165, 167,* 175, 201, *204*
Randrup, A., 279, *304*
Rapoza, N. P., 108, 113, 114, 123, 124, *133*
Rapp, F., 44, 45, 48, *78,* 115, 123, 125, *131, 133*
Rapp, H. J., 277, 278, *303, 304*
Rapp, R., 48, *78*
Rawls, W. E., 533, *537*
Raymer, W. B., 14, 21, *41,* 436, *478*
Razvjazkina, G. M., 364, *386*

Rdzok, E. J., 93, *103*
Recyko, E., 44, 47, 58, *77*
Reed, G., 115, 125, 126, 127, *132*
Reed, S. E., 126, *133*
Rees, M. W., 351, 352, 358, *360*
Reese, B. R., 376, *385*
Reeves, P., 245, *252*
Reeves, W. C., 108, *133*
Regamey, R. H., 110, *130*
Rehacek, J., 321, *332*
Reichman, M. E., 299, *304*
Reilly, B. E., 225, 226, 229, 230, *250*
Reimann, B. E., 220, *252*
Reimer, C. B., 108, 116, 126, *131*
Restle, H., 261, 265, 269, 273, *300*
Reynolds, D. A., 348, *360*
Rhim, J. S., 123, *133*
Rich, M. A., 83, *103*
Richardson, C. C., 396, *406*
Richardson, J., 364, *386*
Richardson, S., 376, *386*
Richter, W. R., 93, *103,* 266, 267, 281, *302*
Rickard, C. G., 528, *537*
Riggs, J., 126, *133*
Riggs, J. L., 115, 116, *133,* 179, *204*
Rivers, T. M., 173, *204*
Roane, P. R., Jr., 142, 156, 160, *167*
Robert, P., 189, 192, *205*
Roberts, D. W., 171, 172, 173, 175, 176, 177, 180, 183, 189, 191, 193, 197, 201, 203, *204, 205*
Robertson, D., 218, 220, 230, 236, 237, 238, 239, *251*
Robertson, J. S., 493, *506*
Robinson, D. M., 64, *78*
Robinson, L. K., 280, *304*
Robinson, W. S., 71, *78,* 413, 419, 420, *430, 431,* 517, *536, 537*
Robl, M. G., 84, *104*
Roger, M., 227, *253*
Roizman, B., 137, 138, 139, 140, 141, 142, 143, 144, 145, 146, 147, 148, 149, 150, 151, 152, 153, 154, 155, 156, 157, 158, 159, 160, 161, 162, 163, *164, 165, 166, 167, 168,* 277, 278, *303, 304*
Romano, M., 107, 108, *133*
Rondhuis, P. R., 108, *133*
Rongey, R., 46, *78*

Subject Index

A

Abortion virus (equidae), classification of, 9

Acrobasia zelleri RAG, poxvirus, main characteristics, 193

Adeno-associated virus-1 (AAV-1), 58
 comparative properties of, 59
 DNA of, 66, 67, 76
 morphology of, 48–50, 54, 56, 59
 immunological properties of, 73–75

Adeno-associated virus-2 (AAV-2), 58
 comparative properties of, 59–62
 immunological properties of, 73–75
 morphology of, 49–51, 59

Adeno-associated virus-3 (AAV-3), 58
 comparative properties of, 59, 60, 62
 DNA of, 67, 68
 immunological properties of, 73–75
 morphology of, 49–52, 59

Adeno-associated virus-4 (AAV-4)
 comparative properties of, 59
 hemagglutination by, 56, 58
 immunological properties of, 73–75
 morphology of, 48–50, 55, 56

Adeno-associated viruses (AAV), 44
 antitumor activity of, 74
 biological properties of, 46
 classification of, 45
 comparative properties of, 59
 DNA of, 65–71
 hemagglutination by, 56, 58
 immunological properties of, 72–75
 morphology of, 48–52, 54–56

Adenoviruses (Adenoviridae), 105–134
 –cell interactions, 121–127
  cytopathology, 122–124
 comparative virology of, 127–129
 definition of, 106–107
 DNA of, 106, 129
  doublet pattern, 37
 DNA-containing oncogenic type, 511, 512
 evolutionary differentiation of, 128
 family characteristics, 5, 6, 7, 8
 genetic relatedness of, 129
 hemagglutinating activity of, 111–115
 helper virus for, 123
 host cell range of, 123–125
 immunological properties of, 115–121
 list of, 108
 oncogenic activity of, 125–128
 structural characteristics of, 109–111, 520–521
 T antigen of, 115–116, 525
 transformation properties of, 521–525
 zoonotic infections from, 129

Adenovirus-2, DNA of, 70, 71, 524

Adenovirus 12
 as DNA oncogenic poxvirus, 512
 morphology, 513
 transformation by, 524

Adenovirus 18, as DNA oncogenic poxvirus, 512

Adoxophyes reticulana (Hb.) nuclear polyhedrosis virus, morphology of, 481

561